THE ROUTLEDGE HANDBOOK OF MECHANISMS AND MECHANICAL PHILOSOPHY

Scientists studying the burning of stars, the evolution of species, DNA, the brain, the economy, and social change, all frequently describe their work as searching for mechanisms. Despite this fact, for much of the twentieth century philosophical discussions of the nature of mechanisms remained outside philosophy of science.

The Routledge Handbook of Mechanisms and Mechanical Philosophy is an outstanding reference source to the key topics, problems, and debates in this exciting subject and is the first collection of its kind. Comprising over thirty chapters by a team of international contributors, the *Handbook* is divided into four Parts:

- Historical perspectives on mechanisms
- The nature of mechanisms
- Mechanisms and the philosophy of science
- Disciplinary perspectives on mechanisms.

Within these Parts central topics and problems are examined, including the rise of mechanical philosophy in the seventeenth century; what mechanisms are made of and how they are organized; mechanisms and laws and regularities; how mechanisms are discovered and explained; dynamical systems theory; and disciplinary perspectives from physics, chemistry, biology, biomedicine, ecology, neuroscience, and the social sciences.

Essential reading for students and researchers in philosophy of science, the *Handbook* will also be of interest to those in related fields, such as metaphysics, philosophy of psychology, and history of science.

Stuart Glennan is the Harry T. Ice Professor of Philosophy and Associate Dean of the College of Liberal Arts and Sciences at Butler University, USA.

Phyllis Illari is Senior Lecturer in Philosophy of Science in the Science and Technology Studies Department at University College London, UK.

ROUTLEDGE HANDBOOKS IN PHILOSOPHY

Routledge Handbooks in Philosophy are state-of-the-art surveys of emerging, newly refreshed, and important fields in philosophy, providing accessible yet thorough assessments of key problems, themes, thinkers, and recent developments in research.

All chapters for each volume are specially commissioned, and written by leading scholars in the field. Carefully edited and organized, *Routledge Handbooks in Philosophy* provide indispensable reference tools for students and researchers seeking a comprehensive overview of new and exciting topics in philosophy. They are also valuable teaching resources as accompaniments to textbooks, anthologies, and research-orientated publications.

Recently published:

The Routledge Handbook of Epistemic Injustice
Edited by Ian James Kidd, José Medina and Gaile Pohlhaus

The Routledge Handbook of Philosophy of Pain
Edited by Jennifer Corns

The Routledge Handbook of Brentano and the Brentano School
Edited by Uriah Kriegel

The Routledge Handbook of Metaethics
Edited by Tristram McPherson and David Plunkett

The Routledge Handbook of Philosophy of Memory
Edited by Sven Bernecker and Kourken Michaelian

The Routledge Handbook of Evolution and Philosophy
Edited by Richard Joyce

The Routledge Handbook of Mechanisms and Mechanical Philosophy
Edited by Stuart Glennan and Phyllis Illari

THE ROUTLEDGE HANDBOOK OF MECHANISMS AND MECHANICAL PHILOSOPHY

Edited by
Stuart Glennan and Phyllis Illari

LONDON AND NEW YORK

First published 2018
by Routledge
2 Park Square, Milton Park, Abingdon, Oxon OX14 4RN

and by Routledge
52 Vanderbilt Avenue, New York, NY 10017

First issued in paperback 2020

Routledge is an imprint of the Taylor & Francis Group, an informa business

British Library Cataloguing-in-Publication Data
A catalogue record for this book is available from the British Library

Library of Congress Cataloging-in-Publication Data
Names: Glennan, Stuart, editor. | Illari, Phyllis McKay, editor.
Title: The Routledge handbook of mechanisms and mechanical
    philosophy / edited by Stuart Glennan and Phyllis Illari. Other titles:
    Handbook of mechanisms and mechanical philosophy
Description: 1 [edition]. | New York : Routledge, 2017 |
Series: Routledge handbooks in philosophy | Includes bibliographical
    references and index.
Identifiers: LCCN 2017001422| ISBN 9781138841697 (hardback : alk.
    paper) | ISBN 9781315731544 (e-book)
Subjects: LCSH: Mechanical movements—History. | Mechanical
    engineering—History.
Classification: LCC TJ15 .R68 2017 | DDC 621—dc23
LC record available at https://lccn.loc.gov/2017001422

ISBN 13: 978-0-367-57341-6 (pbk)
ISBN 13: 978-1-138-84169-7 (hbk)

Typeset in Bembo
by Swales & Willis Ltd, Exeter, Devon, UK

CONTENTS

Contents

Contents

CONTRIBUTORS

Adele Abrahamsen is Project Scientist in the Center for Research in Language at University of California, San Diego. In addition to work in philosophy of science on mechanistic explanation and interdisciplinary relations, she has done cognitive science research on mental representations of meaning, syntactic processing, language–cognition relations, and the onset of the symbolic function.

Marshall Abrams studies probability and modeling in biological and social sciences. After a Ph.D. in philosophy from the University of Chicago, he worked at Colgate, Duke, George Washington University, and the University of Alabama at Birmingham, where he is currently an Associate Professor.

Garland E. Allen is Professor Emeritus of Biology at Washington University in St. Louis. He received his Ph.D. in history of science from Harvard University. His research focuses on the history of biology, particularly genetics and its relationships to evolutionary theory and embryology in the twentieth century. He is author of, among other works, *Life Science in the Twentieth Century* (Cambridge University Press, 1978), *Thomas Hunt Morgan: The Man and His Science* (Princeton University Press, 1978), and *Scientific Process and Social Issues in Biology Education* (Springer, 2016).

Holly Andersen is Associate Professor in the Philosophy Department at Simon Fraser University, in Burnaby, British Columbia. She works in philosophy of science and in metaphysics and epistemology broadly construed. Much of her work relates to causation: causal explanation, application of causal methodology to case studies in philosophy of science, problems related to mental causation, and the metaphysics of causation.

Tudor M. Baetu is Lecturer in Philosophy of Science at the University of Bristol. His research interests are in philosophical issues concerning explanation, modeling, and experiment in biology and medicine.

William Bechtel is Distinguished Professor of Philosophy and a member of the Center for Circadian Biology and the Interdisciplinary Program in Cognitive Science at University of California, San Diego. His research has focused on mechanistic explanation, modeling, and discovery in cell and molecular biology; cognitive science; and most recently, systems biology.

Ingo Brigandt is Canada Research Chair in Philosophy of Biology at the University of Alberta, Canada. His research on evolutionary developmental biology, molecular biology, and systems biology concerns the topics of explanation, reduction and integration, concept change, and the role of values.

Brendan Clarke is Lecturer in History and Philosophy of Medicine at the Department of Science and Technology Studies, University College London. Originally qualified in medicine, his research now concentrates on the intersection between epistemology and clinical medicine.

Carl F. Craver is Professor of Philosophy in the Philosophy-Neuroscience-Psychology Program at Washington University in St. Louis. He is the author of *Explaining the Brain* (Oxford University Press, 2007) and coauthor (with Lindley Darden) of *In Search of Mechanisms: Discoveries Across the Life Sciences* (University of Chicago Press, 2013).

Lindley Darden (Ph.D., conceptual foundations of science, University of Chicago) is Professor of Philosophy, Affiliate in History and Biological Sciences, and Distinguished Scholar/Teacher at the University of Maryland, College Park. She was President of the International Society for History, Philosophy, and Social Studies of Biology (ISHPSSB) from 2001–3, and was elected Fellow of the American Association for the Advancement of Science in 1995. Her most recent book, with Carl F. Craver, is *In Search of Mechanisms: Discoveries Across the Life Sciences* (University of Chicago Press, 2013). Her current work is on the discovery of disease mechanisms.

Lane DesAutels is Assistant Professor in the Department of Philosophy and Religion at Missouri Western State University in St. Joseph, Missouri. His primary research interests are history and philosophy of science (especially biology), metaphysics, and value theory.

Justin Garson is Associate Professor of Philosophy at Hunter College of the City University of New York. His interests are in the history and philosophy of science and, in particular, the intersection of biology and mind. He is the author of *The Biological Mind: A Philosophical Introduction* (Routledge, 2015) and *A Critical Overview of Biological Functions* (Springer, 2016). He is also a co-editor of *The Routledge Handbook of Philosophy of Biodiversity* (Routledge, 2016).

Stuart Glennan is the Harry T. Ice Professor of Philosophy and Associate Dean of the College of Liberal Arts and Sciences at Butler University (Indianapolis, USA). His research focuses on mechanisms, causality, modeling, and scientific explanation. He is the author of *The New Mechanical Philosophy*, forthcoming from Oxford University Press.

William Goodwin is Associate Professor in the philosophy department at the University of South Florida. He is a philosopher of science focused on applied and constructive sciences, such as chemistry and climatology. Much of his work investigates how standard philosophical topics such as the nature of explanation, mechanism, and modeling look different in the context of these philosophically under-investigated scientific fields.

Sara Green is a postdoctoral fellow at the Department of Science Education, University of Copenhagen, Denmark. She works on the epistemic and social implications of research strategies in systems biology and systems medicine.

Marta Halina is University Lecturer in the Department of History and Philosophy of Science at the University of Cambridge. She works in the areas of philosophy of biology, philosophy of cognitive science, and general philosophy of science.

Phyllis Illari is Senior Lecturer in Philosophy of Science in the Science and Technology Studies Department at University College London, and, from July 2017, joint Editor-in-Chief (with Federica Russo) of the *European Journal for Philosophy of Science*. Her current interests are mechanisms, causality, and information, and how they impact on evidence assessment in biomedical sciences.

Stavros Ioannidis is a philosopher of biology and currently a postdoctoral researcher at the University of Athens, Greece. His research focuses on causation and explanation in biology, particularly on mechanistic explanation and developmental explanations of evolution, as well as on more general issues in the metaphysics of science.

Marie I. Kaiser is Assistant Professor at the University of Bielefeld, Germany. Her main research interests are the philosophy of biology, the general philosophy of science, and the metaphysics of biological practice. In particular, her work focuses on the concept of reductive explanation, mechanisms, part–whole relations, causal modeling, complex systems, biological individuality, and the methodology of philosophy of science.

David Michael Kaplan is Senior Lecturer in the Department of Cognitive Science, an Associate Investigator in the Australian Research Council (ARC) Centre of Excellence in Cognition and its Disorders, and an Associate Investigator in the Perception and Action Research Centre at Macquarie University, Australia.

Meinard Kuhlmann, Professor of Philosophy at Mainz University in Germany, received dual degrees in physics and philosophy and has worked at the universities of Oxford, Chicago, and Pittsburgh. His main fields of research are the philosophical aspects of quantum field theory and complex systems.

Daniel Little is Chancellor and Professor of Philosophy at the University of Michigan-Dearborn and Professor of Sociology at the University of Michigan, Ann Arbor. He is the author of many books on philosophy, social science, economic development, and China. His most recent book is *New Directions in the Philosophy of Social Science* (Rowman & Littlefield, 2016). He also maintains an academic blog at UnderstandingSociety.blogspot.com.

Alan C. Love is Associate Professor of Philosophy at the University of Minnesota and Director of the Minnesota Center for Philosophy of Science. His research concentrates on concepts, methods, and reasoning in developmental and evolutionary biology with a special focus on interdisciplinary explanation.

Caterina Marchionni is Academy Research Fellow at the Academy of Finland Centre of Excellence in the Philosophy of the Social Sciences (TINT), University of Helsinki. Her research mainly concerns scientific modeling, explanation, and interdisciplinarity in economics and the social sciences.

Lucas J. Matthews is a philosopher of biology who received his Ph.D. in 2016 from the University of Utah. He is currently a postdoctoral research associate in a behavioral genetics lab in the Department of Psychology at the University of Virginia.

John Matthewson received his Ph.D. at the Australian National University in 2012, and teaches philosophy at Massey University in Auckland, New Zealand. His research focuses on scientific modeling, mechanistic explanation, and the medical sciences.

Marcin Miłkowski is Associate Professor in the Institute of Philosophy and Sociology, Polish Academy of Sciences (Warsaw, Poland). For his *Explaining the Computational Mind* (MIT Press, 2013), he received the Tadeusz Kotarbiński Prize from the Polish Academy of Sciences and the National Science Center Award. He received the Herbert A. Simon Award for significant contributions to the foundations of computational neuroscience.

Maureen A. O'Malley is a philosopher of biology at the University of Bordeaux, France. She specializes in philosophy of microbiology, including microbial systems biology.

Viorel Pâslaru is Associate Professor of Philosophy at the University of Dayton. He has published on topics in philosophy of science and of ecology, such as mechanistic and causal explanation, causation, laws, and nomological explanation in ecology.

Gualtiero Piccinini (Ph.D., University of Pittsburgh, history and philosophy of science, 2003) is Professor of Philosophy and Associate Director of the Center for Neurodynamics at the University of Missouri-St. Louis. His book, *Physical Computation: A Mechanistic Account*, was published in 2015 by Oxford University Press.

Tiberiu Popa is Associate Professor of Philosophy at Butler University (Indianapolis, USA). His recent publications deal, among other things, with Aristotle's science and philosophy of science and with topics of philosophical interest in the Hippocratic Corpus.

Mark Povich is completing his Ph.D. in the Philosophy-Neuroscience-Psychology Program at Washington University in St. Louis. He works on the philosophy of scientific explanation, and has publications in *Philosophy of Science* and *The British Journal for the Philosophy of Science*.

Stathis Psillos is Professor of Philosophy of Science and Metaphysics at the University of Athens, Greece and a member of the Rotman Institute of Philosophy at the University of Western Ontario (where he held the Rotman Canada Research Chair in Philosophy of Science). He is the author or editor of seven books and of more than 120 papers and reviews in learned journals and edited books, mainly on scientific realism, causation, explanation, and the history of philosophy of science. He is a member of Academia Europaea.

Sophie Roux is Professor at the École Normale Supérieure, Paris. Her main research is in the history of early modern thought (Descartes, Galileo, mechanics, mechanical philosophy). She has also published in philosophy of science (thought experiments, mathematization) and history of philosophy of science (Pierre Duhem, Louis Couturat).

Federica Russo is Assistant Professor at the philosophy department, the University of Amsterdam. Her research covers causality, evidence, and modeling in the social, biomedical, and policy sciences, as well as the relation between science and technology.

Benjamin Sheredos received a joint Ph.D. in philosophy and cognitive science from University of California, San Diego in 2016. In addition to philosophy of science, he works in the history of philosophy and has developed a novel understanding of Brentano and Husserl's accounts of the origins of intentionality in mental acts. As a postdoc in UCSD's Center for Circadian Biology, he also applies his research in a team advancing web-based science education.

Catherine Stinson is Postdoctoral Fellow in Philosophy of Neuroscience at the Rotman Institute of Philosophy. Her research covers mechanistic (and non-mechanistic) explanation, computational models, animal models, the metaphysics of abstraction, classification in psychiatry, and body perception.

Jacqueline Sullivan is Associate Professor of Philosophy at the University of Western Ontario. She has published numerous articles on topics in philosophy of neuroscience and philosophy of psychiatry, and co-edited *Classifying Psychopathology: Mental Kinds and Natural Kinds* (MIT Press, 2014).

James Tabery is Associate Professor in the Department of Philosophy and a member of the Department of Pediatrics and the Department of Internal Medicine at the University of Utah.

Emma Tobin is Senior Lecturer at University College London. She completed her Ph.D. at Trinity College Dublin in 2006. She worked at Bristol University with the Arts and Humanities Research Council Metaphysics of Science project. She has authored a number of articles on the topic of natural kinds.

Dingmar van Eck is Postdoctoral Researcher in Philosophy of Science at the Centre for Logic and Philosophy of Science, Ghent University. Most of his current research is on issues related to scientific explanation in the life sciences and engineering sciences.

Petri Ylikoski is Professor of Science and Technology Studies at the Department of Social Research in the University of Helsinki and Visiting Professor at the Institute for Analytical Sociology in Linköping University. His research interests include theories of explanation and evidence, science studies, and social theory. His current research focuses on the integration of findings from biological sciences (neurosciences, genetics, and evolutionary biology) into the social sciences and the foundations of mechanism-based social science. His papers have been published in journals such as *Annual Review of Sociology*, *Philosophical Studies*, *Erkenntnis*, and *Synthese*.

Carlos Zednik is Assistant Professor at the Otto von Guericke University in Magdeburg, Germany. His work concerns the explanatory norms and practices in cognitive science and neuroscience, as well as the nature of embodied and situated cognition.

FOREWORD

My aim in this Foreword is to give some background to the New Mechanism, which has arisen with sources in biology and philosophy, primarily at Chicago, where I started teaching in 1969. About 15 years ago, Carl Craver referred to us as "The Chicago Mechanists"—comprising Peter Machamer (who left before I arrived), Lindley Darden, Bill Bechtel, Bob Richardson, Stuart Glennan, and me. And to some extent, the seeds for these developments emerged in my classroom.

My education in the early 1960s was framed by two things in tension: three years in engineering physics and a year in industry as a designer of mechanical devices, and by contrast, the formal approaches of Rudolph Carnap and Carl Hempel of the Deductive-Nomological model of explanation and the law-based account of reduction of Ernest Nagel. My father, a classical biologist, talked about tissue structures and mechanisms, but I supposed (as did most philosophers who thought about it) that the action of mechanisms could be captured by the intersection (often temporally distributed) of multiple causal laws and boundary processes, leaving laws as fundamental.

My first step toward mechanism came in 1972 when, in analyzing function, I classified as functional new modifications that caused an increase in probability of survival and reproduction of a system under selection. But this new functional trait wasn't a law—it was one in an evolutionary series of causal factors. In a postscript, I saw resonance with Wesley Salmon's new account of explanation in terms of statistically relevant factors, which he contrasted with law-based accounts.

In this period, two sources moved me toward a mechanistic account of explanation. The first was Herbert Simon's classic 1962 paper, "The Architecture of Complexity," which talked about the evolution of complex systems through the aggregation of stable sub-assemblies, and also introduced the idea of near-decomposability as a way of dynamically decomposing a complex system into its parts in terms of relative strength of interaction.

The second and even more direct source for seeing biological explanations as explaining phenomena in terms of mechanisms was Stuart Kauffman's 1971 paper, "Articulation of Parts Explanations in Biology and the Rational Search for Them." Kauffman argued that we must first find a way of decomposing a system into parts (there might be multiple such ways), then constructing a "cybernetic model" using one of these decompositions for how the parts articulated to cause the behavior, and then discovering how these elements were instantiated in the biological system. Kauffman roughed out that paper in our faculty-student philosophy of biology discussion group at Chicago in 1970, and presented his paper at PSA-1970. That group included Lindley Darden and Nancy Maull as students, and evolutionary biologists Richard Lewontin and Richard Levins, Ken Schaffner, molecular biologist Arnold Ravin, and me among others as faculty. I count Kauffman's paper as the first analytical account of mechanistic explanation in biology. In my PSA-1972 paper, I combined Kauffman's ideas of multiple decompositions of a system into parts with Simon's idea of near-decomposability to ask what happened when strongly interacting variables or parts from different partial perspectives of the system were required for a given explanation.

In 1974, mechanisms came to the fore, in a symposium at the PSA meetings on reduction in biology, with Ken Schaffner, David Hull, and Michael Ruse. The focus of the other papers (pro or con) was the Nagel-Schaffner model of reduction, but I puzzled that:

> At least in biology, most scientists see their work as explaining types of phenomena by discovering mechanisms, rather than explaining theories by deriving them from or reducing them to other theories, and *this* is seen by them as reduction, or as integrally tied to it.
>
> *(Wimsatt 1976, p. 671)*

I went on to argue for a mechanistic account, arguing that discovering a mechanism as a relatively stable and manipulable articulation of causal factors better fit the activity of biologists than a search for laws. This notion of mechanism was developed correlatively with a realist account of the nature of levels of organization, and argument against a "nothing-but-ist" eliminative reductionism.

I taught the Simon and Kauffman papers as well as my PSA papers regularly in classes, and Bill Bechtel and Bob Richardson were influenced, I think, by all of them, as well as by a 1970 paper on complexity by Richard Levins, where he distinguished between aggregate, engineered, and evolved systems. Levins' account of engineered systems fit the description of paradigmatic mechanistic explanations of the behavior of a system in terms of interactions between its parts. Evolved systems had parts boundaries that were less clear because they had evolved together, and required the multiple perspectives and parts boundaries of my 1972 paper for their analysis. Simon's work on near-decomposability probably came through most strongly in Bechtel and Richardson's groundbreaking 1993 book, *Discovering Complexity*, which involved decomposition, localization, and re-synthesis (of parts as articulated mechanisms) as explanatory strategies. I continued to interact with Bob and Bill, and to encourage them with their complex and long-maturing book.

My own work turned to other matters until the late 1980s when Stuart Glennan turned up. I was then deeply involved in simulation modeling with Jeffrey Schank. He and Glennan were both very talented programmers, with interests intersecting mechanistic explanation. They had independently discovered the framework and inspiration that object-oriented programming provided for mechanistic thinking and modeling, and I encouraged them to write about it. Glennan chose in his dissertation to develop an account of the nature of mechanisms, and put it to use in developing a novel account of causation. While he had initial troubles getting it accepted, the last chapter of his dissertation became his now well-known 1996 paper, "Mechanisms and the Nature of Causation." The mechanisms literature really took off when Machamer, Darden, and Craver responded to Bechtel, Richardson, and Glennan in their groundbreaking 2000 paper, "Thinking about Mechanisms."

I feel fortunate to have been present as an increasing awareness of the role of mechanisms as elements of scientific explanation emerged, and to have contributed to that development.

Early ideas like Simon's and Kauffman's on interlevel articulatory explanations, and of my own on levels, robustness, aggregativity, multiple complementary decompositions, and emergent complexities, helped to fuel more detailed and focused articulations by Bechtel, Richardson, Machamer, Darden, Craver, and Glennan. Our work also involved two important turns with philosophical import. The first is to take the work and claims of scientists seriously, and to look at what they can bring to philosophy rather than to suppose that the primary mission of philosophers is to bring edification to scientists. The second is a commitment to realism, which tempers the investigations of working scientists and should illuminate ours.

The new mechanistic approach has now become a dominant view across a range of debates in philosophy of science. The chapters in this Handbook analyze the historical, metaphysical, and epistemological dimensions of mechanistic science, and document the spread of philosophical work on mechanism through multiple disciplines. These analyses range from chemistry and (ironically!) physics on up through the biological and social domains where ontological genocide was once promised in the service of eliminative reductionism. Collectively, they show how mechanistic approaches better fit the kinds of investigations actually pursued in these diverse sciences, while explaining why the search for laws in these areas has for the most part been fruitless.

I welcome the multi-dimensional expansion of mechanistic philosophy of science, and the varieties of practice in the compositional sciences, from chemistry to the life and social sciences, that it illuminates. This Handbook is a rich, systematic, and illuminating encyclopedia of what mechanism has wrought. I am proud to be able to introduce it.

William C. Wimsatt
Peter B. Ritzma Professor of Philosophy, University of Chicago, and
Winton Chair of Liberal Arts, University of Minnesota, USA

Reference

Wimsatt, W. C. (1976) Reductive explanation: a functional account, in A. C. Michalos, C. A. Hooker, G. Pearce, and R. S. Cohen, eds., *PSA-1974 (Boston Studies in the Philosophy of Science, volume 30)*. Dordrecht: Reidel, pp. 671–710, reprinted in my (2007) *Re-Engineering Philosophy for Limited Beings: Piecewise Approximations to Reality*. Cambridge, MA: Harvard University Press.

ACKNOWLEDGMENTS

A Handbook like this does not come to be without the work and support of many people, and we would like to thank them for their help. We are very grateful to Tony Bruce, senior publisher at Routledge, who approached Stuart with the idea for a Handbook and who enthusiastically supported his request to bring Phyllis on as co-editor. We are also indebted to Adam Johnson, who was our primary contact at Routledge and did a great deal to advise and support us as we brought this project to completion. Additionally, we would like to thank those who gave us feedback on early plans for the Handbook, including three anonymous reviewers. We also received valuable advice from Bill Bechtel, Carl Craver, and Lindley Darden.

Our biggest debt, of course, is to our contributors. We are pleased to have assembled such an accomplished group of scholars, forty-one men and women from all over the world and at all different stages of their careers. They were attentive to our schedules and responsive to our feedback, and many of them communicated with each other to maximize the coordination and impact of their chapters. They have been, without exception, a joy to work with, and if this book is a success, it is because of them. Also, we thank Bill Wimsatt, who has been a mentor to many in this book, for writing a Foreword to the volume.

We could not have completed this book at all without the support of those close to home. Stuart is grateful to his wife Lesley and son Elliot for their encouragement, and for their understanding during the nights and weekends taken up with editorial work. He also appreciates the support of his Butler colleagues, particularly those in the Dean's office who covered for him while he slipped into cyberspace for meetings with Phyllis. Phyllis is grateful to her husband David, parents Sylvia and Ian, and her colleagues in Team Philosophy for their endless support, and for enduring many mutterings about "the book."

Finally, the editors would like publicly to acknowledge each other. We collaborated closely on every stage of this project, and the result is far better than either of us could have achieved on our own, and we had a lot of fun doing it. Philosophy is always best done together.

1

INTRODUCTION

Mechanisms and mechanical philosophies

Stuart Glennan and Phyllis Illari

Mechanical philosophy is of ancient origin. For philosophers and scientists in many epochs, thinking about mechanisms has proven to be a fruitful way to understand nature. Although mechanical philosophy receded for much of the twentieth century, it is again resurgent; that is the occasion of this Handbook.

The re-emergence of mechanical philosophy has come chiefly from two directions. The first, what has come to be called the New Mechanism, has its origins in the work of a number of philosophers of science (Bechtel & Richardson 1993; Glennan 1996; Machamer et al. 2000; Wimsatt 1976; Thagard 1999) working in the life sciences – including biology, medicine, and cognitive and neuroscience. The second is a growing body of literature on mechanisms and mechanistic explanation in the social sciences, including sociology, political science, economics, and history (Elster 1989; Hedström & Swedberg 1996; Little 1998). For lack of a better term, we will call this approach social scientific Mechanism. Although they arose largely independently, New Mechanism and social scientific Mechanism were both motivated by dissatisfaction with the philosophical image of logical empiricism. In both the life and social sciences, it seemed more plausible to construe scientific inquiry as a search for mechanisms than a search for laws of nature, and to see scientific explanation as causal and mechanistic rather than as a matter of subsuming phenomena under general laws.

The two strands of mechanist thought also share much in common in their conception of what mechanisms are. Mechanisms in both cases are conceived to be complex systems or processes that are "real and local" (Illari & Williamson 2011). They have parts whose activities and interactions are responsible for the phenomena that scientists study. Mechanisms are conceived to have both a "vertical" (constitutive) and "horizontal" (causal) dimension: mechanisms as wholes do what they do because of the activities of their parts. In the social sciences, these parts might be individual people, families, or political parties. In the life sciences, they might be proteins, cells, or organisms. Mechanistic accounts are reductionist in the sense that they take it that the properties and activities of parts underlie the properties and activities of wholes, but it is not nothing-but reductionism. Mechanists take seriously the reality of complex things. They recognize that distinctive scientific domains have distinctive kinds of entities, which engage in distinctive kinds of activities. For instance, social interactions are not the same as chemical interactions. Also, in explaining how mechanisms work, one must look to the organization of a mechanism's parts, and the context in which a mechanism is embedded. As Bechtel has put it, mechanistic explanation must look "down, around, and up."

This Handbook is concerned with both mechanisms and mechanical philosophies; that is to say, it is concerned both with what mechanisms are as things in the world, and with philosophical or scientific theories or approaches in which mechanisms or mechanistic methods figure prominently. Since the word "mechanism" is used to refer both to mechanisms and mechanical philosophies, we will signal this distinction by capitalizing the term when referring to a philosophical or scientific approach, and using lower case when referring to mechanisms themselves. So, for instance, we will speak of New Mechanism or Cartesian Mechanism, but of economic mechanisms or the mechanism of protein synthesis.

What, then, is a mechanism? Within both the New Mechanism and social scientific Mechanism literature there have been quite a number of attempts to offer a working definition or characterization, but we believe that a common core of assumptions can be succinctly captured in a formulation we call *minimal mechanism*:

> A mechanism for a phenomenon consists of entities (or parts) whose activities and interactions are organized so as to be responsible for the phenomenon.

This formulation, which comes from (Glennan forthcoming) and which closely resembles a proposal from (Illari & Williamson 2012), is minimal in two different senses. First, it is a generic and permissive definition – one by which a great many things will count as mechanisms. Second, it represents the common denominator in a set of proposals from new mechanists and social scientific mechanists.[1] We offer minimal mechanism as a characterization not to suggest that questions about the nature of mechanisms are settled – they are not – but rather to provide a framework for organizing discussions and disagreements about what mechanisms are and what role they play in the scientific enterprise. Also, we should emphasize that within this broad category of minimal mechanism there are many kinds of mechanisms, and an informative philosophical analysis of mechanisms must attend to the important differences between the many species of mechanisms.

Just as there are many kinds of mechanisms, there are many kinds of mechanical philosophies. Most broadly we can distinguish them by whether they are concerned primarily with ontological and metaphysical questions, or with epistemic and methodological questions. Metaphysical mechanists are concerned with questions about the constituents and organization of the natural world. They see mechanisms as part of nature and seek to understand questions about what they are and how they are organized. They likely see mechanisms as central to understanding some traditional metaphysical questions, e.g. about the nature of objects and properties, parts and wholes, causal relations and laws of nature. Metaphysical mechanists operate within a tradition of natural philosophy that dates to Democritus. Methodological mechanists, on the other hand, are primarily concerned with questions in the philosophy of science. They understand much of the scientific enterprise as being concerned with the search for mechanisms, so they seek to give accounts of how scientists discover mechanisms, and how they represent, refine, and justify their claims about them, as well as how mechanistic knowledge helps us explain and control natural phenomena.

A second dimension with respect to which we can classify mechanical philosophies is their scope. Some, who argue for a narrow conception of what mechanisms are, see mechanistic science as a particular kind of science that is powerful, but applicable only in limited domains. Others, who argue for a broader conception, see mechanistic methods as applicable to a far wider range of scientific domains. We should emphasize that both of these classifications are idealizations. Metaphysical and methodological questions are of course not that easy to separate, and there is no simple ordering of the scope of mechanical philosophies. Nonetheless, it can be quite helpful in exploring contemporary and historical debates to attend to where philosophers' views fit within this framework.[2]

This Handbook is divided into four parts. The first part delves into the history of scientific and philosophical thinking about mechanisms. The second is focused on the nature of mechanisms themselves; these are areas in which mechanical philosophy addresses basic ontological and metaphysical questions. The third part turns to methodological and epistemological questions raised by the mechanistic approach to natural and social sciences. The final and longest part offers a view of mechanisms and mechanistic approaches in a variety of scientific fields. In the remainder of this chapter, we will survey the contents of these parts and suggest some important connections between them.

Part I: Historical perspectives on mechanisms

As we noted above, mechanical philosophy has a long history. The chapters in Part I of this book recount some important episodes in this history. Understanding these episodes is important in part because they are all major debates and developments within the history of science. But, equally importantly, they allow us to understand both continuities and discontinuities between contemporary and historical mechanistic approaches to nature and science.

In "Mechanisms: ancient sources," Tiberiu Popa begins with a discussion of the ancient sources of mechanistic thought. These commence with Democritus and the ancient atomists and are taken up by the Epicureans. Perhaps most interestingly, Popa shows how Aristotle, despite his criticisms of the atomists, nonetheless adopted – within certain domains – forms of explanation that we would recognize as mechanistic. Sophie Roux, in "From the mechanical philosophy to early modern mechanisms," examines the origins and meanings of the term "mechanical philosophy." She draws on recent historical scholarship that shows that the various metaphysical and scientific projects carried out in the name of mechanical philosophy were quite diverse. While she agrees with critics that many mechanistic explanations from the period do not stand up to scrutiny, she argues that others, especially in the biological realm, are more empirically credible, and bear noticeable resemblances to contemporary-style mechanistic explanations. To illustrate this, Roux examines some of Descartes' uses of mechanism diagrams, and her points show remarkable parallels between early modern and contemporary techniques for visually representing mechanisms – techniques addressed in the contemporary mechanisms literature (see Chapters 18 and 19).

William Goodwin's "The origins of the reaction mechanism" pays welcome attention to a discipline not often covered by new mechanists: chemistry. Goodwin examines the development of the chemical reaction mechanism, including Lapworth's work decomposing reactions into elementary steps. He shows how these decompositions were used to explain kinetic data. In a complementary chapter, "Mechanism, organicism, and vitalism," Garland Allen traces the history of debates about Mechanism, organicism, and vitalism in biology, focusing on developments in the nineteenth and twentieth centuries. Allen situates this debate within broader metaphysical debates about materialism and its limits – showing the interplay between experimental and theoretical work and broader cultural and political developments. Finally, Marcin Miłkowski's "Mechanisms and the mental" covers the long history of mechanistic models of the mental, beginning once again with Descartes but now sweeping all the way to cybernetics and contemporary artificial intelligence. Miłkowski explores the interplay between attempts to build mechanical minds and robots with attempts in cognitive science to understand human and animal cognition mechanistically.

While these chapters are far from sufficient to paint a full portrait of the history of mechanistic philosophy and science, they do show that mechanistic approaches go a long way back, and that from the first they were applied to explain a wide variety of natural phenomena, not just those within the realm of physics. Moreover, these historical episodes show a continued and lively interplay between metaphysical debates about the nature of the material and living world and more concrete empirical explorations that developed and used mechanistic approaches.

Part II: The nature of mechanisms

Part II in this book focuses on some core questions and debates about the nature of mechanisms that have arisen over the last twenty-five years. These chapters flesh out the features that new mechanists point to in their accounts of mechanisms. The part begins with the editors' contribution to the Handbook, "Varieties of mechanisms." Given that mechanisms in the minimal sense form a very broad genus, our aim in this chapter is to provide some taxonomic principles for identifying different species of mechanisms. Investigation into specific varieties of mechanisms allows us to understand methodological differences in different areas of mechanistic science, as well as patterns of similarity that cut across very different scientific domains.

One point of almost universal agreement among contemporary advocates of Mechanism is that mechanisms are identified by their phenomena; that is, they are identified – at least partially – by what they do or produce or how they act. Justin Garson's chapter, "Mechanisms, phenomena, and functions," explores just what these phenomena are, and how they are used to individuate mechanisms. As Garson observes, a mechanism's phenomenon can often be construed as a function, and mechanisms, construed functionally, have important additional characteristics – like a distinction between proper function and malfunction. Garson explores some practical implications of this fact, like the ways in which diseases can be understood as breakdowns of normally functioning mechanisms.

As the definition of minimal mechanism makes clear, mechanisms have two fundamental ingredients – the entities (or parts, or components) of which they are composed, and the activities and interactions that these entities engage in. But while contemporary Mechanists agree about this basic point, there are many questions about just what these constituents are and how they are related. These questions are taken up in Marie Kaiser's chapter, "The components and boundaries of mechanisms." A very basic question about these constituents is whether we should take them to be real things, as opposed to explanatory constructs. And while many mechanists favor a realist interpretation, that realism must be cognizant of the fact that how one identifies the constituents of mechanisms depends upon the phenomenon one seeks to understand. There is, for instance, no single way to carve a multicellular organism into parts. Kaiser also discusses the relative priority of entities and activities. She asks: can one exist without the other, or is one category ontologically primary?

Several chapters in Part II explore the relationship between mechanisms, causation, and laws of nature. Lucas Matthews and Jim Tabery, in "Mechanisms and the metaphysics of causation," situate new mechanist views about the metaphysics of causation within the broader background of recent responses to Humean puzzles about causation. They argue that four different approaches to causation that have been connected with the mechanisms literature – from Anscombe, Lewis, Salmon, and Glennan – all respond to Hume, but with very different desiderata. They suggest that understanding these different desiderata can help untangle a number of debates about the relation between mechanisms and causes. In "Mechanisms, counterfactuals, and laws," Stavros Ioannidis and Stathis Psillos take up the specific question of the relationship between mechanistic and counterfactual conceptions of causation, two approaches that have been quite popular over the last fifteen years. They argue that all ways of understanding mechanisms depend in some way on a prior notion of law or counterfactual. Holly Andersen, in "What would Hume say? Regularities, laws, and mechanisms," takes up the relationship between mechanisms, laws, and regularities, critically discussing the debate, and arguing that there are at least two roles for laws that mechanisms cannot subsume. Together these chapters show how thinking about mechanisms has both informed and been informed by recent debates about causation and laws of nature.

In his "Probability and chance in mechanisms," Marshall Abrams maps out the various ways that the mechanisms studied in the sciences involve activities and outcomes that can be characterized by probabilities. As Abrams notes, it is widely acknowledged that many mechanisms studied in biology and other disciplines are in some sense stochastic, but little has been done to spell out the various kinds of probabilities that are appropriate to describing them. Using the example of the stochastic processes involved in bacterial chemotaxis, Abrams identifies the kinds of probabilities that can be used to characterize the activities of a mechanism's parts. He also discusses how probabilities can be measured and put to use in models and simulations of mechanisms.

In "Mechanistic levels, reduction, and emergence," Mark Povich and Carl Craver turn from questions of causation to constitution. They examine the concept of mechanistic levels, explaining how they are related to other kinds of levels, and how they are different from realization relations. They explore the implications of mechanistic levels for our understanding of reduction and emergence, and mechanistic explanation.

Finally, Emma Tobin, in "Mechanisms and natural kinds," shows how the popular view that natural kinds are property clusters sustained by homeostatic mechanisms requires a clearer conception of what counts as a homeostatic mechanism, and at the same time that there is a possible circularity, because individuation of mechanisms may require appeal to kinds. Both mechanisms and natural kinds prove hard to define and individuate because of boundary issues, raising questions about whether these boundaries are real or conventional.

Collectively, the chapters in this part of the book show how the philosophical debate moved toward an increasingly serious engagement with the metaphysical implications of Mechanism. They do so both by exploring more carefully just what mechanisms are, and by showing how a better understanding of the nature of mechanisms can give us purchase on many traditional questions about the structure of the natural world.

Part III: Mechanisms and the philosophy of science

Part III of the book turns from primarily ontological questions about the nature of mechanisms to epistemological and methodological questions that arise in the activities of mechanism-centered science. The part begins with Marta Halina's chapter on "Mechanistic explanation and its limits." Halina sets out the core features of recent models of mechanistic explanation, and considers the advantages of mechanistic explanations over earlier approaches. She then explores recent challenges to the new mechanist approach, discussing arguments that suggest that knowledge of mechanisms is not necessary for genuine explanations, as well as claims that mechanistic approaches to explanation do not do justice to the centrality of abstraction and idealization in explanation, and do not account for the generality of some kinds of scientific explanations.

John Matthewson's "Models of mechanisms" explores the question of what it is to be a model or representation of a mechanism, situating his account within the broader context of recent work in the philosophy of science on the nature of modeling. Matthewson shows how this can both illuminate our conception of mechanistic explanation and add to the literature on modeling. Abrahamsen, Sheredos, and Bechtel's "Explaining visually: mechanism diagrams" focuses on a specific kind of mechanism representation. Drawing primarily on examples from recent research on circadian rhythms, they explore the many techniques with which scientists use diagrams to efficiently represent the parts and operations in mechanisms, as well as the spatial, temporal, and causal organization of mechanistic processes. They also discuss how diagrams complement other forms of mechanism representation.

Lindley Darden's contribution to this volume is "Strategies for discovering mechanisms." Darden shows how mechanism discovery fits within Norwood Russell Hanson's account of models of discovery. In her account, she emphasizes the iterative character of the discovery process as scientists move back and forth between the phenomena to be explained and the entities and activities that may be responsible for those phenomena.

Part III concludes with David Kaplan's contribution, "Mechanisms and dynamical systems." Kaplan addresses many questions that have been raised about the relations between mechanisms and mechanistic explanation and dynamical models. While some have claimed that dynamical systems theory (DST) provides an alternative framework that can explain phenomena in a number of domains (including systems biology and neuroscience), Kaplan argues that DST, while an important tool for describing the behavior of certain systems, cannot be said to explain them unless the models provide information about the mechanisms responsible for that behavior.

A prominent feature in the chapters in Part III is that their analyses are based upon substantial engagement with contemporary scientific practice and literature, especially in the biological sciences. This, no doubt, is one reason why this research has provided such a promising new lens on traditional issues in the philosophy of science.

Part IV: Disciplinary perspectives on mechanisms

Our final part is by far the longest, with fourteen chapters. Here, we aim to give space for authors to show how debates about mechanisms, however they arise, take place in different parts of the natural, life, social, and engineering sciences. Our selections are tilted toward the biological sciences, reflecting the concentration of research in this area, but we have aimed to balance this with discussions of mechanisms in the social sciences, and in other areas less explored.

The first chapter in this part is Meinard Kuhlmann's "Mechanisms in physics." As Kuhlmann notes, recent philosophical discussions of mechanisms have largely avoided physics. The new mechanistic approach was born in discussions of the special sciences, and many who have advocated for it see it as a repudiation of earlier philosophical approaches to understanding scientific theory and practice that are rooted in physics as a paradigm science. In light of this, Kuhlmann focuses on two questions: first, to what extent do the entities, structures, and processes studied by physics involve mechanisms? Second, how, if at all, are certain features of the fundamental physical structure of the world compatible with the ontological presuppositions of mechanistic approaches in the special sciences?

Lane DesAutels, in "Mechanisms in evolutionary biology," looks at whether the mechanistic framework that has been successfully applied in other areas of biology can be helpful in understanding biological evolution. DesAutels sets out what he takes to be core metaphysical features of mechanisms, and examines whether, in this light, natural selection, drift, and mutation count as mechanisms. He suggests that there may be strategic benefits to thinking about evolution mechanistically, regardless of whether there are any mechanisms of evolution.

In "Mechanisms in molecular biology," Tudor Baetu shows how molecular biology helped shape New Mechanism, and considers how recent developments in the field are leading to reconsideration of the nature of molecular mechanisms. Earlier idealizations that characterized mechanisms, like those responsible for genome replication and expression, as quasi-deterministic molecular machines have given way to a more nuanced picture in which there are stochastic elements and noise. Additionally, increased understanding of the organization of cells and their molecular constituents has shown that assumptions about the modularity of these systems can be false, and also that the intracellular environment has considerable structure which determines the rates of molecular interactions. These features imply that understanding these mechanisms requires new tools, like those provided by systems biology.

In their contribution, "Mechanisms and biomedicine," Brendan Clarke and Federica Russo explore the many ways in which "mechanisms are sought, formulated, and used in medicine." They do so by reflection on six episodes from past and contemporary biomedicine, ranging from the evolving understanding of the various effects and applications of aspirin to the mechanisms connecting asbestos to lung cancer, and to Semmelweis' work on puerperal fever in nineteenth-century Vienna. Their analysis of these episodes shows the epistemic role of evidence of mechanisms, and suggests the importance of a pluralistic approach to evidence in medicine.

Alan Love studies developmental mechanisms, particularly focusing on the relationship between lower-level molecular-genetic mechanisms and higher-level cellular-physical mechanisms. Love uses the distinction between these two classes of mechanisms to elucidate the multi-level processes involved in development. Among other issues, he explores the trade-off between the explanatory generality gained by investigating molecular-genetic mechanisms that operate across taxa and the explanatory completeness gained by integrating them with cellular-physical mechanisms to understand more fully morphological outcomes in specific species.

Viorel Pâslaru's "Mechanisms in ecology" extends the new mechanistic approach from its more established areas toward mechanisms responsible for the distribution and abundance of organisms in their environment. Pâslaru uses research on the shrub *Lonicera maackii* (Amur honeysuckle), applying the minimal account of mechanisms to elucidate individual-level mechanisms that are generally accepted and used in ecology. He also speculates on the challenges associated with the analysis of population-level ecological mechanisms.

Rounding off the chapters on biology widely construed is Ingo Brigandt, Sara Green, and Maureen A. O'Malley's "Systems biology and mechanistic explanation." The authors give a succinct account of the emerging field of systems biology, with an emphasis on its novel techniques for representing and explaining complex biological processes. They show that these explanatory techniques can be used both to extend the range of mechanistic explanation and to construct complementary explanations that are non-mechanistic.

The next two chapters are broadly concerned with mental mechanisms. Catherine Stinson and Jacqueline Sullivan's "Mechanistic explanation in neuroscience" explores the nature of the mechanisms studied in neuroscience by tracing the development of theories of learning and memory. They show that the mechanisms scrutinized by neuroscientists are multi-level and that understanding them requires the integration of models and theories from multiple scientific fields. Carlos Zednik, in "Mechanisms in cognitive science," interprets the best-known model of explanation in cognitive science, David Marr's three-levels account, as an early articulation of a mechanistic approach to explanation. Zednik argues that interpreting Marr's levels mechanistically resolves ambiguities in the original account and allows it to be fruitfully extended to cover research programs in cognitive science that have emerged since Marr's time.

Moving to the social realm, Petri Ylikoski's "Social mechanisms" draws attention to the recent work on mechanism-based explanation in the philosophy of the social sciences. Ylikoski shows how mechanistic explanation has proved to be a useful tool for criticizing existing research practices and metatheoretical views on the nature of the social scientific enterprise. Daniel Little, in "Disaggregating historical explanation: the move to social mechanisms," extends the social mechanisms approach to historical explanation. Little argues that the mechanistic approach provides a far superior model for causal explanations in history than Hempel's covering law model. He suggests that the mechanistic approach fits naturally with current thinking about the character of causal narratives, and can account for the inevitable difficulties of making large historical predictions.

In "Mechanisms in economics," Caterina Marchionni characterizes how mechanisms are conceived in economics in comparison with the other social sciences, and examines how mechanisms are appealed to in philosophical debates about methodological individualism, causal inference, and the extrapolation of causal claims.

In "Computational mechanisms," Gualtiero Piccinini charts the changing understanding of the relationship between mechanisms and computation over the last century. Influenced by the power of the mathematical theory of Turing machines, many philosophers saw Turing machines as providing a precise model of mechanistic processes. But in the past twenty years, the situation has reversed, as philosophers have used an account of mechanisms associated with New Mechanism to identify the particular features of mechanisms that can perform physical computations. As Piccinini puts it, it has been a shift from computation explicating mechanism to mechanism explicating computation.

In the last chapter in our collection, "Mechanisms and engineering science," Dingmar van Eck applies the ontological and explanatory framework of New Mechanism to conceptual problems in engineering science. Van Eck argues that to properly understand mechanistic explanation in engineering science, one must attend to the different kinds of function with which engineers are concerned. He then uses this analysis to explore the interplay between the explanatory work of reverse engineering and the developmental work of engineering design.

The chapters that comprise Part IV of our volume are far from comprehensive, but they give some inkling of the diversity of domains in which the methods of mechanistic science are fruitful, as well as the challenges to and limits of those methods. These investigations, both of the distinctive features of mechanisms and mechanistic explanation within a given domain, and of the surprising points of contact across domains, will, we hope, demonstrate the value of thinking about mechanisms across the sciences.

★

This Handbook is the first reference work on contemporary philosophical research on mechanisms and mechanical philosophy, and we hope that it will serve as a resource to students, philosophers, and scientists. The chapters in this volume provide an entry into a large and diverse literature, but also break some new ground and point to unresolved questions and avenues for further research.

As Sophie Roux (Chapter 3) reminds us, when Robert Boyle first popularized the term "mechanical philosophy" 350 years ago, he lumped together under this moniker a number of distinct and sometimes conflicting philosophical positions and scientific projects. The same no doubt can be said of the mechanistic projects of scientists and philosophers today. But now, as then, we think it safe to say that those projects are worthy of pursuit.

Notes

1 See Illari & Williamson (2012) for a detailed discussion of the prominent new mechanist formulations, and an argument for a generic account like minimal mechanism. Hedström & Ylikoski (2010) provide a survey of proposed definitions from both the new mechanist and social mechanist literatures, and similarly argues for a great deal of commonality across the projects.
2 There have been a number of attempts to identify varieties of mechanisms and mechanical philosophies, among them Andersen (2014a, 2014b); Kuorikoski (2010); Levy (2013). See our further discussion in Chapter 7.

References

Andersen, H. (2014a) "A Field Guide to Mechanisms: Part I." *Philosophy Compass*, 4: 274–83.
——. (2014b) "A Field Guide to Mechanisms: Part II." *Philosophy Compass*, 9(4): 284–93.
Bechtel, W., & Richardson, R. C. (1993) *Discovering Complexity: Decomposition and Localization as Strategies in Scientific Research*. Princeton, NJ: Princeton University Press.
Elster, J. (1989) *Nuts and Bolts for the Social Sciences*. Cambridge: Cambridge University Press.
Glennan, S. S. (1996) "Mechanisms and the Nature of Causation." *Erkenntnis*, 44(1): 49–71.
——. (forthcoming) *The New Mechanical Philosophy*. Oxford: Oxford University Press.
Hedström, P., & Swedberg, R. (1996) "Social Mechanisms." *Acta Sociologica*, 39: 281–308.
Hedström, P., & Ylikoski, P. (2010) "Causal Mechanisms in the Social Sciences." *Annual Review of Sociology*, 36(1): 49–67. DOI: 10.1146/annurev.soc.012809.102632.
Illari, P. M., & Williamson, J. (2011) "Mechanisms Are Real and Local." In P. McKay Illari, F. Russo, & J. Williamson (eds) *Causality in the Sciences*, pp. 818–44. New York: Oxford University Press.
——. (2012) "What Is a Mechanism? Thinking about Mechanisms across the Sciences." *European Journal for Philosophy of Science*, 2(1): 119. DOI: 10.1007/s13194-011-0038-2.
Kuorikoski, J. (2010) "Two Concepts of Mechanism: Componential Causal System and Abstract Form of Interaction." *International Studies in the Philosophy of Science*, 23(2): 1–19.
Levy, A. (2013) "Three Kinds of New Mechanism." *Biology & Philosophy*, 28: 99–114.
Little, D. (1998) *Microfoundations, Method and Causation: On the Philosophy of the Social Sciences*. New Brunswick, NJ: Transaction Publishers.
Machamer, P., Darden, L., & Craver, C. F. (2000) "Thinking about Mechanisms." *Philosophy of Science*, 67(1): 1–25.
Thagard, P. (1999) *How Scientists Explain Disease*. Princeton, NJ: Princeton University Press.
Wimsatt, W. C. (1976) Reductive Explanation: A Functional Account. In A. C. Michalos, C. A. Hooker, G. Pearce, & R. S. Cohen (eds) *PSA-1974 (Boston Studies in the Philosophy of Science, volume 30)*, pp. 671–710. Dordrecht: Reidel.

PART I

Historical perspectives on mechanisms

2

MECHANISMS

Ancient sources

Tiberiu Popa

1. Introduction

A search for possible precursors of our concept of mechanism in ancient texts is potentially rewarding—that is, if we remain mindful of the original methods and aspirations that informed those works and we handle our own terminology with caution. At the beginning of an important article on "Ancient Automata and Mechanical Explanation" (2003), Sylvia Berryman addresses the use of "mechanical" and "mechanistic" in connection with purely material explanations based on contact action. The use of these words, she points out, can be sometimes more opaque than illuminating. One of Berryman's goals—in that article, in her 2009 book on the "mechanical hypothesis," and elsewhere—is to reveal the usefulness of such terminology with specific reference to "a method of investigating the natural world through terms and principles drawn from the discipline called 'mechanics'" (2003: 344). I agree with her assessment that a promising direction of research was largely ignored, while teleology and materialism were regarded by many and for too long as the only positions in ancient natural philosophy worth studying.

At the same time, I believe that, as long as we mark the semantic scope of these terms with reasonable clarity, we should be able to apply them profitably and legitimately to ancient thought from several angles. And that includes the search for early notions—emerging from questions or concerns which were often significantly different from ours—that might correspond in some measure to contemporary understandings of mechanism. The stipulative and composite definition (see also Glennan 2013; Craver and Tabery 2015; Chapter 1 of this volume) I am going to rely on is that a mechanism is a system of entities whose interactions, organization, and specific activities are responsible for its overall stability or for producing regular changes. Philosophically interesting treatments of systems—thus understood—were often instrumental in ancient efforts to articulate topics as diverse as the relations between chance, necessity, and teleology, self-organization in biological contexts, and the sort of causality governing mental processes. The relevance of mechanisms to these topics remained central to early and later modern debates.

As I attempt to sketch a prehistory of the idea of mechanism, I need to be quite parsimonious in my selection. Late antiquity will have to be mostly omitted, given the nature of the sources for that period, which are less amenable to a brief historical survey. I shall focus on the claims and arguments of several philosophers and philosophically minded scientists rather than on more purely scientific studies on mechanics, which might yield less insight into general, theoretical approaches to causation and to what we would call mechanisms. Much of my discussion is devoted to comprehensive views of nature that mobilize causal explanations in terms of interactions between component parts and their operations either universally (Democritus and the Epicureans) or within the confines of some fairly pervasive types of phenomena (Aristotle's inquiry into the properties and operations of the two exhalations; pseudo-Aristotle's study of the lever principle in *Mechanics*). I also evaluate the significance of analogies and technological models in a handful of examples (Aristotle's biology, Lucretius' philosophical poem, and the anonymous *On the Cosmos*).

2. Early atomism

In studies on ancient philosophy, "mechanistic" most frequently describes the worldviews of Democritus and of the Epicureans. Are there features, beyond the synonymy of "mechanistic" and "materialist" in such studies, that warrant a comparison with our concept(s) of mechanism? Let us start with a look at early atomism. For Democritus of Abdera[1] (c. 460–c. 370 BCE), each atom is in important ways much like Parmenides' one being: ungenerated, imperishable, homogeneous. Unlike in Parmenides' metaphysical doctrine, however, plurality and change are not self-contradictory notions. The atoms remain intrinsically the same, but can change their spatial positions with respect to each other. Indeed, all changes are reducible to (re)combinations of these material principles (67A9, 68A58),[2] which are always in motion (67A16, 67A18). Besides, they are perfectly solid or "full," in sharp contrast to the void or "the empty" (67A1, 67A7).

The indefinitely many uncuttables differ from each other with regard to shape and size (and possibly weight, 68B168). Their shapes allow some of them to form clusters or entanglements more easily, following their collisions—if, for example, they have hook-like appendages or some are concave and others convex (68A37). Different atomic shapes can also underlie perceptible qualities (heat is caused by especially small and sharp atoms, 67A14; round ones are liable to be perceived as "sweet," and atoms with many edges can produce the impression of roughness, 68A129), but it is mainly the structures of the aggregates they form that ultimately explain observable properties (e.g. compressibility, 67A19).

We may find the overall spirit of Democritus' explanations quite appealing, but he was often a target for Aristotle, who insisted on the importance of goal-directedness in nature, especially in the realm of life. He takes Democritus to task for failing to furnish a convincing explanatory apparatus for both the complexity of organisms and for the regularity that can be found within species, and for confusedly enlisting spontaneity (*to automaton*) and chance (*tuchē*) in his cosmogonic and cosmological accounts. In *Physics* II.4 (196a24–35; cf. *Parts of Animals* 641b15–23), Aristotle objects that "some"—quite clearly the atomists, chiefly Democritus—suppose inconsistently that our world, much like other worlds in this totality or universe, has come about through the agency of a vortex which was generated spontaneously, whereas they do not take animals and plants to be produced by chance or spontaneously. The surviving fragments and testimonia (e.g. 68A66, 68A68) do not convey a clear picture of the relationship between chance and necessity in Democritus, but the two concepts are probably complementary. As Glennan (2010: 621; cf. Johnson 2009 and Gregory 2013: 464–68) puts it,

> Democritus . . . insists that every event happens from necessity, in the sense of having sufficient antecedent conditions to bring about the event; but that event can still be, from another perspective, a chance event, in that it does not happen for a purpose.

Democritus seems to have tried to mitigate the potential discrepancy between our experience (we can easily notice the repeatability and predictability of many kinds of phenomena) and his insistence on *chance* encounters between atoms, by claiming that a sort of self-selection of atoms with comparable or corresponding shapes and sizes is possible. Just as irrational (*alogōn*) animals tend to seek the company of animals of the same kind, inanimate (*apsuchōn*) things are sorted by size and shape, as when we use a sieve or a winnowing basket to sort grains or as when the waves gather long pebbles and round ones in different places. Fragment 68B164 does not spell out the details of this "mechanism" for selection, but the apparent mutual "attraction"[3] of the atoms is likely due to, say, hooked atoms coming together by chance and clinging to each other or convex and concave atoms happening to collide with one another (68A37, cited above), their congregations producing an enormous number of phenomena, including the operations of our minds.

If our interpretive standard is adherence to the principles of mechanics as a scientific domain, Democritus' physical theory is hardly mechanistic, as Berryman rightly argues (2009: 34–9); indeed, mechanics proper was yet to come into existence. Still, his understanding of causation—based on assumptions about the nature of solid particles, their collisions (*plēgai*) and vibration (*palmos*), and their entanglements—looks decidedly mechanistic in a more general sense of the word. It bears mentioning that his explanations apply at various scales, from the formation and disintegration of entire worlds (*kosmoi*) to physiological and psychological processes. Some causal mechanisms appear thus to be nestled within larger ones, according to hierarchies that are transparently hinted at in the extant fragments or testimonia, although they are not explicitly analyzed there.

Among the phenomena caused by interactions between atoms and compounds of atoms, psychological processes are of special interest. In his survey of earlier theories about the soul as a motive force (*On the Soul* I.2, 405a5–13 = 68A101), Aristotle notes that, for Democritus, the mind (*nous*) or the soul (*psuchē*) is an assemblage of round atoms, their fine grain and their roundness ensuring extraordinary mobility. It is not implausible that such a conglomeration of round atoms might be the result of a process of self-selection, like the one described in the passage that mentions the sieve and the waves. In the next chapter of *On the Soul* (I.3, 406b15–22), Aristotle brings up Democritus again and attributes to him the view that what causally explains the movement of the soul, namely the continuous motion of its round atoms, also accounts for the movements of the body (which is set in motion by the soul). To stress the weakness of this argument, Aristotle writes that Democritus' explanation is reminiscent of the way in which—according to Philippus, an author of comedies—Daedalus made a wooden statue of Aphrodite move as if by itself by pouring quicksilver into it. The problem with Democritus' approach is that it is logically vulnerable (how can *rest* depend on those moving atoms?) and also that we seem able to move sometimes deliberately and as a result of reflection, not merely because of chance happenings. The fact that Aristotle ridicules the atomists' view of self-motion is clear enough and important in itself, but the very comparison between the workings of the soul in Democritus' theory and an automaton, such as an automaton fashioned by a mythical figure, suggests, I think, that even in classical times Democritean explanations were deemed "mechanistic" in the sense that an organized assemblage of parts (of atoms and of structured collections of atoms) and its activities could produce an observable phenomenon.

Democritus' direct and indirect influence was to be profound and lasting, from Epicurus (341–270 BCE) to Gassendi and beyond. The avatars of his doctrine include the famous Epicurean qualification to the Democritean absolute necessity: the uncaused swerve, one of the most distinctive features of Epicurus' doctrine.[4]

3. The Epicureans

We have seen that the first atomists explained psychological processes and states by appeal to atomic motions and the shapes of atoms, in keeping with their dominant metaphysical dogmas. The Epicureans, while embracing the spirit of such accounts, made a number of original contributions, most notably with respect to their treatment of will. Indeed, Lucretius' (who wrote around the middle of the first century BCE) memorable section on the swerve[5] makes it clear that Epicurus' innovation is meant partly to salvage the concept of (free) will. It is also presumably crucial to understanding how aggregates of atoms can ever form, since the atoms tend to "fall" along parallel trajectories and do so, importantly, at equal speed: there is no resistance in void (see Epicurus, *Letter to Herodotus*, 61), as there would be for a body falling through air or water. The speed of bodies falling in air or water varies in proportion to their weights and to the resistance of the medium in which they move (II.225–39). Hence, no atom could possibly catch up with other atoms in a void. The fact that there *are* compounds of atoms, however, compels Epicurus—and his follower, Lucretius—to posit the swerve, a slight and uncaused deviation of the atoms from their parallel trajectories that allows them to collide with each other.

When such swerves take place in the (material) soul, they can cause chain reactions which propagate through the body and instigate our actions. Significant though this is for Epicurean physics in general—the formation of conglomerates of atoms—and for elucidating general traits of animal behavior, this explanatory device was, again, probably meant mainly to justify any reference to intentional action and moral responsibility.[6] What is noteworthy for our purpose is that, in Lucretius' celebrated passage in Book II, psychological aspects like will (*voluntas*) and desire (*voluptas*) and their manifestations (e.g. our going against the movement of a crowd which surrounds us, II.277–83) are tightly connected with notions that were pivotal to theoretical and applied mechanics—weight, collision, air- or water-resistance, etc. Accordingly, speaking of mechanistic explanations in the case of this evolved version of atomism is legitimate for the reason just stated as well as for its general compatibility with our conceptions of mechanisms—even when no explicit technological models come into play.

There is one analogy with a "machine" in Lucretius, by the way, that is rather well known and worth a glance. In Book V he reminds us that the enormous mass, *moles*, and machinery of the world, *machina mundi* (96), with its threefold constitution (sea, earth, sky), is not permanent, despite appearances and false expectations. The formulation seems intended to emphasize at the same time vastness and order—and maybe complexity.[7] Berryman (2009: 38–9) finds this formulation unimpressive: "The meaning of *machina* here could be little more than a vague sense of an arrangement or system, perhaps in contrast to mere undifferentiated mass; there is no suggestion of technology here." I think it is helpful, however, to take into account some of the preceding lines (73–90 in Book V; cf. Epicurus' *Letter to Herodotus* 81, which that passage may reflect), where Lucretius raises the possibility that people's feeble understanding of the laws (*leges, foedera*) of nature might lead them to embrace superstition and to believe that the heavenly bodies, with their wondrously regular revolutions, are divinities (or at least that they are steered by divinities). He appears to mark there a distinction between animate, divine celestial bodies, and a view of the universe as a largely orderly and dynamic mechanism. Incidentally, Lucretius' insistence on the order and regularity of nature is one of the most prevalent motifs in this philosophical poem. A contrast between (a special sort of) living beings moving spontaneously and a functioning mechanical device is thus probably implied here. Finally, and more to the point of the present chapter, this analogy is very much in keeping with the whole Lucretian poem and Epicurean outlook: all phenomena, from the revolutions of the stars to our most intimate thoughts, are explainable through the causal relations between and within the vast realms of nature, and are reducible to interactions between indivisibles and to the "laws" that govern their behavior.

4. Aristotle on the two exhalations

The atomists argued that the world is an infinite swarm of discrete minimal particles moving in "the empty." Aristotle (384–22 BCE) begged to differ. For him, the finite universe is a plenum consisting of five simple stuffs, one of them, the *aithēr*, moving in circular fashion; the other four, which make up the sublunary world, having naturally a rectilinear motion, toward or away from the center of the universe. Perceptible bodies are *continua*, as are movement and time. He also decried the sheer absence of anything like his natural teleology in early atomism, an error that, he thought, inevitably led to grave contradictions. In light of his demolition of Democritus' physics,[8] among other things, one can see why the term "mechanistic" is not generally applied to Aristotle's worldview. Yet, we can ask, are at least *some* aspects of Aristotle's scientific works or *some* of his models mechanistic in a sense that accords in part with the notion of mechanism explored in this Handbook? I start with his "meteorology," a domain where natural teleology is marginal or absent and which may rather aptly be considered to accommodate systematic appeals to and analyses of mechanisms, to use a deliberate anachronism, and then I tackle a few controversial passages from his biological writings.

Aristotle's interest in mechanics proper is no doubt limited (whatever mechanics, *mēchanikē*, might have meant for him),[9] but he does dwell at length on what we might call mechanisms in a more comprehensive sense. He devotes three books, *Meteorology* I–III, to a quasi-classification, description, and, especially, causal explanation of "meteorological" processes. They include what we would recognize as meteorological phenomena, but their scope extends far beyond rain, frost, or wind to include earthquakes, comets, and the Milky Way, all of them occurring, as he thought, in the sublunary sphere. The unifying principle of this inquiry is the theory according to which rainbows and mock suns, rain, and the formation of minerals etc. are caused by the inherent nature of and also by the interaction between two exhalations (*anathumiaseis*) present in the sublunary world: one is dry and smoke-like and emerges when the soil is sufficiently heated by the sun; the other is moist and vaporous and is due chiefly to the evaporation of water. The former rises to become a layer in the proximity of the first celestial sphere and can be easily ignited by it. The latter is situated generally below the dry exhalation and is cooler. The two are normally combined, but one can predominate to various degrees in different regions.

Their behavior is described notably in mechanical terms,[10] often by reference to objects which are ejected under intense pressure. Hot air can be expelled from a certain region when the surrounding cooled air contracts, as one hurls a projectile (4.341b37–342a3, 8); it moves much as a fruit stone is ejected when squeezed between fingertips (4.342a10–11). To take another example from ch. 4 in the first book, to bolster his demonstration of what causes a shooting star, Aristotle wonders whether that process might not be comparable to an exhalation coming from below two lamps, one placed on top of the other. The exhalation causes "the lower lamp to be lit from the flame of the upper; the speed at which this happens is astonishing and is more similar to a projection than to fires being lit in succession" (342a2–7). If this was an experiment carried out by Aristotle, which is not unlikely but is impossible to ascertain, the two lamps and the interaction between them and their properties can be regarded as a rudimentary mechanism (and a sort of testing model for a natural phenomenon) which produces and explains the second flame, in the lower lamp. If so—or even if this is just a thought experiment, used as a model for what we might consider a mechanism—the natural process itself which is supposed to be thus illustrated and clarified, the production of shooting stars, is a kind of mechanism, with identifiable parts and specific activities. Aristotle did not use a technical term for such systems, but his analysis in *Meteorology* I–III of phenomena (the explananda) and of the objects and processes responsible for their generation and defining attributes seems to share some general concerns with modern treatments of mechanisms. His conceptual framework was decidedly different from ours; yet, his effort in that treatise to establish the common and fundamental causal principles of natural systems which could produce extraordinarily different phenomena is an exploit worthy of our admiration, even if many of his conclusions are spectacularly wrong.

Indeed, throughout the first three books of the *Meteorology* we find explanations based on the two exhalations and resorting consistently to thermal and mechanical processes. Cooling, condensation, increased pressure, combustion, and rapid ejection are the main elements of Aristotle's accounts, for instance, of comets and shooting stars (take, e.g., *Meteorology* I.7, where the movement and appearance of shooting stars are compared to what happens when one throws a lit torch into a large amount of chaff, 344a25–8). The quantity of dry or moist exhalation present in a certain region, and its shape and dynamics, are crucial to the production of phenomena which involve the display of particular colors, movements, and shapes rather than others. The Milky Way, therefore, does not look and behave like a comet, although the causes of the two phenomena are related, as we read in *Meteorology* I.8. Generally, modified conditions will be responsible for correspondingly altered phenomena. He makes this point plainly at I.4.341b23 ff. (see also *Meteor* II.9.370a29–33) where he notes that, under the most favorable conditions, a particular kind of smoky exhalation he calls "fire," *pur*, is ignited by the movement of the innermost celestial sphere. The exact outcome differs (*diapherei*), depending on the position (*thesin*) and on the quantity (*plēthos*) of the flammable mixture. If that amount of "fire" extends both broadwise and lengthwise, what we often see is a flame similar to the one produced when the stubble on a farmland is set on fire. If, on the other hand, the combustible material only extends lengthwise, the result will be "torches," "goats," and "shooting stars" (he then proceeds to distinguish the respective characteristics of these phenomena).

The behavior of the two sorts of exhalation constitutes a mechanism not just because of the nature of his analogies with various artifacts, but also and principally because of his explicit and consistent reliance on what will later become core elements of mechanics, including ballistics and pneumatics: trajectories of projectiles, relation between forces (or powers), the effects of increased pressure, etc. Besides, I would venture to suggest that his causal explanations can also be said to involve mechanisms roughly in the sense explored in this volume, even if their components (masses of exhalations, positioned, shaped, and interacting in specific ways and productive of corresponding phenomena) do not form very stable and conspicuously organized systems.

5. Aristotle's biology

Given the pride of place claimed by qualitative changes and also by formal and final causes in many[11] of his works on the science of animals, the rarity of analogies between the workings of living organisms and mechanical devices is no surprise. Such analogies do exist, however, and are relevant here. In ch. 7 of *On the Movement of Animals*, Aristotle invokes a number of *dunameis* or capacities of the soul, including sense-perception, desire, and *phantasia* (very roughly, capacity for mental representation), to explain how movement is initiated in animals. As a result of the operations of those *dunameis*, certain alterations (*alloiōseis*, 701b18) or *qualitative* changes (not reducible to rearrangements of particles, as the atomists would have held) take place and lead to movement. The original impetus for movement is an almost imperceptible change in the region of the heart, which can become warmer or cooler as a result of our desires or of our perception of imminent danger, etc. At 7.701b2–10, he compares the movement of animals to little carriages[12] ridden by children and to automatic mechanical devices (in Greek *automata*) which start moving as a result of a slight change but can subsequently sustain their movement in virtue of their internal organization. This is the case, we read, when some strings are released and wooden pegs in it strike one another. The analogy continues with a more detailed comparison between organic parts and components of automata (and perhaps of the toy carriages): the sinews and the bones correspond to the wooden pegs and the pieces of iron, while the sinews, which can be relaxed to allow movement, are analogous to strings, which can be loosened by a puppeteer. This analogy, then, works with respect to the interaction between (bodily or mechanical) parts, the initiation of movements by a slight original change responsible for a series of consecutive changes, and the similarity between genuine self-motion and the seemingly autonomous movement of automata. One should not lose sight, however, of the fact that Aristotle urges us to also keep in mind the striking difference between the two comparanda: mechanical gears do not need to undergo qualitative changes to perform their respective functions.

Another notable analogy is found in the second book of Aristotle's *Generation of Animals*, in chapters 1 and 5. Some of the difficulties contemplated and solved in the first chapter have to do with whether the semen itself contains any actual parts of the future embryo (the answer is categorically negative, since the matter for the embryo is contributed by the mother) and whether it is the father or rather the seed produced by the father that is responsible for informing the matter. (For current views on mechanisms and developmental biology, see Chapter 25.) Just as the father "moves" the sperm, the latter moves, or informs, the matter. To clarify this, Aristotle musters again the imagery of mechanical devices and notes that:

> Their parts, even while at rest, have in them somehow or other a capacity (*dunamin*), and when something from outside moves the first part, then immediately the next part comes to be in actuality. Hence, as with the automata, in one way that [external mover] moves it, not by being in contact with it anywhere now, but by having at one time been in contact with it; so too that from which the semen originally came, or that which made the semen [moves it], namely not by being in contact with it still, but by having once been in contact with it at some point. In another way, it is the movement within [which moves it], just as the activity of building causes the house to get built.
>
> (*Generation of Animals II.1.734b11–18*, trans. Peck, modified)

The automatic puppets (*ta automata tōn thaumatōn*; *thaumata* means literally "wonders" or "marvels") seem to be regarded as suggestive analogs, then, in so far as they can sustain their movement, following an initial impetus, and can, moreover, move in ways that differ significantly from the nature of that impetus. According to Henry's reading, which I find plausible, what is explained here is not the development of the embryo, but the moving capacity of the sperm. The analogy with an automaton is intended "to show how the sperm can continue to move once it is no longer in contact with the father *and* how the father can still be said to fashion the matter without being in contact with it" (2005: 31).

In ch. 5 (741b7–15) Aristotle turns his attention to the development of the fledgling organism. His main point is that the parts that will become distinguishable in the embryo are present in matter *potentially* and they begin to emerge when a principle of motion (*archē kinēseōs*), present in the semen, comes into play. When the matter has been "moved" in this manner, one stage follows another in a continuous process comparable with the movements of automatic wonders or marvels (*automatois thaumasi*). The coming about of the *actual* parts of the embryo, beginning with the heart, involves successive qualitative changes (741b12). The mention of changes with respect to differentiae such as softness, hardness, and color alerts us to the fact that the preceding analogy (signaling spatial, not qualitative, changes) should not be pushed too far, lest it become more misleading than elucidating. I take this implicit concern to reflect Aristotle's caveat in *On the Movement of Animals* 7, although there is also a potentially significant difference between the two texts with regard to his handling of analogies. Henry argues (2005: 38) that, unlike in *On the Movement of Animals*, the internal motion of the sort of automata brought up here, in *Generation of Animals*, "is the actualisation of a single potential rather than a causal sequence passing through a series of mechanical gears." A (hypothetical)[13] self-moving automaton would thus be a more appropriate analog for a developing embryo than a mechanical puppet whose activity is generated by an external principle and consists of a series of distinct events (causes and effects), not of different stages in a continuous change.

Aristotle's use of this analogy was to prove inspiring, if controversial, and was often discussed in antiquity and later on, but his attitude toward analogies with mechanical devices is quite ambivalent. They can call attention to the relation between an original impetus and self-motion and to the functions of and interactions between the parts of an organism, but they can also be misleading in so far as models based on mechanical processes only convey part of the story, and arguably not the most important one.[14]

6. Mechanica

The *Mechanics* (or *Mechanical Problems*, in Greek: *Mēchanika*; *Mechanica* in its Latin version) has been traditionally included in the Aristotelian corpus. While its authorship is dubious—if not spurious, it is not incompatible with the Stagirite's natural philosophy (as Heath suggested as early as 1921, vol. 1: 344–6). It seems safe to ascribe it to the Peripatetic school in the late fourth or early third century BCE, possibly to Strato of Lampsacus. What makes the *Mechanics* especially interesting is its application of geometrical explanations to physical phenomena and to their consequences for us (e.g. he points out that larger balances indicate weight with greater accuracy, 849b23, which is explainable in the final analysis by appeal to abstract models, such as concentric circles). *Mechanics* frames a host of problems—related to the lever (and, in smaller measure, to pulleys and windlasses, which turn out to illustrate the same principle)—in a way that emphasizes their mutual similarities and the author's belief that they can be expressed in a simple, unifying mathematical language. He stresses the dependence of mechanics both on mathematics (with respect to the "how," the method) and on "physics" or rather natural philosophy (with regard to the object of study, "that about which," 847a24–9). The manifest and sustained interest in tracking down what is common to a variety of physical phenomena and in revealing the philosophically significant points of this investigation makes this treatise a landmark in the history of mechanics, which has only some timid, unsystematic forerunners, judging by the extant evidence.

What distinguishes this text from some of the later and, in some cases, more impressive contributions to advances in mechanics is that, owing to the author's avowed interest in natural philosophy, this constitutes an early self-conscious attempt to articulate a theoretical treatment of what a mechanism (of a particular but very common kind, the lever) is essentially. The main topic of this work and some of the main questions it answers are summarized at the outset of ch. 3. The central paradox which needs to be elucidated in the *Mechanics* is that great weights can be moved by using a comparatively small force.[15] Here is the author's concise description of how a lever (*mochlos*) works:

Since under the same weight the greater radius from the center moves more rapidly, and there are three elements in the lever—the fulcrum, that is the cord[16] or center, and the two weights, the one which causes the movement, and the one that is moved—the ratio of the weight moved to the weight moving it is, then, the inverse ratio of the distance from the center. Now the greater the distance from the fulcrum, the more easily it will move. The reason has been given before that the point further from the center describes the greater circle, so that by the use of the same force, when the mover is farther from the lever, it will cause a greater movement.

(850a37–b7, translation by Hett (1936), with slight modifications)

As a mechanical principle, a lever can be instantiated, according to this work, in the arm of a balance, in the movement of an oar, in an aggregate of wheels in contact with each other, and, more abstractly, in the radius of a circle, or the radii of a set of concentric circles. Note, however, that this passage deliberately ignores the nature of the lever as an instrument which is meant to transmit some force; instead, the reader's attention is focused on the three aspects which, irrespective of the specific type of lever used and the specific conditions under which the phenomenon is produced, can reduce the apparent paradox stated earlier to ratios that are easy to grasp and to formulate in generic fashion. This principle governs an immense range of seemingly incongruent types of processes, from the functioning of scales in markets (chs. 1–3) and the force imparted by oars (chs. 4–5) to the ratio between the height of the yard-arm on a ship and the ship's speed (assuming a wind of constant strength, ch. 6), and from the movement of the potter's wheel (ch. 8) to the distance covered by a missile launched with a sling. Some of the illustrations offered there are about as delightfully or cringingly colorful as mechanics can ever aspire to get (ch. 21: Why is it easier for dentists to remove teeth with the help of a forceps than with their bare hands?). Even certain natural phenomena—for instance, pebbles being shaped by the waves—conform to the same basic explanatory scheme, which can be formulated in discarnate mathematical terms (the relation between circles, etc.). In short, this is a superbly clear and very influential[17] account of the fundamental (interacting) components of a type of mechanism that can be found under the guise of many categories of artificial and natural processes.

7. On the Cosmos

The two most pervasive analogies involving ancient references to technology are those pertaining (A) to the nature of the entire cosmos (see also my comments on Lucretius), and (B) to the structures and functions of living organisms, to their parts and generation (we have seen some early examples in Aristotle's biological works).[18] Such analogies became more frequent with the acceleration of progress in technology in Hellenistic and Roman times.[19]

Let me mention one final example—belonging to the first category and quite typical of post-classical thought in its systematic blend of tenets that can be traced back to different, not always compatible, doctrines. The short exhortative treatise entitled *On the Cosmos* was assigned to Aristotle, but was probably written at some point during the first century CE. It deals succinctly with the structure of the heavens, various regions of the earth, and especially with the relation between the world and its divine keeper (as well as his attributes). Its substance seems to be based on earlier scientific compendia and betrays an eclectic philosophical background, borrowing especially from Aristotle's cosmology and "meteorology."[20] The universe as a whole is imperishable (4.397b8), but the region around the earth is eminently a domain of change. In ch. 5, the author claims that the overall harmony of the world depends on the mutual balancing of opposites, an idea he explicitly attributes to Heraclitus. The interactions between the powers of the elements as well as between the exhalations are responsible for the, admittedly imperfect and fluid, order of the sublunary world. This order, however, is ultimately preserved by a god who acts remotely—from the outermost celestial sphere—through successive powers, *dunameis*, which relay his plans with diminishing degrees of accuracy; the earth is, accordingly, a far less glorious example of harmony than the heavens.

We can find in this theology and teleology echoes, distorted though they might be, from Aristotle, among others. To make the divine portrait more vivid and to justify this god's aloofness, while also arguing for his causal efficacy, the author uses several analogies. After comparing the divinity with the Persian Great King, he embarks on a double analogy of the universe with a piece of machinery (Furley 1955: 390–1, note a, argues it might be a catapult or a ballista; cf. Thom 2014: 117) and with a puppet (an analogy rather reminiscent of several passages in Aristotle's biological corpus—see above). The divinity is able

> to produce all kinds of forms with ease and a simple movement, just as indeed the engineers (*megalotechnoi*) do, producing by means of the single release mechanism of an engine of war many varied activities. In the same way puppeteers, by pulling a single string, make the neck and hand and shoulder and eye and sometimes all the parts of the creature move with a rhythmical movement.
>
> *(Trans. Thom 2014: 45)*

The image of a mechanical device is also evoked by the more dynamic notion of the transmission of change[21] along an assemblage of components and conveying powers, which seem to be semi-autonomous once the ruler and begetter of all things (399a31) provides his initial impulse. What is vigorously emphasized in *On the Cosmos*, therefore, is that the divine keeper does not have to tend to every single component of the cosmic system and to direct every stage in its operation. We have here the picture of a world that, while imbued with divine rationality, is capable of producing its own phenomena, even if the long and complex causal chains behind them lead ultimately beyond nature itself.

8. Conclusion

Recent debates about the essential features of mechanisms and about the relation between mechanisms and laws of nature cannot find, of course, any close antecedents in the texts and theories just surveyed in this chapter. We can track down, nonetheless, a few attempts to causally and comprehensively explain natural events in terms of the interplay among the component parts of some natural system and the activities inherent in it. For the authors discussed here, the discovery of causes and their handling of explanations follow sometimes radically different methods. They investigate systems—mechanisms *avant la letter*—of varying degrees of complexity (instantiations of the lever principle, for instance, are simpler than the formation, structure, and operations of organisms).

Some treatments of explanatory mechanisms are quite self-conscious and crisply articulated; others are vaguer and fairly tentative. Attitudes concerning the appeal to technological models and analogies meant to represent what we would call mechanisms were determined, sometimes in complicated ways, by various philosophical doctrines (natural or providential teleology; distinct approaches to more particular problems, such as the nature of processes characteristic of embryogenesis, etc.). A host of other, more elusive, factors too must have contributed to the shaping of those attitudes: evolving concepts, shifting cultural contexts, polemical stance, and even literary tastes.

There is no robust unity, then, that we can hope to find among these possible and distant precursors of our conception(s) of mechanisms. A few general common points, however, can be identified. The texts examined here tend to mark the phenomena to be explained[22] before distinguishing entities (e.g., uncuttables, exhalations, simple and complex parts of organisms) and activities (collision, uncaused swerve, friction, compression, expulsion, slight "movements" in the region of the heart, relaxation or tightening of sinews, etc.) which, when organized in the right ways, are conducive to changes, often to regular changes. New phenomena can be generated if the conditions—i.e. the organization of the entities making up a mechanism and their activities—change. The atomists' natural philosophy and Aristotle's "meteorology," for example, richly illustrate this aspect. Most of those accounts tend to elucidate: (1) the relation between some original impulse and genuinely or apparently autonomous activity either in the case of individual organisms or of automata or even of the entire cosmos (whose order is ensured remotely by a transcendent divinity); (2) implicitly, the transmission and persistence of the force impressed by that impetus with a higher or lower degree of accuracy; and (3) the interaction between parts: minimal material particles and aggregates of particles, continuous masses of exhalations, celestial spheres and the material constituents of the sublunary world—all this under conditions that hinge on mechanical processes (foreshadowing or following central tenets in ballistics and pneumatics).

Beyond this, we would be hard pressed to find a consensus on what a mechanism is in general. Such dissonance or rather polyphony, however, should be no cause for disappointment. After all, the search for full clarity and consensus is still underway, as this whole volume bears witness.[23]

Notes

1 Democritus probably followed Leucippus quite faithfully with regard to his physical doctrine, which he further developed; Aristotle and several late ancient Aristotelian commentators, among others, often mention Leucippus and Democritus in the same breath.

2 My references to fragments from and testimonia about Democritus follow the Diels-Kranz numbering system.

3 Taylor (1999: 194) believes that there is some evidence that Democritus invoked several types of—attractive and repulsive—forces. Gregory (2013: 449–54) raises a number of serious objections to this interpretation.

4 Epicurus also departed from Democritean physics by rejecting the notion that the atoms are of infinitely many sizes and shapes.

5 *Clinamen*, translating the Greek *paregklisis*; II. 216–93 in Lucretius' *De rerum natura*.

6 Epicurus rejects the notion that everything is ruled by fate in his *Letter to Menoeceus*, 133–4.

7 Bailey (1966: 1335–6): "[A]gain a careful expression; the *mundus* is not only a mass of matter, *moles*, but a complex construction, *machina*."

8 Aristotle also virulently criticizes Plato's "geometric atomism," marred by absurdities like assigning weight to bodies analyzable into planes (the two types of triangles posited in *Timaeus* 53c ff.). Plato's image of the cosmos as a whole is that of a living being (a reflection of its transcendent model, the sum total of ideas), rather than of a mechanical device.

9 There is a mention in *Posterior Analytics* (I.13.78b37) of mechanics as a scientific domain subordinate to stereometry, but he does not elaborate on it. Modern references to Aristotelian "mechanics" generally point to his discussion of pushing, pulling, and the bodies' resistance to such actions and forces (especially the ship-hauling demonstration and the rare quantitative formulas conveying the ratio between velocity and force in *Physics* VII.5), and to his comments on impetus, the movement of a projectile after its release and the presumed role of the medium in sustaining that movement.

10 For a detailed analysis of the quasi-technical terminology in *Meteorology* I–III, see Wilson's section on "Aristotle's lexicon of mechanics" (2013: 65–70).

11 Many, not all: his natural teleology is virtually absent in treatises like *History of Animals*, where he studies morphological and other differentiae ("the fact") to prepare the terrain for inquiries into the *causes* of those differentiae ("the reason why," the cardinal feature of other works, including *Parts of Animals* and *Generation of Animals*).

12 The toy carriages tend to move in a circle because the wheels on one side are smaller than on the other side. Nussbaum (1978: 347) notes that "The difference between the two examples seems to be primarily one of emphasis: the puppet example underlines the generation of a whole series of motions from a single initial motion, the cart example the change in character or a motion because of the nature of the functioning mechanism."

13 Henry (2005: 38–40) believes that the two passages in *Generation of Animals* present a thought experiment and could not refer to an actual device.

14 Recently, there have been some very interesting attempts, however, to demonstrate that Aristotle's biology relies to a more significant extent on the application of mechanical principles: De Groot (2014 (especially 107–62)) and Johnson (forthcoming).

15 Or: power; I am not implying any unduly modern connotations here.

16 I.e. the cord on which, for instance, the beam of a balance is suspended.

17 Its impact is evident not just in later ancient works on mechanics, but also at the dawn of modern science, when this work was widely read.

18 On these two categories of analogies, see, among others, De Solla Price 1964 and Berryman 2009. A few remarkable examples of analogies between biological processes and the functioning of automata can be found e.g. in Galen's *On Seed* I.5 and in his *On the Development of the Fetus* 6.3–4; for useful clarifications of these passages and of their connections with Aristotle's own analogies (discussed in this chapter), see Preus 1977 and Berryman 2002.

19 We have both reports of geared planetaria, complex water-clocks, automata meant to simulate not just the appearance, but also aspects of the behavior of human beings and of non-human animals, and tantalizing archeological evidence (the Tower of Winds in Athens, the Antikythera mechanism, etc.). Nonetheless, the fragmentary nature of the evidence hampers any attempt to outline a coherent history of ancient mechanisms, to properly reconstruct many of them and to attribute their invention or production with reasonable confidence. Berryman argues that, unlike mythical or imaginary examples (e.g. the statues of Daedalus), "real devices offer evidence of the kinds of results that can be achieved by mechanical craftsmanship. Because it seemed that technology could replicate some of the kinds of features taken to be distinctive of living things, the argument that there is a distinction in kind between living and nonliving things, and in the *explanantia* required for the former, faced a formidable challenge" (2003: 365–6).

20 See his discussion of the two exhalations (*anathumiaseis* 4.394a9 ff.). In fact most of ch. 4 is a précis of long tracts of Aristotle's *Meteorology* I–III.

21 The author points out that the same impulse can lead to different outcomes, depending on the constitution of the thing being moved: "It is as if one would throw a sphere, a cube, a cone and a cylinder from a vessel at the same time—for each of them will move according to its own shape—or if one would have a water animal, a land animal and a bird in the folds of one's cloak and throw them out at the same time." Trans. Thom (2014: 47).

22 A clear statement of the necessity of proceeding from "what" to "the reason why" can be found e.g. in Aristotle's *Meteor.* III.2.371b18–22.

23 I would like to express my gratitude to Phyllis Illari, Stuart Glennan, Andrew Gregory, and Monte Johnson for their helpful comments on an earlier version of this chapter.

References

Bailey, C. (1966, reprinted) *Titi Lucreti Cari De rerum natura libri sex*, vols. I–III, Oxford: Clarendon Press.

Berryman, S. (2002) "Galen and the Mechanical Philosophy," in *Apeiron: A Journal for Ancient Philosophy and Science*, 35, 235–53.

—— (2003) "Ancient Automata and Mechanical Explanation," *Phronesis*, XLVIII, 4, 344–69.

—— (2009) *The Mechanical Hypothesis in Ancient Greek Natural Philosophy*, Cambridge: Cambridge University Press.

Craver, C. and Tabery, J. (2015) "Mechanisms in Science," in the *Stanford Encyclopedia of Philosophy*: plato.stanford.edu/entries/science-mechanisms.

De Groot, J. (2014) *Aristotle's Empiricism: Experience and Mechanics in the 4th Century BC*, Las Vegas: Parmenides Publishing.

De Solla Price, D.J. (1964) "Automata and the Origins of Mechanism and Mechanistic Philosophy," *Technology and Culture*, V, 1, 9–23.

Furley, D.J. (1955, reprinted) *Aristotle, On the Cosmos*, Cambridge, MA: Harvard University Press.

Graham, D.W. (2010) *The Texts of Early Greek Philosophy*, Part I, Cambridge: Cambridge University Press.

Glennan, S. (2013) "Mechanisms," in Martin Curd and Stathis Psillos eds., *The Routledge Companion to Philosophy of Science*, London: Routledge, 420–8.

Gregory, A. (2013) "Leucippus and Democritus on Like to Like and *ou mallon*," *Aperion*, 46, 4, 446–68.

Heath, T. (1981 [1921]) *A History of Greek Mathematics*, vol. 1 and 2, New York: Dover Publications.

Henry, D. (2005) "Embryological Models in Ancient Philosophy," *Phronesis*, L, 1, 1–42.

Hett, W.S. (1936) *Aristotle: Minor Works*, London: Harvard University Press.

Johnson, M.R. (2009) "Spontaneity, Democritean Causality and Freedom," *Elenchos*, XXX, 1, 5–52.

Johnson, M.R. (forthcoming) "Aristotelian Mechanistic Explanation," in J. Rocca ed., *Teleology in the Ancient World: Philosophical and Medical Approaches*, Cambridge: Cambridge University Press.

Nussbaum, M. (1978) *Aristotle's* De Motu Animalium, Princeton, NJ: Princeton University Press.

Preus, A. (1977) "Galen's Criticism of Aristotle's Conception Theory," *Journal of the History of Biology*, X, 1, 65–85.

Sambursky, S. (1962) *The Physical World of Late Antiquity*, Princeton, NJ: Princeton University Press.

Taylor, C.C.W. (1999) *The Atomists: Leucippus and Democritus, Fragments*, Toronto: University of Toronto Press.

Thom, J.C. (2014) *Cosmic Order and Divine Power. Pseudo-Aristotle, On the Cosmos*, Tübingen: Mohr Siebeck.

Wilson, M. (2013) *Structure and Method in Aristotle's* Meteorologica: *A More Disorderly Nature*, Cambridge: Cambridge University Press.

3

FROM THE MECHANICAL PHILOSOPHY TO EARLY MODERN MECHANISMS

Sophie Roux

Early modern natural philosophers put forward the ontological program that was called "mechanical philosophy" and they gave mechanical explanations for all kinds of phenomena, such as gravity, magnetism, the colors of the rainbow, the circulation of the blood, the motion of the heart, and the development of animals. For a generation of historians, the mechanical philosophy was regarded as the main alternative to Aristotelian orthodoxy during the so-called Scientific Revolution and mechanical explanations were presented as paving the way for the use of experiments and mathematics in the understanding of natural phenomena.

However, the historical category of mechanical philosophy was later criticized as being too broad, while early modern mechanical explanations were condemned by more epistemologically oriented minds for being incompatible with, or at least not necessarily connected to, the use of experiments and mathematics. In the last ten years, just as the new mechanistic literature emerged in philosophy of science, there has been a reevaluation of early modern mechanical explanations in a domain that had been until then considered peripheral to the so-called Scientific Revolution, namely the domain of biology, anatomy, physiology, and medicine. Although they were neither confirmed nor predictive, some early modern explanations in these domains appear to have a cognitive value similar to the value of contemporary mechanistic explanations.

1. Establishing mechanical philosophy

Boyle is often said to have coined the term "mechanical philosophy." It would be more exact to say that he was the first to use this term to advertise the general program of explaining all natural phenomena by matter and motion alone. There were indeed some earlier uses of the term. After he read *Meteors* and *Dioptrics*, Libert Froidmont complained in a letter from 1637 that Descartes fell too often into the "coarse and somewhat bloated (*ruda & pinguiscula*)" Epicurean physics; he noted in particular that the reduction of the Aristotelian elements to small parts of various figures was "too gross and mechanical (*nimis crassa & mechanica*)." In his answer to Froidmont, Descartes allowed his philosophy to be qualified as "mechanical," but he reversed the negative connotations associated with this adjective and insisted that his physics manifested a certitude similar to the certitude of mathematics only in as much as it was mechanical. Froidmont had only to see the numerous problems that Descartes was able to solve to understand that there was absolutely no reason to condemn his "bloated and mechanical philosophy (*pinguiscula & mechanica philosophia*)" (Descartes, 1964–74, vol. I, pp. 402, 406, 420–1, discussed in Roux, 1996, pp. 15–17, Gabbey, 2004, pp. 18–20, and Roux, 2004, pp. 32–4).

Just a few years after Descartes' death, Henry More dubbed him "the great Master of this Mechanical Hypothesis" (1653a, p. 44). Later on, More recommended "that admirable Master of Mechaniks *Des-Cartes*" to be taught in universities to make the students see the limits of mechanical explanations and to enable them to beat hollow the "pretender to Mechanick Philosophy" (1659, n.p., discussed in Gabbey, 1982, pp. 220–2). These occurrences of the term "mechanical philosophy" are interesting, but they should not be equated with the establishment of mechanical philosophy. Descartes had a positive program for reducing all natural phenomena to matter and motion alone (Descartes, 1964–74, vol. VIII, pp. 314–23, vol. XI, p. 47, *passim*), but he was not ready to include in this program those who had other conceptions of matter and motion than his own. As for More, he did not consider mechanical philosophy as a program that would have delineated how to frame physical explanations in the years to come; for him, Descartes' explanations were essentially incomplete, and their gaps gave evidence that some kind of immaterial substance should be introduced, whether it be God or the Spirit of Nature (1653b, pp. 42–7 and 1659, pp. 193–204, discussed in Gabbey 1990b, pp. 22–5 and 30–1).

The story is different in *Some specimens of an attempt to make chymical experiments useful to illustrate the notions of the corpuscular philosophy* that Robert Boyle wrote in the late 1650s and published in his *Certain Physiological Essays* (1661). In the Preface, Boyle insisted that, notwithstanding the differences between their philosophies, Descartes and Gassendi agreed upon two things. First, contrary to Aristotelians and to chemists, they intended to "deduc[e] all the Phenomena of Nature from Matter and Local Motion." Second, they proposed to stand up for the defense of Christian religion. At this point, Boyle needed a term to baptize the program that both Descartes and Gassendi would have shared. He actually proposed several ones: it could be called "Corpuscular," because it explains natural phenomena through corpuscles, "Phoenician," because it its inventor is supposed to be the Phoenician Moschus, or, still, "Mechanical," because it gives an account of phenomena by motion, which is "obvious and very powerfull in Mechanical Engines" (1999–2000, vol. II, p. 87, discussed in Roux, 1996, pp. 18–20). This ontological program was an important program, even if it was only one of Boyle's two programs for natural philosophy, the other one being "experimental philosophy"; that is, the epistemological program to acquire and secure knowledge of nature through observations and experiments (1999–2000, vol. II, p. 14, vol. III, p. 12, *passim*; see Sargent, 1994, 1995; Chalmers, 1993, 2012; Gaukroger, 2006, pp. 352–99; Anstey and Vanzo, 2012, 2016; Anstey, 2014). Boyle was not sparing his words; to designate this ontological program, he spoke also of the "New," the "Real," or the "Atomical" philosophy (1999–2000, vol. XI, p. 292). But his most common terms are "the corpuscular (or corpuscularian) philosophy (or hypothesis)" and "the mechanical philosophy (or hypothesis)," the first emphasizing the kind of entities that appear in explanations (corpuscles), the second the kind of activities that these entities are engaged in (motions).

In *The Origin of Forms and Qualities according to the Corpuscular Philosophy* (1666), publicized by Henry Oldenburg in the *Philosophical Transactions* as "a kind of *Introduction* to the Principles of the *Mechanical Philosophy*" (1666, p. 191), Boyle identified positively the eight tenets that are constitutive of this ontological program:

1)   There is only one universal matter, which is extended, divisible and impenetrable.
2)   The diversity between bodies comes from various affections of matter, the main of these affections being local motion.

3) Motion divides matter into parts of various sizes and figures.
4) Between several parts of matter, there are relations of order and situation, which determine the texture of a body.
5) Parts of matter in motion make some impressions on the senses of animals.
6) The sensible qualities that are associated to these impressions are nothing but the effects of matter in motion on the senses.
7) Substantial forms are nothing but the names given to aggregates of sensible qualities.
8) Generation, alteration and corruption are nothing but the names given to transformations of matter in motion. (1999–2000, vol. V, pp. 305 *sqq.*)

Referring to those that Boyle himself designated as the two founding fathers of mechanical philosophy, to wit Descartes and Gassendi, will make clear that the mechanical philosophy was large enough to include various interpretations of most of these tenets.

Regarding 1, contrary to Aristotelian matter, which is a pure power, matter is a true substance (Descartes, 1964–74, vol. XI, p. 33; Gassendi, 1658, vol. III, p. 636b). Being "universal," matter is everywhere one and the same: it is a "homogeneous" or "uniform" substance (Descartes, 1964–74, vol. III, pp. 211–12, vol. VIII, p. 52, vol. XI, p. 17). Its only properties are extension, divisibility, and impenetrability. That bodies are extended was granted by all natural philosophers; the only question was if extension was the only essential property of matter. By mentioning "divisibility," Boyle seems to adhere to Descartes' view, according to which there are no atoms because any part of matter can be divided further away (1964–74, vol. III, pp. 191, 213–14, 477, vol. VI, pp. 238–9, vol. VIII–1, pp. 51–2, 58–9). Still, Descartes' view was that matter is *infinitely* divisible, while Boyle claimed to remain neutral with respect to this question (1999–2000, vol. VIII, pp. 103–4). In fact, since Boyle admitted that there are *minima* or *prima naturalia* that are not naturally divided, though they could be divided in thought or by God (1999–2000, vol. V, pp. 325–6), except for the words that he uses, he subscribes to Gassendi's view, according to which there are atoms that have parts but that cannot be divided by any natural force (1658, vol. I, p. 256b). Similarly, putting on a par extension and impenetrability (that is, the property that a body has of excluding any other body from its location) is truer to Gassendi than to Descartes: Gassendi presented, in addition to gravity or weight, extension and impenetrability as two equally fundamental properties (1658, vol. I, pp. 55a, 267a), while Descartes asserted that impenetrability is to be derived from extension (1964–74, vol. V, pp. 269, 341–2, vol. XI, p. 33).

Regarding 2 and 4, because of its homogeneity and uniformity, matter is not sufficient in itself to account for the variety we see in natural phenomena. This variety comes from the different shapes, sizes, and motions of parts of matter and from their relationships (Descartes, 1964–74, vol. VIII–1, pp. 52–3, vol. XI, p. 34; Gassendi, 1658, vol. I, pp. 366a–71b).

Regarding 3, here, Boyle adheres to Descartes' view, according to which motion is prior to shape and size because it is responsible for slicing matter into corpuscles of various sizes and shapes (1964–74, vol. XI, p. 34). A related point is that, for Descartes, motion does not belong to matter in itself, while, for Gassendi, matter is not inert but active, weight being a principle of motion internal to atoms as such (1658, vol. I, pp. 276b, 335b). In that respect, Boyle goes along with Descartes: against Gassendi and the Ancient atomists, he opposes the opinion that motion is essential to matter, because it would pave the way to atheism (1999–2000, vol. V, p. 306).

Regarding 5 and 6, since there are only corpuscles in motion in the world, the various qualities that animals perceive cannot but be caused by these corpuscles in motion. While Descartes did not try to give specific names to combinations of corpuscles, Gassendi and Boyle suggested that there are intermediate levels between the lowest atomic level and the highest level, which are accessible through observations and experiments; they called these intermediate levels respectively *semina rerum* or *molecula* (Gassendi, 1658, vol. I, pp. 282b, 472a, 335b) and textures or primary concretions (Boyle, 1999–2000, vol. V, pp. 333–4).

Regarding 7 and 8, mechanical philosophy is an eliminative reductionism that aimed at replacing the Aristotelian ontology. Considering the variety of ways in which the previous tenets were interpreted, the rejection of Aristotelian ontology was probably the only common denominator between early modern mechanical philosophers.

As he himself acknowledged, Boyle did not create this ontological program out of nothing: in the first half of the seventeenth century, there had been attempts to replace the dominant Aristotelian natural philosophy with a natural philosophy formulated in terms of matter and motion alone. But, possibly because of his own "reconciling Disposition," or because of a larger Latitudinarian context, Boyle was one of the first authors to propose to put aside the differences that were important to the members of the former generation and to offer to the members of the generations to come to create a common program in which to inscribe their works (Garber, 2013). Thus, the resemblances between Descartes and Gassendi were brought to the forefront and their differences were pushed into the background. These were metaphysical differences that concerned God and the soul, most of which were violently expressed in the controversy that went through Descartes' *Meditationes*, Gassendi's *Fifth Objections*, Descartes' *Fifth Replies*, Gassendi's *Disquisitio metaphysica*, and finally Descartes' letter to Clerselier of 12 January 1646 (Lennon, 1993; Lolordo, 2007; Osler, 1994; Roux, 2008). There were also differences concerning the very principles of natural philosophy, as we just noted (Lennon, 1993; Roux, 2000).

More important for us here, there were what we perceive today as epistemological differences between types of explanations (Roux, 2009, 2012). Consider, for example, bodies that fall on the Earth. This had to be explained, because gravity or attraction were considered as mere words, action at a distance being proscribed in favor of action by contact. According to Gassendi, the fall of heavy bodies is caused by the conjunction of two external forces, on one hand the pushing force of the air from above, on the other the force of some magnetic corpuscles emitted by the Earth that pull bodies down to the Earth as small insensible ropes or hooks would do (1658, vol. III, pp. 489–96). According to Descartes, the fall of heavy bodies follows from the impossibility of the void and from his third law of nature, according to which every body, when moving circularly, tends to recede from the center of its motion, a tendency which is greater for swifter and smaller bodies. Void being impossible, the smaller and swifter bodies would not recede from the center if the bigger and slower bodies did not approach the center, which means that these bigger and slower bodies go down (1964–74, vol. VIII, pp. 213–17; the explanation is slightly different in *Le Monde* (vol. XI, pp. 72–80), but its kingpin is the same law of nature and the impossibility of the void).

Let us call Gassendi's explanation "corpuscular," in so far as the burden of explanation lies on the properties of insensible magnetic corpuscles, and Descartes' explanation "nomological," because the burden of his explanation lies on a law of nature. The distinction between corpuscular and nomological explanations does not exhaust the different types of mechanical explanations, though. There were also explanations that consisted in exhibiting the causal interactions of parts of more or less complex structures. This is the type of explanation that was put in place when animals were compared with machines that can be decomposed in parts that act on each other. For example, Perrault suggests that the influx of animal spirits into a muscle causes its release, which itself causes the contraction of the antagonistic muscle, just as the release of the guy supporting a mast causes the slackening of the opposed guy (Perrault, 1680, pp. 75–7, discussed in Des Chene, 2005; Roux, 2012). Considering the importance that comparison with machines had for this third type of explanation, it could be called "machinical."

To avoid neologisms, but also to comply with the distinction proposed by Ernan McMullin (1978, discussed below), let us call these "structural" explanations.

Partly because of Boyle's position at the Royal Society and partly because the ontological program of explaining natural phenomena in terms of matter and motion alone was at this time already well admitted by a number of natural philosophers, the terms "corpuscular philosophy" and "mechanical philosophy" were successful, but not always with the same connotations. Among the members of the newly founded Royal Society, the adjectives "mechanical" and "experimental" were paired together, as if each of them contributed to make the meaning of the other more precise. Henry Power presented himself as an "experimental and mechanical philosopher" (1664, p. 193); Robert Hooke contrasted "the *real*, the *mechanical*, the *experimental Philosophy*" and "the Philosophy of discourse and disputations" (1665, Sig.a2r); Samuel Parker explained that he preferred "the Mechanical and Experimental Philosophy" to the Aristotelian philosophy (1666, p. 45). In contrast, on the Continent, even those who thought that everything occurs mechanically did not often use the term "mechanical philosophy"; when they did, it referred to the demand that physical explanations are formulated in intelligible terms and connected to the mathematical sciences.

It was only by reference to Boyle and the Royal Society that in this period the young Leibniz happened to speak of "corpuscularians" and of "corpuscular philosophy" (1923, vol. VI–1, pp. 489–90, VI–2, pp. 325, 327), a term that he used at least once as equivalent with "mechanical philosophy" (1923, VI–2, p. 325). Still, he already called "mechanical" what does not rely on the supposition of fictitious entities, but on clear and simple terms (1923, vol. II–1, pp. 266, 284, 287, 372, 379, 393). From the end of the 1670s onward, Leibniz continued to occasionally use "corpuscular philosophy" (1923, vol. II–2, pp. 396, 845, VI–4, p. 477), but the more frequent terms from his pen were "mechanical philosophy" and related words like "mechanism," that he used in association with laws and mathematics (1923, vol. II–2, p. 172, VI–4, pp. 485, 1559, 1566, 2009, 2118, 2342). The association of the mechanical with what is clear and intelligible, if not with mathematics, is also pregnant in the eulogies that Bernard Le Bovier de Fontenelle wrote as secretary of the *Académie des sciences*. Speaking of a chemist, he rejoiced that "the sound philosophy . . . undertook to reduce this mysterious chemistry to the simple corpuscular mechanics" and he defined the corpuscular philosophy as "the one where only clear ideas are admitted, that is figures and motions" (1740, vol. II, pp. 444, 250, 322).

Thus, because of the variety of these connotations, "mechanical philosophy" changed from a term for an already large ontological program into a kind of a broad umbrella term covering several programs associated with the erosion of Aristotelian natural philosophy and with the development of the new sciences (Roux, 1996). The first historians who promoted mechanical philosophy as a key component of the Scientific Revolution took over this broad umbrella. In her influential paper from 1952, Marie Boas identified, among many others, Descartes, Boyle, and Newton as the three main early modern mechanical philosophers who were able to "put atom to work" (1952, p. 540). She suggested a progression from each to the next one: Descartes would have offered a theory of matter accounting for many unexplained physical properties, Boyle would have expanded it thanks to his brilliant experiments, Newton would have put the emphasis on the mathematized law of attraction and, if he "inclined towards the non-mechanical one for want of experimental evidence," "he would have preferred a mechanical explanation" (1952, p. 519). For her, the early modern mechanical philosophy in its full form consisted in adapting the ancient atomism to the new science of mechanics, characterized by the use of experiments and mathematics (1952, pp. 414, 426, 520–1).

Another example of these first promoters of the historiographical category of the mechanical philosophy is Eduard Jan Dijksterhuis. According to the opening words of his monumental

Mechanization of the world picture, first published in Dutch in 1950 and then translated into several European languages, the emergence of "the conception of the world usually called mechanical or mechanistic" had more profound and far-reaching effects than any other conception of the world (1961, p. 3). In a word, these historians described the ontology of matter and motion as necessarily implying a certain access to nature (that is, through observations and experiments) as well as a certain development of knowledge (thanks to mathematics). For them, the mechanical philosophy was thus essentially linked to the two most eminent characteristics of the Scientific Revolution and, beyond that, of modern science; that is, experimentation and mathematization.

2. Challenging mechanical philosophy

Still, questions about the delineation of mechanical philosophy emerged and the cluster of associated programs began to come apart. Of course, some of these programs were *sometimes*, in *some* places, and in *some* respects associated, but a natural philosopher engaged in one of them did not necessarily engage in all the others. Already in *The Mechanization of the world picture*, Dijksterhuis made a distinction between mechanics as the science of motion and mechanics as the theory of machines; in his conclusion, he pointed out that what led to the mechanization of the world picture was not the theory of machines, but rather a science of motion that should be characterized as a "mathematism" rather than as a "mechanism" (1961, pp. 4, 498). In his short textbook, *The Construction of Modern Science, Mechanisms and Mechanics*, which has achieved an enduring success since its first publication in 1971, Richard Westfall dissociated two traditions that had been conflated by Boas: the mechanical philosophy, which "conceived of nature as a huge machine and sought to explain the hidden mechanisms behind phenomena" and the Platonic-Pythagorean tradition, which developed in the mathematized science of mechanics. According to him, the tension between these two conflicting traditions was resolved only with Newton, which is to grant some significance to mechanical philosophy (1977, pp. 1, 36, 42, 120). But when Westfall comes to the point of explaining what exactly the mechanical philosophy brought to sciences such as optics, chemistry, and biology, he presents it as a language, an idiom, a facade, a robe, or even a puppet regime (1977, pp. 41–2, 50, 56, 73, 77, 94, 104). This amounts to saying that mechanical philosophy was something external to the sciences. Introducing Gassendi, Westfall went as far as speaking of "the occupational vice of mechanical philosophers, the imaginary construction of invisible mechanisms to account for phenomena" (1977, p. 41, see also pp. 1, 56).

It was only, however, in the 1990s that the category of mechanical philosophy was systematically debunked from a historical point of view. It was pointed out that to regard the mechanical philosophy as the principal alternative to Aristotelian orthodoxy was a significant oversimplification of the historical situation. Learned studies emphasized that the great authors owed more to the Scholastics than they had been willing to admit, and for the lesser authors, that they had at times taken such complex positions that the division between the mechanical and the non-mechanical philosophies became impossible to draw in practice, all the more so given that the Aristotelian doctrine of *minima naturalia* helped the development of some corpuscularianisms (Leijenhorst, 2002; Lüthy, 1997, 2001a, 2001b; Murdoch, 2001). It had been known for twenty years that various magical and hermetic philosophies were available at the time (Rhigini Bonelli and Shea, 1975; Westman and McGuire, 1976), but only in the 1990s were the alchemical roots and active principles in some early modern corpuscularianisms recognized (Clericuzio, 1990, 2000; Henry, 1986, 1989; Newman, 1994, 2001, 2006; Principe, 1998). In addition, it was pointed out that the first generation of the so-called mechanical philosophers did not identify themselves as belonging to a common program, but rather as competitors in the search for new philosophies intended to replace the philosophy of the schools (Gabbey, 2004; Garber, 2013).

In short, the category of mechanical philosophy came to be thought of as lacking any real historical significance, except when applied to authors like Boyle who explicitly defined it. It was particularly feared that it represented a bad partition of seventeenth-century natural philosophies, ill-suited for drawing out their inexhaustible richness and their fast transformation. While I concur with the historical criticisms that have been addressed to the obviously too broad category of mechanical philosophy, in this chapter I want to focus on the epistemological grounds that have been used to scrutinize mechanical explanations. For the sake of clarity, I will distinguish a *de jure* evaluation that bears on mechanical explanations in general from a *de facto* evaluation that affects some mechanical explanations only.

The *de jure* evaluation was based on a normative notion of science, according to which something deserves to be positively evaluated if and only if it complies with certain well-established standards; that is, in the case of science, the use of mathematics and the recourse to experimental proofs. For example, Gaston Bachelard famously said that Descartes' physics "should be left in its historical solitude," because it would be a physics "where objects are not measured, a physics without equations, a geometrical representation without scale, without mathematics" (1951, p. 35). Similarly, with regards to the recourse to experimental proofs, Alan Chalmers argued that Boyle's successes were "achieved in spite, rather than because, of his allegiance to mechanical philosophy" (1993, p. 541; and again, 2012). Contrary to the first historians who promoted the category of mechanical philosophy, Bachelard and Chalmers described the link between mechanical explanations on the one hand and mathematization or experimentation on the other as a kind of union against nature: as such, mechanical explanations would stay outside of proper science—they would even constitute an obstacle to the achievement of scientific norms. If mechanical philosophy was saved, it was not for its explanatory successes, but only because it functioned as a unifying discourse that legitimized scientific practices (Gaukroger, 2006, pp. 253–5, 260, 397–406).

There were also *de facto* criticisms addressed to mechanical explanations. In this case, the objector did not say that they were flawed in principle and that they were doomed to be false, but in a more descriptive way noted that some of them happened to be vacuous. Ernan McMullin characterized structural explanations as obtaining when the properties or behavior of a complex entity are explained by the structure of that entity; that is, by "a set of constituent entities or processes and to the relationships between them." Moreover, he noted that, by contrast with nomological explanations, such explanations are causal, the *explanans* being a structure that causes the *explanandum* (McMullin, 1978, pp. 139, 145–7). Relying on the notion of structural explanation, Alan Gabbey proposed to consider mechanical explanations as a kind of structural explanations. (This is obviously a restriction; if mechanical explanations are defined as those that are formulated in terms of matter and motion alone, they also include corpuscular and nomological explanations, as was noted in the first part of this chapter.) This led him to formulate a nuanced judgment on mechanical explanations.

Some of them were indeed successful when the phenomenon at stake was represented by a working model which instantiated a physical law or property or still a more familiar reality (the distinction between the two cases is presented as one of degree in Gabbey, 2001, p. 454). It may happen that the working model at stake leads to a falsification of the explanation— for example, Newton falsified the explanation that Descartes proposed of the motion of the planets around the Sun by comparison with the motion of small bodies in a basin of swirling water—but this, being only the usual game of science, does not mean that the explanation was flawed in principle (Gabbey, 1985, pp. 11–12; 1990a, pp. 274–86; 2001, p. 453).

But, continues Gabbey, not all mechanical explanations were successful in this sense: most notably, the explanations that pretended to account for sensations in terms of "adequate" or "appropriate" corpuscles and motions failed because they turned out to be tautological or circular. For example, Walter Charleton writes that the "Odours . . . can never be explained, but by assuming a certain *Commensuration*, or Correspondency, betwixt the Particles amassing the Odour, and the Contexture of the Olfactory Nerves" (quoted and discussed in Gabbey, 1985, pp. 12–14, 1990a, pp. 278–82, 2001, pp. 461–2; on the "incommensurability" between a sensation and its mechanical explanation, see Meyerson, 1951, pp. 334–7).

When they proposed such explanations, mechanical philosophers did not progress one step beyond Molière's doctor. When this doctor said that opium causes sleep because it has a dormitive virtue, he referred to an empirical property, since opium has indeed the property to make one's sleep, but redoubles the *explanandum*—this empirical property—by a vacuous *explanans*—a dormitive virtue; asserting, as Cartesians did, that opium has a corpuscular structure that is appropriate to make one sleep is no better, because the *explanans*—the "appropriate" corpuscular structure of opium—only redoubles the *explanandum*—the fact that one sleeps when we swallow opium (Gabbey, 2001, p. 462). As to the reason why mechanical philosophers proposed such tautological explanations, it was because of their situation of competition with the Aristotelians: Aristotle's worldview, to which they intended to offer a plausible alternative, continued to shape their agenda (Gabbey, 1985, pp. 13–14, 1990a, p. 279; see also Clarke, 1989, p. 189).

Although Gabbey had a nuanced judgment concerning the value of mechanical explanations, his papers were perceived as reinforcing the *de jure* criticisms of mechanical explanations. When these were studied at all, they were not considered as an epistemologically legitimate practice, but as an outdated phenomenon to be studied from a contextualized point of view— much like the remnants of a lost civilization, the meaning of which we cannot perceive. This contextualized treatment of mechanical explanations was all the more striking given that, at about the same time, historians of science claimed that practices that had been until that point neglected and even rejected as irrational, for example alchemical practices, had a rationality of their own (Principe, 1998; Newman, 2001, 2006). In the last ten years, however, while the new mechanism literature emerged in philosophy of science, there was a reevaluation of early modern mechanical explanations in a domain that had been until then considered peripheral, namely the domain of biology and biology-related disciplines. Comparing early modern explanations with the new mechanism literature will bring to light a number of points of epistemological interest.

3. Reevaluating mechanical explanations

As we have seen, the *de jure* negative evaluation of mechanical explanations was grounded on a specific picture of science, according to which the science *par excellence* would be a physics relying on mathematical laws; not surprisingly, when McMullin and Gabbey began to evaluate them more positively, it was through the notion of structural explanation, which was introduced by contrast to the notion of nomological explanation. Similarly, if one sets aside some forerunners (Haugeland, 1978; Salmon, 1984; Glennan, 1996, 2002), the mechanistic literature of the 2000s began with the observation that the standard philosophy of science, being modeled after parts of physics that rely on mathematized laws, did not say a word about mechanisms, notwithstanding the fact that the use of mechanisms is more than frequent in contemporary biology and biology-related disciplines (Machamer, Darden, and Craver, 2000, pp. 7–8, 23; Bechtel and Abrahamsen, 2005, pp. 422–3; Bechtel and Richardson, 2010, pp. xvii–xviii; Bechtel, 2011, pp. 533–4, 537, 539). Such a similarity suggests that we draw a parallel between contemporary mechanistic explanations and early modern mechanical explanations.

This parallel is all the more justified given that the word "mechanism" itself was a neologism first coined by the very members of the Royal Society who, in the 1660s in England, began to use the term "mechanical philosophy." In the *Immortality of the Soul*, More opposes Descartes' assertion that we may blink without the intermeddling of the soul,

> for if one . . . can keep himself from the fear of any hurt . . . he may easily abstain from winking: But if fear surprise him, the Soule is to be entitled to the action, and not the meer Mechanisme of the Body.
>
> *(1659, p. 103)*

Similarly, Power claimed that the microscope would allow "the illustrious wits of the Atomical and Corpuscularian Philosophers" to see "the curious Mechanism and organical Contrivance of those Minute Animals, with their distinct parts, colour, figure and motion" (1664, p. 5), the *Philosophical transactions* reported that Thomas Willis' *Pharmaceutice rationalis* exposes the "mechanism and power" by which "medicaments do their works," the "mechanical way" being contrasted to "specifique vertues" (1673, pp. 6166–7), and Nehemiah Grew presented as a "mechanism" a membrane that functions as a bow (1682, p. 13). In his *Micrographia*, Hooke used often the word "mechanism" in association with "curious," "stupendious," and "excellent" to designate an exquisitely framed structure or contrivance which is apt to accomplish certain functions (1665, pp. 91, 95, 102, 134, 152, 154, 165, 170–1, 173). There are differences between the philosophical backgrounds of More, Power, Willis, and Hooke, but under their pen, "mechanism" referred, most notably with regards to the anatomy and physiology of plants, animals, and human beings, to a delicate material structure that accomplishes a function without the intervention of any internal vital or spiritual principle (Bertoloni Meli, 2011, pp. 12–16, to whom I owe some of the preceding occurrences). Here again Leibniz can be considered as a go-between in as much as he introduced the word on the Continent (1923, vol. II–1, p. 713, II–3, pp. 452, 713). However, in French "la mécanique" was for a long time preferred to "le mécanisme" for designating, especially in the case of animals, a structure that accomplishes a function (Fontenelle, 1740, vol. I, pp. 175, 249, vol. II, pp. xii, xxi). It was only in Stahl's *Disquisitio de mechanismi et organismi diversitate* and in his ensuing controversy with Leibniz that mechanism began to be opposed to organism (1737).

Before drawing a parallel between early modern mechanical explanations and contemporary mechanistic explanations, a few caveats are necessary. First, it should not cover up the differing intellectual contexts in which the two enterprises were formulated (Theurer, 2013, pp. 913–14). The mechanical philosophy was a reaction against Aristotelian natural philosophy and various Renaissance natural philosophies drawing a sharp distinction between artificial and natural beings; against them, the mechanical philosophy insisted on the homogeneity of nature and on the universality of its laws. The contemporary mechanistic literature is directed against the 1960s philosophers of science who prioritized physics over the other sciences and consequently put the stress on a nomological account of the sciences; against them, this mechanistic literature insisted that one should take into account mechanisms to capture the specificity of biology with regard to the other sciences.

Second, although the contemporary mechanistic literature includes sometimes some references to the early modern period (Glennan, 1996, pp. 50–1; Bechtel, 2011, pp. 535–6; Nicholson, 2012; Theurer, 2013) and, conversely, although some historians of early modern natural philosophy were involved in this literature (Des Chene, 2005; Hutchins, 2015), the development of a mechanistic literature in the philosophy of science and the reevaluation of early modern mechanical explanations in the history of science were largely independent.

Third, it is out of the question to dwell on the details of the proliferating contemporary mechanistic literature. It is all the more out of the question given that this literature is not completely unanimous on what is a mechanism, since it includes two distinct traditions, the first one focusing on mechanical systems, the second one on causal processes (for other attempts of disambiguation of the notion of mechanism, see Glennan, 2002; Nicholson, 2012; Theurer, 2013). To overcome this problem, I will focus primarily on an especially relevant paper of the first tradition (Bechtel and Abrahamsen, 2005).

Bechtel and Abrahamsen give the following definition of mechanism:

> A mechanism is a structure performing a function in virtue of its component parts, component operations, and their organization. Moreover:
>
> - The component parts of the mechanism are those that figure in producing a phenomenon of interest.
> - Each component operation involves at least one component part.
> - Operations can be organized simply by temporal sequence.
> - Mechanisms may involve multiple levels of organization.
>
> *(2005, p. 423)*

It is a definition that immediately makes sense if one thinks about mechanical devices like clocks and automata that were constructed from the medieval period to early modern times. It is indeed by reference to machines that Bechtel and Richardson first introduced the notion of mechanistic explanation, albeit noting that this notion was extended when the technological context evolved (Bechtel and Richardson, 2010, pp. 17–18; Bechtel, 2011, pp. 534–6; but note that Craver and Darden, 2013, pp. 15–16 suggest we should distinguish machines and mechanisms). Thus it is important to ask the question of what stays of the initial inspiration when the notion is extended. While Machamer, Darden, and Craver answer this question by associating mechanism with entities and activities in general (2000, pp. 3–4), Bechtel and Abrahamsen say, not incompatibly but more precisely, that a mechanism is a system composed of parts that interact to perform functions.

Discovering a mechanism is to bring to light various relations thanks to different cognitive strategies. First, a strategy of decomposition that reveals the mereological relation between the parts and the whole structure, a relation which is relative, since something which is a part with respect to the whole structure can be considered as a whole with respect to its component parts. Second, there is a strategy of localization, since these parts are arranged in specific spatial relationships in order to interact, while the operations they perform follow successive temporal orderings. Finally, there are relations of causality between parts and operations as well as between structure and function: parts produce operations, while the whole structure performs a function. Discovering a mechanism implies decomposing a structure in parts, identifying their localizations and interactions, and describing them as causes of the phenomenon at stake.

Since mechanisms are said to be composed of localized parts that perform operations in time, it is only natural to represent them in diagrams that make use of spatial relations to convey information both on the situation of parts in space and on the succession of operations in time. Bechtel and Abrahamsen notice that, considering what our cognitive faculties are, the information that is to be found in diagrams representing mechanisms could not be conveyed verbally, or at least could not be conveyed so easily, inference procedures being different whether one starts with a proposition or with a diagram. Moreover, they add that, to reason about diagrams, the cognitive agent probably refers to former perceptions to mentally animate the diagram and imagine how a part may act on another part. When looking at the different parts of the diagram and following the temporal order, usually indicated by arrows, she will be able to simulate the sequences of operations—provided, of course, that the mechanism at stake is simple enough (Bechtel and Abrahamsen, 2005, pp. 426–31; Bechtel and Richardson, 2010, xix; Chapter 18, this volume; Machamer, Darden, and Craver, 2000, pp. 8–11 observe that mechanisms are represented in diagrams as well, but they cannot easily justify it because of their larger notion of mechanism).

When confronted with such a notion of mechanism, the historian of early modern science could react negatively and say that, when the new mechanists refer to the early modern period, they do it crudely. But the contemporary notion of mechanism and its correlation with diagrams may nonetheless illuminate the epistemic strategies at stake in some early modern mechanical explanations. Considering that the contemporary notion of mechanism appeared in the philosophy of biology, the obvious move is to turn to biological writings; and considering that Descartes played the role of the mechanical philosopher *par excellence*, it is natural to focus in particular on Descartes' biological writings. While the mechanical philosophy has been for a long time censured as irrelevant to biology (Westfall, 1977, pp. 94–104), it is indeed in this domain that historians of early modern thought began to see mechanical explanations with a new eye.

Spirits and Clocks, the book that Dennis Des Chene devoted to Descartes' biology, was a starting point in this respect. Des Chene focused on machines as they were represented in the so-called Theaters of machines, huge books displaying pictures of machines that the early modern engineers boasted they were able to build. It has been often noted since Jurgis Baltrusaïtis and Geneviève Rodis-Lewis that Descartes may have referred to these books, more especially to Salomon de Caus' *Des raisons des forces mouvantes*, when, at the beginning of his treatise *On Man*, he alluded to the marvelous machines that were found in the Gardens of the Greats (De Caus, 1615; Descartes, 1964–74, vol. XI, pp. 130–2; Baltrusaitis, 1955; Rodis-Lewis, 1956). And it has been consequently argued that he may have been inspired by these books to think that animals are only more complex machines than those machines that we construct.

Des Chene did much more than pick up on some comparisons between animal bodies and machines: he showed that Descartes explained and represented organisms as the engineers explained and represented machines in Theaters of machines. De Caus and other engineers considered machines as spatial structures that can be decomposed into parts, each part producing an operation and all the operations being coordinated to perform a function. According to Des Chene, Descartes explained the human body in the very same way: he isolated parts (the heart, the circulatory system, the lungs), analyzed the effect produced by operations of a localized part (the heart) on the neighboring parts (the circulatory system and the lungs), and explained the functions that this succession of operations composes for the organism as a whole (circulation of the blood, respiration). Des Chene insisted also that it is essential for the machines displayed in these Theaters to be visible and to exhibit through pictures how their component mechanisms function. Likewise, Descartes' biological treatises not only described with words the sequences of operations that occur in human bodies, but also made use of pictures following the same pictorial conventions as De Caus (2001, pp. 71–89). One cannot but think that De Caus and Descartes bring into play a notion of mechanism and a use of pictures that is similar to what Bechtel and Abrahamsen highlighted.

To push this further, considering that the new mechanist program appeared because the nomological and reductionist account of the sciences was inadequate in the case of biology, it can be asked what Descartes makes of laws and corpuscular reduction. Des Chene noted that

Descartes' anatomical descriptions in terms of mechanisms were independent from any quantitative assessment of forces and motions (2001, pp. 83–4), but he maintained that, in principle, the reduction to simpler mechanisms should end up in descriptions formulated in terms of laws applied to extended things (2001, pp. 71–2, 154). In a recent paper, Barnaby Hutchins argued further that the reductionist ontology of the mechanical philosophy does not show up when Descartes analyzes the operations of the human body; systems of mechanisms would be the only entities at play. Hutchins' general argument is that, if ontological commitments do not intervene in practice when a phenomenon is explained, it is as if they did not exist. He has a more specific argument: the explanation of the beating of the heart in the *Description of the Human Body* would be both systemic and non-reductionist. Systemic because the explanation of the beating of the heart calls for the explanation of other associated functions (circulation of the blood and respiration), which themselves involve explanations of other bodily functions still (nutrition and assimilation). Non-reductionist because, in the explanation of the beating of the heart, Descartes has recourse not only to corpuscles, but also to higher-level entities (flesh, pores, fibers) and higher-level operations (the lengthening of the heart). The lengthening of the heart is especially important: Descartes, hypothesizing that it is because of this lengthening that some corpuscles of blood are expulsed, does not reduce the higher level to the lower level, but, on the contrary, makes the higher level intervene on the lower level (Hutchins, 2015).

The new literature on mechanisms and the history of early modern thought converge also in the case of diagrams. As we have seen, Bechtel and Abrahamsen insist that there is an affinity between mechanisms and diagrams that helps us understand what would be difficult to capture in verbal propositions alone, and Des Chene showed that the same pictorial conventions were implemented by De Caus for representing machines and by Descartes for representing organs. More generally, the numerous pictures that are to be found in Descartes' writings have recently been taken more seriously than before. Christoph Lüthy was one of the first scholars to recognize Descartes' innovation when he introduced pictures in books of natural philosophy, given that there were none in scholastic textbooks. However, he judged severely their function; according to him, Descartes used pictures to bridge the gap between logical deduction and rhetorical persuasion, or between his general program of natural philosophy and his particular explanations: they appeared when logical demonstrations failed (Lüthy, 2006, pp. 103, 110–11).

Although Lüthy is correct that a picture does not make a demonstration, it is still worth following the logic of the pictures and understanding their cognitive function. Not only were the *Essays* densely illustrated, but Descartes took great care of their pictures. He was a poor painter himself, but he followed the advice of his friend Constantyn Huygens who favored woodcuts over copperplates and preferred to place each picture in front of the corresponding text. Huygens recommended an engraver who was "rather a philosopher" and had "the understanding as quick as the chisel" and the choice fell on Franz Schooten the Younger (Descartes 1964–74, vol. I, pp. 589, 607, 614).

Huygens anticipated that, among the pictures of the *Essays*, Figure 3.1, taken from *Météores*, would be the most difficult to draw (Descartes, 1964–74, vol. I, pp. 607, 614, discussed in Zittel, 2009, pp. 209–13). The first difficulty was probably that the picture had to convey the idea that there are corpuscles of water that have an elongated figure, although nobody has ever seen them. The trick of this picture is to include realistic elements, like the big rock on the right side or the cloud above it, that are so to speak able to transfer by contiguity some of their reality to the invisible corpuscles that they stand alongside. The second difficulty was that the idea of a transformative motion should be conveyed through a static picture, since Descartes thought corpuscles of water are progressively transformed into vapors, and vapors into clouds. The reader is invited to see this transformation that happens in time with the help of the letters

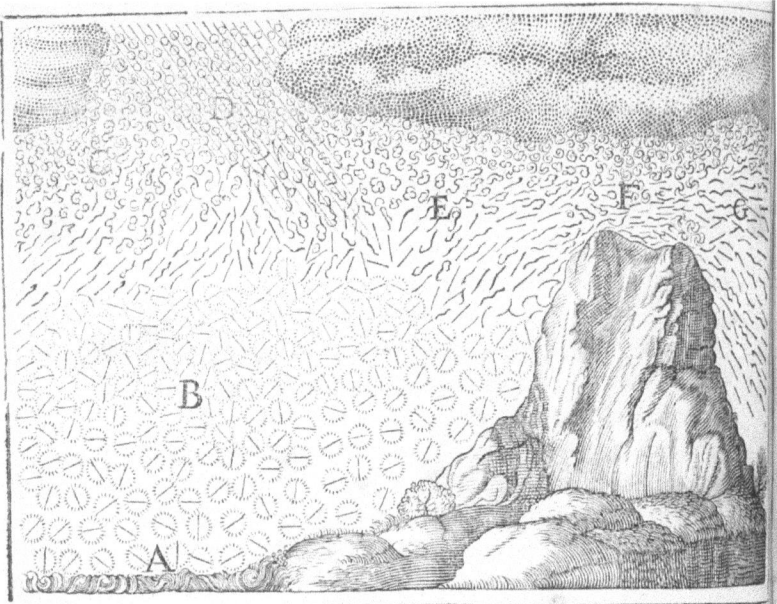

Figure 3.1 *Météores*, in Descartes, 1964–74, vol. VI, p. 242

that guide her gaze from the water corpuscles above (A) to the vapors in which these corpuscles occupy more space because they rotate on themselves (B), and finally to different species of clouds that move at the top of the pictures (C, D, E, F, G). It goes without saying that the existence of such corpuscles was never experimentally confirmed. Still, such a picture functions like a diagram that has the cognitive function of helping the reader to see nature as composed of corpuscles that are transformed by motion.

The pictures of the posthumously published treatise *On Man* (discussed in Wilkin, 2003; Des Chene, 2001, pp. 74, 84–6; Zittel, 2009, pp. 306–6, 2011) will help us to make a further point. Because Descartes left only two illustrations while his treatise continuously referred to pictures, his executor Claude Clerselier began to look for an engraver in 1657, then in 1659 launched a "call for pictures" that happened to stay open for almost five years until he received at last proposals from both Louis de La Forge and Gérard von Gutschoven (Clerselier, 1659 and 1664, in Descartes 1964–74, vol. V, p. 764, vol. XI, pp. xiii–xvii). In between Frans Schuyl had published in 1662 a Latin translation of the treatise, with other pictures still. Schuyl's copperplates were much more detailed, realist, and expressive than the diagrammatic woodcuts by La Forge and Gutschoven that were published in Clerselier's edition.

Figure 3.2 gathers two pictures from Clerselier's edition representing the brain while awake and while asleep, the corresponding picture from Schuyl's edition being Figure 3.3. Or again, compare the representation of the pineal gland by Gutschoven (Figure 3.4) with its representation as a small organ in the middle of the brain by Schuyl (Figure 3.5), where, as Wilkins (2003) observed, the realism went so far as representing the pineal gland on an independent bit of paper that can be moved by the reader. Although Claus Zittel has recently contested the pre-eminence given by all editors and commentators to Clerselier's pictures over Schuyl's pictures (2009, pp. 306–46, 2011), one does not need to decide who to support to understand that two conceptions of pictures were at stake. Clerselier, who published his edition two years

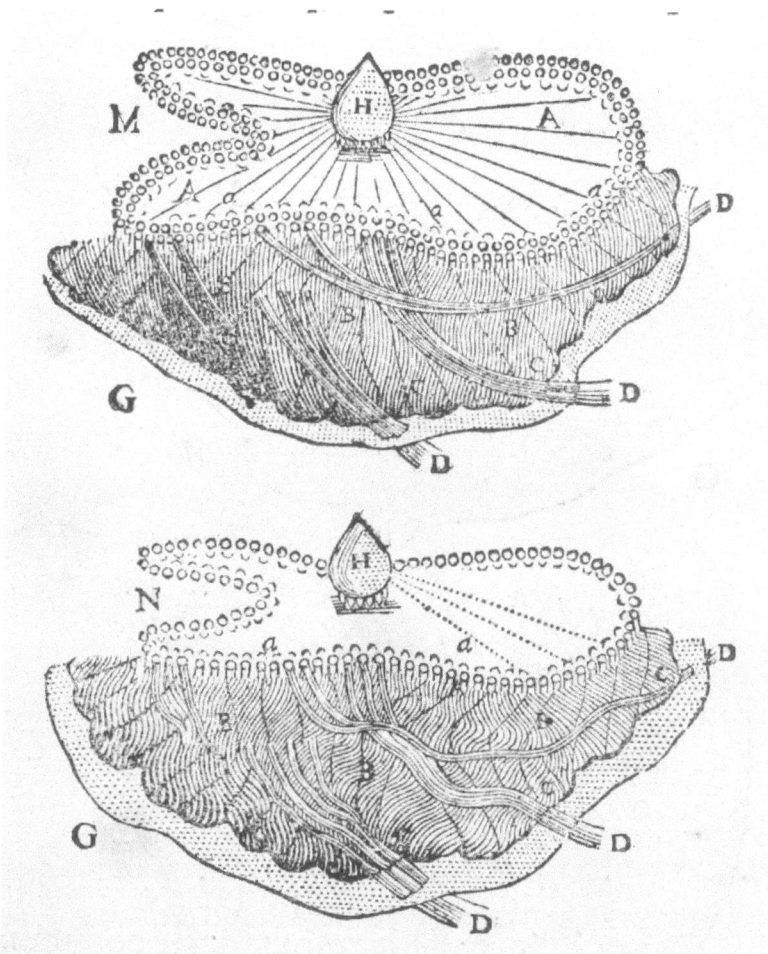

Figure 3.2 The brain awake (above) and asleep (below), Clerselier's edition, in Descartes, 1964–74, vol. XI

after Schuyl's edition and was probably embarrassed by being so late, justified his own mode of representation. He could not but recognize that, as far as the engraving and the printing are concerned, Schuyl's pictures were better than those of La Forge and Gutshoven, but he added that they were nevertheless not so good to make things intelligible.

> One should not be astonished if these figures [of Clerselier's edition] bear no resemblance to nature, because the purpose was not to make a book of anatomy . . . , but only to explain through these figures what M. Descartes advanced in his book, where he speaks more often of things that are not to be sensed, but that had to be rendered sensible in order to be more intelligible. But there is nothing more easy than to put them back in nature and to conceive them as they are, after having considered them other than as they are.
>
> *(Clerselier, 1664; this passage is not reproduced in Descartes, 1964–74, vol. XI)*

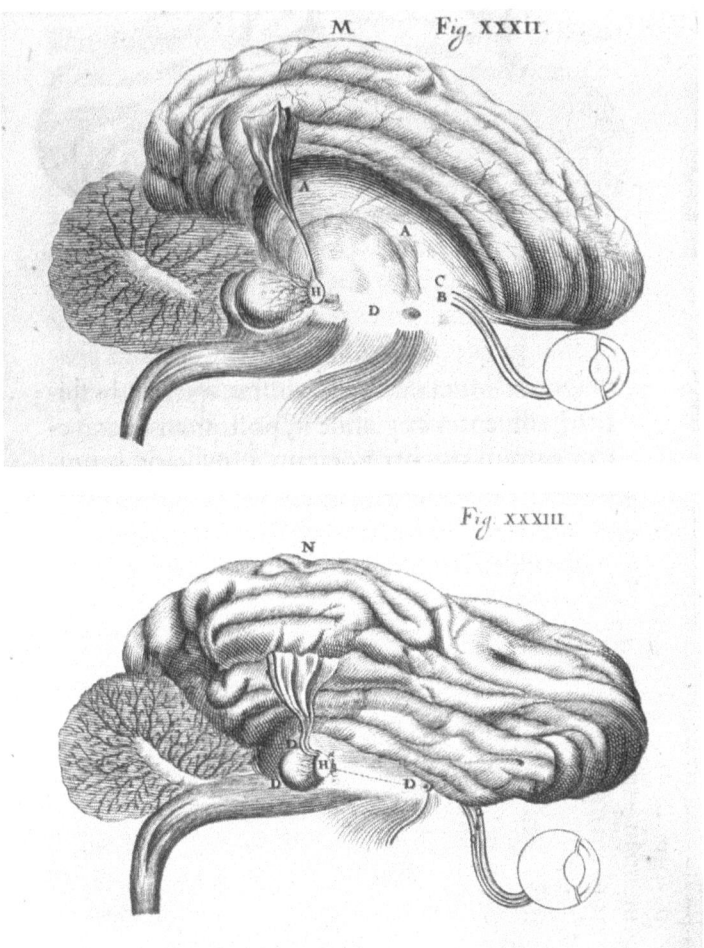

Figure 3.3 The brain awake (above) and asleep (below), Schuyl's edition, in Descartes, 1662, pp. 77–8

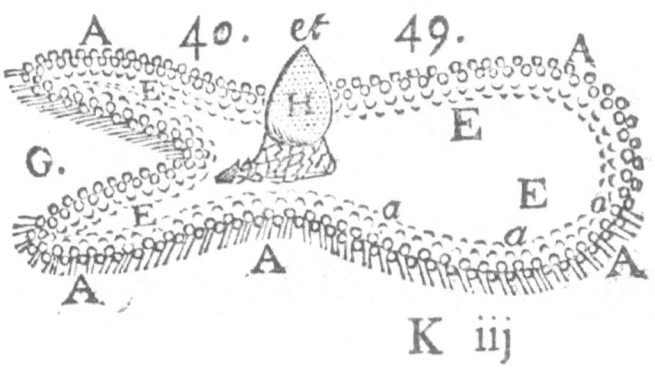

Figure 3.4 The pineal gland, Clerselier's edition, in Descartes, 1964–74, vol. XI

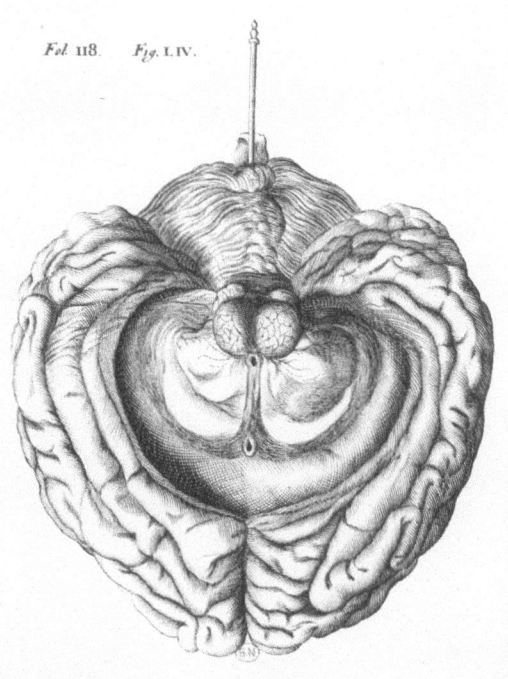

Fol. 118.   *Fig.* L.IV.

Figure 3.5   The pineal gland, Schuyl's edition, in Descartes, 1662, p. 1188

Thus, while Schuyl had recourse to the conventions used in anatomy books because he considered that Descartes' explanations of animal functions would be best conveyed if the reader was first persuaded that the pictures would conform to nature, Clerselier ended up defending more schematic diagrams because he thought that one should first make Descartes' explanations intelligible, which required depicting parts and operations that are unseen. While Schuyl favored empirical adequacy, Clerselier preferred intelligibility, which he equated to visibility.

4. Conclusion

When they turn to history, some new mechanists suggest that mechanistic thinking began in the seventeenth century with a "restrictive ontology" and an "austere worldview" and that this view was progressively diversified and enriched (Machamer, Craver, and Darden, 2000, p. 15; Craver and Darden, 2005, p. 234; Craver and Darden, 2013, pp. 4–5). As has been shown, the mechanical ontology was probably more diversified in the early modern period than is usually thought. But the most important lesson to draw from this chapter is that this ontology should be distinguished from the search for mechanical explanations (Des Chene, 2001, p. 14, 2005, p. 246 put forward a similar distinction between mechanism as a doctrine and mechanism as a method). The mechanical ontology was crucial to the early modern sciences because it preliminarily excluded Aristotelian and Renaissance doctrines that seemed to be incompatible with decomposition and localization. But, when the matter at hand was the explanation of specific phenomena, even Descartes the arch-mechanist made the search for particular mechanisms the primary focus; and, after him, natural philosophers or physicians did not hesitate to search for mechanisms without being committed to a mechanical ontology (Des Chene, 2005; Bertoloni Meli, 2011). In that respect, focusing on specific mechanical explanations while drawing inspiration from the new mechanistic literature is probably a more fruitful perspective on the early modern period than trying to spell out the inexhaustible varieties of the mechanical philosophy.

References

Primary sources

Boyle, R. (1999–2000) *The Works of Robert Boyle*, ed. M. Hunter and E.B. Davis, 14 vols. London: Pickering and Chatto.

Caus, S. de (1615) *Les raisons des forces mouvantes, avec diverses machines tant utiles que plaisantes. Aus quelles sont adionts plusieur desseings de grotes et fontaines.* Frankfurt: J. Norton.

Charleton, W. (1654) *Physiologia Epicuro-Gassendo-Charltoniana: or a Fabrick of Science Natural, Upon the Hypothesis of Atoms, Founded by Epicurus, Repaired by Petrus Gassendus, Augmented by Walter Charleton.* London: T. Newcomb for T. Heath.

Clerselier, C. (1659) "Préface," *Lettres de Monsieur Descartes. Tome second*, ed. C. Clerselier. Paris: C. Angot.

Clerselier, C. (1664) "Préface," *L'Homme de René Descartes et la formation du foetus avec les remarques de Louys de La Forge*, ed. C. Clerselier. Paris: C. Angot.

Descartes, R. (1662) *Renatus Des Cartes de homine, figuris et latinaitate donates a Florentio Schuyl*, ed. F. Schuyl. Leiden: F. MOyardus et P. Leffen.

Descartes, R. (1664) *L'Homme de René Descartes et la formation du foetus avc les remarques de Louis de La Forge*, ed. Claude Clerselier. Paris: C. Angot.

Descartes, R. (1964–74) *Œuvres de Descartes*, ed. C. Adam and P. Tannery (1897–1913), new presentation by B. Rochot and P. Costabel, 13 vols. Paris: Vrin.

Fontenelle, B. Le Bovier de (1740) *Éloges des académiciens, avec l'Histoire de l'Académie royale des sciences*, 2 vols. La Haye: I. van der Kloot.

Gassendi, P. (1658) *Petri Gassendi Opera Omnia*, 6 vols. Lyon: L. Anisson and J.B. Devenet.

Glanvill, J. (1671) *A Praefatory Answer to Mr Henry Stubbe, the doctor of Warwick wherein the malignity, hypocrisie, falshood of his temper, pretences, reports, and the impertinency of his arguings & quotations in his animadversions on Plus ultra are discovered.* London: J. Collins.

Grew, N. (1682) *The Anatomy of Plants, with an Idea of a Philosophical History of Plants, and Several other Lectures, Read before the Royal Society.* London: W. Rawlings.

Hooke, R. (1665) *Micrographia or Some physiological descriptions of minutes bodies made by magnifying glasses: with observations and inquiries thereupon.* London: J. Martyn and J. Allestry.

Leibniz, G.W. (1923) *Leibniz. Sämtliche Schriften und Briefe*, ed. Preussische (then Deutsche) Akademie der Wissenschaften, Darmstadt (then Leipzig, then Berlin).

More, H. (1653a) *An antidote against atheisme, or An appeal to the natural faculties of the minde of man, whether there be not a God by Henry More.* London: R. Daniel.

More, H. (1653b) *An antidote against atheisme, or, An appeal to the natural faculties of the minde of man, whether there be not a God.* London: R. Daniel.

More, H. (1659) *The Immortality of the Soul.* London: W. Morden.

Parker, S. (1666) *A Free and Impartial Censure of the Platonick Philosophy.* Oxford: W. Hall for R. Davis.

Perrault, C. (1680) *La Mécanique des animaux*, in *Essais de physique. Tome III.* Paris: J.-B. Coignard.

Philosophical Transactions of the Royal Society of London. London: J. Martyn.

Poli, M. (1706) *Il Trionfo degli Acidi Vendicati dalle Calunnie di molti Moderni.* Roma: G. Placo.

Power, H. (1664) *Experimental philosophy, in three books containing new experiments microscopical, mercurial, magnetical: with some deductions, and probable hypotheses, raised from them, in avouchment and illustration of the now famous atomical hypothesis.* London: J. Martyn and J. Allestry.

Stahl, G.E. (1737) *Theoria medica vera. Physiologiam & Pathologiam, tanquam doctrinae medicae partes vere contemplativas, e naturae & artis veris fundamentis, intaminata ratione, & inconcussa experientia sistens.* 2nd edition (1st edition, 1708). Halle: Orphanotropheum.

Secondary sources

Anstey, P. (2014) "Philosophy of Experiment in Early Modern England: The Case of Bacon, Boyle and Hooke," *Early Science and Medicine*, vol. 19, no. 2, pp. 103–132.

Anstey, P., and Vanzo, A. (2012) "The Origins of Early Modern Experimental Philosophy," *Intellectual History Review*, vol. 22, no. 4, pp. 499–518.

Anstey, P., and Vanzo, A. (2016) "Early Modern Experimental Philosophy," in Sytsma, J. (ed.) *A Companion to Experimental Philosophy*, Malden: Blackwell, pp. 87–102.

Bachelard, G. (1951) *L'activité rationaliste de la physique contemporaine*. Paris: Presses universitaires de France.

Baltrusaitis, J. (1955) *Anamorphoses ou perspectives curieuses*. Paris: O. Perrin.

Bechtel, W., and Richardson, R.C. (2010) *Discovering Complexity. Decomposition and Localization as Strategies in Scientific Research*. 2nd edition (1st edition, 1993). Cambridge, MA: The MIT Press.

Bechtel, W., and Abrahamsen, A. (2005) "Explanation: A Mechanist Alternative," *Studies in History and Philosophy of Biology and Biomedical Sciences*, vol. 36, pp. 421–441.

Bechtel, W. (2011) "Mechanism and Biological Explanation," *Philosophy of Science*, vol. 78, no. 4, pp. 533–557.

Bennett, J.A. (1986) "The Mechanics' Philosophy and the Mechanical Philosophy," *History of Science*, vol. 24, no. 1, pp. 1–28.

Bertoloni Meli, D. (2011) *Mechanism, Experiment, Disease: Marcello Malpighi and Seventeenth–Century Anatomy*. Baltimore: The Johns Hopkins University Press.

Boas, M. (1952) "The Establishment of the Mechanical Philosophy," *Osiris*, vol. 10, pp. 412–541.

Chalmers, A. (1993) "The Lack of Excellency of Boyle's Mechanical Philosophy," *Studies in History and Philosophy of Science*, vol. 24, no. 4, pp. 541–564.

Chalmers, A. (2012) "Intermediate Causes and Explanations: The Key to Understanding the Scientific Revolution," *Studies in History and Philosophy of Science*, vol. 43, no. 4, pp. 551–562.

Clarke, D.M. (1989) *Occult Powers and Hypotheses: Cartesian Natural Philosophy under Louis XIV*. Oxford: Oxford University Press.

Clericuzio, A. (1990) "A Redefinition of Boyle's Chemistry and Corpuscular Philosophy," *Annals of Science*, vol. 47, no. 6, pp. 561–589.

Clericuzio, A. (2000) *Elements, Principles and Corpuscles: A Study of Atomism and Chemistry in the Seventeenth Century*. New York, Dordrecht, Boston, London: Kluwer Academic Publishers.

Clerselier, C. (1659) "Préface," in *Lettres de Monsieur Descartes. Tome second*, ed. Claude Clerselier. Paris: Charles Angot.

Craver, C., and Darden, L. (2013) *Thinking about Mechanisms: Discovery across the Life Sciences*. Chicago: University of Chicago Press.

Des Chene, D. (2001) *Spirits and Clocks. Machine and Organism in Descartes*. Ithaca, NY: Cornell University Press.

Des Chene, D. (2005) "Mechanisms of Life in the Seventeenth Century: Borelli, Perrault, Régis," *Studies in History and Philosophy of Biological and Biomedical Sciences*, vol. 36, pp. 245–260.

Dijksterhuis, E.J. (1961) *The Mechanization of the World Picture*. Oxford: Oxford University Press.

Gabbey, A. (1982) "Philosophia Cartesiana Triumphata: Henry More (1646–1671)," in Lennon, T., Nicholas, J.M. and Davis, J.W. (eds.), *Problems of Cartesianism*, Kingston, Ontario: McGill-Queen University Press, pp. 171–250.

Gabbey, A. (1985) "The Mechanical Philosophy and its Problems: Mechanical Explanations, Impenetrability, and Perpetual Motion," in Pitt, J.C. (ed.), *Change and Progress in Modern Science*, Dordrecht, Boston, Lancaster: D. Reidel Publishing Company, pp. 9–84.

Gabbey, A. (1990a) "Explanatory Structures and Models in Descartes' Physics," in Belgioioso, G., Cimino, G., Costabel, P. and Papuli, G. (eds.), *Descartes: il Metodo e I Saggi*, Roma: Istituto della Enciclopedia Italiana, pp. 273–286.

Gabbey, A. (1990b) "Henry More and the Limits of Mechanism," in Hutton, S. (ed.), *Henry More (1614–1687): Tercentenary Studies*, New York: Kluwer Academic Publishers, pp. 19–35.

Gabbey, A. (2001) "Mechanical Philosophies and their Explanations," in Lüthy, C., Murdoch, J.E. and Newman, W.R. (eds.), *Late Medieval and Early Modern Corpuscular Matter Theories*, Leiden: Brill, pp. 441–466.

Gabbey, A. (2004) "What was 'Mechanical' about 'The Mechanical Philosophy?'," in Palmerino, C.R. and Thijssen, J.M.M.H. (eds.), *The Reception of the Galilean Science of Motion in Seventeenth-Century Europe*, New York: Kluwer Academic Publishers, pp. 11–24.

Garber, D. (2013) "Remarks on the Pre-history of the Mechanical Philosophy," in Garber, D. and Roux, S. (eds.), *The Mechanization of Natural Philosophy*, New York: Kluwer Academic Publishers, pp. 3–26.

Gaukroger, S. (2006) *The Emergence of a Scientific Culture: Science and the Shaping of Modernity, 1210–1685*. Oxford: Oxford University Press.

Glennan, S.S. (1996) "Mechanisms and the Structure of Causation," *Erkenntnis*, vol. 44, pp. 49–71.

Glennan, S. (2002) "Rethinking Mechanistic Explanation," *Proceedings of the Philosophy of Science Association*, vol. 3, pp. S342–S353.

Haugeland, J. (1978) "The Nature and Probability of Cognitivism," *The Behavioral and Brain Science*, vol. 2, pp. 215–260.

Henry, J. (1986) "Occult Qualities and the Experimental Philosophy: Active Principles in Pre-Newtonian Matter Theory," *History of Science*, vol. 24, no. 66, pp. 335–381.

Henry, J. (1989) "Robert Hooke, the Incongruous Mechanist," in Hunter, M. and Schaffer, S. (eds.), *Robert Hooke. New Studies*, Woodbridge, UK: Boydell, pp. 149–180.

Hutchins, B.R. (2015) "Descartes, Corpuscles and Reductionism: Mechanism and Systems in Descartes' Physiology," *The Philosophical Quarterly*, vol. 65, no. 261, pp. 669–689.

Leijenhorst, C. (2002) *The Mechanisation of Aristotelianism: The Late Aristotelian Setting of Thomas Hobbes' Natural Philosophy*. Leiden, Boston and Köln: Brill.

Lennon, T. (1993) *The Battle of the Gods and Giants: The Legacies of Descartes and Gassendi, 1615–1755*. Princeton, NJ: Princeton University Press.

Lolordo, A. (2007) *Pierre Gassendi and the Birth of Early Modern Philosophy*. Cambridge: Cambridge University Press.

Lüthy, C. (1997) "Thoughts and Circumstances of Sébastien Basson: Analysis, Micro-History, Questions," *Early Science and Medicine*, vol. 2, pp. 1–73.

Lüthy, C. (2001a) "An Aristotelian Watchdog as Avant-Garde Physicist: Julius Caesar Scaliger," *Monist*, vol. 84, no. 4, pp. 542–561.

Lüthy, C. (2001b) "David Gorlaeus's Atomism, or: The Marriage of Protestant Metaphysics with Italian Natural Philosophy," in Lüthy, C., Murdoch, J.E. and Newman, W.R. (eds.), *Late Medieval and Early Modern Corpuscular Matter Theories*, Leiden: Brill, pp. 245–290.

Lüthy, C. (2006) "Where Logical Necessity Becomes Visual Persuasion: Descartes' Clear and Distinct Illustrations," in Maclean, I. and Kusukawa, S. (eds.), *Transmitting Knowledge: Words, Images and Instruments in Early Modern Europe*, Oxford: Oxford University Press, pp. 97–133.

Osler, M.J. (1994) *Divine Will and the Mechanical Philosophy: Gassendi and Descartes on Contingency and Necessity in the Created World*. Cambridge: Cambridge University Press.

Machamer, P., Darden, L., and Craver, C.F. (2000) "Thinking about Mechanisms," *Philosophy of Science*, vol. 67, pp. 1–25.

McMullin, E. (1978) "Structural Explanations," *American Philosophical Quarterly*, vol. 15, no. 2, pp. 139–147.

Meyerson, É. (1951) *Identité et réalité*. Paris: Vrin.

Murdoch, J.E. (2001) "The Medieval and Renaissance Tradition of *minima naturalia*," in Lüthy, C., Murdoch, J.E. and Newman, W.R. (eds.), *Late Medieval and Early Modern Corpuscular Matter Theories*, Leiden: Brill, pp. 91–141.

Newman, W.R. (1994) "The Corpuscular Theory of J. B. Helmont and Its Medieval Sources," *Vivarium*, vol. 31, no. 1, pp. 161–191.

Newman, W.R. (1996) "Boyle's Debt to Corpuscular Alchemy," in Hunter, M. (ed.), *Robert Boyle Reconsidered*, Cambridge: Cambridge University Press, pp. 107–118.

Newman, W.R. (2001) "Experimental Corpuscular Theory in Aristotelian Alchemy: From Geber to Sennert," in Lüthy, C., Murdoch, J.E. and Newman, W.R. (eds.), *Late Medieval and Early Modern Corpuscular Matter Theories*, Leiden: Brill, pp. 291–331.

Newman, W.R. (2006) *Atoms and Alchemy: Chymistry and the Experimental Origins of the Scientific Revolution*. Chicago, London: The University of Chicago Press.

Nicholson, D.J. (2012) "The Concept of Mechanism in Biology," *Studies in History and Philosophy of Biological and Biomedical Sciences*, vol. 43, pp. 152–163.

Principe, L.M. (1998) *The Aspiring Adept: Robert Boyle and his Alchemical Quest*. Princeton, NJ: Princeton University Press.

Rhigini Bonelli, M.L. and Shea, W.R. (eds.) (1975) *Reason, Experiment and Mysticism in the Scientific Revolution*. New York: Science History Publications.

Rodis-Lewis, G. (1956) "Machinerie et perspectives curieuses dans leur rapport avec le cartésianisme," *XVIIe siècle*, vol. 32, pp. 461–474.

Roux, S. (1996) *La philosophie mécanique (1630–1690)*. Unpublished PhD, EHESS, 2 vols, p. 808.

Roux, S. (2000) "Descartes atomiste?", in Gatto, R. and Festa, E. (eds.), *Atomismo e continuo nel XVII secolo*, Napoli: Vivarium, pp. 211–274.

Roux, S. (2004) "Cartesian Mechanics," in Palmerino, C.R. and Thijssen, J.M.M.H. (eds.), *The Reception of the Galilean Science of Motion in Europe*, Dordrecht: Kluwer Academic Publishers, pp. 25–66.

Roux, S. (2008) "Les *Recherches métaphysiques* de Gassendi: vers une histoire naturelle de l'esprit," in Taussig, S. (ed.), *Gassendi et la modernité*, Turnhout: Brepols, pp. 105–140.

Roux, S. (2009) "À propos du colloque *The Machine as Model and as Metaphor*," *Revue de synthèse*, vol. 30, no. 1, pp. 165–175.

Roux, S. (2012) "Quelles machines pour quels animaux? Jacques Rohault, Claude Perrault, Giovanni Alfonso Borelli," in Gaillard, A., Goffi, J.-Y., Roukhomovsky, B. and Roux, S. (eds.), *L'automate. Machine, métaphore, modèle, merveille*. Bordeaux: Presses universitaires de Bordeaux, pp. 69–113.

Salmon, W. (1984) *Scientific Explanation and the Causal Structure of the World*. Princeton, NJ: Princeton University Press.

Sargent, R.-M. (1994) "Learning from Experience: Boyle's Construction of an Experimental Philosophy," in Hunter, M. (ed.), *Robert Boyle Reconsidered*, Cambridge: Cambridge University Press, pp. 57–78.

Sargent, R.-M. (1995) *The Diffident Naturalist: Robert Boyle and the Philosophy of Experiment*. Chicago: The University of Chicago Press.

Theurer, K. (2013) "Seventeenth-Century Mechanism: An Alternative Framework for Reductionism," *Philosophy of Science*, vol. 80, no. 5, pp. 907–918.

Westfall, R.S. (1977) *The Construction of Modern Science: Mechanisms and Mechanics*, 2nd edition. Cambridge: Cambridge University Press.

Westman, R.S., and McGuire, J.E. (eds.) (1976) *Hermeticism and the Scientific Revolution. Papers read at a Clark Library Seminar, March 9, 1974*. Los Angeles, William Andrews Clark Memorial Library: University of California at Los Angeles.

Wilkin, R. (2003) "Figuring the Dead Descartes: Claude Clerselier's *Homme* de René Descartes (1664)," *Representations*, vol. 83, pp. 38–66.

Zittel, C. (2009) *Theatrum philosophicum: Descartes und die Rolle ästhetischer Formen in der Wissenschaft*. Berlin: Akademie-Verlag.

Zittel, C. (2011) "Conflicting Pictures: Illustrating Descartes' Traité de l'homme," in Dupré, S. and Lüthy, C. (eds.), *Silent Messengers: The Circulation of Material Objects of Knowledge in the Early Modern Low Countries*, Berlin, Lit-Verlag, pp. 217–260.

4

THE ORIGINS OF THE REACTION MECHANISM

William Goodwin

1. Introduction

Chemists have characterized reaction mechanisms in a variety of different ways, which vary by sub-discipline and time period. For the purposes of this chapter, however, I will consider a chemical reaction mechanism to be "a specification, by means of a sequence of elementary chemical steps, of the detailed process by which a chemical change occurs" (Lowry and Richardson, 1987, p. 190). These elementary chemical steps may themselves be described by providing "a picture of the participating species at one or more crucial instants during the course of the reaction" (Gould, 1959, p. 127). These characterizations of mechanism are drawn from relatively contemporary texts in theoretical organic chemistry, which limits how robustly applicable they are. However, it is appropriate, in a chapter such as this, to consider the notion of mechanism applicable in this subfield because it was the field in which reaction mechanisms first emerged in modern chemistry.

Before going on to highlight some of the major episodes in the development of the chemical reaction mechanism, it will be worth considering the extent to which this notion of mechanism fits characterizations of mechanism in the philosophical literature. Minimally, "[a] mechanism for a phenomenon consists of entities (or parts) whose activities and interactions are organized so as to be responsible for the phenomenon" (Introduction to this volume). In the case of the organic reaction mechanism, the phenomenon for which a mechanism is provided is a chemical reaction, typically characterized by a chemical equation. A chemical equation represents a reaction by providing a (stoichiometrically balanced) description of the reactants and the products of the transformation. The reactants and products are themselves represented by chemical formulas of one form or another. In the organic case, these are typically structural formulas, which model[1]—in varying levels of detail—the composition, bond connectivity, and spatial distribution of atoms in a molecule.

A mechanism may be provided for a chemical reaction by decomposing the original chemical equation into parts, each of which is itself a chemical equation (see Figure 4.1). In such cases, a sequence of chemical equations, each of which represents a step in the original transformation, collectively add up to produce the original transformation. Some products of the intermediate steps in such a system of chemical equations, which are represented with their own structural formulas, are more or less stable chemical entities that then function as reactants for subsequent steps.

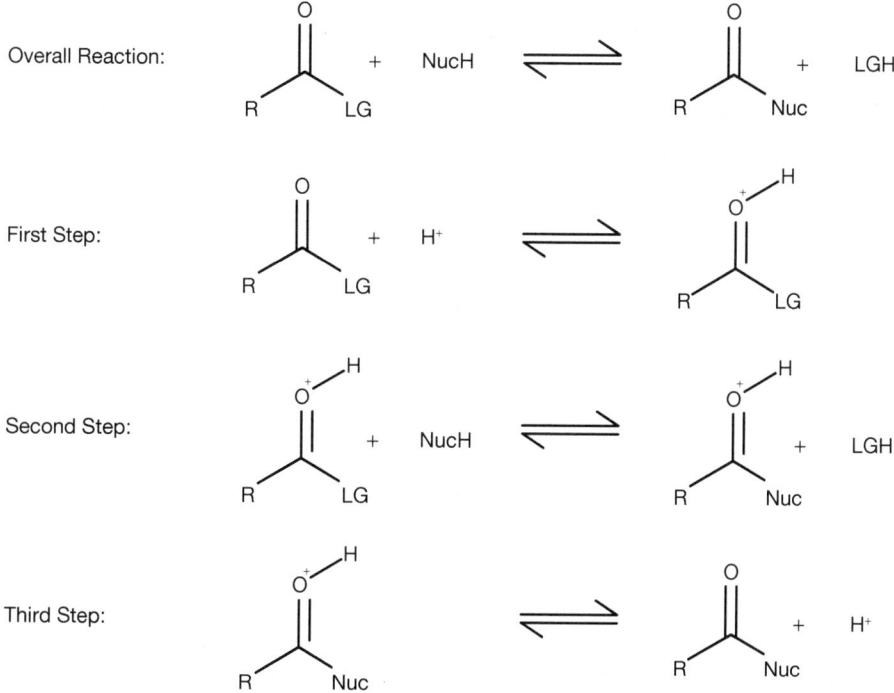

Figure 4.1   The decomposition of a chemical reaction into steps. This mechanism depicts an acid-catalyzed Addition–Elimination reaction on a carboxylic acid derivative. H⁺ is a catalyst, while the other charged species in the second and third steps are intermediates. Nuc represents a nucleophile, LG a leaving group, and R a generic group

These are the intermediates of the reaction, and though they don't occur in the original chemical equations, their structures often do a lot of the explanatory work of the reaction mechanism. Additionally, some chemical species may be reactants in some earlier step of the system of equations, but products later on, thereby canceling out and again not occurring at all in the original chemical equation. These are catalysts of the reaction, and they too may play a crucial role in the explanations provided by the mechanism. So a mechanism of this sort provides a decomposition of a chemical reaction into a sequence of steps that include descriptions of the intermediates and catalysts of a chemical transformation by providing their structural formulas. The sequence of steps and their relative rates can be crucial to the mechanism because they allow for the identification of the bottleneck step, which determines the rate at which the overall process takes place. Additionally, the entities (intermediates and catalysts) occurring in these elementary steps can be subject to structural analysis to explain important features of the original chemical reaction, often including its relative rate and product distribution.

Not all organic reaction mechanisms proceed by decomposing the original reaction into parts, in the sense of a sequence of sub-reactions. Many chemical reactions, including, typically, the elementary reactions into which more complex reactions are decomposed, proceed in one step, without forming any more or less stable intermediates. Still, reaction mechanisms are sought for these reactions. The entities invoked in such mechanisms are quasi-structural descriptions of certain crucial points in the transition between reactants and products including, typically, the highest-energy configuration along the reaction coordinate.[2] And, much as with the intermediates in the more complex case, the representations of the crucial points of the transition can be subjected to structural analysis, allowing the chemist to explain important facts about the chemical reaction. Many contemporary reaction mechanisms combine both strategies for describing the process by which a chemical reaction occurs: first, they decompose the reaction into its sub-reactions, and then they provide structural characterizations of the transition states of at least some of those sub-reactions (most crucially the bottleneck, or rate-determining step of the reaction system).[3]

Chemical reactions usually involve the breaking and/or making of bonds: certain bonds in the reactants must be broken so that the bonds in the product can be made. Mechanisms must therefore describe stages in this process of bond breakage and formation. As a result, mechanisms generally presuppose or invoke some account of chemical bonding. Furthermore, to explain why bonds break and form between particular atoms, a mechanism typically relies on some account of inter and/or intramolecular interactions. Chemists' understanding of both chemical bonding and the interactions between or within molecules underwent significant changes during the same period of time that the reaction mechanism emerged as a useful theoretical tool in organic chemistry. As a result, there is no simple answer to the question of what sorts of activities and interactions are attributed to the entities invoked in chemical mechanism. Early mechanisms invoke accounts of chemical bonding and interaction that look quite different from our modern accounts, though they are often recognizable in retrospect as approximations to these more modern accounts.

A reaction mechanism is a model of a chemical transformation. It is a model in just about all of the relevant philosophical senses of the term: it is a primarily non-linguistic partial representation, intermediate between theory and observation, that allows for both the concrete application of theory and the refinement of novel explanatory concepts.[4] In the first place, it is a staged description of a chemical transformation that allows for the theoretical apparatus of reaction kinetics or Transition State Theory to be applied, either quantitatively or qualitatively, to the transformation. Second, it leaves out as much chemical detail as is possible (including, in many cases, the solvent in which the reactants are dissolved), given the phenomena that it hopes to explain. Third, it is a depiction of the microscopic actors in a macroscopic process, so that at best it characterizes one of the innumerable ways that such transformations actually occur at the microscopic level. And lastly, the depictions of these microscopic actors are themselves structural models of the chemical species that they depict. Chemists are generally quite explicit about the provisional and instrumental character of reaction mechanisms. For instance, in the foundational textbook of physical organic chemistry, the author claims:

> A mechanism . . . is a scientific tool by which to obtain verifiable relationships between measureable quantities; it is to be judged by its utility in correlating known facts and predicting new ones, not by its agreement with some unknowable absolute truth.
>
> *(Hammett, 1940, p. 98)*

Thus, reaction mechanisms are theoretical tools for explaining and predicting chemical phenomena, not primarily attempts to describe in some realistic way the actual processes by which chemical reactions take place.

2. Setting the stage for mechanisms

Reaction mechanisms, in something like their modern form, first appeared in the organic chemistry literature at the beginning of the twentieth century (Hammett, 1966). By the end of the nineteenth century, organic chemistry had developed an extremely successful theory—structure theory—that allowed it to individuate, classify, and rationalize much of the chemical behavior of organic compounds. The goal of structure theory, as articulated by Aleksandr Butlerov, was to have one structural formula for each distinct compound and then "when the general laws governing the dependence of chemical properties on chemical structure have been determined, this formula [would] express all of those properties" (Brock, 2000, p. 256). Structural formulas depicted organic compounds in terms of the "bonds" between the elements (or atoms) that composed them. Some structural chemists interpreted chemical structure realistically, as something like a picture or model of the arrangements of atoms in organic molecules. Others, particularly early on, resisted this realistic interpretation, opting instead to think of chemical structures as abstract symbols that encoded the possible transformations of chemical kinds (see Duhem, 2002; Nye, 1993, ch.6). This ambiguity was facilitated, at least in part, by the fact that organic chemists had no consensus account of the nature of chemical bonding; thus the very connections depicted by a chemical structure were a black box. Gradually, perhaps because of the emerging awareness that the newly discovered electron had a role in chemical bonding, not only were chemical structures interpreted increasingly realistically, but organic chemists began to aspire to provide causal accounts of the processes by which one chemical structure was transformed into another.

In 1908, we find in an annual report on the state of organic chemistry the following claim:

> The great advances made in recent years in our knowledge of the structure of organic compounds have directed attention all the more vividly to the gaps which remain. This is especially the case with regard to reactions. The chemical equation only represents, as a rule, the initial and final stages, the mechanism by which the result is obtained remaining obscure, and only to be filled by the assumption of hypothetical intermediate products, the nature of which it is often very difficult to define.
>
> *(Desch and Morgan, 1909, p. 74)*

This quote clearly brings out what was at this time a quite legitimate worry about the doubly hypothetical "intermediate products" that might be invoked to fill in the mechanisms of chemical reactions. These intermediates are hypothetical not only in the sense that they are postulated merely to explain features of the reaction and are not measureable or observable on their own, but also because the very terms in which they are described are conditioned upon some hypothetical account of chemical bonding and interaction. The speculative character of accounts of reaction mechanisms was tolerated because of their potential payoff. The annual report of the following year makes it clear what that payoff was:

> The application of physico-chemical methods to the study of organic compounds is largely responsible for the increasing interest shown in the relation of physical properties to chemical constitution, and in the mechanism of reactions. Whilst the older structural organic chemistry, aiming principally at the synthetical formation of compounds and the determination of constitution, was mainly concerned with the final products of a given reaction, more and more attention is now being devoted to the dynamical aspects of reactions, embracing the quantitative study of velocity, etc., as well as the molecular mechanism by which the interchange is effected.
>
> *(Desch and Lapworth, 1910, p. 57)*

The recently developed methods of what would later be called physical organic chemistry had provided new kinds of information about both organic molecules and chemical transformations.

It was in an effort to explain these newly recognized features of chemical reactions that organic chemists early in the twentieth century resorted to speculation about the mechanisms of reactions. As suggested in this quote, it was the rates of chemical reactions that primarily motivated speculations about mechanisms.

The most fundamental result from physical chemistry that is relevant to the rates of organic reactions is what is sometimes called the generalized Law of Mass Action:

> In dilute systems the rate of every chemical reaction is proportional to the product of the concentrations of the substances which are reacting, and is independent of the concentrations of all other substances and the presence or absence of all other reactions.
>
> *(Hammett, 1966, p. 465)*

Physical chemists accumulated evidence for this principle over the course of the 1890s so that, by the first decade of the twentieth century, "the generalized Law of Mass Action [was] accepted without reservation or hesitation and [was] used with complete confidence as the basis for judgments about . . . mechanism" (Hammett, 1966, p. 465). To see how this law could be used to underwrite speculations about reaction mechanisms, I will describe one of the first studies on organic reaction mechanisms that helped to establish "the whole technique of the kinetic investigation of reaction mechanism on a firm basis" (Hammett, 1966, p. 466).

3. Lapworth's kinetic studies of reaction mechanism

Hydrogen cyanide (HCN) will add to the carbon-oxygen double bond in many aldehydes (RCOH) and ketones (RCOR') to form a cyanohydrin ($R_2C(OH)(CN)$). In 1903, Lapworth published an account of the mechanism of this addition reaction supported by kinetic studies. The great innovation in his account was to work backwards from an apparent violation of the generalized Law of Mass Action to a multistep mechanism for the reaction that reconciled the observed rate behavior with this general principle.

The stoichiometrically balanced equation for cyanohydrin formation is as follows:

$$R_2CO + H^+ + CN^- \rightarrow R_2C(OH)CN$$

And according to the generalized Law of Mass Action, this would mean that the rate of reaction should be proportional to the products of the concentrations of the reactants. But this was not what Lapworth had found. Instead he reported that adding acid (H^+) to the reaction reduced its rate, contrary to the increase expected by the general principle. Adding bases, which don't occur among the reactants at all, actually increased its rate. Furthermore, adding cyanide ions (CN^-) to the reaction mixture led to dramatic increases in rate. Since hydrogen cyanide is a weak acid, addition of acid would reduce the amount of HCN that dissociated to form cyanide ions, while addition of base would promote cyanide ion formation. The upshot was that "the reaction velocity depended mainly on the concentration of cyanogen ions [CN^-]" (Lapworth, 1903, p. 997).

To accommodate this rate data, Lapworth proposed that the reaction actually took place in two steps. First the cyanide ion united, in a reversible process, with the neutral carbonyl carbon, which was polarized because the oxygen possessed "residual affinity." The result was a complex ion, which then went on to react with hydrogen ions (H^+) to form the cyanohydrin (see Figure 4.2). If it were supposed, in addition, that the second step was "very rapid in comparison with the first" (Lapworth, 1903, p. 1001)—that is, that the first step was what is

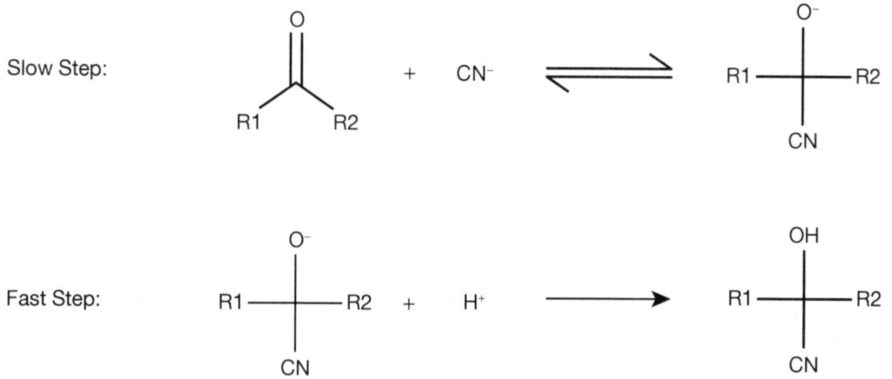

Figure 4.2 Lapworth's decomposition of cyanohydrin formation. In the first, slow step, a cyanide ion combines with a ketone to form a complex ion. In the second, faster step, the complex ion combines with a hydrogen ion to form the cyanohydrin. The overall rate should be proportional to the cyanide ion concentration, but not to the hydrogen ion concentration

now called the "rate-determining step"—then the prediction of the generalized Law of Mass Action would be borne out by the rate data. When applied to the first equation, this general principle entails that the rate should be proportional to the concentration of CN^-, which is just what the rate results had indicated.

Following up on his proposal, Lapworth tested his theory on a variety of different compounds. He speculated that in certain circumstances, it should be possible to run the reaction not with hydrogen cyanide, but instead with a cyanide salt, which would supply the CN^- required in the first step but not the hydrogen ions required in the second. In such cases it might be possible to produce and isolate the potassium salt of the complex ion that Lapworth speculated was the product of the first step of the reaction. He reports, "[i]t was found that crystalline products could be prepared by the direct action of potassium cyanide on benzaldehyde and camphor quinone, and that these products have approximately the composition of hydrated potassium salts of the corresponding cyanohydrins" (Lapworth, 1904, p. 1207). Lapworth produced several other kinetic studies of this general sort, each proposing a multistep mechanism proceeding though organic ions, that reconciled measured rate data with the generalized Law of Mass Action.[5]

There are some important things to note about the mechanism that Lapworth provides. First, though he proposes and defends his account of the mechanism on the basis of rate data, he is concerned to show that the organic ion that he proposes as the product of the first step of the reaction is a stable entity by finding a way to isolate it as a salt. Perhaps because the role of organic ions in the transformations of organic compounds was still controversial, it was evidently advantageous for the species appealed to in a mechanism to be stable and isolable. More modern researchers, however, do "not find it necessary to isolate an intermediate in order to establish a mechanism" (Hammett, 1966, p. 466). Lapworth, furthermore, made no attempt to speculate about any transitory, unstable intermediate stages in the transformations of one organic species into another. This is just to say that Lapworth's mechanism decomposes a chemical reaction into parts, where those parts are sequentially arranged sub-reactions, but there is no attempt to look inside those sub-reactions themselves to describe transitory structures, such as the transition state.

Additionally, Lapworth is clearly working with some conception of both how organic reactions happen and why ions unite with organic compounds at some particular places rather than others. This is to say, he has theoretical views about bonding and interaction that inform his account of mechanisms. In fact, Lapworth endorsed the general view that most changes in organic compounds were due to "electrolytic dissociation" and thus proceeded through organic ions of the sort proposed in this mechanism. In addition, he later developed a broader theory that assigned "relative polarities" to the atoms in an organic molecule by first identifying a "key atom" (what we would now identify as the most electronegative atom in the molecule) and then "alternating polarities" of the atoms as they were further and further removed from the key atom. Finally, he classified reagents as "anionoid" or "cationoid," and these reagents could be expected to attack atoms in an organic molecule of the opposite polarity. In the case of the attack of cyanide ions on a carbonyl carbon, the cyanide ion would be an anionoid reagent and thus directed at atoms with a positive polarity. The carbon in a carbonyl compound would have a positive polarity because it was one atom removed from the oxygen, the key atom, with a negative polarity.

Lapworth's speculations about reaction mechanisms clearly show how rate data can be used to decompose a chemical reaction into elementary reactions which themselves are consistent with the guiding principle of reaction kinetics. He managed to do this while insisting on "an electrochemical point of view at the molecular level" (Robinson, 1947, p. 995), which was eventually recast and subsumed within the electronic theory of valence (the idea that covalent chemical bonds are the result of the sharing of pairs of electrons, recognized around 1920). The electronic theory of valence itself was eventually recast in turn within quantum mechanical accounts of chemical bonding in the early 1930s. Still, Lapworth's speculations about how to think about and rationalize chemical reactions "forged a necessary link in the chain of theory" (Robinson, 1947, p. 995) which grounds modern accounts of chemical mechanism. He is generally credited with having helped to establish both the central role of ions in chemical change, and the importance of polarization in understanding the interactions of chemical species (Schofield, 1995; Russell, 1971, pp. 287–91). Oddly, however, Hammett (1966, p. 466) reports that after Lapworth's early work on mechanisms:

> [A] remarkable thing happened; the whole development stopped almost completely for twenty years or more. There were a few kinetic investigations, but they were isolated, out of the main stream of chemical activity and unnoticed by the great mass of chemists.

Indeed reaction mechanisms don't return to center stage in the annual reports on progress in chemistry until 1938, when the author reports (Watson, 1938):

> The study of the mechanisms of chemical reactions, which has not received much attention in recent Reports, continues to provide a fruitful field of research, and our knowledge of many changes in which organic compounds participate has been enriched by the results of a large number of recent investigations.

This leads to the interesting question of why the study of mechanisms was effectively stopped for twenty years, but, following Hammett, I will not try to answer this question. Instead I will concentrate on the sorts of developments that allowed organic chemists to begin to look inside the elementary reactions, and thus to expand the power of the reaction mechanism as an explanatory tool. After briefly discussing some of these developments, I will go on to provide an example of the extended explanatory power of this second coming of reaction mechanisms.

4. Moving inside the elementary reaction

As has already been suggested, one important change in the twenty years between Lapworth's early work and the work in the 1930s that ushered in the contemporary reaction mechanism was in theories of chemical bonding and the description of the "activities" and "interactions" of the species described in a chemical mechanism. Both the electronic theory of the chemical bond and its quantum mechanical reinterpretations helped organic chemists to coherently and consistently rationalize both the bonding and interactions of these species. For instance, Lapworth's "anionoid" or "cationoid" reagents came to be called nucleophiles and electrophiles, according to a "predominating constitutional affinity either for atomic nuclei or for electrons" (Ingold, 1953, p. 198). These reagents would then interact with the polarized covalent bonds of organic substrates, which had uneven distributions of electrons ranging over the positive atomic centers. Ingold, who was a central figure in the renewed interest in reaction mechanisms, developed a system for classifying the sorts of intra and intermolecular interactions between organic species. This system attempted to take into account both the electronic nature of chemical bonding and the delocalization of electrons (eventually) described by quantum mechanics. These interactions could then be used to explain not only why a particular reaction occurred at a particular place on a molecule, but also relative differences between the "energies" of such interactions. Something like this description of the internal dynamics of chemical bonding and interaction had been present in Lapworth's accounts of the interactions between ions and polarized organic molecules, but "[i]t was only when the picture of a chemical bond became something electrical and hence potentially fluid that fruitful correlations were sought and found between electronic displacements affecting bond character and modes and ease of reaction" (Bartlett, 1956, p. 10). By the time of the second coming of the reaction mechanism, Ingold and others had "rationalized a large body of organic experience in terms of polarizations transmitted through the molecule by essentially two mechanisms," basically "inductive displacement of electrons" and what would later (after the quantum mechanical reinterpretation of the chemical bond) be called resonance (Bartlett, 1956, p. 10). Though organic chemists' tools for thinking about bonding and interaction have been articulated in the meantime, these are still fundamental concepts for understanding organic reaction mechanisms.

A second crucial series of developments that allowed chemists to finally move inside the elementary reactions was the development of theories of reaction rates. There were two distinct theories of this sort developed in the 1930s. Both of them were attempts to model the temperature dependence of reaction rates in a way that drew upon the newly developed mathematical machinery of chemical thermodynamics, canonically presented in (Lewis and Randall, 1923). The Collision Theory, as (Hammett, 1940, p. 112) calls it, was principally applicable to reactions in the gas phase.[6] The Transition State Theory, on the other hand, was taken up by organic chemists who are principally interested in reactions that take place in solution. The Transition State Theory, described in (Hammett, 1940, pp. 115–18) and (Ingold, 1953, pp. 43–52), is an extension or articulation of the thermodynamics of chemical equilibria onto the problem of reaction rates. To appreciate its significance, it is first important to see how organic chemists were able to use chemical thermodynamics, in conjunction with their newly refined understanding of chemical bonding, to understand or rationalize the effects of changes in chemical structure or medium on a chemical equilibrium.

According to chemical thermodynamics, the eventual balance between reactants and products in a chemical reaction depends on their difference in a thermodynamic quantity known as the "free" energy. Subject to some important caveats and exceptions, it is often possible in the sorts of circumstances relevant to the organic chemist to make predictions (typically qualitative or relative) about how this free energy difference will change as a result of changes in either the chemical structures of the species in equilibrium or the medium in which the equilibrium takes place. So, for instance, by appealing to Ingold's inductive effect, one might predict that replacing one atom in a reactant structure with a more electronegative atom would result in products that are, relatively speaking, of even lower energy. This would result in changes to the overall free energy change in the reaction, thereby changing the eventual balance between reactants and products in an anticipatable way. In other words, chemical thermodynamics, in concert with the updated accounts of chemical bonding and interaction, allows for a structural analysis of the eventual balance between reactants and products. This is a further step on the path of fulfilling Butlerov's ambition for structural chemistry in that features of a chemical reaction (the eventual balance between reactants and products) can be explained in terms of the structures of the reacting species.

Unfortunately, this sort of thermodynamics "has no concern with the rate of reaction, which depends on reaction mechanism: it has no concern with mechanism" (Ingold, 1953, p. 40). Only by a creative application of this thermodynamic machinery to a model of chemical transitions could the sort of structural analysis facilitated by thermodynamics be used to think about reaction rates and therefore mechanisms. The model of chemical transitions behind Transition State Theory abstracts from the actual collisions that are a "necessary part of the process of reaction" (Hammett, 1940, p. 116) and instead focuses on the changes in energy (in the first instance, potential energy) that accompany an idealized reaction trajectory. The idealized reaction trajectory is something like the least-energy path by which the reactants of an elementary reaction can come together, breaking some old bonds and making new ones in the same process, and then leave as newly formed products. If the potential energy of the reacting chemical system is plotted versus a measure of progress along this reaction trajectory (typically called the reaction coordinate), then there must be an energy maximum corresponding to some point in the idealized reaction trajectory. This highest-energy intermediate configuration, which would typically be at some midpoint in the simultaneous breaking and forming of bonds, is called the transition state, and the energy difference between the reactants and the transition state is called the activation energy.

The transition state is not an intermediate in the traditional sense because it is not stable and is extremely short lived as a result; furthermore, the transition state does not have a typical chemical structure because some of the bonds in it are partially broken while others are only partially formed. Still, the trick of the Transition State Theory is to treat the reactants and the transition state as if they were in chemical equilibrium. Following this strategy, the eventual balance between reactants and transition state is governed by the "free" energy difference between them. And after some additional assumptions and approximations, the reaction rate can be taken to be proportional directly to "the number of systems in the transition state" (Ingold, 1953, p. 47). The free energy difference which, according to this approach, determines the rate of the reaction can be related—again, subject to some important caveats and exceptions—to the activation energy of the reaction.

The upshot is that the same sort of structural analysis that was possible in the case of chemical equilibria can now be applied to questions about the rates of reactions, and thereby to mechanisms, by considering how changes in structure or medium will affect the activation energies of a class of reactions. As Ingold puts it, "we can often profitably apply the ideas of the [transition state] theory in qualitative discussions of effects of molecular structure, or of solvent, on the heat and entropy of activation, and therefore on rate of reaction" (Ingold, 1953, p. 48). Because of these developments in the theory of reaction rates, therefore, the organic chemist "could treat the effect of structure, of medium, and the like on rates and on equilibrium in the same terms" (Hammett, 1966, p. 467) and thereby extend even further Butlerov's vision of the explanatory powers of chemical structure.

5. Explaining the Walden inversion

One of the great triumphs of the second coming of the reaction mechanism was to finally bring clarity to a stereochemical puzzle first identified in the 1890s by Paul Walden.[7] Stereochemistry refers to the arrangement of atoms in space, and most particularly to the "handedness" of chemical compounds. When a carbon atom is bound to four distinct groups, those groups are arranged about the carbon as if they were at the vertices of a tetrahedron with the carbon at its center. As a result, there are two distinct such arrangements possible—these are non-super-impossible mirror images of one another called enantiomers. Enantiomers have very similar chemical structures and so their chemical and physical behavior is also very similar.

However, there are some chemical differences between them (some enzymes, for instance, will only react with one enantiomer) and they have different optical activities; they rotate plane-polarized light in opposite directions. Walden had shown that it was possible, by a series of reactions, to convert an optically active substance into its enantiomer. Chemists at the time had assumed that substitution reactions of the sort Walden employed would proceed in a way that either kept the spatial arrangement the same, or perhaps randomized the spatial arrangement. Walden showed that at some point during his series of reactions, the groups around carbon had to systematically shift from one spatial arrangement to another. In the forty years after the discovery of Walden's inversion, many additional inversions were discovered; however, not all substitution reactions on optically active starting products led to inversion, some resulted in the retention of configuration, and others led to a mix of enantiomeric products.

Prior to the mechanistic approach described below, attempts to explain the Walden inversion "were generally quite unsatisfactory since they were unable to predict what conditions or reagents might produce the inversion" (Ramsay, 1981, p. 109). For our purposes, we can think of the mechanistic explanation of the Walden inversion, worked out by about 1940 by Ingold and Hammett and others, as consisting of three parts. First, the description and justification of a substitution mechanism that would result in the inversion of stereochemical configuration. Second, the description and justification of a substitution mechanism that would result in the mixing (and possible retention) of configurations. And lastly, a scheme for deciding either which mechanism a particular substitution was likely to undergo or, failing that, how changes to the reaction setup were likely to influence the mechanism of a substitution.

The general form of a substitution or displacement reaction,[8] which is "by far the most important type of reaction in organic chemistry" (Hammett, 1940, p. 131), is:

A + B–C → A–B + C   where A, B, and C are atoms or groups of atoms.

There are, prima facie, three ways that such a substitution could occur, corresponding to the stages in which bonds are broken and formed. Either the bond between A and B forms first, followed by the rupture of the bond between B and C, or the bond between B and C ruptures first, followed by the formation of the bond between A and B, or lastly, the bond breakage and formation might take place simultaneously. It turns out that for the kind of substitutions that are relevant to the Walden inversion, only the second two occur. Furthermore, one of these possible mechanisms results in inversion of configuration, while the other results (most typically) in a mixture of configurations.

The synchronous mechanism, named S$_N$2 by Ingold, is an elementary reaction. It takes place in one step that proceeds through a transition state in which the bond between A and B is being formed at the same time that the bond between B and C is breaking. If one assumes that A and C repel one another, then (if A, B, and C all stand for single atoms) it stands to reason that the lowest-energy path for this transformation would have A approaching B–C along the line of the bond but on the opposite side of B. Then, as the new bond forms, C would exit along this same line on the opposite side from A. This would result in a transition state that minimizes repulsion, is lowest in energy, and therefore is the most likely reaction path. The case of substitution on an asymmetric carbon[9] by the S$_N$2 mechanism is similar except that an asymmetric carbon has three other groups to which it stays linked during the substitution process.

The behavior of these three linkages during the substitution "is like that of the ribs of an umbrella in a gale" (Hammett, 1940, p. 159); they invert, resulting in a change of configuration during the process of converting to the product (see Figure 4.3). That this "backside attack" mechanism would result in the lowest-energy transition state for a simultaneous substitution was supported by several arguments resting upon the newly available electronic and quantum mechanical accounts of chemical bonding and interaction. Furthermore, one of the first significant uses of isotopic labeling in chemical kinetics (see Ingold, 1953, pp. 377–9) established not only that could such inversion occur, but also that it was universal for substitution of this sort.

The sequential mechanism, named S$_N$1 by Ingold, proceeds in two steps. In the first step, which is rate determining, the bond between the asymmetric carbon and the leaving group, with some help from the solvent, breaks, resulting (typically) in an intermediate ion (much like Lapworth's mechanism). In the second step the new group forms a bond with this intermediate resulting in the final product. Because the intermediate in this process is a (solvated) carbon ion in which the carbon is bound to only three other groups (the ones that don't change during the substitution), this intermediate no longer has a "handedness." The prior configuration is lost, and when the new group adds to this intermediate it can do so in two different ways, each resulting in a different enantiomer. Thus when a substitution occurs by this mechanism, it will typically result not in an inversion, but rather in a mixture of the possible enantiomers.[10]

The last element in "solving" the Walden inversion was to develop a scheme for deciding by which mechanism a particular reaction had proceeded or would proceed. Ingold and co-workers showed how to distinguish whether a substitution occurred by the synchronous or the sequential route by kinetic criteria (measuring how the rate of the reaction varied with the changing concentrations of certain reactants). They then went on to explore the structural features and reaction conditions that promoted one or the other of these mechanistic routes. Basically, which mechanism occurs is decided by which transition state (the one in the synchronous reaction or the one in the rate-determining step of the sequential reaction) is lower in energy. This can be analyzed in structural terms, considering things such as the energetic cost of breaking the initial

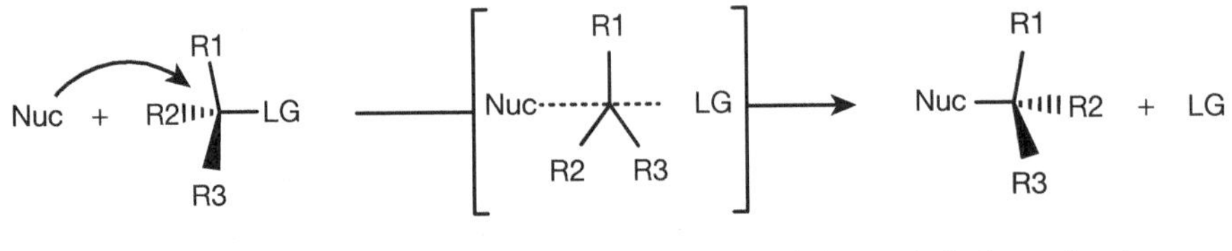

Figure 4.3 The "backside attack" mechanism, in which a nucleophile (Nuc) displaces a leaving group (LG) from an asymmetric carbon (originally bonded to groups R1, R2, and R3 in addition to LG), thereby resulting in an inversion of configuration. The structure in brackets depicts the transition state

bond, the steric environment of the transition states, and the potential stabilizing effects of the proposed solvent, etc. The upshot was that one could plan substitution reactions that should, on structural grounds, go by one or the other of these mechanisms (typically in synthesis one wants to employ the stereospecific S_N2 mechanism because it gives a single, predictable product); and furthermore, one could tell, based on the kinetics of the actual reaction, whether the reaction had gone as planned. As a result of this second coming of the reaction mechanism, then, rational planning of stereospecific synthesis first became possible.[11]

I hope to have shown, in this example, how moving inside the elementary reactions allowed organic chemists to explain new features of chemical reactions. Theories of chemical bonding and reaction rates, along with the techniques of reaction kinetics, allowed for the solution of a recalcitrant structural puzzle. Furthermore, the extrapolation of structural analysis into the theory of reaction rates, made possible by the Transition State Theory, allowed not only for structural accounts of why a reaction proceeds by a particular mechanism, but also rational prediction of how novel reactions were likely to proceed.

6. Conclusion

Attempts to explain the relative rates of chemical reactions initially led chemists to introduce a new theoretical tool—the reaction mechanism. Initially, these were decompositions of a chemical reaction into sequential steps containing hypothetical intermediates. Eventually, facilitated by developments in theories of chemical bonding and reaction rates, chemists moved inside these elementary steps, describing the transitory structures theoretically relevant to the rates of chemical reactions. With this move, the explanatory power of the reaction mechanism dramatically increased. Structural features of the products of the most important types of reaction in organic chemistry were now explicable in mechanistic, and therefore structural, terms. Furthermore, this new mechanistic understanding could be leveraged to facilitate the rational planning of novel chemical syntheses, and this has always been the coin of the realm for the organic chemist.[12]

Notes

1 See Goodwin (forthcoming) for an account of structural formulas as the primary models of organic chemistry.

2 The reaction coordinate is the lowest-energy path from reactant to products, over the course of which some bonds in the reactants are broken, while new bonds in the product are formed. The highest-energy point along this path is called the transition state.

3 Bartlett (1956, pp. 9–10) says, "If in any organic reaction one could write reliable structural formulas for the *n* intermediates and the *n*+1 transition states, the problem of its mechanism might be considered solved."

4 See Goodwin (forthcoming) for the roles of models in chemistry.

5 See Hammett (1940, pp. 96–9) for an account of Lapworth's reasoning as applied to the halogenation of ketones.

6 Hammett (1966, p. 466) speculates about some of the ways that work in the kinetics of gas reactions helped trigger the "renaissance in the study of kinetics and mechanisms of reactions in solution," but I am not able to discuss these here.

7 My description of the Walden inversion and its mechanistic explanation draws on the accounts in Ramsay (1981, pp. 107–15), Ingold (1953, pp. 372–400), and Hammett (1940, pp. 157–83).

8 It is nucleophilic substitutions that are relevant to the Walden inversion. There are also electrophilic substitutions, which I will not discuss in this chapter.

9 An asymmetric carbon is one with four distinct groups bound to it, which would generally make a molecule containing such a carbon optically active.

10 It is, not surprisingly, substantially more complicated than this. There are some circumstances in which the S_N1 mechanism will result in retention of configuration and others when it favors inversion, but these exceptions are also rationalizable in mechanistic terms.

11 This is not meant to suggest that the only significant role of mechanistic studies is in synthetic design. As a helpful referee pointed out, mechanistic understanding of organic chemistry has also facilitated rational drug design, which is increasingly replacing discovery by trial and error.

12 See Goodwin (2011) for more on how mechanisms contribute to synthesis.

References

Bartlett, Paul D. (1956). "Reaction Mechanisms" in *Perspectives in Organic Chemistry,* ed. Sir Alexander Todd. Interscience Publishers, New York, pp. 9–27.

Brock, W. H. (2000). *The Chemical Tree: A History of Chemistry*. Norton, New York.

Desch, C. H. and Morgan, G. T. (1909). "Organic Chemistry." *Annual Reports on Progress in Chemistry 1908* 5: 74.

Desch, C. H. and Lapworth, A. (1910). "Organic Chemistry." *Annual Reports on Progress in Chemistry 1909* 6:68.

Duhem, Pierre. (2002). *Mixture and Chemical Combination*. ed. and trans. Paul Needham. Dordrecht, Kluwer Academic Publishers.

Goodwin, W. (2011). "Mechanisms and Chemical Reaction" in *Handbook of the Philosophy of Science: Philosophy of Chemistry*, ed. A. Woody, R.F. Hendry, and P. Needham. Elsevier Science, the Netherlands, pp. 309–27.

Goodwin, W. (forthcoming). "Models of Chemical Structure" to appear in the *Springer Handbook of Model-Based Science*.

Gould, E. (1959). *Mechanism and Structure in Organic Chemistry*. Henry Holt and Company, New York.

Hammett, Louis P. (1940). *Physical Organic Chemistry: Reaction Rates, Equilibria, Mechanisms*. McGraw-Hill Book Company, New York.

Hammett, Louis P. (1966). "Physical Organic Chemistry in Retrospect." *Journal of Chemical Education* 43(9): 464–9.

Ingold, C. K. (1953). *Structure and Mechanism in Organic Chemistry*. Cornell University Press, Ithaca, NY.

Lapworth, A. (1903). "Reactions Involving the Addition of Hydrogen Cyanide to Carbon Compounds." *Journal of the Chemical Society* 83: 995–1005.

Lapworth, A. (1904). "Reactions Involving the Addition of Hydrogen Cyanide to Carbon Compounds. Part II. Cyanohydrins Regarded as Complex Acids." *Journal of the Chemical Society* 84: 1206–1214.

Lewis, G. N. and Randall, M. (1923). *Thermodynamics*. McGraw-Hill, New York.

Lowry, T. H. and Richardson, K. S. (1987). *Mechanism and Theory in Organic Chemistry*, Third Edition. Harper and Row, New York.

Nye, Mary Joe. (1993). *From Chemical Philosophy to Theoretical Chemistry: Dynamics of Matter and Dynamics of Disciplines 1800–1950*. University of California Press, Berkeley.

Ramsay, O. B. (1981). *Stereochemistry*. Heyden, London.

Robinson, R. (1947). "Arthur Lapworth, 1872–1941." *Journal of the Chemical Society* 989–996.

Russell, C. A. (1971). *The History of Valency*. Humanities Press, New York.

Saltzman, Martin (1972). "Arthur Lapworth: The Genesis of Reaction Mechanism." *Journal of Chemical Education* 49(11): 750–752.

Schofield, K. (1995). "Some Aspects of the Work of Arthur Lapworth." *Ambix* 42(3): 160–186.

Watson, H. B. (1938). "Organic Chemistry." *Annual Reports on Progress in Chemistry 1938* 35: 208.

5

MECHANISM, ORGANICISM, AND VITALISM

Garland E. Allen

1. Introduction

The term "mechanism" has been used widely in the life sciences at least since the seventeenth century. Embodied in the "mechanical philosophy" of Thomas Hobbes (1588–1679), Pierre Gassendi (1592–1655), René Descartes (1596–1650), and Robert Boyle (1627–91), among many others, it came to dominate both the epistemology and ontology of virtually all modern western science (Durbin, 1988). In this sense "Mechanism" refers to a world-view or philosophical system that sees nature from a materialist perspective and to varying degrees, sometimes literally, sometimes figuratively, marked with analogies to machines or machine-like processes. It is also still referred to as the "mechanical philosophy" or "metaphysical Mechanism," since it is "concerned with questions about the constituents and organization of the natural world" (Glennan and Illari, Chapter 1, this volume). However, as philosophers of science know, the term "mechanism" has a second, related, and overlapping meaning designated as *operative* or *explanatory mechanism* (hereafter, "operative mechanism").

Operative mechanisms are processes in which a specific set of localized interactions of component parts leads to some regular or predictable outcome, as in the *mechanism* of enzyme action or the *mechanism* of genetic recombination. According to "new mechanist" philosophers such as Peter Machamer, Lindley Darden, and Carl Craver, these mechanisms involve a set of entities, activities, and spatial-temporal interactions that produce and explain the phenomenon in question (Machamer, Darden, and Craver, 2000; Craver and Darden, 2013; see also Bechtel and Abrahamsen, 2005). Following the editors' format laid out in Chapter 1, when the word "mechanism" refers to a philosophical or world-view, it will be capitalized, while when it refers to operative or local mechanisms it will not. This chapter will focus on the historical development of philosophical Mechanism beginning with the seventeenth century, though focusing primarily on the nineteenth and twentieth centuries.

While focusing primarily on the philosophical aspects of Mechanism in the life sciences, it is also important to emphasize that this world-view was not merely an abstract philosophical discourse, but one that evolved in various social and economic contexts, and was influenced not only by changing technologies available in other sciences, but also by economic, social, and political developments in society at large. Its uses and meaning have always been very much context-dependent.

2. Mechanism and materialism in the life sciences

As pointed out in Chapter 1, Mechanism derives from the mechanical philosophy of the sixteenth and seventeenth centuries, and in various forms has been one of the dominant philosophical and methodological approaches in the life sciences up to the present day (Boas, 1952; Dijksterhuis, 1961: 3, 431; Durbin, 1988: 179; Shapin, 1996). Mechanism is just one of several forms of a broader philosophy of materialism and encompasses, though is not necessarily limited to, the associated concepts of atomism and matter-in-motion tracing back to the writings of Democritus and Lucretius in the classical world (Durbin, 1988). Throughout its history "Mechanism" has been used in both an ontological and epistemological sense. For the most part, however, working biologists have been less prone to engage in extended ontological discussions about whether organisms *really* are nothing more than complex machines or aggregates of chemical reactions, preferring to use mechanical or chemical analogies as a heuristic device to pose questions and test hypotheses experimentally.

While differing in focus, earlier Mechanisms and new Mechanism have overlapping commitments. Deriving from the machine analogy, both expect that any process can be described in terms of material, physical components that work together in an organized way harmoniously and consistently. For example, just as the cogs in a gear-shift mechanism fit together to produce the motion of a vehicle, so during enzyme-mediated catalysis a substrate fits into the active site of its enzyme in such a way that specific chemical bond rearrangements are facilitated, yielding a new end-product.

Although Mechanism has tended to dominate much of western science, especially biology, over the past 350+ years, periodically voices have been raised in protest against what seemed like mechanists' over-simplistic way of viewing organisms. The alternative has been one or another forms of what has been referred to as "holism" or "organicism" that claimed living organisms could never be fully understood as, or reduced to, simple mechanical or chemical processes, but had to be looked at as "wholes," as entities in themselves, not merely as an additive aggregation of parts. In the period 1700–1850 many of these alternative views were mostly couched in mystical terms, as the result of some immaterial, "vital," or "directive" force ("vitalism") that had no counterpart in the non-living world. Many vitalistic theories were considered to be serious alternatives to standard mechanistic explanations in that they posited law-like activities, but just not the sort that arose from ordinary physical or chemical processes.[1] These mechanist-vitalist debates punctuated the history of biology, producing a pendulum swing at times away from Mechanism, at times back toward it, with each swing, however, altering in various ways the mechanistic research programs that were under dispute. The periods of the most blatant promotion of mechanistic thinking include the mid-to-late seventeenth and eighteenth centuries, the mid-nineteenth century, and much of the twentieth century.

Mechanism is a form of materialism that gained ascendancy during the scientific revolution of the seventeenth century as the Mechanical Philosophy, or later, mechanistic materialism (Boas, 1952; Westfall, 1971; Shapin, 1996). Materialism is the view that all phenomena in the universe depend upon the interactions of some sort of material particles (derived from Greek "atomism") in continuous motion, whose collisions generate events/phenomena in the world. It also embodies the assumption, or metaphysical claim, that our ideas about phenomena derive from interaction with the material world through our senses. This means that matter is primary and our ideas about it are secondary, or derivative (in contrast to non-materialist, or "idealistic" philosophies that claim ideas are primary, or exist as abstract categories, and the material world is derived from these pre-existing and non-material categories). The mechanistic world-view also claims that nature can best be understood as a mosaic of separate parts, a detailed description of which, when combined together, yields a complete description/understanding of the phenomena in question.

From the mechanistic materialist point of view, the proper way to study any system is to take it apart (analysis) and determine the characteristics of the individual, isolated parts under as controlled a set of conditions as possible. For some mechanists, such as biochemists studying enzyme kinetics, the lowest level of analysis might be a solution of purified enzyme and its substrate. For others, such as cell biologists studying respiration, the lowest level of organization might be a group of molecules attached to a specific cell membrane system. To understand a given biological process, the mechanistic materialist strategy has been to start with isolating those parts by some analytical procedure and identifying them structurally and functionally. Mechanistic materialists generally follow some kind of reductionist strategy, that is, to investigate higher-level processes by reducing them to their lower levels of organization: for example, cells in terms of molecules, organs in terms of cells, organisms in terms of organ-systems, and the like. In this formulation mechanistic materialism is downward looking, proceeding ultimately to the most general and basic material-level explanation as possible. By treating living processes in material, molecular, or atomistic terms, reductionism brings biology into complete accord with physics and chemistry.

In discussing the relationship between parts and wholes, it is sometimes assumed that mechanistic materialists have no appreciation of complexity and no interest in how components in a system interact to produce an overall effect. This is not the case for most mechanists. For example, new mechanists identify the components of a complex system (muscle cells, for example) to understand how they work together in relation to the whole (coordinated muscle contraction). For mechanistic materialists, investigating the components separately reveals their individual characteristics more precisely than is possible when studying them as part of a complex whole. However, a tacit assumption of most mechanistic materialists is that once the characteristics of each component are known, their relationship to each other and to the whole will become apparent. It has been a cardinal principle of Mechanism that the properties of living systems can be understood in terms of the laws of physics and chemistry (a view sometimes referred to as "physicalism"). Thus, the lowest level of organization to which mechanists have tended to reduce complex phenomena has been that of atoms and molecules (Gilbert and Sarkar, 2000; Bechtel and Richardson, 1993; Schaffner, 1967, 1976; Allen, 1975).

In their attempt to model biology after the physical sciences in the later nineteenth and early twentieth centuries, mechanists emphasized the importance of *experimentation*, which allowed the biologist to distinguish between alternative hypotheses; hypotheses that were incapable of experimental test—for example, that living systems were organized by a non-chemical, non-physical "vital" force—were considered of no scientific value. Experimentation thus served epistemologically as a corrective against unbridled speculation and idealistic metaphysics.

Standing in opposition to mechanistic materialism in the later nineteenth and early twentieth centuries were a variety of views that can be clustered under the general term "holism" (also called "organicism" or "emergentism") which, as the names imply, were concerned with how complex systems function organically, as a unified "whole." Holistic biologists sought to provide an approach that differed from what they saw as the piecemeal, oftentimes naïve and simplistic views promoted by zealous mechanists such as Carl Vogt or Jacques Loeb (see below). In the period starting around 1880, two categories of holistic thinking emerged, especially in Europe: materialist and non-materialist, though the two were often mistakenly conflated. Among the former are included holistic materialism and dialectical materialism, and among the latter vitalism and some versions of organicism. Holistic and dialectical materialism share a materialist epistemology, seeking to account for living processes as functioning wholes within the framework of known physical and chemical laws. Vitalism, on the other hand, claims that living organisms defy description in purely physico-chemical terms, because organisms possess some non-material, non-measurable forces or directive agents that account for their complexity. Vitalism was regarded by mechanistic materialists as fuzzy-minded and subjective, offering no real guidelines for practical investigation.

Gilbert and Sarkar (2000) have pointed out that even though other versions of holistic thinking in the early twentieth century shared with Mechanism a strictly materialist epistemology, they were nonetheless largely ignored or rejected because they were perceived as keeping "very bad company"—either with vitalism of one sort or another, Nazi organicism, or Marxist dialectical materialism—all of which advanced some kind of holistic/organicist view (Gilbert and Sarkar, 2000: 4–5). Yet, in substance, holistic materialism, and the more formally developed subcategory, dialectical materialism, stand in sharp contrast to vitalism in not postulating *any* forces that cannot be understood in terms of the known laws of physics and chemistry. What united all forms of holism was the clear recognition that living organisms were capable of activities that had no counterpart in the machine world: self-replication, purposeful or ordered response to stimuli, elaborate self-regulatory capabilities, and the incredible efficiency of their energy transduction. To understand these complex interactive processes, it was necessary to get beyond the individual parts to look at the whole system or process. In the period 1900–40, finding methods—conceptual and experimental—to accomplish these investigations was a major goal of holistically oriented materialist biologists (including the embryologists Hans Spemann, Ludwig von Bertalanffy, Joseph Needham, Paul Weiss, and the geneticist Richard Goldschmidt).

As part of their opposition to mechanistic thinking, holistic materialists emphasized the importance of distinguishing between levels of organization in a complex system (in studying organisms this could include the atomic, molecular, cellular, tissue, organ-system, organismic, population, or ecosystem levels), proceeding to investigate each level on its own terms. The holistic approach does not preclude starting with a reductionist, analytical breakdown of a complex system into its component parts, but it does emphasize that this is not enough. For example, in studying the function of the mammalian kidney, holistic materialists might first determine all of the tissue and cell types of which the kidney is composed. These might be studied initially in isolation, using methods of histology, cytology, and cell physiology. However, holistic materialists argue that it is also necessary to study the kidney in the context of the whole, functioning organism where its response to variables such as blood pressure, ion concentration, and hormones could be revealed. Isolating and studying separately the ten or so individual tissue types that make up the kidney could not be expected to provide a full picture of how the kidney functions *as a whole* within the intact organism.

The nephron, the main filtration site in the kidney, depends for its function completely on its interaction with cells in different regions of the kidney, from the cortex, where the first major filtration of numerous ions occurs, to the medulla, where selective reabsorption takes place. Each level of organization within the organ—cellular, tissue, and organ—has its own properties and characteristics that cannot be understood only by examining them separately. In addition, these tissues types are organized anatomically in such a way that their position within the kidney is a crucial aspect of their function. It is in their appreciation of the concept that each level of organization in a complex system has its own special properties, and that these must by studied by techniques appropriate for *that level*, that holists have differed in one significant way from mechanists in the past.

For holistic biologists, complex systems (even relatively simple ones) show emergent properties that are the product of the individual parts *plus* their interactions (what we call today *synergistic effects* are an example of emergent properties). To those holistic thinkers committed to materialistic explanations, emergent properties were not mystical, since they resulted from the *interaction* of the parts within the whole. But emergent properties could not be predicted from knowing only the individual components making up a given system. Explanations relevant to a given level had to be derived from studying that level itself and not merely by isolating, describing, and then extrapolating upward to higher levels.

A more formalized version of holism, known in the twentieth century as dialectical materialism, became the official philosophy of science (and social science) in the Soviet Union after 1917 and the People's Republic of China after 1949. Dialectical materialism shares all the basic characteristic of holistic materialism: concern with levels of organization, interaction of parts, and emergent properties, but it added several important features. The most unique and defining of these is the dialectical insistence itself: that all processes can be best understood in terms of the interaction of opposing forces, or agents within a system, and between any system and its external environment. A classic example of dialectical materialist thinking applied to biology has been Darwinism, where the dynamics of evolutionary change is constantly moved forward by the interaction of two opposing tendencies: heredity (non-variable reproduction) and variation (variable reproduction), adjudicated by natural selection (Prenant, 1943: 138).

With either process by itself there is no evolution; with both present, evolution becomes inevitable. A chief feature of dialectical materialism, its advocates point out, is that it provides a way to investigate and understand the dynamic change that characterizes all systems in the universe. Random events may of course affect the way in which any system changes, but more constant factors driving change lie at a deeper level internal to the system itself. Further discussion of the characteristics of dialectical materialism take us too far afield from the discussion of mechanistic materialism. The point of mentioning it here is to indicate that from the 1860s onward another version of holistic thinking existed that was overtly and self-consciously materialistic and devoid of mystical or vitalistic overtones. That it was not more consciously pursued was a result of the political climate following the Bolshevik revolution and the ensuing gulf between Soviet and western science and philosophy (Graham, 1986).

3. Mechanistic philosophy in the nineteenth and twentieth centuries

Early mechanistic approaches to understanding living organisms in the seventeenth and eighteenth centuries were highly simplistic and literally cast in terms of familiar mechanical devices and processes. They were a way to explain biological processes in terms and models that were familiar. The use of such mechanistic analogies grew in part out of the increasing presence of machinery in daily life (water pumps, mining hoists, and the like). Organisms, like the Newtonian universe, operated like machines, with precision, regularity, and predictability, features that life scientists wanted to extend to living organisms in the same way physicists had done for planetary systems. Even in popular culture of the later eighteenth century, the mechanistic view found expression in a fascination with mechanical models of animals, such as Vaucanson's Duck, a wind-up model that could feed, drink, digest, and defecate. Mechanistic models were expanded and made more sophisticated as the nineteenth century progressed

Historians have dubbed the mid- and later nineteenth century as the "age of materialism," in part because of the rise of industrial capitalism, which emphasized the production and distribution of material commodities, a concern for the acquisition and movement of capital, and the increased presence of machines in the workplace. It was in this period that Marx worked out the laws of capitalist development, placing them in the context of a materialist theory of history. This widespread materialist culture also had its own versions of mechanistic theories in biology, particularly in medicine and physiology.

Medical materialism developed in the mid-nineteenth century, particularly in France and Germany, in the wake of the revolutions of 1848. Ideologically, materialism was associated with the attempt to eliminate the Cartesian dualism between body (viewed mechanistically) and the mind/soul (viewed idealistically, non-materially), and as part of the growing anti-clericalism of the period (Temkin, 1946: 324). In France, Henri Dutrochet (1776–1847) investigated the passage of materials across membranes as a purely physical process and Francois Magendie (1783–1855) pioneered experimental studies of nutrition (using animal models) and the structure and function of the nervous system. Magendie's successor at the College de France, Claude Bernard (1813–78), pursued a quasi-mechanistic research program in his studies on the glycogenic function of the liver and the maintenance of the *milieu intérieur*, or internal environment. He did, however, leave some room for vitalistic processes that could not be explained solely by physical or chemical means (Temkin, 1946: 326).

In Germany, too, physiology was at the center of a new mechanistic biology, especially embodied in the influential Berlin medical materialists of the late 1840s and 1850s: Ludwig von Helmholtz (1821–94), Emil Du Bois-Reymond (1818–96), Carl Ludwig (1816–95), and Ernst von Brücke (1819–92). Interested particularly in the electrical properties of nerve and muscle tissue, this group pioneered a thorough physical and chemical research program that explicitly made no distinction between organic and inorganic matter. In his "Investigations concerning Animal Electricity" (1848) Du Bois-Reymond looked forward to the day when "physiology would dissolve completely into biophysics and biochemistry" (Temkin, 1946: 326). It was in this same vein that Karl Vogt (1817–95) could write (1845): "thoughts have about the same relation to the brain as bile has to the liver or urine to the kidney" (quoted in Temkin, 1946: 324).

Toward the end of the century mechanistic explanations found clear expression in another area of life sciences: embryology. Wilhelm His (1831–1904) attempted to account for the folds that occurred in the various germ layers in vertebrate development by mechanical processes that he illustrated with the folding properties of a series of rubber sheets and tubing. Slightly later (1881) Wilhelm Roux (1850–1924) advanced his mosaic theory of embryonic development (also known as the Roux-Weismann theory, after August Weismann, 1834–1914, who had come up independently with a similar view). Their focus had been on the mechanism of differentiation during embryogenesis. Roux claimed with each cell division hereditary units were parceled out in such a way that each cell generation received increasingly specialized particles, so that by the time differentiation was complete, each cell type (muscle, nerve, skin) contained only the particles determining that cell's specific characteristics. To test his theory, Roux killed one of the first two blastomeres of the frog egg with a hot needle. True to his prediction, he got some half-embryos that developed through an early embryonic (the gastrula) stage. From this work Roux developed an entire research program, known as *Entwicklungsmechanik*, aimed at promoting an experimental and mechanistic understanding of embryonic differentiation.

However, in 1891 a young German embryologist, Hans Driesch (1867–1941), working at the Naples Zoological Station, carried out a similar series of experiments using the sea urchin. Instead of killing one of the blastomeres, Driesch separated them by vigorously shaking the cultures. Driesch expected half-embryos also, but found that the separated blastomeres each developed into a complete, though smaller, embryo. Driesch, who was also initially a mechanistic materialist, interpreted the results as contradicting Roux's strictly mechanical model. No machine could reconstruct the whole out of an incomplete set of individual parts. The sea urchin embryo acted as what Driesch called a "harmonious equipotential system," and for several subsequent years he carried out experiments to study the characteristics of self-regulation and adjustment of embryos to altered conditions. Eventually, by the early 1900s, however, he despaired of finding a mechanistic, physico-chemical solution to the problem, and adopted an increasingly vitalistic interpretation (see below).

4. The early twentieth century: Jacques Loeb and the mechanistic conception of life

In the twentieth century Mechanism found its most explicit and forceful exponent in a German émigré to the United States, physiologist Jacques Loeb (1858–1924). Loeb's work illustrates more clearly than many others how one form of mechanistic materialism was put into practice as a scientific research program.

Born in Germany in the same year as the publication of the influential essay "The Mechanistic Conception of Life" by Rudolf Virchow (1821–1902), Loeb was, in the words of historian Donald Fleming, "a child of his time" (Fleming, 1964: xi). Virchow's views were exemplars of an explicitly materialistic interpretation of the natural and social worlds (Allen, 1992). Interested in the problems of human volition and the expression of the will, Loeb received an M.D. degree from the University of Strasbourg, working subsequently in the laboratory of Adolf Fick (1829–1901), who had been a student of the Berlin medical materialists and espoused their view that life was nothing more than an expression of physical and chemical laws. Embedded in this mechanistic view was the assumption, as in physics, of a strictly deterministic universe where, if one knew all the inputs, it would be theoretically possible to predict the precise outcome of any process, including human behavior. At Würzburg Loeb also came under the influence of plant physiologist Julius Sachs (1832–97), who gave him an appreciation for the importance of experimentation and an interest in the specific problem of plant and animal movements, or tropisms. It was only after his emigration to the United States in 1891 that Loeb's mechanistic philosophy became explicit and took its most concrete shape. Loeb's research on artificial parthenogenesis (the development of an unfertilized egg into an adult organism) provides a useful illustration of what Mechanism meant as a biological philosophy.

No process in late nineteenth- or early twentieth-century biology remained more of a bastion for metaphysical explanations than that of fertilization of the egg by a sperm and the subsequent developmental events that it triggered. Once the sperm has penetrated the egg, a cascade of events takes place, rapidly in most cases, leading to the first cleavage, in which the single-celled zygote becomes a two-celled embryo (the blastomeres in Roux's experiment). This remarkable series of events raised many questions: How does the entrance of sperm into the egg cytoplasm initiate division? Does the point of penetration determine the plane of division, and if so how does the plane of the first division affect the future axis (anterior-posterior, dorsal-ventral) of the organism? What role, if any, does the egg cytoplasm play in the course of development? And, of course, there is the fundamental question of how differentiation takes place.

Although a strong proponent of Roux's *Entwicklungsmechanik* program, Loeb at first saw no contradiction between Driesch's results and a mechanistic interpretation. He, too, had experimented with sea urchin eggs at Woods Hole, and found that fragments of embryos, or even eggs, could be stimulated to produce full embryos under the right conditions. It was when Driesch declared himself an unabashed vitalist that Loeb parted ways, since he saw vitalism as an unacceptable metaphysics that served no function other than stifling research.

Loeb focused on the problem of fertilization, since nothing seemed to be more basic to the distinction between living and non-living matter than the moment in the initiation of a new life. It was also known, however, that under natural conditions, the eggs of some species (ants, bees, wasps, aphids) undergo normal development *without* being fertilized, i.e., natural parthenogenesis (Loeb, 1900). Loeb asked whether parthenogenetic development could be induced by physical or chemical means, which, if possible, could provide a new mechanistic approach to a fundamental, vital process.

Loeb's experiments involved placing sea urchin eggs in sea-water of varying osmolarities (concentration of ions), and of varying types of salt combinations. The ideas behind these experiments came from Sachs, who had studied the effects of varying salt concentrations on water uptake, transpiration, and photosynthesis in plants, and from the physical chemist Svante Arrhenius (1859–1927), whose theories had convinced Loeb that biological phenomena were controlled by precise ionic concentrations. Living phenomena such as fertilization thus could be approached from the standpoint of physical chemistry. He hoped to *replicate* that normal process by known physico-chemical agents (Loeb, 1912 [1964]: 6).

When Loeb placed unfertilized sea urchin eggs in hypertonic sea-water, he found that on returning them to normal sea-water, the eggs started to divide (though at first none developed beyond early blastula stage) (Loeb, 1899: 326–7). By manipulating variables such as types and concentrations of various ions or the length of time the eggs were immersed, Loeb determined the proper mixture and concentration of ions and immersion time (2 hours) that would not only yield a large number of cleavages, but would also support development through the swimming larval stage (Loeb, 1899: 329). As he wrote:

> I consider the chief value of the experiments . . . to be the fact that they transfer the problem of fertilization from the realm of morphology [which meant at the time descriptive and often speculative work] into the realm of physical chemistry.
>
> *(Loeb, 1899: 332)*

Others soon showed that purely physical stimuli, such as pricking the egg with a needle, could also initiate development. These were all positive confirmations that life could be understood by employing the concepts of physics and chemistry, and testing them empirically by controlled experiments. Loeb's interest in this work was not merely in finding the operative *mechanism* by which fertilization actually triggered development, but rather in demonstrating that living processes followed the laws of physics and chemistry, and through proper experimental investigation could be brought under human control. The "Mechanistic Conception of Life" represented progress for Loeb precisely because it led to control, to the engineering of nature (Pauly, 1987: 114).

Loeb's mechanistic view was cast in the reductionist language of physics and chemistry. Loeb's form of reductionism sought to explain complex processes by examining their increasingly lower levels of organization. What Loeb meant by reductionism (and he did use the term) was to trace higher-level processes down to particular physical and chemical reactions. As he wrote in 1888:

> Whatever appear to us as innervations, sensations, psychic phenomena, as they are called, I seek to conceive through reducing them—in the sense of modern physics—to the molecular or atomic structure of the protoplasm, which acts in a way that is similar to (for example) the molecular structure of an optically active crystal.
>
> *(Pauly, 1987: 38)*

Loeb's commitment to Mechanism was part of a larger program he shared with many colleagues to make biology a hard science, and thus to place it on an equal professional footing with the physical sciences.

5. Mechanistic approaches in the later twentieth century

Loeb's mechanistic philosophy was highly influential in a variety of areas of biology outside of general physiology. It became part and parcel of Thomas Hunt Morgan's (1866–1945) research program in genetics, and was codified in the "beads-on-a-string" representation of genes as atomistic parts of chromosomes (Morgan et al., 1915: 60). Others saw genes as the biologist's equivalent to the atoms or molecules of physics and chemistry, and the organism as a mosaic of traits determined by the individual, particulate gene. Genetics became an exemplar of the mechanistic approach for the period 1915–50. Mechanistic thinking was also highly influential in the population genetics of the founders of the evolutionary synthesis, particularly in the writings of Ronald A. Fisher (1890–1962) in 1930. Building on work in laboratory genetics by Morgan and others, Fisher saw evolution as the process of selection acting on discrete Mendelian genes, not the organism as a whole. Fisher's claim, in his path-breaking *The Genetical Theory of Natural Selection* (Fisher, 1930), was that by reducing populations of organisms to individual, freely recombining genes, he could accomplish for evolutionary biology what the kinetic theory of gases had accomplished for physics and chemistry.

Neurobiology continued the mechanistic trend started by the Berlin medical materialists of the previous century with studies of neuronal conduction not as the flow of electrons along a wire, but of ions across the axon membrane down a concentration gradient to produce a self-perpetuating action potential. Conduction or inhibition across synapses between neurons could also be reduced to molecules released by the pre-synaptic and absorbed by receptors on the post-synaptic membrane. Even in psychology, in the first half of the century, mechanistic thinking was prominent, especially in the behaviorism of John B. Watson (1878–1958) and B.F. Skinner (1904–90). Their views of operant conditioning were often couched in machine-like language, and saw the organism and its behavior as the result of a set of input-output processes that was both reductionistic and deterministic.

The rise of biochemistry and molecular biology in the middle and second half of the century provided a new stimulus to mechanistic approaches. Biochemists in the 1940s and 1950s referred to cells as "bags of enzymes" while elucidation of the molecular structure of deoxyribonucleic acid (DNA) provided one of the most far-reaching revolutions in mechanistic thinking in recent times. The fact that similar mechanisms for DNA replication, protein synthesis, and gene control seemed to apply to all organisms was exemplified in the claim by molecular biologist Gunther Stent (1924–2008) that "What is true of *E. coli* [as exemplar of all bacteria] is true of the elephant." The implication is not only that there is a unity among diverse living organisms on earth, which biologists already recognized, now extended to the molecular level, but also that multicellular organisms can be viewed simply as large collections of unicellular organisms. This brand of Mechanism ignored the emergent properties that characterize the different levels of organization in complex organisms.

6. Beyond mechanistic materialism: (w)holism and organicism

Since the rise of mechanistic materialism in the seventeenth century, there have been reactions to what seemed like the overly simplistic machine models of living organisms. Initially, these were expressed as some form of vitalism in which living beings were thought to possess non-material, non-measurable energies or forces that were qualitatively different from those described in non-living entities. Echoes of this vitalism can be found in the writings of French anatomist and physiologist Xavier Bichat (1771–1802), who saw the ability to experience sensations and thought as evidences of vital properties very different from those of the inorganic world.

However, by the mid-nineteenth century, Driesch notwithstanding, vitalism as such had lost much of its significance, as it offered no prospects for a meaningful research program.

However, one aspect of the vitalistic perspective that did attract attention, especially in the early twentieth century, was the holistic or organismic movement, which focused attention on organisms, populations, and even ecosystems as "wholes"; that is, not as machines atomized into discrete and separable components. Traces of holistic or organismic thinking are apparent in the *Naturphilosophie* movement in Germany in the late eighteenth and early nineteenth centuries, which emphasized the importance of looking at organisms whose properties were something more than just a sum of their individual parts. A concern with holistic thinking remained a part of many aspects of German biology well into the twentieth century, although it had some proponents in Europe and North America as well.

Historian Anne Harrington has suggested that holism in biology in the early twentieth century grew out of not only reaction to the biological Mechanism of Roux, Loeb, and others, but also as a reaction to the cultural fragmentation associated with "modernism," World War I, urbanization, industrialization, and the perceived intrusion of machines into everyday life. The fragmentation was most painfully obvious in Germany following the harsh political and financial terms forced on it by the Versailles Treaty after World War I. In combination with disillusionment and embittered national pride, Germany produced a romanticized nationalism and nostalgia for a past in which it was believed that humans were more harmoniously integrated into the natural world (Harrington, 1996: 19–33). Thus the holistic view was intimately tied up with a variety of scientific and cultural trends that set it against the mechanistic world-view that many thought had led to an undermining of morality, the social fabric and "man's place in nature."

In the period between 1900 and 1950, holistic philosophy included a wide spectrum of approaches with widely differing views on both the philosophical meanings of holism and the practical applications of holistic thinking to scientific research programs. Driesch was, of course, the major and most extreme exponent. His claim that embryonic development was guided by an "entelechy," an organizing, directive force that consumed no energy, was immaterial, but it was the factor that distinguished living from non-living matter. There were many others, however, including Jakob von Uexküll (1864–1944), Ludwig von Bertalanffy (1901–72), and Henri Bergson (1859–1941), who espoused similar views. In one way or another each of these individuals advocated an anti-mechanistic, holistic approach, replacing the machine analogy with an organicism that repudiated the reduction of all living processes to simple matter-in-motion (Harrington, 1996).

But in the twentieth century there were two sorts of holistic approaches: those that still advocated some sort of vitalistic property of living systems, and those that claimed that living systems obeyed all the laws of physics and chemistry, yet because of their various levels of organization, displayed properties that were greater than the sum of their parts. Holists of either type argued that even if ions in solution in the cell cytoplasm, or in blood plasma, obeyed Arrhenius' laws of electrolytic dissociation, that was not what made an organism "alive." What was important was that organisms displayed the wholeness of the organic process, characterized by their organizational relationships. For all holists the whole was always greater than the sum of the parts, thus showing emergent properties that were not inherent in any of the parts by themselves.

Holists with a vitalistic tinge, such as von Uexküll, interpreted emergent properties as an expression of a vital property that could never be understood from a purely analytical approach. In a tribute to Driesch and his concept of the embryo as a "harmonious equipotential system," von Uxeküll wrote:

Driesch succeeded in proving that the germ cell does not possess a trace of machine-like structure, but consists throughout of equivalent parts. With that fell the dogma that the organism is only a machine. Even if life occurs in the fully organized creature in a machine-like way, the organization of the structureless germ into a complicated structure is a power *sui generis*, which is found only in living things and stands without analogy. . . . It is not to be denied that vitalists are the victors all along the line.

(Quoted in Harrington, 1996: 51)

To mechanists, this interpretation represented a resurgence of German romanticism, a new *Naturphilosophie* that was not only philosophically backward-looking but from a pragmatic point of view had no significant research potential. At the same time, however, a new form of holistic thinking, what came to be called *holistic materialism*, was emerging. Holistic materialism attempted to approach biological systems with an appreciation of their emergent properties in materialist terms that provided the basis for an experimental research program. However, so powerful were the forces of mechanistic thinking in the first half of the twentieth century that attempts at holistic alternatives, even when expressed in materialistic, non-mystical ways, were easily misunderstood or viewed with considerable suspicion.

Among those who tried to develop a holistic materialist, experimental framework was German embryologist Hans Spemann (1869–1941) and his students in the 1920s–40s. Trained, like Loeb, at Würzburg in morphology, Spemann took up experimental work in embryology as an alternative to the purely descriptive and speculative approach of his mentors. Although appreciating Driesch's motivations to treat the embryo holistically, Spemann could not agree with his overtly vitalistic interpretation. Starting in the late 1890s through the mid-1920s, Spemann developed the concept of embryonic induction to account for the sequence of events in tissue differentiation. His approach was holistic, materialistic, and experimental, based on removing tissues during different stages of development and transplanting them to various regions of younger or older embryos. His initial example was the induction of the lens in the developing vertebrate (frog) eye. Spemann noted that the lens began to form when the optic vesicle, an outgrowth of the underlying brain tissue, first made contact with the overlying ectodermal tissue. Spemann showed experimentally that if the optic vesicle were removed, the overlying ectoderm did not differentiate into a lens; conversely, if the optic vesicle were transplanted from the anterior to the flank or posterior region of the embryo, it produced lenses in these otherwise eyeless regions. Spemann hypothesized that induction required physical and (presumably) chemical contact between the inducing and the induced tissue. For Spemann this process became a paradigm for how differentiation could be viewed in the embryo as a whole: as a series of inductive cascades of increasingly greater specificity (Hamburger, 1988: 18).

Spemann then took the process of induction to even earlier embryonic stages: to the gastrula stage after the blastula has begun to invaginate (push inward) to create a two-layered embryo. The point of invagination was known as the blastopore. Spemann, and his graduate student Hilde Mangold (neé Proescholdt, 1898–1924), found that transplanting tissue from the dorsal (upper) side of the blastopore into the ventral (underside) of an older embryo induced the formation of a whole secondary embryo. The dorsal lip tissue appeared to be the primary inducer, or "organizer" as it came to be called, of the entire vertebrate embryonic process. Working out details of factors influencing the function of organizers became the focus of work in his laboratory for over a decade (1924–35). (For this work Spemann was awarded the 1935 Nobel Prize in Physiology or Medicine.)

As a holistic materialist, Spemann focused his experimental work only at the tissue and organ-system levels of organization. He was not interested in reducing the process of induction to the chemical, physical, or even genetic level. Spemann's view was similar to Driesch's concept of the embryo as a "harmonious equipotential system" (a term that Spemann himself adopted and used), but not in a vitalistic way. Spemann thought that induction was a process based on interaction of material components and that while molecules were certainly involved, reduction to the molecular level would lose the essence of the process itself. For Spemann, the most interesting processes could only be revealed at those higher levels at which their emergent properties occurred. The basis of the organizer was to be found in its wholeness: inducing tissue had no meaning except in its interaction with inducible tissue. The whole, then, was greater than the sum of the parts in isolation.

Other embryologists followed suit. Spemann's student, Viktor Hamburger (1900–2001), carried on his mentor's holistic approach, but with a greater appreciation for the necessary interplay between mechanistic and reductionist approaches. From the mid-1930s onward, Hamburger pioneered work on development of the nervous system in the chick, noting that while some inductive process triggered neurons to begin growing out from the spinal cord toward developing limb buds, if the limb buds were removed nerve growth was significantly reduced (this work led to the later discovery of nerve growth factor, or NGF). Both tissues were involved in a reciprocal interaction and thus led to the emergence of fully innervated and functional limbs. Other developmental biologists who were greatly influenced by Spemann in the 1940s–70s included Paul A. Weiss (1898–1989) and Conrad H. Waddington (1905–75). Weiss developed a holistic view of the embryo as a series of "fields" in which various inductions and interaction took place at different times during embryogenesis. Waddington made significant advances in integrating genetics with development in a metaphorical "landscape" in which both systems co-evolved through the action of natural selection. More than in most other fields, embryologists promoted a holistic materialism that went beyond older mechanistic conceptions of life.

7. Conclusion

This chapter has focused on characterizing the mechanistic materialist philosophy that has been at the center of attempts to understand living systems from the early seventeenth through the end of the twentieth century, and to distinguish it from various emerging holistic or organismic philosophies, often associated with some form of vitalism. Looking beyond the latter years of the twentieth century, however, it is apparent that there is a flowering of holistic materialist and even dialectical materialist views in biology that is taking the life sciences in new directions. Gilbert and Sarkar have provided an exemplary survey of organicist thinking in widely different fields, from developmental biology, evo-devo (evolutionary developmental biology), ecology, neurobiology, and other interdisciplinary attempts to see biological systems as composed of multiple levels of organization with emergent properties at each new level of organization (Gilbert and Sarkar, 2000).

Richard Burian has argued that the older Mendelian concept of the gene as a discrete atomistic unit was grossly oversimplified in light of new work in molecular genetics, where segments of DNA are not functionally or even structurally discrete, and thus the Mendelian gene cannot be reduced to the molecular gene. In this sense, perhaps nowhere has the shift from a mechanistic approach and mechanistic research strategy between 1920 and 2000 been more apparent than in genetics as a whole. Yet nowhere in this changing philosophical approach has there been any tendency to return to forms of vitalism or non-material processes. The old mechanist-vitalist debates of the 1890s through the 1930s have been purged from biology at every level.

There has also been a small but significant resurgence of formalized dialectical materialist views in the life sciences, particularly in the work of evolutionary biologists such as Richard Lewontin (1929–) and Richard Levins (1930–2016). Their elaboration of a dialectical philosophy in various fields of evolutionary biology, population genetics, and ecology in the book *The Dialectical Biologist* suggests how a dialectical materialist approach can be highly fruitful as a way to understand the dynamics of living processes (Levins and Lewontin, 1985). They also suggest that if you really observe nature holistically, you cannot help but develop a dialectical materialist approach, even if not expressed as a formal philosophical system. This point has been made with respect to Darwin, who was not very philosophically inclined but couched his whole theory of evolution by natural selection in terms of the interaction of opposing forces, or processes (Allen, 1989, 1991, 1992).

Today, with the advent of computer technologies and fields such as bioinformatics with large data sets, systems, and "complexity" theory, holistic thinking is no longer so suspect in a variety of fields. As biology has begun to supplant physics as the pre-eminent science of the day, its philosophical underpinnings are not only becoming more sophisticated, but also less rigid. Like organisms themselves, biological approaches to understanding their functions require a variety of approaches, and must be flexible depending on what system is being studied, what questions are being asked, and what methods are available for research. Both analytical (Bechtel and Richardson's "decomposition") and synthetic, holistic approaches are necessary and complementary. New mechanists such as Craver and Darden argue that the search for mechanisms is not reductionist in the classical sense, as mechanisms can be studied at various different levels of organization. These new approaches are less monolithic and more flexible. As evolutionary biologist and historian Ernst Mayr (1904–2005) argued, biology is not bound to try to subsume all its theories under general laws, but can be free to develop its generalizations in a statistical sense, allowing for variation not only in the organisms being studied themselves but also in the messy world of lab and field observation and experimentation.

Note

1 As a movement in the philosophy of science, vitalistic claims should not be confused with religious arguments such as Special Creationism or Intelligent Design, which also appeal to metaphysical explanations, but couch them in terms of an individual, unknowable creator or deity.

References

Allen, G. (1975) *Life Science in the Twentieth Century*, New York: Cambridge University Press.
Allen, G. (1989) "Dialectical Materialism in Modern Biology," *Science and Nature* 3: 43–57.
Allen, G. (1991) "Mechanistic and Dialectical Materialism in 20th Century Evolutionary Theory: The Work of Ivan I. Schmalhausen," in L. Warren and H. Koprowski (eds), *New Perspectives on Evolution*, New York: John Wiley, 15–36.
Allen, G. (1992) "Evolution and History: History as Science and Science as History," in M. Nitecki and D. Nitecki (eds), *History and Evolution*, Albany, NY: State University of New York Press, 211–239.
Allen, G.E. (2001) "The Classical Gene: Its Nature and Its Legacy," in L.S. Parker and R. Ankeny (eds), *Mutating Concepts, Evolving Disciplines: Genetics, Medicine and Society*, Dordrecht: Kluwer Academic Publishers, 11–41.
Ash, M. (1995) *Gestalt Psychology in German Culture, 1890–1967: Holism and the Quest for Objectivity*, Cambridge: Cambridge University Press.
Bechtel, W. and A. Abrahamsen (2005) "Explanation: A Mechanistic Alternative," *Studies in History and Philosophy of Biology and Biomedical Sciences* 36: 421–441.
Bechtel, W. and R. Richardson (1993) *Discovering Complexity*, Princeton, NJ: Princeton University Press.
Boas, M. (1952) "The Establishment of the Mechanical Philosophy," *Osiris* 10: 412–541.

Clark, A. (1997) *Being There. Putting Brain, Body and World Together*, Cambridge, MA: MIT Press.

Commoner, B. (2002) "Unraveling the DNA Myth," *Harper's Magazine* (February): 39–47.

Craver, C. (2001) "Role Functions, Mechanisms and Hierarchy," *Philosophy of Science* 68: 53–74.

Craver, C. and L. Darden (2013) *Thinking about Mechanisms: Discovery across the Life Sciences*, Chicago: University of Chicago Press.

Dijksterhuis, E.J. (1961) *The Mechanization of the World Picture*, Oxford: Oxford University Press.

Durbin, Paul T. (1988) *Dictionary of Concepts in the Philosophy of Science*, New York: Greenwood Press.

Engels, F. (1940 [1935]), *The Dialectics of Nature*. "Preface" to the English translation by J.B.S. Haldne, New York: International Publishers.

Fisher, R.A. (1930) *The Genetical Theory of Natural Selection*, Oxford: Clarendon Press.

Fleming, D. (1964) "Introduction," in Jacques Loeb, *The Mechanistic Conception of Life*, Cambridge, MA: Harvard University Press, vii–xlii.

Gilbert, S. and S. Sarkar (2000). "Embracing Complexity: Organicism for the 21st Century," *Developmental Dynamics* 219: 1–9.

Graham, L. (1986) *Science, Philosophy and Human Behavior in the Soviet Union* (2nd ed.), New York: Columbia University Press.

Hamburger, V. (1984) "Hilde Mangold, Co-discoverer of the Organizer," *Journal of the History of Biology* 17: 1–11. Also reprinted as an Appendix to Hamburger, 1988: 173–180.

Hamburger, V. (1985) "Hans Spemann, Nobel Laureate 1935," *Trends in NeuroSciences* 8: 385–387.

Hamburger, V. (1988) *The Heritage of Experimental Embryology*, New York: Oxford University Press.

Harrington, A. (1996) *Reenchanted Science: Holism in German Culture from Wilhelm II to Hitler*, Princeton, NJ: Princeton University Press.

Holtfreter, J. (1933) "Nachweis der Induktionsfähigkeit abgetöter Keimteile." *Roux' Archiv für Entwicklungsmechanik* 128: 584–633.

Levins, R. and R. Lewontin (1985). *The Dialectical Biologist*, Cambridge, MA: Harvard University Press.

Loeb, J. (1890) *Der Heliotropismus der Tiere und seine Übereinstimmung mit dem Heliotropismuus der Pflanzen*. Würzburg: Georg Hertz. Translated in Loeb, J. (1905) *Studies in General Physiology*, Chicago: University of Chicago Press, 1–88.

Loeb, J. (1899) "On Some Facts and Principles of Physiological Morphology," *Biological Lectures Delivered at the Marine Biological Laboratory of Woods Hole, Summer, 1893*, Boston: Ginn & Company, 37–61. Reprinted in Maienschein (1986).

Loeb, J. (1900) "On the Nature of the Process of Fertilization," *Biological Lectures Delivered at the Marine Biological Laboratory of Woods Hole, Summer, 1899*, Boston: Ginn & Company, 273–282. Reprinted in Maienschein (1986).

Loeb J. (1903) "Sind die Lebenserscheinungen wissenschaftlich und vollständig erklärbar?" *Die Umschau* 7: 21; presented also as a lecture at Berkeley, as "Phenomena of Life," and reported in *The Daily Californian*, March 6, 1903. Quoted in Pauly, 1987.

Loeb, J. (1904) "The Recent Development of Biology," *Science* 20: 781.

Loeb, J. (1909) "On the Nature of Formative Stimulation (Artificial Parthenogenesis)," in D. Fleming (ed.), *The Mechanistic Conception of Life*, Cambridge, MA: Harvard University Press.

Loeb, J. (1912) "The Mechanistic Conception of Life," *Popular Science Monthly* (January). Reprinted in Fleming, D., ed. (1964) *The Mechanistic Conception of Life*, Cambridge, MA: Harvard University Press, 5–34.

Machamer, P., L. Darden, and C. Craver (2000) "Thinking about Mechanisms," *Philosophy of Science* 67: 1–25.

Maienschein, J. (1996) *Defining Biology: Lectures from the 1890s*, Cambridge, MA: Harvard University Press.

Morgan, T.H., A.H. Sturtevant, H.J. Muller, and C.B. Bridges (1915) *The Mechanism of Mendelian Heredity*, New York: Henry Holt.

Oparin, A.I. (1964 [1938]) *Life: Its Nature, Origin and Development*. Translated by Ann Syng, New York: Academic Press.

Pauly, P. (1987) *Controlling Life: Jacques Loeb and the Engineering Ideal in Biology*, Chicago: University of Chicago Press.

Prenant, M. (1943) *Biology and Marxism*. Translated by C. Desmond Greaves, New York: International Publishers.

Provine, W. (2001) *The Origins of Theoretical Population Genetics*, Chicago: University of Chicago Press. Reprint of the 1972 edition, with a new "Afterword."

Schaffner, K. (1967) "Approaches to Reductionism," *Philosophy of Science* 34: 137–147.

Schaffner, K. (1976) "Reductionism in Biology: Prospects and Problems," in R.S. Cohena and A. Michalos (eds), *Proceedings of the 1974 Meeting of the Philosophy of Science Association*, Dordrecht: D. Reidel, 613–632.

Schmalhausen, I. (1949) *Factors of Evolution*, Philadelphia: Blakiston; Reprint ed., Chicago: University of Chicago Press (1986).

Shapin, S. (1996) *The Scientific Revolution*, Chicago: University of Chicago Press.

Spemann, H. (1923) "Zur Theorie der tierischen Entiwcklung," *Rektoratsrede* 1–16. Freiburg im Breisgau: Speyer und Kaerner.

Spemann, H. (1927) "Neuen Arbeiten über Organisatoren in der tierischen Entwicklung," *Naturwissenschaften* 15: 946–951.

Spemann, H. (1938) *Embryonic Development and Induction*, New Haven, CT: Yale University Press.

Stein, G. (1988) "Biological Science and the Roots of Nazism," *American Scientist* 76: 50–58.

Temkin, O. (1946) "Materialism in French and German Physiology of the Early Nineteenth Century," *Bulletin of the History of Medicine* 20: 322–327.

Uexküll, J. (1937) "Die neue Umweltlehre. Ein Binderglied zwischen Natur- und Kulturwissenschaften," *Der Erziehung: Monatschrift für den Ausammenhang von Kultur und Erziehung im Wissenschaft und Leben* 13: 185–199.

Westfall, R. (1971) *The Construction of Modern Science*, New York: John Wiley.

6

MECHANISMS AND THE MENTAL

Marcin Miłkowski

1. Introduction: the notion of mechanism evolves with the understanding of the mental

In this chapter, I sketch the history of mechanistic models of the mental, as related to the technological project of trying to build mechanical minds, and discuss the changing use of these models. In section 2, I introduce the Cartesian notion of mechanism, which shaped the debate in the centuries that followed. Early mechanistic proposals are also connected with early attempts to formulate the computational account of thinking and reasoning, upheld notably by Hobbes and Leibniz. In section 3, associationist and behaviorist models of the mind are sketched, along with attempts to understand the neural system in terms of connections and associations. Early robotic models, built mostly by behaviorists and other students of animal behavior, are also introduced. In section 4, the focus is on early computational and cybernetic models of the mind. In section 5, I deal with computational models of mental mechanisms as proposed by students of artificial intelligence and cognitive science. The history of uses of mechanistic models sheds light on different kinds of explanations of the mental.

The notion of mechanism has played an immense role in the history of attempts to scientifically explain mental phenomena. On the one hand, it has always been related to the philosophical mind–body problem, and, on the other hand, to the possible scope of mechanistic explanations. Mechanization of thinking is also one of the classical problems in philosophy of mathematics, and the notion of mechanism, in a much more abstract form, appears in the history of mathematics as well.

The philosophical mind–body problem is the question of how the mental and the physical are related. One way to defend a naturalistic solution to the mind–body problem is to build a mechanistic model of the mind. Such a model might show, even as a proof of an in-principle possibility, that thinking can be instantiated in a physical mechanism. Conversely, the proof that thinking may not be mechanized hints at the possibility that the mind is not physical, after all.

However, one may also contend that mechanistic models of thinking, even if successful, do not decide the mind–body question at all. In other words, even if thinking is just the exercise of computational mechanisms, there still could be some non-mechanistic aspects of thought or perception that completely evade the mechanistic explanation. Indeed, this is the position taken by computational dualists, from Leibniz to David Chalmers. Yet the existence of limits to mechanization may be interpreted not as evidence that human minds transcend mechanism but that they are also subject to the same limits, as the debate over limiting theorems in mathematics shows (see Chapter 33).

In short, the development of technological mechanisms has affected the ways one conceived of possible explanations of the mental, and mechanists tried to describe, build, and elucidate mechanisms that exhibit at least some cognitive capacities, including mathematical reasoning and conceptual thought. Early mechanistic models were just proofs of possibility of mechanization; however, current mechanistic explanations strive to go beyond the question of how it is possible to display such-and-such a mental phenomenon, and (at least partially) explain actual mental processes.

2. Early mechanists: Descartes and Hobbes to Leibniz and mechanical ducks

René Descartes is known for his defense of mind–body dualism, but he was also a staunch proponent of mechanistic explanations, which paved the way for modern science. His understanding of mechanisms was strongly influenced by the technology available in his age, in particular fountains and clocks. A *mechanism* for Descartes is a spatial system whose operation depends on local interactions among its parts. These interactions can be further described geometrically. Notably, mechanisms can be explained without recourse to teleology, and indeed, the Cartesian account of the mechanistic explanation is strongly opposed to the Aristotelian and medieval models of explanation, which used to appeal to final causes, theological considerations, angelology, basic elements, and Pythagorean principles, all at the same time.

The significance of Descartes does not stop here. He also offered early mechanistic accounts of the nervous system and psychological phenomena (see Chapter 28). The nervous system was conceived of in hydraulic terms of flows of animal spirits. These are produced in the blood and may exert influence over bodily parts. The basic building block of the nervous system, according to Descartes, is the *reflex arc*: the stimulus pulls tiny wires of the nervous system, which in turn open little valves in the brain, releasing animal spirits to hollow nerve tubes that lead to appropriate muscles (see Figure 6.1). This general outline of the reflex arc as the basic operation of the nervous system has remained immensely important, even if subsequent research rejected hydraulic metaphors. The modern notion of *reflex* itself was introduced by a medical doctor Thomas Willis in his treaty *De motu musculari* (Willis 1694). Willis was influenced by Descartes, but his mechanism was framed in optical terms (reflection being an optical phenomenon); the reflex is a pure reaction to external influence rather than something that relies on internal resources (as in Descartes' model (cf. Canguilhem 1955)).

However, Descartes argued that there are limits to mechanistic explanation of the mind. He thought that the physiological machine just described can have functions such as

> the reception by the external sense organs of light, sounds, smells, tastes, heat and other such qualities, the imprinting of the ideas of these qualities in the organ of the "common" sense and the imagination, the retention or stamping of these ideas in the memory, the internal movements of the appetites and passions, and finally the external movements of all the limbs.
>
> *(Descartes 1664/1985, 1: 100)*

But in *Discourse on the Method*, he claimed that no mechanism can be built that could "use words, or put together other signs, as we do in order to declare our thoughts to others" (Descartes 1637/1985, 1: 140). According to him, it is not conceivable that "a machine should produce

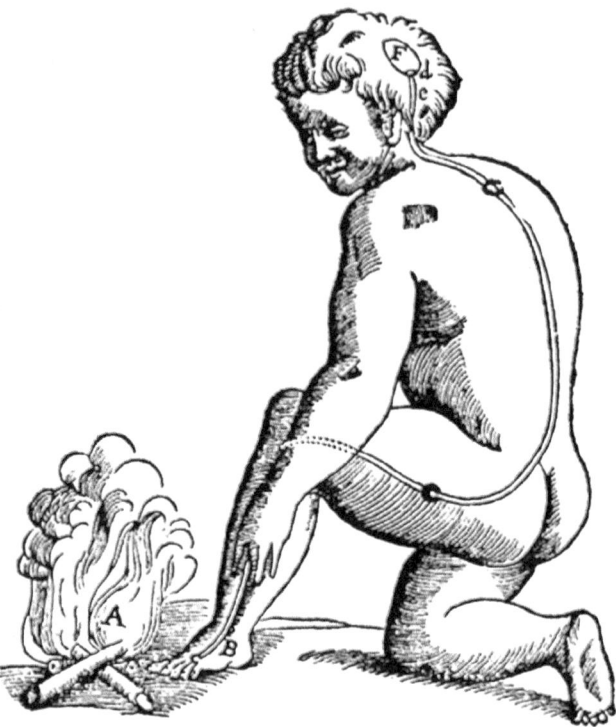

Figure 6.1   The reflex arc of the painful stimulus according to Descartes. The fire A is a stimulus afflicting the skin (B) and moving the fine thread (C), which goes to the valve (d, e). The valve opens the cavity (F), from which the animal spirit is released, and this in turn makes the head turn and move the hand and the foot

different arrangements of words so as to give an appropriately meaningful answer to whatever is said in its presence." Moreover, no machine could work for a variety of purposes, and, in contrast, reason is a universal instrument. A human being's rational capacities must therefore come from a non-physical rational soul, which Descartes believed to be connected with the material brain through the pineal gland (Descartes 1649/1985, 1: 340). The Cartesian challenge is still alive today (Wheeler 2008).

In contrast to Descartes, Hobbes was probably the first proponent of a thoroughly mechanistic and computational theory of mind; as he claims, ratiocination is just computation: "to compute is, either to collect the sum of many things that are added together, or to know what remains when one thing is taken out of another" (Hobbes 1655/1839, 3). However, Hobbes failed to deliver even a sketchy description of a mechanism that could indeed think by computing. But soon Leibniz envisaged that any philosophical argument could be simply settled by computing in a universal mathematical language (Leibniz 1666). The seventeenth century witnessed important developments in the construction of various machines, including Blaise Pascal's arithmetic machine, but none of them was fit to perform complex logical computations. Further development in engineering led to even more sophisticated machines, the Digesting Duck, created in 1739 by Jacques de Vaucanson, being probably the most prominent. The mechanical duck appeared to be able to digest food; however, it only collected food in one container and produced artificial feces from another one. Another imaginary

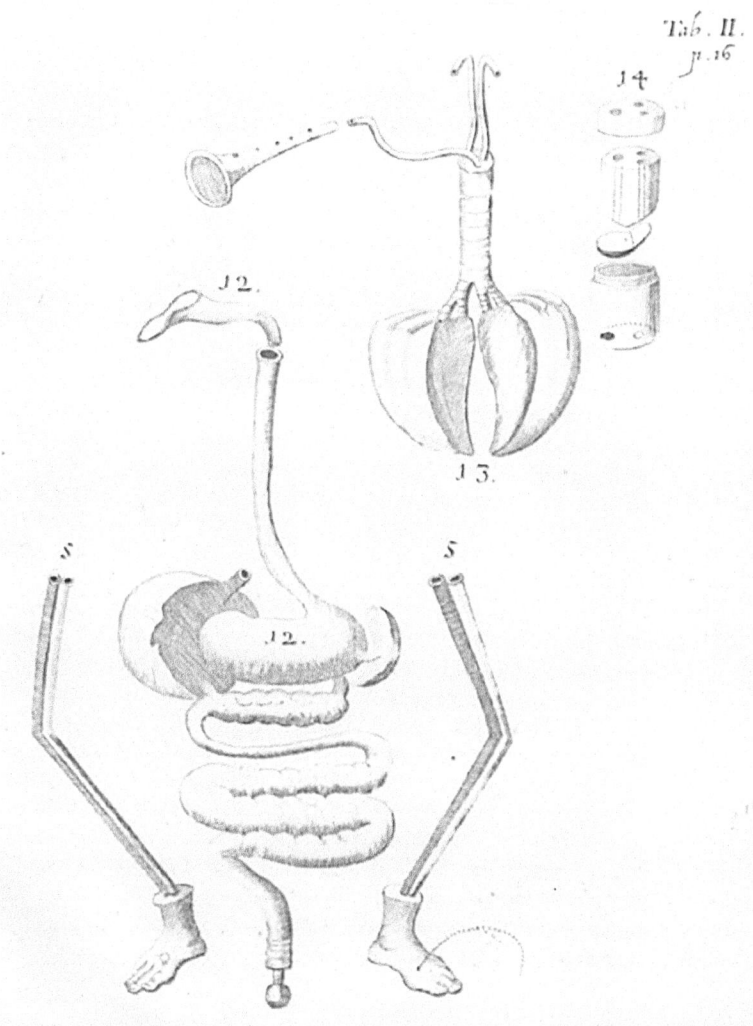

Figure 6.2 *Statua Humana Circulatoria*, Salomon Reisel's imaginary model of respiration

model of human respiration, *Statua Humana Circulatoria*, conceived by Salomon Reisel, is depicted in Figure 6.2. At the same time, such contraptions aroused the imagination of proponents of mechanical reductionism.

The mechanical reductionism is mostly associated with Julien de La Mettrie, who offered a purely mechanical account of the mental in his *Man a Machine* (La Mettrie 1748/1912). Note, however, that La Mettrie did not espouse the Cartesian notion of the mechanism; he claimed that the matter is capable of movement and sensibility only when appropriately *organized*. To him, organization is not just a matter of spatial arrangement, as in Descartes, but also of inner complexity that may be quantified. In this, La Mettrie embraced a notion of mechanism closer to the one defended by New Mechanists than the Cartesian one. At the same time, he admitted that it was an open question what kind of organization contributed to sensation.

3. Associationistic models of the neural system and early robotic models

The most influential criticism of the Cartesian account of thought came from John Locke, who criticized the theory of innate ideas defended by Descartes, and claimed that all thought came from the senses. Yet for the development of the mechanistic understanding of the mental, another empiricist claim is much more important: namely, that ideas form thoughts by association. The associationist theory of ideas has roots as deep as in Plato's *Phaedo*, and early formulations can also be found in Hobbes. But it was David Hume, having identified three principles of association—similarity, contiguity of time and place, and cause and effect—who had the greatest influence on subsequent theorizing about mental mechanisms. For example, having experienced similar impressions when touching fire, a stove, and hot metal, we may associate all of them with the notion of *something hot*.

Soon after, associationism was linked together with hydraulic models of the mind. For example, Sigmund Freud's speculations about neurological mechanisms underlying psychological phenomena relied on associationist and hydraulic hypotheses (compare Bilder and LeFever 1998). In particular, a difficult question was how new associations were formed. Associationism suggested that there had to be a certain mechanism, and as late as at the turn of the twentieth century, physical models of association learning were offered (for a comprehensive review of such models, see Cordeschi 2002).

In the second half of the nineteenth century, the first labs in experimental psychology were formed by Wilhelm Wundt in Germany and William James in the US. Wundt's work had deep roots in psychophysical research, whose champions included physicists Herman Helmholtz (who also posited unconscious inference) and Ernst Mach. However, Wundt's goal was to explain psychological phenomena with lawlike generalizations, just like in psychophysics, rather than by appeal to organized mechanisms. His experimental paradigm, related largely to introspection, did not uncover a large body of well-confirmed laws. In opposition to Wundt's introspective psychology and to its American competition in the work of Titchener, the behaviorist movement started. Interestingly, behaviorists, such as E. Thorndike, also tried to discover general laws of learning that would hold true of all animals.

Behaviorism is committed to a broadly associationist view of the learning organism. Hence, at the turn of the twentieth century, there was a large interest in understanding the underlying principles of association. This was also driven by new discoveries in neurobiology; in particular, Sir Charles Sherrington proposed that elementary building blocks of the nervous systems, the neurons, communicate via synapses (see Chapter 28). At the time, Santiago Ramón y Cajal and Camillo Golgi debated over the general anatomy of the nervous tissue. Cajal argued that the nervous tissue is composed of multiple cells while Golgi claimed that the brain is a single continuous network, called the *reticulum*. Sherrington's proposal of the synaptic communication, related to his detailed study of reflexes, was one of the tipping points in the debate (Burke 2007). Moreover, he also refined the Cartesian notion of the reflex arc as the basic operational blueprint of the nervous system. The notion of the reflex arc—as composed of a receptor, conductor, and effector—was compatible with the behaviorist psychology, which focused on stimulus-response pairs.

Such is the background for the early connectionist theory defended by Edward Thorndike (1911). According to him, "the animal mind is, by any definition, something intimately associated with his connection system or means of binding various physical activities to various physical impressions" (Thorndike 1911, 16). The "connection system" underlies the capacity of animals to learn and consists of connections between stimuli coming from the external environment and collected by a "receptor system," and motor responses.

Subsequently, artificial models of learning mechanisms were proposed. For example, Bent Russell built hydraulic networks that demonstrated effects of learning over time, supported by experimental results of reinforcement on animals (Russell 1913). This approach was then continued by a prominent behaviorist, Clark Hull, in his robotic models of learning in the 1920s and 1930s. Hull wanted to build robots to prove the possibility that complex forms of behavior were not limited to living organisms. Hull's models were no longer hydraulic but electrochemical. His model of the conditioned reflex was supposed to free "the science of complex adaptive mammalian behavior from the mysticism whichever haunts it" (Krueger and Hull 1931, 267). Importantly, it was not an exact replica of animal behavior but a *conceptual model* that proved that a number of features of behavior are mechanically replicable. This feature of simulative models remained important throughout the rest of the twentieth century.

However, the time was not yet ripe for simulative research and later Hull shied away from further robotic simulations, as they were not treated seriously. For Edward Tolman, with his hypothetical robotic "schematic sowbug," a simplified artificial animal that illustrated his theory that all behavior was entirely stimulus driven (Tolman 1939), robotic models were not essential in studying behavior either. Not only behaviorists built robots at the time; one particularly important exception was Alfred Lotka (1925), known for his work on mathematical predator/prey models. His simple mechanical animal model—a "correlating apparatus"—was an important predecessor of later cybernetic robots for two reasons. First, Lotka approached the evolution of complex systems in terms of reciprocal equilibrium. No longer does the nervous system work merely as a reflex arc; correlating actions with the environment is "essentially cyclic in character: It has its origin in the external world, which becomes depicted in the organism, provokes a response, the terminal step of which is usually, if not always, a reaction upon the external world" (Lotka 1925, 340). Second, his model uses the notion of information as controlling the behavior of an animal, and his simplistic model of representation (called "depiction" by Lotka) predates later work on cognitive representations. This notion allows talk of how the animal's behavior can be influenced by future events—represented thanks to the antennas mounted on the mechanical robot to detect the edge of a table—without appealing to the notion of the final cause (Cordeschi 2002, 128).

4. Computation and cybernetics: from Babbage and Turing to Ashby

Another crucial development in the nineteenth century was the blueprint for a general-purpose computational machine. Charles Babbage argued that his Difference Engine (see Figure 6.3) could serve as a simple model of the geological laws (Babbage 1837; Dolan 1998). This was one of the first uses of numerical computation for simulation purposes, but the implications of the Difference Engine do not end here. It was the predecessor of the Analytic Engine, the first general-purpose programmable machine to be designed, as Babbage himself and Lady Ada Lovelace observed. And a general-purpose mechanism was exactly what Descartes thought was impossible.

It was only Alan Turing (1937) who brought the consequences of the general-purpose computing machines to light. Having formalized the notion of a computing machine, called later a *Turing machine*, he defined a universal machine that was able to operate like any other Turing machine just by operating on the description of that machine on its input tape (for more, see Chapter 33). However, universal Turing machines can only compute effectively computable procedures, so there are limits to their capacities. Turing referred to the important work by Kurt Gödel (1933) who proved that in the first-order predicate logic with elementary arithmetic,

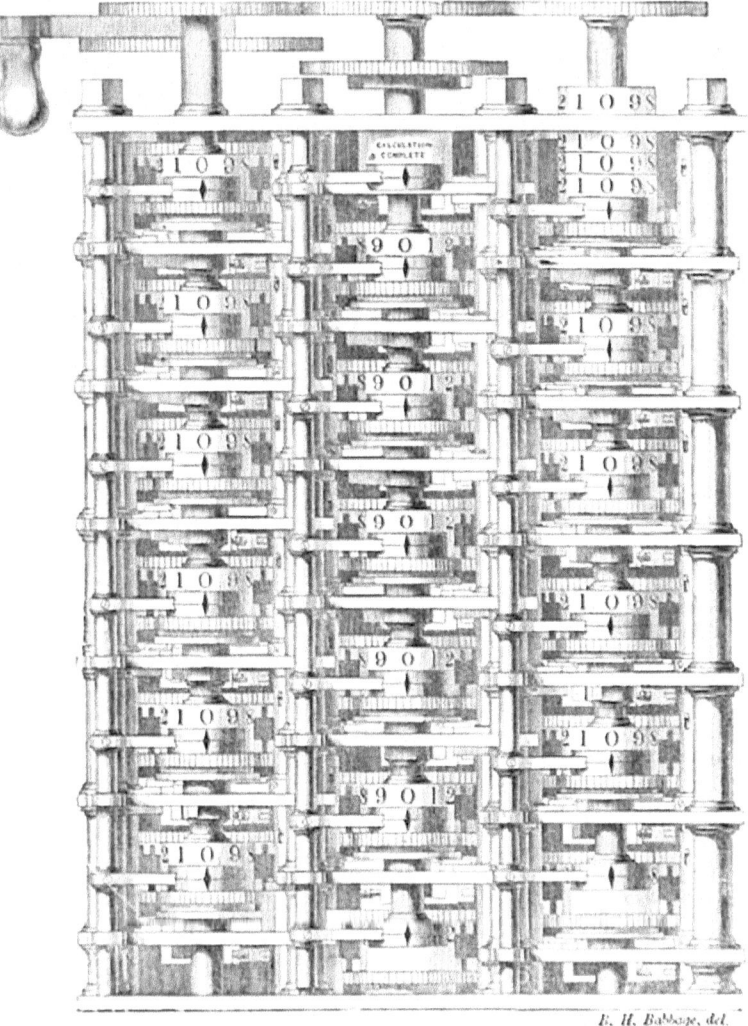

Figure 6.3 Difference Engine by Charles Babbage. Fragment of the original blueprint

there are always true statements that cannot be proved in that very logic. This later led to an antimechanistic argument that no machine could simulate a human mathematical mind (Lucas 1961; Penrose 1989). However, these arguments need to presuppose that the human mind is logically consistent, which also cannot be decided using an effective procedure (Putnam 1960). (The consistency presupposition is unjustified, and the proof used by Lucas leads to a contradiction (Krajewski 2007).)

Turing suggested that artificial computational machines can be thought to be generally intelligent as soon as they could converse with people without being detected as machines (Turing 1950). In this conversational test of intelligence—called later *the Turing test*—the criterion of thinking is what Descartes considered impossible.

Even more importantly, Turing was not alone when he thought machines could be programmed to display intelligence. It was a common assumption among British and American defenders of *cybernetics*—a general theory of control and communication in the animal and the machine, as Norbert Wiener defined it in the eponymous title of his book (Wiener 1948).

Cybernetics made several important contributions to the development of mechanistic models of the mental (Pickering 2009; Johnston 2008). First, it offered a new conceptual framework for natural teleology, i.e. for describing the goals of physical systems without recourse to any non-physical factors (Rosenblueth, Wiener, and Bigelow 1943). The notion of feedback, fundamental to cybernetics, was used to elucidate the relation of the mechanism to its environment, showing how the environment can control the behavior of the mechanism. The *feedback* occurs when the output of the system is directed back to the input of the system, which effectively creates a loop. There are two kinds of feedback: negative and positive. The *negative feedback* occurs when some function of the output of the system is fed into its input to stabilize the output. It promotes stability and is frequently used for control purposes (Sluckin 1954). The *positive feedback*, on the other hand, occurs when a small disturbance in the output of the system increases the magnitude of the disturbance. It can lead to chaotic oscillations and exponential growth. Note that machines with feedback had been invented much earlier than the 1940s. There were other regulators or governors before (for example, the Watt governor used to control the steam engine), but there was no precise mathematical theory of control.

Second, cybernetics used the newly developed *theory of information* (Weaver and Shannon 1964). Claude Shannon defined a mathematical notion of information as uncertainty of the receiver of the information as to what information it will receive over the noisy channel from a sender (for introductions to the theory of information, see Floridi 2010; Burgin 2009). It has become possible to quantify the information sent. However, the open question was how to relate this theory to meaningful (semantic) information. Semantic information is *about* something, in contrast to purely structural, probabilistic, or syntactic information, which boils down to the existence of mere regularities. A number of theorists proposed their own approaches to semantic information (MacKay 1969; Bar-Hillel 1955; Dretske 1981).

Third, the focus on complexity of control allowed the hypothesis that the main function of the brain is to control—in a stable way—the overall behavior of the animal. One particularly influential notion was that of *homeostasis* or keeping the internal conditions of the system relatively constant and stable. W. Ashby built a model of a homeostatic machine—Homeostat—that was supposed to show how complex systems adapt to the environment (Ashby 1960). Even if the Homeostat does not seem particularly exciting (it has no effectors to move it), it shows what Ashby called "ultrastable behavior": it adjusts its internal organization to react appropriately to the outside disturbance. As such, Homeostat goes beyond simple reflex arc models and simple animal stimulus-driven behaviors.

Fourth, cybernetics promoted interdisciplinary research and asked integrative questions about mechanisms of control and communication in animals and machines. This promoted building robotic models of animals, and the use of cybernetic notions in biology and psychology. For example, W. Grey Walter built robots (Walter 1950a, 1950b, 1953)—called tortoises for their looks—as models of animal behavior: environmentally induced exploration (in a robot jokingly called *Machina speculatrix*); and conditional reflexes (in *Machina docilis*). These electromechanical tortoises were milestones in the history of biorobotics (the discipline that explains biological behavior with robots; compare Webb 2002) and bionics (the discipline that builds biologically inspired robots). The technological obstacles to building more complex robots could not, however, be easily surpassed, and robots gave way to computational modeling in information-processing psychology. Yet Grey Walter's research remained influential to defenders of classical cognitive research (Miłkowski 2016), and later, in the 1980s, inspired behavioral robotics (Brooks 1999).

5. Computational models and artificial intelligence

Cybernetics, information theory, and the advent of digital computers made it possible to consider the development of computer simulation of mental processes. While in the nineteenth and early twentieth century, Charles Sherrington, William James, and Alfred Lotka would talk of the telephone switchboard as the model of the brain, in the second half of the twentieth century the preferred mechanistic model of the mental was the programmable digital computer. There were four reasons given for the general analogy between computers and minds (Apter 1970). First, both are general-purpose machines. Second, both process information. Third, both can create *models* of reality. In other words, one of the basic functions of the brain is to build models of the environment; this was particularly stressed by Kenneth Craik (1967). Fourth, both computers and brains have a large number of similar elementary building blocks that perform simple operations, which, taken together, contribute to complex behavior.

In 1958, Herbert Simon and Allen Newell predicted that in ten years most psychological theories will take the form of "computer programs, or of qualitative statements about the characteristics of computer programs" (Simon and Newell 1958, 7–8). Their view was extreme, and many cognitive psychologists did not share their enthusiasm (Baars 1986; Neisser 1963). Nevertheless, psychologists used information-theory notions to design and analyze significant experimental results (Sperling 1960; Miller 1956; Broadbent 1958). Even in 1967, almost ten years after Simon and Newell made their initial prediction, psychological references to computer programs were highly metaphorical and lacked experimental evidence (Miller, Galanter, and Pribram 1967). In the 2000s, however, the prediction holds true: over 80 percent of articles in theoretical journals focus on computational models (Busemeyer and Diederich 2010).

What made the prediction true? One factor was the new interdisciplinary field formed at the end of the 1970s that came to be called *cognitive science* (Arbib et al. 1978). It may be a stretch to claim that it was a scientific revolution in the strict sense (Kuhn 1970), as some initially claimed (Gardner 1985): first, it does not seem incommensurable with previous psychological research (for example, Tolman's (1948) claim that rats use cognitive maps to navigate mazes is easily translatable into cognitive terminology); and second, it's deeply connected with multiple disciplines that already coalesced around cybernetics. Cognitive science made significant contributions to mechanistic models of the mental even if some would deny that all cognitive models are mechanistic.

In the 1950s, another important field was formed: Artificial Intelligence (AI). In AI, two basic approaches were dominant: connectionist machine learning, initially in the Perceptron machine (Rosenblatt 1958), and the rule-based approach, defended by Simon and Newell. The Perceptron was an electromechanical, artificial neural network (ANN) with extremely simplified neurons. It could learn by the so-called delta rule that allowed researchers to adjust the weights of connections between neurons. However, it was soon discovered that using the delta rule, Perceptrons could not learn elementary logical operations such as exclusive disjunction, or XOR (Minsky and Papert 1969). This has led ANN research to stall until the 1980s. In their place, rule-based approaches dominated, but they also failed to quickly deliver the promised results.

These two approaches have their counterparts in cognitive science. The rule-based approach, also called the *symbolic* approach because of its reliance on a particular conception of cognitive representation, was the mainstream approach until the early 1980s. For example, Allen Newell and Herbert Simon (1972) simulated problem-solving processes with rule-based systems, which were then evaluated by comparing the simulation with the behavioral data (such as verbal reports of subjects and eye-tracking data). Newell and Simon stressed that their methodology was ill-suited for motor and perceptual processing.

The connectionist approach has its roots in associationism, and it became much more sophisticated in the 1980s when new methods of ANN learning appeared. Notably, the backpropagation algorithm allowed the training of networks with multiple layers of hidden neurons, and the ANNs were used to simulate parallel cognitive processing. Significant progress has been made in transcending the limitations of associationism: the association relation is symmetrical while many cognitive classical structures are ordered (McClelland, Rumelhart, and PDP Research Group 1986; Rumelhart and McClelland 1986; for a general assessment, see Bechtel and Abrahamsen 2002). However, defenders of the symbolic approach claimed that ANNs either do not offer a full account of cognitive structures, with such features as compositionality or systematicity, or they are mere implementations of symbolic approaches (Fodor and Pylyshyn 1988).

All this criticism notwithstanding, ANN modeling has matured significantly. While the early ANNs were extremely simplified and idealized, the current models use much more biologically plausible models of neurons, sometimes with a large variety of neuron types. ANNs are also routinely described using control theory, which allows using dynamical systems theory to design their architecture (Eliasmith and Anderson 2003). They are also used in computational neuroscience, for example to build large brain models (Eliasmith 2013). The large models strive for biological plausibility (in the mechanistic terminology, they are how-plausibly models; see Craver 2007), and while they are heavily idealized, they are able to offer novel predictions. In other words, they are no longer just proofs of possibility that a mechanism implementing a mental capacity may be built, but idealized models of how actual brains may work.

Because of the heavy stress on computational models, general objections against cognitive science were raised. The first one is related to the notion of mental representation as used in cognitive research. In most explanatory models, it was assumed that it's sufficient to simulate the syntactic structure of representations for the semantic properties to appear; as John Haugeland quipped: "if you take care of syntax, the semantics will take care of itself" (Haugeland 1985, 106). John Searle raised an important objection against this formalist principle in his Chinese Room thought experiment (Searle 1980): one can easily imagine a person with a special set of instructions in English who could manipulate Chinese symbols and answer questions in Chinese without understanding it at all. Hence, understanding is not reducible to syntactic manipulation. While the discussion around this thought experiment is hardly conclusive (Preston and Bishop 2002), the problem was soon reformulated by Stevan Harnad as the "symbol grounding problem" (Harnad 1990). How can symbols in computational machines mean anything? Or, to spell this question out a little more generally, what mechanism is able to mean anything?

Formulated this way, Searle's question is yet another version of the mind–body problem as related to intentionality, or aboutness of mental processes or states. In the nineteenth century, Franz Brentano claimed that intentionality distinguishes the mental from the physical (Brentano 1900; Chrudzimski 1999). In this, he followed Descartes: no mechanism could entertain thoughts. However, with the advent of digital computers, a new argument was put forward: thought and content are causally relevant in the physical world just because syntactic entities are computed over. Searle argues that syntax is not sufficient for intentionality, and in this, most theorists of mental representation agree. The most influential position, teleosemantics, defended by Ruth Garett Millikan (1984) and Fred Dretske (1997), is that to represent is, roughly, to have the biological function of making information available to cognitive and motor processes. This position remains controversial (Ramsey 2007; Hutto and Myin 2013; Fodor and Pylyshyn 2015), especially for the followers of the cybernetic approach to cognition in terms of control. They claim that the appeal to (cognitive) representation in control mechanisms plays no particular explanatory role (Van Gelder 1995; Chemero 2009), and that computational explanations may be replaced by dynamical explanations (essentially similar to the ones offered by Ashby in his account of Homeostat). However, most contemporary criticisms of computational cognitive science focus on two other properties of mental processes: qualities of experience and the ability to use common sense.

It was argued that no computational mechanism can produce the qualitative states of consciousness (called *qualia*) and that consciousness is not physical (Jackson 1986; Chalmers 1996). These arguments are usually put forward in the following form: no amount of physical knowledge (for example, about computational structures of the brain) is sufficient to know the qualitative consciousness. The two most frequent replies to such arguments are the following: first, it is claimed that qualitative knowledge of conscious states is not knowledge-that but knowledge-how (Ryle 2009), which is in principle not available intersubjectively (Churchland 1985); and second, it is argued that it's just the lack of imagination of philosophers trying to understand the epistemic situation of a person with all the relevant physical knowledge that creates the illusion that there is something to qualia that is beyond all physical description (Dennett 2005).

The last important argument against computational cognitive science is that it cannot explain how people smoothly cope with their everyday tasks, or to understand their common sense (Dreyfus 1979). Indeed, it is still a Holy Grail of artificial intelligence to build a machine that can make commonsensical inferences and act accordingly. The defenders of AI claim that the actual scope of common-sense knowledge is vaster than critics of AI imagine.

6. Conclusion

The history of mechanistic models of the mental reflects theorizing about mechanisms as such. First, mechanisms were considered to be entities in space, whose changes could be described geometrically, and whose interactions could be only local. Then, they were considered causal and temporal as well. La Mettrie subsequently stressed that mechanisms are organized systems. At the same time, the understanding of the mental in terms of mechanisms of information processing progressed. Descartes proposed a general outline of the nervous system in terms of a reflex arc but considered important mental faculties to be exempt from the purview of mechanism. He supposed that there could be no mechanism for thinking and drawing inferences. Soon, however, Leibniz sketched a proposal of a universal, logical, and computational calculus of thought. As soon as computing mechanisms were built, it was suggested that they could be used to build not only rough approximations of the mind, but also precise causal models, including also cyclical interactions with the environment. Subsequently, the gap between mathematical, computational simulations of the mental and the neurobiologically plausible models began to close.

The role of mechanistic models has also changed. Here, it is useful to distinguish mechanistic simulations from theoretical descriptions of mechanisms. Until the second half of the twentieth century, mechanistic simulations, usually owing to limits of technology, were mere proofs of possibility that the mind might be, in spite of appearances, mechanical. Some modelers were overly enthusiastic about the capacities of their artifacts; for example, Walter would claim that his tortoises were capable of self-recognition, while they simply responded to light sources, and just because they had lamps mounted on them, they would react to their own image in the mirror. Theoretical descriptions of mechanisms were not bounded by technological limitations, and they could offer some insight into how the mental processes actually occur. Hence, already the model of the reflex arc was not just a mere how-possibly model that would prove that the mechanical mind is possible (and Descartes would even retort that the mechanical mind is a contradiction in terms). It was a how-plausibly model of the nervous system (not the mind) whose role was to explain at least a large number of muscle movements. Only in the second half of the twentieth century have mechanistic simulations been built for similar explanatory purposes. Yet technological and theoretical limitations have not all been overcome, and they are still far from complete models of how actual mechanisms work. But the underlying assumption of modelers in cognitive neuroscience today is that such models not only can but also should be built (Piccinini and Craver 2011; Miłkowski 2013; Boone and Piccinini 2016).

Different technological artifacts have been treated as models of the mental, from a wax tablet, through clocks and the automated telephone switchboard, to a digital computer. However, as Margaret Boden stresses, the computer is not just another metaphor of the mind; it might also be the last metaphor because it may offer a precise causal model of the mind (Boden 2008). Most mechanistic models of the mental are no longer controversial. Even self-proclaimed dualists rarely believe that it is impossible to understand intentionality mechanistically. The only remaining bastion of anti-mechanists is qualitative consciousness, although in recent years, some progress has been made in terms of measuring consciousness (Seth et al. 2008) and building more plausible causal models of its processes (Boly et al. 2013). We still do not know how to model consciousness mechanistically, but it is no longer obvious that we will never know.

References

Apter, Michael. 1970. *The Computer Simulation of Behaviour*. London: Hutchinson.

Arbib, Michael A., Carl Lee Baker, Joan Bresnan, Roy G. D'Andrade, Ronald Kaplan, Samuel Jay Keyser, Donald A. Norman, et al. 1978. *Cognitive Science, 1978*: *Report of The State of the Art Committee to The Advisors of The Alfred P. Sloan Foundation*. New York: The Alfred P. Sloan Foundation. http://csjarchive.cogsci.rpi.edu/misc/CognitiveScience1978_OCR.pdf.

Ashby, W Ross. 1960. *Design for a Brain: The Origin of Adaptive Behavior*. New York: Wiley.

Baars, Bernard J. 1986. *The Cognitive Revolution in Psychology*. New York: Guilford Press.

Babbage, Charles. 1837. *The Ninth Bridgewater Treatise. A Fragment*. London: J. Murray.

Bar-Hillel, Yehoshua. 1955. "Information and Content: A Semantic Analysis." *Synthese* 9 (1): 299–305. doi:10.1007/BF00567416.

Bechtel, William and Adele Abrahamsen. 2002. *Connectionism and the Mind*. Oxford: Blackwell.

Bilder, Robert and F. Frank LeFever, eds. 1998. *Neuroscience of the Mind on the Centennial of Freud's Project for a Scientific Psychology*. New York: New York Academy of Sciences.

Boden, Margaret A. 2008. "Information, Computation, and Cognitive Science." In *Philosophy of Information: Volume 8*, edited by Pieter Adriaans and Johan Van Benthem, 8: 741–61. Elsevier B.V. doi:10.1016/B978-0-444-51726-5.50023-6.

Boly, Melanie, Anil K. Seth, Melanie Wilke, Paul Ingmundson, Bernard J. Baars, Steven Laureys, David Edelman, and Naotsugu Tsuchiya. 2013. "Consciousness in Humans and Non-Human Animals: Recent Advances and Future Directions." *Frontiers in Psychology* 4. Frontiers. doi:10.3389/fpsyg.2013.00625.

Boone, Worth and Gualtiero Piccinini. 2016. "The Cognitive Neuroscience Revolution." *Synthese*, June 193 (3): 1509–1534. Springer Netherlands. doi:10.1007/s11229-015-0783-4.

Brentano, Franz. 1900. *Psychologie Vom Empirischen Standpunkt*. Hamburg: F. Meiner.

Broadbent, D. E. 1958. *Perception and Communication*. Oxford: Pergamon Press.

Brooks, Rodney A. 1999. *Cambrian Intelligence: The Early History of the New AI*. Cambridge, MA: The MIT Press.

Burgin, Mark. 2009. *Theory of Information: Fundamentality, Diversity and Unification*. Singapore: World Scientific Publishing Co.

Burke, Robert E. 2007. "Sir Charles Sherrington's the Integrative Action of the Nervous System: A Centenary Appreciation." *Brain: A Journal of Neurology* 130 (Pt 4): 887–94. doi:10.1093/brain/awm022.

Busemeyer, Jerome R. and Adele Diederich. 2010. *Cognitive Modeling*. Los Angeles: Sage.

Canguilhem, Georges. 1955. *La Formation Du Concept de Réflexe Aux XVIIe et XVIIIe Siècles*. Paris: Presses universitaires de France.

Chalmers, David J. 1996. *The Conscious Mind: In Search of a Fundamental Theory*. New York: Oxford University Press.

Chemero, Anthony. 2009. *Radical Embodied Cognitive Science*. Cambridge, MA: The MIT Press.

Chrudzimski, Arkadiusz. 1999. "Die Theorie Der Intentionalität Bei Franz Brentano." *Grazer Philosophische Studien* 57 (July): 45–66. doi:10.5840/gps1999574.

Churchland, Paul M. 1985. "Reduction, Qualia, and the Direct Introspection of Brain States." *The Journal of Philosophy* 82 (1): 8–28.

Cordeschi, Roberto. 2002. *The Discovery of the Artificial. Studies in Cognitive Systems.* Dordrecht: Springer Netherlands. doi:10.1007/978-94-015-9870-5.

Craik, Kenneth. 1967. *The Nature of Explanation.* Cambridge: Cambridge University Press.

Craver, Carl F. 2007. *Explaining the Brain: Mechanisms and the Mosaic Unity of Neuroscience.* Oxford: Oxford University Press.

Dennett, Daniel C. 2005. *Sweet Dreams: Philosophical Obstacles to a Science of Consciousness.* Cambridge, MA: The MIT Press.

Descartes, René. 1985. *The Philosophical Writings of Descartes.* Edited by John Cottingham, Robert Stoothoff, and Dugald Murdoch. Vol. 1. Cambridge: Cambridge University Press.

Dolan, Brian P. 1998. "Representing Novelty: Charles Babbage, Charles Lyell, and Experiments in Early Victorian Geology." *History of Science* 36: xxxvi.

Dretske, Fred I. 1981. *Knowledge and the Flow of Information.* 2nd ed. Cambridge, MA: The MIT Press.

——. 1997. *Naturalizing the Mind.* Cambridge, MA: The MIT Press.

Dreyfus, Hubert. 1979. *What Computers Still Can't Do: A Critique of Artificial Reason.* Cambridge, MA: The MIT Press.

Eliasmith, Chris. 2013. *How to Build the Brain: A Neural Architecture for Biological Cognition.* New York: Oxford University Press.

Eliasmith, Chris and Charles H. Anderson. 2003. *Neural Engineering: Computation, Representation, and Dynamics in Neurobiological Systems.* Cambridge, MA: The MIT Press.

Floridi, Luciano. 2010. *Information: A Very Short Introduction.* Oxford: Oxford University Press.

Fodor, Jerry A. and Zenon W. Pylyshyn. 1988. "Connectionism and Cognitive Architecture: A Critical Analysis." *Cognition* 28 (1–2): 3–71.

——. 2015. *Minds without Meanings: An Essay on the Content of Concepts.* Cambridge, MA: The MIT Press.

Gardner, Howard. 1985. *The Mind's New Science: A History of the Cognitive Revolution.* New York: Basic Books.

Gödel, Kurt. 1933. "Zum Entscheidungsproblem Des Logischen Funktionenkalküls." *Monatshefte Für Mathematik Und Physik* 40 (1): 433–43. doi:10.1007/BF01708881.

Harnad, Stevan. 1990. "The Symbol Grounding Problem." *Physica D* 42: 335–46.

Haugeland, John. 1985. *Artificial Intelligence: The Very Idea.* Cambridge, MA: The MIT Press.

Hobbes, Thomas. 1839. *The English Works of Thomas Hobbes of Malmesbury.* London: J. Bohn.

Hutto, Daniel D. and Erik Myin. 2013. *Radicalizing Enactivism: Basic Minds without Content.* Cambridge, MA: The MIT Press.

Jackson, Frank. 1986. "What Mary Didn't Know." *The Journal of Philosophy* 83 (5): 291–5. doi:10.2307/2026143.

Johnston, John. 2008. *The Allure of Machinic Life: Cybernetics, Artificial Life, and the New AI.* Cambridge, MA: The MIT Press.

Krajewski, Stanisław. 2007. "On Gödel's Theorem and Mechanism: Inconsistency or Unsoundness Is Unavoidable in Any Attempt to 'Out-Gödel' the Mechanist." *Fundamenta Informaticae* 81 (1): 173–81.

Krueger, Robert C. and Clark L. Hull. 1931. "An Electro-Chemical Parallel to the Conditioned Reflex." *Journal of General Psychology* 5: 262–9.

Kuhn, Thomas S. 1970. *The Structure of Scientific Revolutions.* Chicago: University of Chicago Press.

La Mettrie, Julien. 1912. *Man a Machine.* Translated by Gertrude Carman Bussey and Mary Whiton Calkins. Chicago: The Open Court Pub. Co.

Leibniz, Gottfried. 1666. *Dissertatio de Arte Combinatoria, in qua Ex Arithmeticae Fundamentis Complicationum Ac Transpositionum Doctrina Novis Praeceptis Extruitur, & Usus Ambarum per Universum Scientiarum Orbem Ostenditur.* Lipsiae: apud Joh. Simon Fickium et Joh. Polycarp. Seuboldum Literis Spörelianis.

Lotka, Alfred. 1925. *Elements of Physical Biology.* Baltimore: Williams & Wilkins Company.

Lucas, J. R. 1961. "Minds, Machines and Gödel." *Philosophy* 9 (3): 219–27.

MacKay, Donald MacCrimmon. 1969. *Information, Mechanism and Meaning.* Cambridge, MA: The MIT Press.

McClelland, James L., David E. Rumelhart, and PDP Research Group, eds. 1986. *Parallel Distributed Processing: Explorations in the Microstructures of Cognition, Volume 2: Psychological and Biological Models.* Cambridge, MA: The MIT Press.

Miller, George A. 1956. "The Magical Number Seven, Plus or Minus Two: Some Limits on Our Capacity for Processing Information." *Psychological Review* 63 (2): 81–97. doi:10.1037/h0043158.

Miller, George A., Eugene Galanter, and Karl H. Pribram. 1967. *Plans and the Structure of Behavior*. New York: Holt.

Millikan, Ruth Garrett. 1984. *Language, Thought, and Other Biological Categories: New Foundations for Realism*. Cambridge, MA: The MIT Press.

Miłkowski, Marcin. 2013. *Explaining the Computational Mind*. Cambridge, MA: The MIT Press.

——. 2016. "Models of Environment." In *Minds, Models and Milieux: Commemorating the Centennial of the Birth of Herbert Simon*, edited by Roger Frantz and Leslie Marsh, 227–38. New York: Palgrave Macmillan. doi:10.1057/9781137442505_13.

Minsky, Marvin and Seymour Papert. 1969. *Perceptrons: An Introduction to Computational Geometry*. Cambridge, MA: The MIT Press.

Neisser, U. 1963. "The Imitation of Man by Machine: The View That Machines Will Think as Man Does Reveals Misunderstanding of the Nature of Human Thought." *Science* 139 (3551): 193–7. doi:10.1126/science.139.3551.193.

Newell, Allen and Herbert A. Simon. 1972. *Human Problem Solving*. Englewood Cliffs, NJ: Prentice-Hall.

Penrose, Roger. 1989. *The Emperor's New Mind*. London: Oxford University Press.

Piccinini, Gualtiero and Carl F. Craver. 2011. "Integrating Psychology and Neuroscience: Functional Analyses as Mechanism Sketches." *Synthese* 183 (3): 283–311. doi:10.1007/s11229-011-9898-4.

Pickering, Andrew. 2009. *The Cybernetic Brain: Sketches of Another Future*. Chicago: University of Chicago Press.

Preston, John and Mark Bishop. 2002. *Views into the Chinese Room: New Essays on Searle and Artificial Intelligence*. Oxford: Clarendon Press.

Putnam, Hilary. 1960. "Minds and Machines." In *Dimensions of Mind*, edited by Sidney Hook, 148–79. New York: New York University Press.

Ramsey, William M. 2007. *Representation Reconsidered*. Cambridge: Cambridge University Press. doi:10.1017/CBO9780511597954.

Rosenblatt, F. 1958. "The Perceptron: A Probabilistic Model for Information Storage and Organization in the Brain." *Psychological Review* 65 (6): 386–408. doi:10.1037/h0042519.

Rosenblueth, Arturo, Norbert Wiener, and Julian Bigelow. 1943. "Behavior, Purpose and Teleology." *Philosophy of Science* 10 (1): 18. doi:10.1086/286788.

Rumelhart, D. E. and James L. McClelland. 1986. *Parallel Distributed Processing: Explorations in the Microstructure of Cognition. Volume 1. Foundations*. Cambridge, MA: MIT Press.

Russell, S. Bent. 1913. "A Practical Device to Stimulate the Working of Nervous Discharges." *Journal of Animal Behavior* 3 (1): 15–35. doi:10.1037/h0070584.

Ryle, Gilbert. 2009. *The Concept of Mind*. New York: Routledge.

Searle, John R. 1980. "Minds, Brains, and Programs." *Behavioral and Brain Sciences* 3 (3): 1–19. doi:10.1017/S0140525X00005756.

Seth, Anil K., Zoltán Dienes, Axel Cleeremans, Morten Overgaard, and Luiz Pessoa. 2008. "Measuring Consciousness: Relating Behavioural and Neurophysiological Approaches." *Trends in Cognitive Sciences* 12 (8): 314–21. doi:10.1016/j.tics.2008.04.008.

Simon, Herbert A. and Allen Newell. 1958. "Heuristic Problem Solving: The Next Advance in Operations Research." *Operations Research* 6 (1): 1–10.

Sluckin, Wladyslaw. 1954. *Minds and Machines*. Harmondsworth: Penguin.

Sperling, George. 1960. "The Information Available in Brief Visual Presentations." *Psychological Monographs: General and Applied* 74 (11): 1–29. doi:10.1037/h0093759.

Thorndike, Edward L. 1911. *Animal Intelligence*. London: The Macmillan Company.

Tolman, Edward Chace. 1939. "Prediction of Vicarious Trial and Error by Means of the Schematic Sowbug." *Psychological Review* 46 (4): 318–36. doi:10.1037/h0057054.

——. 1948. "Cognitive Maps in Rats and Men." *Psychological Review* 55 (4): 189–208.

Turing, Alan. 1937. "On Computable Numbers, with an Application to the Entscheidungsproblem." *Proceedings of the London Mathematical Society* s2-42 (1): 230–65. doi:10.1112/plms/s2-42.1.230.

——. 1950. "Computing Machinery and Intelligence." *Mind* LIX (236): 433–60. doi:10.1093/mind/LIX.236.433.

Van Gelder, T. 1995. "What Might Cognition Be, If Not Computation?" *The Journal of Philosophy* 92 (7): 345–81.

Walter, W. Grey. 1950a. "An Imitation of Life." *Scientific American*. doi:10.1038/scientificamerican0550-42.

——. 1950b. "An Electro-Mechanical Animal." *Dialectica* 4 (3): 206–13. doi:10.1111/j.1746-8361.1950.tb01020.x.

——. 1953. *The Living Brain*. New York: Norton.

Weaver, W. and C. E. Shannon. 1964. *The Mathematical Theory of Communication*. Urbana: University of Illinois Press.

Webb, Barbara. 2002. "Can Robots Make Good Models of Biological Behaviour?" *Behavioral and Brain Sciences* 24 (6): 1033–94. doi:10.1017/S0140525X01000127.

Wheeler, Michael. 2008. "God's Machines: Descartes on the Mechanization of Mind." In *The Mechanical Mind in History*, edited by Phil Husbands, Owen Holland, and Michael Wheeler, 307–30. Cambridge, MA: The MIT Press.

Wiener, Norbert. 1948. *Cybernetics, or Control and Communication in the Animal and the Machine*. New York/Paris: J. Wiley/Hermann.

Willis, Thomas. 1694. *Opera Omnia, Nitidius Quam Unquam Hactenus Edita, Plurimum Emendata*. Coloniae: Sumptibus Gasparis Storti.

PART II

The nature of mechanisms

7

VARIETIES OF MECHANISMS

Stuart Glennan and Phyllis Illari

1. Why explore mechanistic variety?

A common worry about the utility of philosophical analyses of mechanism is that the term "mechanism" is applied to such a staggeringly diverse collection of things that nothing informative can be said about it as a general category. This is especially true for those like us (Glennan forthcoming; see also Chapter 1, Illari and Williamson 2012) who have argued for a conception of mechanism that is expansive enough to allow most anything scientists have called mechanisms to fall under its extension. In response to such worries, we have argued that, while the things scientists (and philosophers) call mechanisms are heterogeneous, there are enough commonalities to allow for an informative analysis of mechanisms. Such an analysis can show what mechanisms are, and distinguish them from things that are not mechanisms.

Even if this analysis is successful, it is important for advocates of expansive conceptions to explore the varieties of mechanisms. The class of mechanisms is a heterogeneous lot, and exploring the nature and scope of these variations will give us insight into both metaphysical and methodological questions about mechanisms. There are (at least) two ways we might explore this variety. One is to look at exemplars of mechanisms that are studied in various scientific domains. The chapters in Part IV collectively exemplify this approach, and they do much to illustrate the special techniques, challenges, and opportunities for mechanistic research within those domains. In this chapter, though, we will take a different approach. Our goal will be to describe a set of taxonomic dimensions along which we can characterize the varieties of mechanisms that exist within the natural and social world. This system of classification will often cut across disciplinary and domain boundaries.

As we explore the varieties of mechanisms, we need to contrast our project with another related project. Our goal in this chapter is focused on the varieties of mechanisms that exist in the world, rather than the variety of philosophical or methodological approaches that have gone by the name of "Mechanism" (or "mechanicism" or "mechanical philosophy").[1] There are certainly varieties of mechanical philosophies just as there are varieties of mechanisms, and some philosophers have offered taxonomies of those varieties (see also Chapters 3 and 5 for historical discussions). It is, moreover, easy to slip between varieties of mechanisms and varieties of mechanical philosophies, because different kinds of philosophical approaches will often assume or privilege certain varieties of mechanisms. We shall return to the issue of the relationship between mechanisms and mechanical philosophies in the conclusion of our chapter.

2. The dimensions of mechanistic variety

Any attempt to provide a scheme for classifying things must begin with a specification of the set of things to be classified. For us, that specification is embodied in minimal mechanism (see Chapter 1):

> A mechanism for a phenomenon consists of entities ·(or parts) whose activities and interactions are organized so as to be responsible for the phenomenon.

We start with this permissive characterization, not because it is better or "more correct" than less permissive ones, but because its permissiveness allows us to treat more restrictive conceptions as varieties of mechanisms.

How might one fruitfully sort out the vast array of things that count as mechanisms under the minimal definition? Our proposal is that a set of classificatory dimensions can naturally be read off the characterization of minimal mechanism.[2] Specifically, a mechanism can be classified according to (1) the kind of phenomena for which it is responsible, (2) the kinds of entities, activities, and interactions that comprise the mechanism, and (3) the way in which these entities, activities, and interactions are organized. To these three dimensions, we add another: (4) the mechanism's etiology, i.e. the way it came to exist and have the properties it does. A mechanism's etiology is logically independent of its current constitution and properties, but there are interesting historical and causal dependencies between how a mechanism came to be, what it is made of, how it is organized, and what it does.

Within each of these four dimensions, there is great variety, and we cannot hope in this short piece to do more than offer a few examples of that variety, but it should be enough to illustrate our main point—namely that attending to the varieties of mechanisms will help illuminate both metaphysical and methodological debates. In identifying these four dimensions as dimensions, we are suggesting that they are to a large extent independent. For instance, mechanisms with very different kinds of entities, activities, and interactions can have similar kinds of organization, or mechanisms with very different etiologies can be responsible for similar kinds of phenomena. On the other hand, our decision to treat entities and activities as lying within a single dimension reflects our understanding that entities or parts have a comparatively narrow range of activities and interactions in which they can engage. In this and other ways, the independence of these dimensions is not complete, and indeed understanding the ways in which, e.g., particular kinds of organization are characteristic of mechanisms with certain sorts of parts is crucial to understanding the taxonomic space.

3. Varieties of phenomena

It is a truism among new mechanists that mechanisms are individuated by their phenomena or behavior. All mechanisms are mechanisms for. This usage is common in scientific and technological contexts. For instance, one speaks of mechanisms for protein synthesis, thermal regulation, or signal amplification. Mechanisms are classified as mechanisms of a certain kind if they produce the same kind of phenomena, so for instance, artificial and natural hearts are both hearts in the sense of being mechanisms for pumping blood within a circulatory system, even if the material constitution and internal operations are quite different.

There is no non-pragmatic standard by which to determine if two mechanisms are responsible for the same kind of phenomena. If two factories produce cars, one Fords and the other Volvos, or one sedans and the other SUVs, do they produce the same kind of thing or different?

Obviously there is no proper answer to this question apart from some criteria of sameness, and such criteria will depend upon one's interests. But while we cannot offer an absolute standard for type-identity of phenomena, we can say much about structural features of phenomena by which we might classify them. These will show us that very disparate phenomena may be, in some respects, of similar kinds, and these similarities will bear on a variety of methodological and metaphysical issues.

Consider, for example, the question of the regularity of the mechanism's phenomenon (see Chapter 12). Machamer, Darden, and Craver (MDC), in their much-quoted definition of mechanism (Machamer et al. 2000), argued that mechanisms produce regular changes from start or set-up conditions to finish or termination conditions. A paradigm of a mechanism of this kind is protein synthesis. Given appropriate start conditions (presence of appropriate substrates, promoter regions, etc.), transcription of DNA will be initiated, leading through a sequence of steps to synthesis of proteins in the ribosome (the termination condition). Such a process is regular in the sense that it happens many times in the life of a cell, but it is far from universal—because various things can interfere with protein production. For instance, gene regulation can inhibit production of certain proteins, or the cellular machinery can become damaged.

The regularity of a mechanism's behavior provides an important way of classifying mechanisms by their phenomena. While early characterizations, like MDC, impose a regularity requirement, our approach is to see that different varieties of mechanisms are different in the degrees and respects of their regularity.

One sense in which mechanisms can be regular is that the phenomena they are responsible for can recur. Recurrence comes in two basic varieties. The same token mechanism can exhibit the phenomenon on multiple occasions, and the phenomenon can be exhibited by multiple tokens of the same type of mechanism. Protein synthesis recurs in both of these ways. The same cellular machinery continually produces proteins over its lifetime, and there are many cells producing the same proteins. In other cases, you can have one kind of recurrence without the other. Apoptosis is a kind of mechanism that is responsible for programmed cell death. Many cells die this way, but each cell dies only once. On the other hand, token mechanisms, like the Old Faithful Geyser in Yellowstone National Park, can repeatedly exhibit the same kind of phenomena. Among mechanisms with recurrent phenomena, we can also distinguish between the more or less regular. For instance, we can characterize mechanisms by the probability with which start-up conditions actually lead to termination conditions (see Chapter 13).

Issues about regularity and recurrence are important in both methodological and metaphysical debates. For instance, one of the epistemological arguments for a regularity requirement on mechanisms is that if mechanisms are not in certain ways regular, then we will not be able to discover them. Experimental techniques in the sciences depend upon the fact that the same phenomena recur across many tokens of the same kind of mechanism. Because of this, for instance, one can study the mechanism of protein synthesis in a few cells of some species of model organism and apply those conclusions to protein synthesis in cells of organisms in a wide variety of taxa. Arguably, the distinction between the historical and experimental sciences is that the experimental, much more than the historical, study recurrent phenomena. Historians (of both natural and human history) are often concerned with the explanation of singular events, and these events are often produced by mechanisms that are non-recurrent, or "one-off." For instance, natural historians may be interested in the causes of the Cambrian explosion, while historians of human civilization will be interested in the causes of distinctive events in human history, like the fall of the Roman Empire or the stock market crash in 1929. Investigation of these "ephemeral mechanisms" require distinctive methodologies (Glennan 2010; see also Chapter 31, this volume).

Regularity and recurrence also play roles in metaphysical debates about the relationship between mechanisms, causes, and laws. Recurrent mechanisms are the foundation of the new mechanist's account of the nature of laws. Laws in the higher-level sciences, if such exist, are thought to be effects—descriptions of the behavior of recurrent and regular mechanisms. At the same time, the idea of the existence of one-off mechanisms is essential if one thinks, as Glennan does, that causal claims are claims about the existence of mechanisms. This is required because not all causally connected events are produced by mechanisms that are regular or recurrent.

Another useful way to classify mechanistic phenomena is by the kind of relationship a phenomenon bears to the entities, activities, and interactions that are responsible for it. Craver and Darden (2013) have suggested that sometimes mechanisms *underlie* phenomena, whereas other times they *produce* them. When a mechanism underlies some phenomenon, the mechanism's phenomenon is constituted by the collective activities and interactions of its parts, and for this reason such phenomena are called *constitutive*. In contrast, when a mechanism produces some phenomenon, that phenomenon is a (causal) result of the activities and interactions of the mechanism's parts, rather than being made of them. For this reason, we call such phenomena *non-constitutive*. The contrast between these two kinds of phenomena can be seen by comparing protein synthesis to the mechanism responsible for muscle contraction. Protein synthesis can be seen as a paradigm of a productive, non-constitutive mechanism, because the proteins are made by, but not made of, the organized activities and interactions of the parts of the mechanism. Muscle contraction seems quite different. Muscles consist of cells called myocytes, which contain within them bundles of fibrils, which can shorten via the sliding of filaments made of proteins. When a muscle contracts, it does this because the cells of which its tissues are made contract, and this in turn happens because of the sliding filaments within the fibrils within the cell. Here it seems that the result of the mechanism is not some product made by the mechanism; the contracting of the muscle just is the contracting of its cells and the fibrils within them. This "just is" relation is a constitutive one.

Attending to the distinction between these kinds of phenomena can be helpful in sorting out a variety of metaphysical and epistemological issues. Metaphysical debates about the nature of causation, production and difference making, and causal chains are concerned chiefly with the "horizontal" production of non-constitutive phenomenon (Bogen 2004; Campaner 2013; Glennan forthcoming, Chapter 7; see also Chapters 10, 11, and 12, this volume; Kincaid 2012; Salmon 1984). On the other hand, mechanisms for constitutive phenomena have been at the center of discussions of "vertical" relations between parts and wholes in science (Craver 2007; Gillett and Aizawa 2016; see also Chapters 9 and 14, this volume; Leuridan 2012; Winther 2009), which offer a new perspective on both metaphysical and methodological versions of debates about reduction and emergence. There has also been discussion of experimental methods for identifying working parts of constitutive mechanisms (Craver and Darden 2013) and of the challenges of integrating accounts at these different levels of mechanisms (Stinson 2016; see also Chapter 28, this volume).

4. Varieties of entities, activities, and interactions

Perhaps the most obvious way to classify kinds of mechanisms is by the constituents of which they are made. According to minimal mechanism, these constituents are of two kinds—first the entities that are working parts of the mechanism, and second the activities and interactions in which these entities engage. The varieties of entities in mechanisms are diverse, ranging from DNA and mRNA in protein synthesis, through people and institutions such as Central Banks in economic mechanisms, to electrons and star cores in astrophysics. The same mechanism can include entities of notably different sizes. For instance, it appears that the mechanism responsible for core collapse in supernovae involves interactions between parts that range in size and mass between the subatomic and stellar cores with masses greater than our sun (Illari and Williamson 2012). Relatedly, the mechanisms responsible for some phenomenon may include entities of quite different kinds. For instance, the mechanisms that account for the diversity and abundance of flora in urban landscapes will involve artifacts in the built environment (like buildings and bridges) and the social and physical actors that build them, as well as human and non-human fauna in the area, the nutrients within the soil, and the sun and rainwater that sustains growth.

Just as mechanisms can be classified by the kinds of entities of which they are made, so too can they be classified by the kinds of activities and interactions in which the parts of mechanisms engage. It is one of the important insights of the new mechanist literature that concepts like "activity," "interaction," and "cause" are abstract. Proper characterization of mechanisms requires one to use more specific concepts like "pushing," "folding," "binding," "trading," and "collapsing" (Bogen 2008; Machamer et al. 2000). These specific activity concepts tell one much more about how and under what conditions mechanisms bring about their phenomena.

It is not enough, though, to distinguish between the abstract category of activities and interactions, and specific sorts of activities and interactions. Rather, activities and interactions can be characterized in a hierarchy of increasingly less abstract ways, corresponding to increasingly concrete and determinate varieties of activities and interactions. For example, many activities and interactions are involved in protein synthesis, but each one is an activity or interaction of a particular kind. Transcription is a very general activity, but any particular transcription is a transcription of a particular coding strand of DNA, into a matched strand of mRNA, which will be more determinate. Similarly, protein folding is the folding of a polypeptide chain—a chain of amino acids—into a functional protein. Specific foldings take many forms, going through at least three stages to form secondary, tertiary, and quaternary protein structure.

Glennan (forthcoming, Chapter 2) has argued that the parallel point needs to be made about entities. The concept of entity is extremely abstract, and classifications of kinds of entities will form hierarchies of increasingly concrete and determinate forms. For instance, mRNA and proteins are different kinds of entities, but each of these kinds have a variety of more determinate kinds that fall under them. The same is true with functionally specified entities: all promoters have features in common that make them the kind of entity they are, but they come in many more determinate forms, promoting the transcription of different coding strands of DNA.

Entity and activity varieties provide us with one way of classifying mechanisms into kinds. We can lump mechanisms by the kinds of entities which are their parts, and also by the kinds of activities that those parts engage in. Some mechanism classifications emphasize similarities in entities, while others emphasize similarities in activities. For example, biochemical mechanisms are largely grouped together as similar based on their entities, and competitive mechanisms are largely grouped based on their activities and interactions, while social mechanisms share a bit of both.

The kinds of entities and kinds of activities involved in a mechanism constrain but do not completely determine each other. This is crucial to discovery methods that involve using knowledge of an entity to identify its activity, or vice versa (see Chapter 19, this volume). Particular activities are activities of particular entities, and a particular entity can only take part in some activities, while a particular activity can only be produced by some entities, but entities don't have to have proper activities. Generally quite different kinds of entities may engage in similar activities, and particular kinds of entities may engage in very different sorts of activities and interactions. Think of all the things that mouths or screwdrivers can do, or of the many different kinds of entities that one can use to break a window or plug a leak.

Indeed, whether something counts as an entity or activity at all is not something that can be answered except locally, in the context of particular phenomena. We take this to be Dupré's point when he argues that nothing is an object *tout court*, but only at a particular timeframe (Dupré 2012). On a long timeframe, Dupré argues, even mountains are processes (which we take to be activity-like), because on a geological timeframe, even mountains are constantly changing. Something similar might be said for at least some activities and entities in mechanisms, which exist at a particular timescale, depending on the phenomenon that the mechanism is for. Cells and tissues are sufficiently stable that particular cells or tissues often count as entities in many mechanisms for bodily phenomena. Nevertheless, they are sufficiently changeable that they do not count in mechanisms for other phenomena, and may count instead as whole mechanisms or even systems—which are in their turn decomposable into *both* entities and activities.

5. Varieties of organization

What a mechanism does, and so how we discover it and use it in explanation, depends not just on the entities and activities of which it is made, but also on how these constituents are organized. Electrical circuits nicely illustrate this fact: the very same resistors or capacitors can exhibit very different resistance or capacitance depending upon whether they are wired in parallel or in series. Similarly, the developmental processes by which fertilized eggs divide and develop into embryos and ultimately mature organisms depend upon variations in concentrations of proteins within different regions of the egg. The fact, for instance, that a mature fruit fly has its wings and body segments where it does depends crucially upon the locations and rates at which gene products express within the developing embryo.

We treat mechanistic organization as a separate dimension of mechanistic variety, because it can vary largely independently of the kinds of entities and activities that constitute the mechanism. Consider, for example, forms or organization like positive and negative feedback loops. In positive feedback loops, a mechanism produces a change in some property of the system, and this change in turn feeds back into the system, amplifying the effect. In contrast, in negative feedback loops, the changing property feeds back in a way that dampens the effect. These and other forms of organization can be found in all manner of mechanistic processes and systems—electrical, molecular and chemical, genetic, climatological, economic, and social. It is the fact that such forms of organization induce predictable patterns of behavior in mechanisms that allows for the development of representational and modeling techniques like dynamical systems theory, control theory, and information theory that are largely independent of the particular kind of entities or activities involved (see Chapter 20, this volume).

These kinds of organizational varieties can be called topological and functional. They show how parts of mechanisms and their activities are arranged—spatially, temporally, and causally. These kinds of features are often captured in mechanism diagrams (see Chapter 18, this volume) or via mathematical formalisms like structural equation models or Bayes nets, which indicate which parts are connected to which, and the functional forms of such dependencies.

The varieties of topological and functional organizations of mechanisms are essentially limitless, and we shall not try to survey them further here. We do, however, want to call attention to a few more basic ways to classify mechanistic organization that are of both methodological and ontological import. To begin, we can classify mechanism by the number of parts they have. Muscular skeletal mechanisms, for instance, have relatively few parts (muscles, bones, cartilage, etc.)—few enough that scientists can identify each of those parts along with its role in the mechanism.

In contrast, in molecular mechanisms like the mechanism of protein synthesis, there are many parts—e.g., many segments of mRNA and many proteins—and operations are completed many times over. Whether mechanisms have few or many parts has important consequences for the techniques scientists use to describe, manipulate, and explain them. For instance, in mechanisms of protein synthesis, diagrams represent token entities and activities of a process that occurs many times over. Description of what the mechanism does will not be simply in terms of producing a protein, but in producing proteins in various concentrations and at various rates. This is very different from a representation of the parts of a circuit in an electronic amplifier, where the wiring diagram explicitly identifies and locates each part.

We can also classify mechanisms by the degree to which their parts are uniform. Some mechanisms are composed of a small set of largely uniform parts, in the way that Lego structures may be made out of a large number of similar blocks. Other mechanisms have a variety of different parts with different capacities and roles. For instance, at the lowest molecular level, the components of DNA, RNA, and proteins are small in number, and each of these building blocks (nucleotides, amino acids) is more or less identical, with the structural and functional diversity of proteins arising from the many different ways in which these simple building blocks can be combined. At higher levels of mechanistic organization in multi-cellular organisms, we often see much more structural and functional diversity in parts. Locomotion in animals, for instance, requires the orchestration of very different types of entities and activities—muscular, skeletal, pulmonary, respiratory, neurological, and so on.

Finally, we can classify mechanistic organization by what brings about and maintains relationships between parts. At one extreme, which we call induced organization, these relationships are imposed by an external agent, like when a host arranges the seating of guests at a dinner party; at the other, which we call affinitive organization, these relationships arise from the affinities that parts bear to each other, like when the guests at the dinner party seat themselves. In such a case, the arrangements of the guests and who interacts with whom will be determined in part by chance, but also by who each guest is disposed to sit with. All mechanistic organization is to some degree affinitive, because of the simple fact that different entities have different capacities to act and interact with each other. However, sometimes, as in many chemical processes, reactions occur largely due to random interactions between molecules with different kinds of affinities, whereas in other cases, like the case where protein concentrations vary from the front to middle to back of a developing egg cell, the organization was induced by prior processes that set the next stage in the developmental mechanism. All of these forms of organization matter to how we mechanistically explain and discover mechanisms, especially since we have become able to recognize and model forms of organization that recur in many mechanisms (see Chapter 20, this volume).

6. Varieties of etiology

Because mechanisms are localized in space and time, they must have etiologies—that is to say, there must be some causal process that led up to their existence. There is a difference between what caused a mechanism to come into being and how it works now, though as we shall show in this section, there are often connections between a mechanism's etiology on the one hand, and what it is made of and how it is organized on the other.

While mechanical philosophy owes much in its origins to analogies with machines of human construction, the etiology of such mechanisms is typically unlike that of mechanisms responsible for naturally occurring phenomena. Mechanical devices like windmills or cars have what we call designed-and-built etiologies. This is to say that an agent, the designer, identified some phenomena they wanted to produce (like the grinding of corn or moving passengers over roads), and then set about to collect and arrange a set of parts so that their activities and interactions would in fact bring about the desired result. It is not just the artifacts of human engineering that can be designed and built. Many social, political, and economic systems are also "engineered" in this way, and there is no requirement that designers and builders must be human agents.

Most mechanisms, however, evolve over time. Evolved etiologies are by no means limited to biological systems. In the most basic sense, evolved mechanisms are those that are built and modified over time, so that the present characteristics of the mechanism have emerged gradually from earlier stages or versions of the mechanism. While the idea of evolved mechanisms brings most immediately to mind biological mechanisms, abiotic systems and processes like stars, volcanoes, and weather systems also have evolved etiologies in our sense, as do many social, legal, and economic mechanisms—like property or commodity markets or systems regulating property, marriage, and child-rearing.

There are a variety of different kinds of processes that can underlie the evolution of mechanisms and mechanical systems. For instance, in stellar evolution, the gradual changes in stars from one stage to another depend mainly upon the star's mass and the concentrations of various elements within the stellar core. As a star consumes hydrogen in fusion, gradually its properties will change. Evolutionary biology is concerned with the etiology of populations of reproducing organisms (and their traits), and there appear to be a number of processes—selection, mutation, migration, drift—which can drive this evolution. We believe that these represent varieties of etiological mechanisms, though whether and in what sense they are mechanisms is a matter of some debate (see Chapter 22, this volume). Similarly, when we speak of developmental mechanisms, we are describing mechanisms by which an organism and its component mechanisms evolve in the developing embryo (see Chapter 25, this volume).[3]

One other way that a mechanism can come to be is by the operation of ephemeral mechanisms. Ephemeral mechanisms are mechanisms in which the relationship between the parts of the mechanism that are responsible for the phenomena are short lived, unstable, and thrown together by happenstance. The mechanisms responsible for one-off events, like forest fires, car crashes, and romantic liaisons, are typically ephemeral, but seldom do enduring mechanisms arise from such processes. In a very different context, Donald Davidson imagined the possibility of lightning striking a tree and somehow miraculously rearranging the tree to create an exact molecule-by-molecule replica of himself, which he called Swampman. Were such miraculous events to occur, the process by which the mechanisms within Swampman came to be would be ephemeral in an extreme way.

While we distinguish these three kinds of etiologies, the etiologies of particular mechanisms typically involve several of these elements in concert. For instance, while biological organisms and the mechanisms that operate in them are the product of evolutionary processes, it is clear that many events that impact the outcome of those processes will be the result of ephemeral mechanisms—for instance, mechanisms responsible for one-off mutations or for environmental changes that lead to extinctions. Many artifacts and technologies have designed-and-built etiologies, but the design and build process can be iterated, so present versions are either evolutionary modifications of earlier versions, as when we upgrade the plumbing in our houses, or newly designed and built tokens of a type, but where the design of the type itself has been modified as the result of an evolutionary process. For instance, my new iPhone was designed and built, but its design is deeply constrained by earlier iterations of iPhones.

While a mechanism's constituents and organization are to some degree independent of its etiology, there are clearly connections. What a mechanism does and how it does are often constrained by its history, and traces of its history are evident in the present mechanism. How we set about discovering the mechanism and using it in explanation and in attempts to control the world by altering the operation of the mechanism can be affected by that history. Perhaps most obviously, evolved and ephemeral mechanisms will tend to rely on affinitive forms of organization, because there is no external agent to impose organization on them. We could not effectively alter the operation of the mechanism without taking this into account. Similarly, if Simon's argument is correct, evolved systems will tend to have "modular" designs (Simon 1996; see Chapter 14, this volume). Also, evolved systems will contain vestiges of earlier versions of the system that have become entrenched; this is obviously true of populations of organisms evolving by natural selection, but it is equally true of my iPhone, or for that matter the organization of Britain's National Health Service. Effective policy or technological change needs to consider this.

7. Upshots

After this too brief exploration of the varieties of mechanism, we would like to reflect on the philosophical significance of this classificatory project, and to discuss how it is related to other attempts to classify mechanisms and mechanical philosophy.

To begin, let us consider whether and in what sense these varieties of mechanisms are real. Our language has been realist. Mechanisms are things in the world, and we can place them into categories according to the different properties they have. Nonetheless, our realism is tempered. The dimensions of mechanistic variety do not allow us to sort mechanism tokens neatly into kinds. The properties by which we sort cross cut. Mechanisms may have similar patterns in the kind of phenomena they produce (e.g., in being regular or constitutive) while involving very different kinds of activities. Or, again, mechanisms made with very different kinds of entities can be organized in similar ways, and mechanisms that produce the same kind of phenomena may have very different etiologies. Another problem for realism is vagueness. How many parts does one need to have to be a many-parted mechanism, or with what probability must a start-up condition lead to a termination condition to have us count a mechanism as regular? A third problem is that mechanisms will often be hybrids involving multiple varieties within each of these dimensions. Consider, for instance, the varieties of entities and activities involved in mechanisms responsible for mood disorders, or the variety of organizational motifs in gene regulatory networks.

However, the fact that there are many ways to carve up the world does not mean that there isn't a world out there that constrains and makes sense of our carvings. We think our account is consistent with what Mitchell calls a "pluralist realist approach to ontology, which suggests not that there are multiple worlds, but that there are multiple correct ways to parse our world, individuating a variety of objects and processes that reflect both causal structures and our interests" (Mitchell 2009, p. 13; see also Wimsatt 1994 and Chapter 15, this volume). While Mitchell mentions "interests," we should also add abilities, since how we parse the world will depend upon the tools we have. Take for instance the distinction between few and many-parted mechanisms. This distinction is not only vague; it is largely determined by our cognitive and computational resources. Many-parted mechanisms require different techniques for discovery, representation, and control—but this is because of our abilities rather than any intrinsic feature of the mechanism. More generally, we think that the way we classify mechanisms into varieties reflects the models we use to represent them, and that the suitability of such models depends both on token mechanisms in the world and upon our interests and epistemic resources (Glennan forthcoming, Chapter 4).

We turn now to a discussion of the relation of our account of mechanistic variety to other taxonomic efforts in the literature. As we noted at the outset, care needs to be taken to distinguish kinds of mechanisms, which are things in the world, from kinds of (capital "M") Mechanisms, which are philosophical and scientific claims or views about the role of mechanisms and mechanistic reasoning. This distinction is easily lost in taxonomic discussions, since mechanistic philosophical projects typically implicitly or explicitly presuppose a conception of what mechanisms are in the world.

Let us begin with Levy's taxonomy, which is of big "M" Mechanisms. He identifies three distinct but related philosophical projects that he calls "Causal Mechanism (CM)," "Explanatory Mechanism (EM)," and "Strategic Mechanism (SM)" (Levy 2013). CM is a metaphysical project intended to give an account of causation in terms of mechanisms. EM is a thesis about explanation to the effect that good explanations must cite information about mechanisms, while SM aims to draw attention to the heuristics used in mechanism discovery. The projects of EM and SM are epistemological and methodological. Quite rightly, Levy observes that Causal Mechanism of the sort proposed by Glennan must embrace a permissive conception of mechanisms in the world (like our minimal mechanism) if it is to have any hope of making the claim that causal connections are all or mostly mediated by mechanisms. Similarly, he observes that Strategic Mechanism, which he associates most prominently with Bechtel and Richardson (1993), adopts a narrower conception of mechanisms, in which they are machine-like, and where decomposition and localization are powerful strategies. Explanatory Mechanism, which he associates most prominently with MDC (2000) and Craver (2007), also focuses on a somewhat narrower conception of mechanism, especially those that are always or for the most part regular, and have hierarchical organization.

Andersen (2014) offers a broadly similar distinction between two projects, which she calls Mechanism$_2$ ("Mechanisms as an ontology of the world"), which aims to use mechanisms to give a complete account of what there is, including an account of causation, and Mechanism$_1$ ("Mechanisms as integral to scientific practice"), which focuses on the practices of discovery, representation, and explanation associated with hierarchically organized mechanisms. Roughly, Andersen's Mechanism$_2$ and Levy's CM identify metaphysical or ontological projects, while Andersen's Mechanism$_1$ and Levy's EM and SM identify explanatory, epistemological, and methodological projects.[4]

Nicholson explores the concept of mechanism in biology, and argues that the word "mechanism" has principally been used to refer to machine-like structures in the world: "machine mechanisms," "a step-by-step explanation of . . . a causal process," and to "Mechanicism": "the philosophical thesis that conceives organisms as machines" (Nicholson 2012, p. 153). The first of the theses refers to a kind of small "m" mechanism, while the last of these is clearly a big "M" Mechanism (see Chapter 5, this volume). The middle of these is neither, but it is associated with what Nicholson calls "the Mechanismic Program," and which most philosophers now call "the New Mechanism" and is close to Andersen's Mechanism$_1$ and Levy's Explanatory Mechanism.

Kuorikoski's (2010) taxonomy distinguishes two kinds of small "m" mechanisms—"computational causal systems" and "abstract forms of interaction." The former he takes to be the sort of mechanisms that have been chiefly of interest in recent discussions of mechanisms in biology (and we surmise in EM, SM, and Mechanism$_1$). The latter is the concept Kuorikoski sees at work in discussions of mechanisms in the social science literature. The former kind of mechanisms are the sorts of mechanisms amenable to decomposition and localization. The latter refer to mechanisms where these strategies fail, but which can be characterized abstractly in terms of certain kinds of organization. He cites as examples selection mechanisms in evolution, various kinds of market mechanisms, crowding out, and self-fulfilling prophecies. (As Kuorokoski notes, abstract form of interaction (AFI) mechanisms are not themselves abstract, but the explanations of such mechanisms are.)

We find that there is much that is informative in these kinds of attempts to taxonomize mechanisms and Mechanisms, but there are clear limitations to these approaches—limitations that are mitigated by more carefully attending to the varieties of mechanisms. These attempts all try to identify a few kinds of mechanisms, and to illustrate how these different kinds of mechanisms show up in different scientific domains or are appropriate for different projects. For instance, we notice a frequent distinction between a "broad" and a "narrow" conception of mechanism, where the narrow conception is "machine-like" and the broad one is not.

We find this approach problematic because it is unclear how narrow a narrow conception should be, and any narrow conception can be broadened in many different directions. In particular, these accounts suppose a clear univocal conception of what a machine is, but machines themselves are massive in their variety. A calculator is a machine, and so is a Watt governor, and how different are their properties!

We also doubt that we can identify some kinds of mechanisms that are important for ontological and metaphysical projects, and others that are the kinds that scientists talk about. As we note above, Levy and Andersen suggest that Mechanism$_2$/Causal Mechanism requires a broad conception of mechanism; while this is true so far as it goes, we think that the distinctions between varieties of mechanisms we make here will be central to creating a broad account of mechanism that still recognizes important variations. For instance, if we do not clearly distinguish between constitutive and non-constitutive phenomena, we will not be able to sort out the relationships between part-whole and causal forms of dependence, which are often important. At the same time, there are other interesting projects that, in spite of being metaphysical, might still focus on rather narrow forms of mechanisms. For instance, we might imagine that distinctions along the dimensions we have offered would be of assistance in sorting out various forms of emergence. Similarly, we can observe that very broad conceptions of mechanisms, like Kuorikoski's AFIs, can be methodologically significant. Illari (2011) also argues for the need for a relatively broad understanding of mechanisms to characterize an important role for evidence of mechanism in causal inference in medicine (see also Clarke et al. 2014).

We can moreover do much to explicate some of the proposed taxonomies of mechanisms by reference to our account of mechanistic varieties. Take, for instance, Kuorikoski's notion of mechanisms as abstract forms of interaction. From our perspective, the sorts of mechanisms Kuorikoski identifies are mechanism varieties that are identified by the kinds of organization they have. Different kinds of markets (e.g., perfect markets, monopolies and oligopolies, and markets exhibiting various kinds of market failures) will be the kind of markets they are in virtue of how they are organized. Monopolies, for instance, have only one seller, and we can make predictions about how that market will behave in light of that fact.

In a recent unpublished paper, Levy and Bechtel have argued that it is time to move to "Mechanism 2.0." They describe structures and processes that biologists are seeking to understand and that appear to be mechanisms (in our sense of minimal mechanism) but for which the tools of discovery, representation, and explanation described by new mechanists of the last 20 years do not quite work. They suggest that it is best to stop worrying about whether these structures and processes are or are not mechanisms by some definition, and instead to explore the various ways in which these biological mechanisms break the narrow mold. We certainly concur, and we hope that our approach provides some tools that will help in that exploration.

Notes

1 As we noted in Chapter 1, the word "mechanism" is used both to refer to worldly mechanisms and to philosophical or methodological theories or approaches that are mechanistic. We will continue to follow the convention we adopted there of referring to the former with lower-case "m" and the latter with upper-case "M."

2 Much of the terminology used in this taxonomy is introduced and elaborated in more detail in Chapter 5 of Glennan (forthcoming).

3 In biology, the word "evolution" is typically used to refer to the change in populations and their characters over time, and contrasted with development, which is responsible for changes in individual organisms from conception to maturity. But both evolutionary and developmental mechanisms are kinds of etiological mechanisms, and the mechanisms that result from both of these processes have evolved (as opposed to designed and built) etiologies.

4 Andersen seems to us to move between big "M" and small "m" mechanisms, presumably because she thinks the philosophical projects presuppose particular conceptions of mechanisms. She provides three additional conceptions of mechanism, which are mechanisms "that are ontologically 'flat,' or at least not explicitly hierarchical in character: equations in structural equation models of causation, causal-physical processes, and information-theoretic constraints on states available to systems" (Andersen 2014, p. 284). These are more clearly small "m" mechanisms (though obviously structural equations are not in the world).

References

Andersen, H. (2014) 'A Field Guide to Mechanisms: Part II'. *Philosophy Compass*, 9(4): 284–93.

Bechtel, W., & Richardson, R. C. (1993) *Discovering Complexity: Decomposition and Localization as Strategies in Scientific Research*. Princeton, NJ: Princeton University Press.

Bogen, J. (2004) 'Analysing Causality: The Opposite of Counterfactual is Factual'. *International Studies in the Philosophy of Science*, 18(1): 3–26.

——. (2008) 'Causally Productive Activities'. *Studies in History and Philosophy of Science Part A*, 39(1): 112–23.

Campaner, R. (2013) 'Mechanistic and Neo-Mechanistic Accounts of Causation: How Salmon Already Got (Much of) It Right'. *Metateoria*, 3(February): 81–98.

Clarke, B., Gillies, D., Illari, P., Russo, F., & Williamson, J. (2014) 'Mechanisms and the Evidence Hierarchy'. *Topoi*, 33(2): 339–60. Springer. DOI: 10.1007/s11245-013-9220-9.

Craver, C. F. (2007) *Explaining the Brain*. Oxford: Oxford University Press.

Craver, C. F., & Darden, L. (2013) *In Search of Mechanisms: Discovery across the Life Sciences*. Chicago: University of Chicago Press.

Dupré, J. (2012) *Processes of Life: Essays in the Philosophy of Biology*. Oxford: Oxford University Press.

Gillett, C., & Aizawa, K. (2016) *Scientific Composition and Metaphysical Ground*. (K. Aizawa & C. Gillett, Eds). London: Palgrave Macmillan UK. DOI: 10.1057/978-1-137-56216-6.

Glennan, S. S. (2010) 'Ephemeral Mechanisms and Historical Explanation'. *Erkenntnis*, 72(2): 251–66. DOI: 10.1007/s10670-009-9203-9.

——. (forthcoming) *The New Mechanical Philosophy*. Oxford: Oxford University Press.

Illari, P. M. (2011) 'Mechanistic Evidence: Disambiguating the Russo–Williamson Thesis'. *International Studies in the Philosophy of Science*, 25(2): 139–57. DOI: 10.1080/02698595.2011.574856.

Illari, P. M., & Williamson, J. (2012) 'What Is a Mechanism? Thinking about Mechanisms across the Sciences'. *European Journal for Philosophy of Science*, 2(1): 119. DOI: 10.1007/s13194-011-0038-2.

Kincaid, H. (2012) 'Mechanisms, Causal Modelling, and the Limitations of Traditional Multiple Regression'. (H. Kincaid, Ed.). *The Oxford Handbook of Philosophy of Social Science*. Oxford: Oxford University Press, 46–64.

Kuorikoski, J. (2010) 'Two Concepts of Mechanism: Componential Causal System and Abstract Form of Interaction'. *International Studies in the Philosophy of Science*, 23(2): 1–19.

Leuridan, B. (2012) 'Three Problems for the Mutual Manipulability Account of Constitutive Relevance in Mechanisms'. *British Journal for the Philosophy of Science*, 63(2): 399–427. DOI: 10.1093/bjps/axr036.

Levy, A. (2013) 'Three Kinds of New Mechanism'. *Biology & Philosophy*, 28: 99–114.

Machamer, P., Darden, L., & Craver, C. F. (2000) 'Thinking about Mechanisms'. *Philosophy of Science*, 67(1): 1–25.

Mitchell, S. D. (2009) *Unsimple Truths: Science, Complexity and Policy.* Chicago: University of Chicago Press.

Nicholson, D. J. (2012) 'The Concept of Mechanism in Biology'. *Studies in History and Philosophy of Biol & Biomed Sci,* 43(1): 152–63. Elsevier Ltd. DOI: 10.1016/j.shpsc.2011.05.014.

Salmon, W. C. (1984) *Scientific Explanation and the Causal Structure of the World.* Princeton, NJ: Princeton University Press.

Simon, H. A. (1996) *The Sciences of the Artificial,* 3rd ed. Cambridge, MA: MIT Press.

Stinson, C. (2016) 'Mechanisms in Psychology: Ripping Nature at Its Seams'. *Synthese,* 193: 1585–614. Springer. DOI: 10.1007/s11229-015-0871-5.

Wimsatt, W. C. (1994) 'The Ontology of Complex Systems: Levels of Organization, Perspectives, and Causal Thickets'. *Canadian Journal of Philosophy,* Supplement: 207–74.

Winther, R. (2009) 'Part-Whole Science'. *Synthese,* 178(3): 397–427.

8

MECHANISMS, PHENOMENA, AND FUNCTIONS[1]

Justin Garson

1. Introduction

A mechanism is always a mechanism for a phenomenon. The phenomenon that a mechanism serves is not somehow incidental to that mechanism, but constitutive of it: mechanisms are identified, and individuated, by the phenomena they produce. Thus, we can ask about mechanisms for blood clots, or demand-pull inflation, or social cohesion in naked mole rats. But it does not make sense to ask, "how many mechanisms are in the human body?" or even, "is the universe a mechanism?" These latter two questions are nonsensical because they do not specify a relevant phenomenon. That each mechanism has a phenomenon has become a platitude in the new mechanism literature in the philosophy of science (e.g., Glennan 1996, 52; Bechtel and Abrahamsen 2005, 423; Craver 2007, 122; Darden 2008, 960; Craver and Darden 2013, 52)—though as it turns out, there are different ways in which a mechanism "has" its phenomenon.

Sometimes, however, philosophers describe mechanisms not as having phenomena but as serving functions. "Mechanisms are identified and individuated by the activities and entities that constitute them, by their start and finish conditions, and by their functional roles" (Machamer, Darden, and Craver 2000, 14). "A mechanism is a structure performing a function in virtue of its component parts, component operations, and their organization" (Bechtel and Abrahamsen 2005, 423). "Mechanisms are systems that produce some phenomenon, behavior or function" (Glennan 2010, 256). "The entities and activities that are part of the mechanism are those that are relevant to that function or to the end state, the final product that the mechanism, by its very nature, ultimately produces" (Craver 2013, 141). Sometimes, only parts of mechanisms are described as having functions; sometimes the mechanism as a whole is described as serving a function. In the following I will mainly focus on this latter use (though see section 3).

The fact that people sometimes describe mechanisms in terms of their *functions*, rather than (or in addition to) their *phenomena*, raises an intriguing question: are these two ways of talking about mechanisms mere terminological variants of one another? Or, when we describe a mechanism as having a function, are we saying something more than, or other than, that it has a phenomenon? If so, what more, precisely, is being said? We can also frame the question from an ontological perspective. Compare the set of all mechanisms and the set of all mechanisms that serve functions. Are these two sets coextensional? Or is the set of mechanisms that have functions a proper subset of the set of mechanisms that have phenomena? In other words, is it that *all* mechanisms have phenomena, but for *some* of those mechanisms, those phenomena happen to be their functions, too?

In the following, I urge the view that there are at least two broad senses of mechanism at play in science and in the philosophy of science. (Technically, these should be thought of as two families of senses of mechanism, since each family may encompass several definitional variants.) According to the first sense of mechanism, which I will call, following Glennan (forthcoming), "minimal mechanisms," mechanisms are defined in terms of their phenomena, but there is no additional implication that these phenomena are in any intuitive sense functions of those mechanisms, where function is aligned with design, purposiveness, goal-directedness, or utility. Using Glennan's example, in this minimal sense, my car is a mechanism for locomotion, but it is also a mechanism for melting chocolate bars, even though melting chocolate bars is not one of my car's functions. Its function is just to get me from place to place. (Perhaps it has other legitimate functions, for example to serve as a status symbol.) In this minimal sense, the aggregation of mis-folded proteins is part of the mechanism for Alzheimer's disease, even though misfolded proteins do not have the function of producing Alzheimer's disease.

The second sense of mechanism is what I call, following Garson (2013), the functional sense of mechanism. Here, mechanisms are identified by the functions they serve, where "function" is understood as having a connotation of teleology, purposiveness, or design—though as we will see, working out precisely what this sense of "function" amounts to is beset with controversy. Cast in an ontological vein, I urge that the class of functional mechanisms is an interesting proper subset of the class of minimal mechanisms. In this sense of mechanism, a car typically would not be described as a mechanism for melting chocolate bars, even though it does so fre-quently enough and even though we can explain how that works. The heart would not usually be considered a mechanism for causing hemorrhages; it is a mechanism for pumping blood, even though it does both of those things and we understand how both processes work.

One virtue of recognizing a distinct, functional sense of mechanism is that it helps us think about the normativity of mechanisms; that is, it helps us understand how mechanisms can break (see section 4). A mechanism breaks when it cannot perform the function that defines it. In other words, the class of functional mechanisms is coextensive with the class of mechanisms that can break. It is not clear whether the minimal sense of mechanism has the resources to explain how a mechanism can break. A second virtue is that it helps us think about the distinc-tion between a mechanism's target and its byproduct (e.g., Craver and Darden 2013, 69), for example locomotion and melting chocolate bars in the case of the car. The target/byproduct distinction maps neatly onto the traditional function/accident distinction. A mechanism's target is just its function; a byproduct is any effect that is incidental to its discharging its function (in the functions literature, an "accident"). A third virtue of this functional sense of mechanism is that it helps us organize biomedical knowledge well (see section 4; also see Chapter 24).

In the following, I will do four things. First, I will provide an overview of the different ways that a mechanism can have a phenomenon (section 2). Second, I will provide a survey of what various philosophers have had to say about a mechanism's function (section 3). Even when philosophers agree that there is some intimate connection between mechanisms and functions, they disagree about the nature of that relationship and about the underlying concept of func-tion. (This is why I say that the functional sense of mechanism is actually a family of senses of mechanism.) Third, I will discuss some benefits of recognizing a distinct, functional sense of mechanism (section 4). In closing, I will return to the problem I started with: what is the rela-tionship between these two ways of talking about mechanisms? The most economical solution is to think of functional mechanisms as constituting an interesting subset of minimal mechanisms.

2. How mechanisms have phenomena

There are two, and arguably only two, ways that a mechanism has a phenomenon. First, a mechanism can have a phenomenon in the sense that it causes the phenomenon to occur. That is, the phenomenon is the effect of the mechanism's operation. Craver and Darden (2013, 65) refer to this as the "producing" relation. Second, a mechanism can have a phenomenon in the sense that the operation of the mechanism somehow constitutes, or realizes, the phenomenon. Craver and Darden (ibid.) refer to this as the "underlying" relation (though there are open questions about what, precisely, this constitution relation amounts to; see, e.g., Couch 2011; Romero 2015; Kaiser and Krickel 2016). The fact that mechanisms can have a phenomenon in one of these two ways gives rise to two different styles of explanation: etiological mechanistic explanation and constitutive mechanistic explanation (see Craver 2001, 69). Kästner (in prep) explicitly makes this connection between the distinction between producing/underlying, on the one hand, and the distinction between etiological/constitutive explanations, on the other. I will describe each of these relations in turn.

Consider the genetic mechanisms involved in cystic fibrosis (for now I will discuss cystic fibrosis as "having" a mechanism or mechanisms, rather than arising from the breakdown of a mechanism). Cystic fibrosis is characterized by debilitating and even fatal respiratory blockages, among other problems. When we describe the genetic mechanism involved in the respiratory blockade characteristic of this disease, we typically do so by describing a certain cause-and-effect sequence. Although there are many biochemical pathways underlying cystic fibrosis, one very common pathway begins with the deletion of three nucleotides in a certain segment of DNA (the delta F-508 mutation in the CFTR gene). This causes the loss of an amino acid, phenylalanine, in the corresponding protein sequence (the CFTR protein). This makes the protein misfold, which, in turn, disrupts the normal passage of chloride across cell membranes. This can lead to a buildup of sticky mucus in the lungs and elsewhere, though there is still some uncertainty about the precise pathways involved. In this case, the delta F-508 mutation is part of a mechanism for the disruption of chloride transportation in this first, "producing" sense. Disruption of chloride transport is a late stage in a long and complex cause-and-effect sequence.

In contrast, consider the mechanism for the patellar (knee-jerk) reflex. A tap to the patellar tendon causes a muscle spindle in the quadriceps muscle to stretch, which sends a sensory signal to the ventral horn of the spinal cord. The sensory neuron synapses directly onto a motor neuron, which sends a command back to the quadriceps to contract, causing the leg to kick. (This reflex is an example of a "monosynaptic" reflex because it only involves a single synapse.) Importantly, this mechanism does not *cause* the patellar reflex. The reflex is not the terminal stage of a cause-and-effect sequence. Rather, the patellar reflex is constituted by the whole sequence of events. When one explains how this mechanism works, one offers a constitutive etiological explanation. (In contrast, if one wants to describe the mechanism that causes the kick, we are back to the producing relation, since kicking is the terminal stage of a causal sequence that begins with tapping the patellar tendon.)

Craver and Darden (2013) consider a third potential way that a mechanism can have a phenomenon, which they call the "maintaining" relationship. This is exemplified by the maintenance of homeostatic set points, like the water level of the blood. Consider the mechanism for dopamine homeostasis. Mesolimbic dopamine neurons originate in the ventral tegmental area of the midbrain and terminate in the nucleus accumbens. They play a role in mediating reward, although the precise function of these dopamine signals is unclear (Berridge and O'Doherty 2013). These neurons use a homeostatic mechanism for maintaining a more-or-less constant level of dopamine in the synapse, usually at concentrations of between 20 and 50 nM

(Grace 2000, 336). To do this, the dopamine neuron's axon terminal has autoreceptors to monitor the level of dopamine outside the cell. When the level rises significantly above this "set point," the autoreceptor sends a signal to decrease the synthesis of dopamine (by inhibiting the enzyme tyrosine hydroxylase). When it drops significantly below this level, the autoreceptor activates that same enzyme and causes more dopamine to be synthesized inside the neuron. Sometimes, scientists wish to know how such "set points" are maintained over the long run in the face of fairly regular perturbations.

On reflection, however, it is not clear that we need to recognize a third, distinct, category for how mechanisms "have" phenomena to make sense of homeostatic phenomena like dopamine homeostasis. Rather, dopamine homeostasis is just another sort of phenomenon that we might wish to give a mechanistic explanation for. When we do provide a mechanistic explanation, we will either identify a producing relation, or an underlying relation, depending on the details of our analysis (Kästner in prep). Plausibly, if you were to ask, "how does the dopamine neuron maintain a fairly steady extracellular dopamine level despite frequent disruptions?" and I tell you about the dopamine neuron's autoreceptors and how they regulate tyrosine hydroxylase, I am describing the mechanism that underlies dopamine homeostasis, and giving a constitutive mechanistic explanation. If, instead, I were to give you a developmental account of how genetic and early environmental factors interact to create dopamine homeostasis, I am providing an etiological mechanistic explanation.

It would be easy to form the impression that when scientists pursue a mechanistic explanation, they first fix the phenomenon clearly and then discover its mechanism. But things are rarely that simple. Rather, there is often a back-and-forth movement between the way we describe a phenomenon, and the way we describe its mechanism. This movement is important for documenting the role of mechanisms in the process of scientific discovery (see Chapter 19). This is what Bechtel and Richardson (1993, 173) call "reconstituting the phenomenon"; Glennan (2005) and Craver and Darden (2013; Chapter 4) also discuss it. Sometimes in the history of science, scientists have a certain conception of the phenomenon they are after, and they start to identify the mechanisms underlying it. What they discover about those mechanisms changes how they think about the phenomenon itself. Consider how the concept of memory has changed over the last 50 years because of advances in psychology and neuroscience (Bechtel 2008). Or consider how the notion of schizophrenia has changed over the last century. Garson (forthcoming) describes how the study of recreational amphetamine use in the United States helped to "reconstitute the phenomenon" of schizophrenia in the 1960s and 1970s. The Swiss psychiatrist Eugen Bleuler described the fundamental phenomenon as a "loosening of associations"; today, most Western psychiatrists would consider this "loosening of associations" as representative of, at best, one subtype of schizophrenia, as a result of ongoing research.

3. Mechanisms and functions

Several philosophers have recognized an intimate connection between mechanisms and functions, where "function" has the connotation of teleology, design, or purpose. Consider a statement like, "the function of zebra stripes is to deter biting flies," or "the function of eyespots on butterfly wings is to deflect attack away from vital organs." Here, "function" connotes teleology or purpose: deflecting attack is in some loose sense what the eyespots are "there for." Philosophers of biology have developed many different theories about what functions are (Garson 2016). The point I wish to make here is that some philosophers and scientists have explicitly defined the notion of mechanism in terms of a rich, teleological notion of function. These philosophers include Craver (2001, 2013), Piccinini (2010),

Piccinini and Craver (2011), Moss (2012), Garson (2013), and Rosenberg (2015), though they disagree about the precise relationship between mechanisms and functions and they disagree about what functions themselves are.

Why would anyone want to tie functions and mechanisms together in this way? One way to motivate this narrower, functional, sense of mechanism is by bald appeal to intuition. In most everyday contexts, it seems natural to say that a car is a mechanism for locomotion. In contrast, it seems strange to say that a car is a mechanism for melting chocolate bars. It seems equally strange to say that a car is a mechanism for spewing carbon dioxide. But why does it seem strange? All three of those phenomena (locomotion, melting chocolate bars, emitting carbon dioxide) are very common, and they are completely explicable in terms of physics and engineering. In everyday contexts, however, only the first description seems natural or normal; the latter two sound unusual or strained. Why?

The same point can be made in the biological context instead of the context of artifacts and tools. In the minimal sense of mechanism, the heart is a mechanism for pumping blood, and it is also a mechanism for making beating sounds that one can listen to through a stethoscope. But we typically say that the heart is a mechanism for pumping blood. It sounds strange to say that it is a mechanism for making beating sounds, even though it does both and both are entirely explicable in terms of basic physics and biology. What is going on here?

One way of articulating this intuition (rather than explaining it) is in terms of the distinction between a mechanism's target and its byproduct (Craver and Darden 2013, 69). In the case of the car, it seems natural enough to identify locomotion as the mechanism's target and the emission of CO_2 as a byproduct. In the case of the heart, it seems natural enough to identify the mechanism's target as pumping blood and the beating sounds as a byproduct. The fact that these attributions are fairly stable in most everyday contexts raises the question of why we speak this way.

This is where functions come in. An obvious difference between locomotion and melting chocolate bars (or spewing CO_2) is that locomotion is the car's *function*. That is what cars are designed to do. Emitting CO_2 is a (rather unfortunate) byproduct of its function. That is not why cars exist. A similar point can be made about the heart. The reason the heart is a mechanism for pumping blood, rather than making beating sounds, is that pumping blood is the heart's function. That is what the heart is for. Perhaps there are some contexts where people are willing to say that the heart is a mechanism for making heart sounds (for example, in a diagnostic context). But to the extent that there is some context-dependency in the way people talk about mechanisms, that might be explicable in terms of the context-dependency in the way people talk about functions. Quite generally, according to this functional sense of mechanism, for X to be (part of) a mechanism for Y, Y must be its function. As I will show in section 4, this construal of mechanism also helps us understand the normativity of mechanisms; that is, how mechanisms break.

Suppose we accept that, in some contexts, mechanisms are defined in terms of the function they serve. That raises an obvious question: what are functions? Philosophers have explored this question for the last 40 years with little agreement (see Garson 2016 for a recent overview). There are three mainstream theories of function on the market: the selected effects (SE) theory, the causal role (CR) theory, and the biostatistical theory (BST), though this list is not exhaustive. Roughly, according to the SE theory, the function of a trait is the reason it evolved by natural selection. The function of the heart is to pump blood, rather than make beating sounds, because it evolved by natural selection for pumping blood. BST characterizes a trait's function in terms of its current-day contribution to survival and fitness, rather than in terms of selection history. Like SE, BST defines function in terms of evolutionary considerations (that is, present-day fitness) but remains neutral about history. According to the CR theory, the function of a system *part* consists in its contribution to some system capacity that an investigator has picked out as especially worthy of attention. Unlike the other views, CR emphasizes the contextual and perspective-dependent aspects of functions.

I am not interested, in this place, in characterizing these theories of function more rigorously or discussing alternatives. Rather, the point I wish to make is that people might think about mechanisms slightly differently depending on how they think about functions. As Piccinini (2010, 286) put the point, "different notions of mechanism may be generated by employing different notions of function." That is why I say that the functional sense of mechanism is really a family of different senses of mechanism. We can talk about the SE-functional sense of mechanism, the CR-functional sense of mechanism, and so on.

For example, one fairly austere way of thinking about mechanisms is to define mechanism in terms of function, and then define functions in terms of selected effects (that is, using the SE theory of function). Some philosophers and scientists have done just that. This is a fairly narrow way of thinking about mechanisms since it implies that for *X* to be a mechanism for *Y*, *X* must have been shaped by natural selection for doing *Y*. We can refer to this as the *SE-functional sense of mechanism*. In this sense, the heart is a mechanism for pumping blood, and not making beating sounds, because it was shaped by natural selection for pumping blood.

In the 1960s, the evolutionary theorist G. C. Williams (1966, 9) recommended this way of talking about mechanisms. As he put it, "the designation of something as a *means* or *mechanism* for a certain *goal* or *purpose* will imply that the machinery involved was fashioned by selection for the goal attributed to it." Here, mechanisms are defined in terms of function ("purpose"), which, in turn, is defined in terms of natural selection. Williams thought there was something deeply counterintuitive about describing something as a "mechanism" for an effect that it was not plausibly selected for. "Should we therefore regard the paws of a fox as a mechanism for constructing paths through snow? Clearly we should not" (13). By the same token, I suppose that Williams would also be loath to say that, "a car is a mechanism for melting chocolate bars," or that, "the heart is a mechanism for causing hemorrhages."

Some of the contemporary evolutionary psychologists have adopted Williams' usage in the way that they characterize psychological mechanisms (see Garson 2013, 322 for discussion). Among philosophers, Rosenberg (2015) seems to endorse this SE-functional sense of mechanism. In (Garson 2013) I endorse the functional sense of mechanism, but do not specify which theory of function is correct. However, I emphasize (p. 319) that function should be defined somewhat narrowly in terms of selection, fitness, or design.

A reservation one might have about the SE-functional sense of mechanism is that it might be *too* narrow; that is, it might fail to account for the wide spectrum of mechanisms out there. For example, sometimes neuroscientists talk about neural "mechanisms" for various activities, even when they do not think those activities are their evolved functions (Craver 2013, 141). It also would not apply to mechanisms outside of the contexts of biology and engineering, such as in physics, chemistry, or geology. As Glennan (2005, 445) points out, one may wish to talk about mechanisms underlying the eruption of geysers, even though geysers have no evolved functions. However, there is a broader sense of function that can encompass these diverse contexts. This is the CR theory of function, to be discussed shortly.

Craver (2001, 2013) explored the connection between mechanisms and functions in substantial detail. He says that mechanisms are defined and individuated in terms of the functions they serve. As he puts it, "the entities and activities that are part of the mechanism are those that are relevant to that function or to the end state, the final product that the mechanism, by its very nature, ultimately produces" (2013, 141). Here, the "function" of a mechanism is equated with its "end state" or "final product." This way of thinking about mechanisms also makes an appearance in Machamer, Darden, and Craver's well-known paper, which states that,

> to the extent that the activity of a mechanism as a whole contributes to something in a context that is taken to be antecedently important, vital, or otherwise significant, that activity too can be thought of as the (or a) function of the mechanism as a whole.
>
> *(2000, 6)*

However, instead of embracing the SE theory of function, Craver accepts the CR theory of function. We can label the whole package of ideas the *CR-functional sense of mechanism*. One distinctive feature of CR functions is that they have an explicitly perspectival, or contextual, character. For Craver, the function of the "uppermost" system in a mechanistic analysis is simply some capacity that a researcher or research team has found especially worthy of attention. To say the system has a function does not imply that it has an intrinsic goal or purpose. To the extent that CR functions can be associated with teleology or purpose, they have a *derived* teleology and *derived* purpose—derived, that is, from the interests and goals of the researchers who investigate them. As Craver puts it,

> This teleological feature of mechanistic description is also implicit in the fact that mechanisms such as the NMDA receptor are bounded: a judgment has been made about which entities, activities, and organizational features are in the mechanisms and which are not.
>
> *(Ibid., 140)*

Craver's view about mechanisms raises complex questions about realism and antirealism about mechanisms (see Chapter 9).

One of the benefits of Craver's expansive way of thinking about mechanisms is that it allows us to make sense of mechanism-talk outside of the contexts of biology and engineering (for more on the topic of mechanisms in engineering, see Chapter 34). Nothing prevents us from giving a CR-functional analysis of, say, El Niño phenomena, or demand-pull inflation, or even the way that atoms aggregate into molecules.

Lenny Moss (2012) also thinks there is a tight connection between mechanisms and functions (or "goals"). He points out that, often in the life sciences, biologists only attribute mechanisms to systems that are in some sense goal-directed: "To count as a biological mechanism the phenomenon in question thus must be perceived as being an expression of the ostensible 'purposiveness' of the living cell or organism" (165). Note that to say that an organism or biological system is "goal-directed" is different from saying that it has functions (see Garson 2016, chapter 2 for discussion). The point here is that Moss agrees with Craver (2013) and Garson (2013) that mechanisms have a teleological dimension. Part of Moss' evidence for this view is that, in many cases, we would refrain from calling something a mechanism if it were merely an artifact of laboratory procedure and not somehow expressive of an organism's goal-directedness:

> if cells stick to tissue culture plastic because of a chemical reaction with the plastic that resulted in the happenstance chemical production of an epoxy resin this too would be registered as an artifact and not as a "mechanism of adhesion" . . . . It is strictly the teleological aspect which makes the difference.
>
> *(Ibid.)*

There is a potential source of ambiguity that comes up when we discuss the relation between mechanisms and functions. There is a difference between attributing a function to a mechanism, qua whole, and attributing functions to the mechanism's parts. Up until now, the philosophers that I have discussed hold that, for X to be a mechanism for Y, Y must be X's function (or at least, X must be part of a goal-directed system, as in Moss' view). In other words, they see the mechanism as a whole as the thing that has a function, purpose, or goal. Other theorists tend to attribute functions, first and foremost, to the parts of mechanisms, rather than (or in addition to) the whole mechanism. This seems to be a fairly common stance. For example, Craver (2001) emphasizes the way that parts of mechanisms have functions. He refers to these part-functions as "mechanistic role functions." At times, Bechtel and Richardson (1993) also use "function" specifically when they are describing the activities of a mechanism's *parts*, rather than a mechanism's phenomenon. For example, they define "mechanistic explanations" as explanations that "account for the behavior of a system in terms of the functions performed by its parts and the interactions between these parts" (17). Piccinini and Craver (2011) also describe the way that mechanisms have functions, but if I understand them correctly, they typically focus on the way that parts of mechanisms have functions, rather than the whole mechanism.

Craver (2001) makes an interesting distinction regarding these mechanistic role functions. He observes that when we attribute a function to a component of a system, we can point to the item's activity without making any reference to the role that it plays in the broader system, or we can emphasize its role, irrespective of the specific activity by which it performs this role. A simple example will clarify the distinction. Consider the function of the heart. On the one hand, we can describe the heart as having the function of *beating*. This just describes what it does irrespective of the contribution it makes to the system (circulating blood). On the other hand, we can describe the heart as having the function of *helping to circulate blood*. This describes the role it plays in the circulatory system without telling us the specific activity it performs (beating). Craver (2001, 65) refers to these two roles as the item's "isolated activity" and "contextual role" respectively. In his view, "a complete description of an item's role would describe each of these."

There is one further source of ambiguity that arises when we think about functions. I suspect that some writers in the new mechanism tradition use the term "function," purely synonymously with "phenomenon." That is, in some cases I suspect that "function" represents nothing more than a terminological variation on "phenomenon" or "behavior," and hence the class of mechanisms that have "functions" in this weak sense is coextensional with the class of what Glennan calls "minimal mechanisms." For example, Glennan (2002, 127, fn. 6) notes that "it is tempting to use 'function' in place of the term 'behavior'." Ultimately, he suggests there that we resist that temptation, because "function" may carry inappropriate connotations of design. Similarly, Bechtel and Abrahamsen (2005, 433) characterize function in terms of the familiar function/structure distinction, which is devoid of connotations of teleology or purpose. However, I see a potential problem here. As I will describe in the next section, we often want to describe a mechanism as "broken," and one good way to make sense of that is by appealing to function, in some appropriately restrained sense of that term.

4. When mechanisms break

Biologists and biomedical researchers employ a colorful vocabulary to talk about how mechanisms break. A mechanism can have a "breakdown." It can be "usurped," "coopted," or "hijacked." It can be "interfered with," "disrupted," "impaired," or "disabled." It can simply "fail." For example, "drugs of abuse can hijack synaptic plasticity mechanisms in key brain circuits" (Kauer and Malenka 2007, 844). "Potentially irreversible impairments of synaptic memory mechanisms in these brain regions are likely to precede neurodegenerative changes" (Rowan et al. 2003, 821). The idea that mechanisms can break plays an important role in the way that biomedical researchers organize knowledge. Many philosophers in the new mechanism tradition have also emphasized how the idea of a mechanism's breaking is crucial for understanding causation, identifying components of mechanisms, and explaining disease (e.g., Bechtel and Richardson 1993, 19; Craver 2001, 72; Glennan 2005, 448; Darden 2006, 259).

So, what is it for a mechanism to break? At first pass, it is tempting to say that a mechanism breaks just when it cannot yield the phenomenon that defines it. My truck is a mechanism for locomotion, but because of a rusty spark plug, it cannot start. Sophia's immune system is a mechanism for fighting off harmful pathogens, but because of a low white blood cell count (leukopenia) due to a virus, it cannot do that (or not as effectively as usual). It is broken.

But this way of speaking raises a deeper question. Suppose my truck cannot start. Why would we say that the truck is a broken mechanism for locomotion, rather than simply that it is not currently (or no longer) a mechanism for locomotion? What license do we have (no pun intended) to say that my truck is a mechanism for locomotion, that is, that locomotion is its phenomenon, even when it cannot actually yield this phenomenon? Why not just say it is no longer a mechanism for locomotion?

Indeed, some of the standard definitions of mechanism, read strictly, do not allow us to speak of broken mechanisms. The set of entities and activities constituting my broken truck are *not* "organized such that they are productive of regular changes [that is, those constituting locomotion]." My busted truck is not a "complex system that produces the behavior [of locomotion] by the interaction of a number of parts." This suggests that we should find some way of loosening up those definitions to allow my truck to be a broken mechanism for locomotion, rather than simply not a mechanism for locomotion at all.

There are a few ways to respond to this situation. One way is to bite the bullet and say that, strictly speaking, there is no such thing as a broken mechanism. For example, in Glennan's minimal sense of mechanism, one would have to say that my broken truck is not a mechanism for locomotion, since it cannot actually perform that activity (though it might still be a mechanism for something else such as melting chocolate bars). This maneuver has the disadvantage that it seems to run against much of biomedical usage, as indicated in the quotations that open this section. Another way of accounting for the possibility of broken mechanisms is to explain it in terms of a purely statistical or epistemic norm. For example, one might hold that, when we say that the truck is a "broken" mechanism, all we mean is that it is acting in a way that is atypical or unexpected. This is a step in the right direction, but we should not equate brokenness with atypical or unexpected behavior. The medulla oblongata is a mechanism for helping us breathe, and it is also a mechanism for initiating the gag reflex. The latter is atypical, and perhaps unexpected, but one would not want to say it is broken when it causes us to gag.

My suggestion here is that a mechanism breaks when it cannot perform the function that defines it. The function of my car is locomotion, but it cannot perform that function because of a rusty spark plug, so it is broken. The function of Sophia's immune system is to fight pathogens, but it cannot do so because of an unusually low white blood count, so it is broken. Perhaps there are other ways of explaining what it is for a mechanism to break; my only claim is that this is an obvious option. I also do not want to restrict functions to selected effects or design (as in artifacts); as I indicated in the last section, one might construe function broadly in terms of the CR theory of function. My point here is that when we say that a mechanism is broken, we seem to imply that it has some function that it is unable to perform, even if we disagree about precisely what functions are.

Why are functions uniquely suited to this role? Because functions, at least on standard analyses, have a normative character. To say that functions are "normative" just means it is possible for a token system to *have* a function that it cannot *perform*. Functions have the remarkable property that they linger, as it were, even in the absence of the corresponding capacity. Standard analyses of function seem to provide a reasonable explication of this normative character, though there is ongoing debate about how exactly the different theories of function should account for normativity. For example, on the standard SE theory, the function of a trait is the reason it evolved by natural selection. So, the function of a trait has to do with its history, rather than its current-day capacities. As a consequence, it is easy to understand how a trait can possess a function (owing to its history) that it is nonetheless unable to perform.

Mechanisms, too, at least in standard biological usage, have a normative character. Mechanisms are the sorts of things that can break. Put simply, it is possible for X to be a mechanism for Y even when X cannot do Y. In my view, the normativity of mechanisms derives from the normativity of functions. On reflection, this is not such a radical philosophical move. Philosophers of biology have often appealed to the normativity of functions to make sense of the normative character of other biological categories. Dretske (1986) and Neander (1995) appeal to the normativity of function to make sense of the normative character of representations. Neander (1991) and Rosenberg and Neander (2009) appeal to the normativity of function to make sense of the way that types of traits are individuated. Appealing to function is a natural, even obvious, way to think about the normative character of mechanisms.

The idea that mechanisms serve functions not only helps us think about how mechanisms break; it is also useful for organizing biomedical knowledge. Suppose we accept that mechanisms serve functions, in some rich, teleological sense of that term. Then, we (generally) would be prohibited from talking about mechanisms for disease. For example, in the functional sense, there is no mechanism for anencephaly. Rather, anencephaly is an explicable result of the breakdown of a mechanism for neurulation. Moghaddam-Taaheri (2011, 608–10) makes a similar point in her discussion of medicine, though she does not situate this insight within a general theory of mechanism. Neander (2017, chapter 3) argues that diseases are generally best explained in terms of deviations from proper function, though she does not specifically relate this point to the concept of mechanism.

Let me clarify the point by contrasting two different ways of talking about diseases. On the first way of talking about disease, all diseases have mechanisms. There is a mechanism for spina bifida, and anencephaly, and encephalocele, among others. Biomedical research would be the organized attempt to explicate the mechanisms for these diverse diseases, along with their spatial, temporal, and hierarchical features. On the second way of talking about disease, diseases generally do not have mechanisms. Rather, there are only mechanisms for functions. Here, there is only one functional mechanism involved, namely the mechanism for neural tube formation. Different diseases result when this mechanism breaks in various ways.

Biomedical researchers often adopt this latter way of speaking, though see Garson (2013) for important qualifications. The reason is that the latter way of speaking is highly informative, and it also provides a useful heuristic for discovery. First, when I describe anencephaly as the result of a breakdown in a mechanism for neurulation, rather than as having its own mechanism, I convey critical information about its etiology. I am guiding you to the root problem, as it were, underlying anencephaly. Second, I set up a heuristic for future biomedical discoveries. Anencephaly results from disrupting neural tube folding at the *anterior* neuropore. What happens if folding is disrupted at the *posterior* neuropore instead? These are the kinds of questions that come up when we frame diseases in terms of broken mechanisms.

In conclusion, I accept a form of mechanism pluralism. There are different concepts of mechanism at play in biology and in philosophy, and my goal is not to argue for the unique correctness or superiority of a single one. In one sense (Glennan's "minimal mechanisms"), the notion of mechanism serves as a way of understanding causation and constitution quite generally, even outside of biology and engineering. For example, it can help us to understand mechanism-talk in areas such as physics, chemistry, and geology, where talk about functions is generally out of place. In another sense (the "functional" sense), mechanisms are defined in terms of the functions they serve, where function is thought of in a suitably rich teleological sense. Mechanisms in this functional sense constitute an interesting subset of mechanisms in this minimal sense. Perhaps there are other senses of mechanism as well (see Chapter 7). This functional sense of mechanism can help us understand certain aspects of mechanisms, such as the distinction between a mechanism's target and byproduct, the normativity of mechanisms, and the role of mechanisms in biomedical research.

Note

1 I am grateful to several people for discussions that helped me develop the ideas in this chapter. These include Carl Craver, Lindley Darden, Lenny Moss, Karen Neander, Gualtiero Piccinini, Anya Plutynski, and Sahotra Sarkar. I am also grateful to Phyllis Illari and Stuart Glennan for inviting me to contribute a chapter to this volume and for their valuable editorial comments.

References

Bechtel, W. (2008) *Mental Mechanisms: Philosophical Perspectives on Cognitive Neuroscience*, New York: Routledge.

Bechtel, W., and Abrahamsen, A. (2005) "Explanation: A Mechanist Alternative," *Studies in the History and Philosophy of Biological and Biomedical Sciences* 36: 412–441.

Bechtel, William, and Richardson, R. C. (1993) *Discovering Complexity: Decomposition and Localization as Strategies in Scientific Research*, Princeton, NJ: Princeton University Press.

Berridge, K. C., and O'Doherty, J. P. (2013) "From Experienced Utility to Decision Utility," in P. W. Glimcher and E. Fehr (Eds.) *Neuroeconomics: Decision Making and the Brain*, 2nd ed. Amsterdam: Elsevier, 335–351.

Couch, M. B. (2011) "Mechanisms and Constitutive Relevance," *Synthese* 83: 375–388.

Craver, Carl F. (2001) "Role Functions, Mechanism, and Hierarchy," *Philosophy of Science* 68: 53–74.

——. (2007) *Explaining the Brain: Mechanisms and the Mosaic Unity of Neuroscience*, New York: Oxford University Press.

——. (2013) "Functions and Mechanisms: A Perspectivalist View," in P. Huneman (Ed.) *Functions: Selection and Mechanisms*. Dordrecht: Springer, 133–158.

Craver, C. F., and Darden, L. (2013) *In Search of Mechanisms: Discoveries Across the Life Sciences*, Chicago: University of Chicago Press.

Darden, Lindley. (2006) *Reasoning in Biological Discoveries*, Cambridge: Cambridge University Press.

——. (2008) "Thinking Again about Biological Mechanisms," *Philosophy of Science* 75: 958–969.

Dretske, Fred. (1986) "Misrepresentation," in R. Bogdan (Ed.) *Belief: Form, Content, and Function*. Oxford: Clarendon Press, 17–36.

Garson, J. (2013) "The Functional Sense of Mechanism," *Philosophy of Science* 80: 317–333.

——. (2016) *A Critical Overview of Biological Functions*, Dordrecht: Springer.

——. (forthcoming) "'A Model Schizophrenia': Amphetamine Psychosis and the Transformation of American Psychiatry," in S. Casper and D. Gavrus (Eds.) *Technique in the History of the Brain and Mind Sciences*. Rochester, NY: University of Rochester Press.

Glennan, Stuart. (1996) "Mechanisms and the Nature of Causation," *Erkenntnis* 44: 49–71.

——. (2002) "Contextual Unanimity and the Units of Selection Problem," *Philosophy of Science* 69: 118–137.

——. (2005) "Modeling Mechanisms," *Studies in History and Philosophy of Biological and Biomedical Sciences* 36: 443–464.

——. (2010) "Ephemeral Mechanisms and Historical Explanation," *Erkenntnis* 72: 251–266.

——. (forthcoming) *The New Mechanical Philosophy*, Oxford: Oxford University Press.

Grace, A. A. (2000) "Gating of Information Flow within the Limbic System and the Pathophysiology of Schizophrenia," *Brain Research Reviews* 31: 330–341.

Kaiser, M. I., and Krickel, B. (2016) "The Metaphysics of Constitutive Mechanistic Phenomena," *British Journal for the Philosophy of Science*. doi: 10.1093/bjps/axv058.

Kästner, L. (in prep) "On the Curious Trinity of Mechanisms."

Kauer, Julie A., and Malenka, Robert C. (2007) "Synaptic Plasticity and Addiction," *Nature Reviews Neuroscience* 8: 844–858.

Machamer, Peter, Darden, Lindley, and Craver, Carl F. (2000) "Thinking about Mechanisms," *Philosophy of Science* 67: 1–25.

Moghaddam-Taaheri, Sara. (2011) "Understanding Pathology in the Context of Physiological Mechanisms: The Practicality of a Broken-Normal View," *Biology and Philosophy* 26: 603–611.

Moss, Lenny. (2012) "Is the Philosophy of Mechanism Philosophy Enough?" *Studies in History and Philosophy of Biological and Biomedical Sciences* 43: 164–172.

Neander, Karen. (1991) "Functions as Selected Effects: The Conceptual Analysts' Defense," *Philosophy of Science* 58: 168–184.

——. (1995) "Misrepresenting and Malfunctioning," *Philosophical Studies* 79: 109–141.

——. (2017) *A Mark of the Mental: In Defense of Informational Teleosemantics* Cambridge, MA: MIT Press.

Piccinini, Gualtiero. (2010) "The Mind as Neural Software? Understanding Functionalism, Computationalism, and Functional Computationalism," *Philosophy and Phenomenological Research* 81: 269–311.

Piccinini, G., and C. Craver. (2011) "Integrating Psychology and Neuroscience: Functional Analyses as Mechanism Sketches," *Synthese* 183: 283–311.

Romero, F. (2015) "Why There isn't Inter-Level Causation in Mechanisms," *Synthese*. doi: 10.1007/s11229-015-0718-0.

Rosenberg, A. (2015) "Making Mechanisms Interesting," *Synthese*. doi: 10.1007/s11229-015-0713-5.

Rosenberg, Alex, and Neander, Karen. (2009) "Are Homologies Selected Effect or Causal Role Function Free?" *Philosophy of Science* 76: 307–334.

Rowan, M. J., Klyubin, I., Cullen, W. K., and Anwyl, R. (2003) "Synaptic Plasticity in Animal Models of Early Alzheimer's Disease," *Philosophical Transactions of the Royal Society of London B* 358: 821–828.

Williams, George C. (1966) *Adaptation and Natural Selection*, Princeton, NJ: Princeton University Press.

9

THE COMPONENTS AND BOUNDARIES OF MECHANISMS

Marie I. Kaiser

Most new mechanists agree that mechanisms are made of two kinds of components (see Chapter 1). The first kind of components are material objects that are variously called entities (Machamer, Darden and Craver (MDC) 2000), parts (Glennan 1996, 2002), or component parts (Bechtel 2006, 2008). Examples include neurotransmitters, muscle fibers, and genes. The second kind of components are variously called activities (MDC 2000), interactions (Glennan 1996, 2002), or operations (Bechtel 2006, 2008). They are what entities do and what produces change (MDC 2000). Examples include binding, contracting, and being transcribed. I will use the term "component" to refer to these two kinds of things that make up or compose mechanisms.[1]

Since the origin of the new mechanical philosophy, metaphysical questions about the components of mechanisms (hereinafter "mechanistic components") have figured prominently in the debate (e.g., Tabery 2004; Psillos 2004; Machamer 2004; Bogen 2008; Torres 2009; Illari and Williamson 2013). MDC (2000) provoked this dispute by defending a dualistic ontology of mechanistic components, according to which mechanisms are made up of components belonging to two distinct ontological categories: entities (having certain properties) and activities. MDC contrast their dualism with Glennan's (1996, 2002) allegedly monistic view that mechanisms consist of entities (having certain properties) and of interactions, which are nothing but occasions on which a property change of one entity brings about a change in properties of another entity.

The dispute between dualists and monists involves two sets of metaphysical questions: First, how can the two kinds of mechanistic components be further characterized? What are their major features and what distinguishes them from each other? How are entities/parts and activities/interactions/operations related to more well-known ontological categories, such as material objects, properties, processes, events, and dispositions? I address these questions in section 1. Second, what is the relation between the two kinds of mechanistic components? Is one more fundamental than the other? Can activities, for instance, be reduced to entities and their properties? In what sense are the two kinds of mechanistic components mutually dependent? I examine these questions in section 2.

The second part of this chapter (sections 3 and 4) is concerned with the individuation of the components of mechanisms and with the boundaries of mechanisms. Mechanisms are always mechanisms *for* specific phenomena or behaviors (Glennan 1996, 2002). Accordingly, the components of a mechanism are individuated with respect to the phenomenon the mechanism is responsible for (see Chapters 1 and 8): Only those entities/parts and activities/interactions/operations that are relevant to the phenomenon count as components of the mechanism. This raises the question of how to spell out the conditions of relevance. What makes entities/parts and activities/interactions/operations relevant to a specific phenomenon? How do we draw the boundary between what belongs to a mechanism and what does not? In section 3, I present different criteria for individuating the components of mechanisms—natural boundaries, robustness/stability, strength of interactions, mutual manipulability, and INUS conditions—and I discuss their merits and limitations.

The claim that the individuation of a mechanism (by individuating all of its components) depends on the characterization of the respective phenomenon invites the question of how real the boundaries of mechanisms are. Mechanistic antirealists, as I call them, argue that mechanisms do not exist as well-delineated entities in nature because it is the scientist who imposes boundaries on certain sets of components while pursuing specific explanatory interests and adopting specific perspectives (Bechtel 2015; see also Wimsatt 1972, 2007). In section 4, I examine whether such an antirealistic move is inevitable and which arguments can be offered to defend the realistic view that mechanisms exist in nature and have real boundaries and a determined set of components.

1. The ontological nature of mechanistic components

The goal of this section is to provide a general characterization of the two kinds of components that make up mechanisms. I will refer to mechanistic components as entities and as activities because entity-activity terminology has prevailed in the debate (even Glennan forthcoming speaks about activities now) and because there are good reasons for this terminological choice: First, the term "entity" is more suitable than the term "part" because, as I argue somewhere else (Kaiser forthcoming), being a component of a biological mechanism is not the same as being a biological part of a biological object in general. The components of a mechanism are individuated with respect to a single behavior, whereas the parts of biological objects are individuated with respect to all of their characteristic behaviors. Moreover, natural boundaries are crucial to the individuation of biological parts but they do not constrain the individuation of mechanistic components in the same way (more on this in section 3). Second, we can drop Bechtel's (2006, 2008) term "operation" because he does not tie any metaphysical claim to this term. Third, against the term "entity" one might object that it is commonly used as a placeholder term for *any* ontological category (including activities) and that we should be more precise and speak about material objects instead (Kaiser and Krickel 2016). But in the context of this handbook it is advisable to stick to the term "entity" in this specific usage because it is an established term in the debate. Fourth, I prefer the term "activity" over the term "interaction" because interactions can plausibly be seen as special kinds of activities. Fifth, against the concept of an activity one might object that it is too restrictive and should be replaced by the broader metaphysical concept of an occurrence (which includes processes, events, and states; Kaiser and Krickel 2016). For the present purposes, however, I stick to the familiar concept of an activity while critically discussing what activities are.

I begin here by considering what entities are (as components of mechanisms). The mechanism for muscle contraction is composed of entities such as actin filaments (i.e., strands of actin molecules), calcium ions, and sarcoplasmic reticula (i.e., cell organelles involved in the storage and release of calcium ions). What the new mechanists refer to as "entities" are concrete *material objects* that are located in space and time. Typically, entities are spatially extended, have certain shapes and sizes, and often are surrounded by characteristic spatial boundaries, such as the membrane comprising the sarcoplasmic reticulum. Entities exist in time and are wholly present at different time points. Contrary to activities, entities are not, themselves, temporally extended in the sense of having temporal parts.[2] A calcium ion in my muscle fiber now is wholly present; it does not consist of the calcium ion now, the calcium ion two days ago, the calcium ion in one minute, etc. Entities can engage in processes, events, or activities which are temporally extended (e.g., diffusing through a channel). But the calcium ion as such is not temporally extended. In metaphysical terms, entities are *continuants* not occurrents, as activities are (Kaiser and Krickel 2016).

Entities are *bearers of properties*, including dispositional properties (or "causal powers"; Glennan forthcoming). Properties allow entities to engage in specific activities (MDC 2000, 3; Craver and Darden 2013, 16). For example, a sarcoplasmic reticulum is surrounded by a membrane that contains different kinds of ion channels and ATP-driven calcium pumps. These properties enable the sarcoplasmic reticulum to engage in the activity of releasing calcium ions.

Psillos (2004, 312) draws our attention to the fact that entities can exist without actually engaging in activities (e.g., if entities do not manifest their dispositions). I agree that entities, in general, can exist without engaging in any activities. But the important point is that entities can only be components of mechanisms if they engage in activities. Without doing something, an entity cannot be relevant to the phenomenon the mechanism is responsible for. All conditions of relevance require something to occur, something to be active (see section 3). Hence, component entities are necessarily *involved in activities*; that is, they are "working entities" (Darden 2008, 961; Craver and Darden 2013, 18). But we need to be cautious at this point. Entities that are components of mechanisms need not be active at all times. For the working of some mechanisms it is important that entities engage in activities only at certain times, not at others. For example, the sarcoplasmic reticulum releases calcium ions when an action potential arrives, not during later stages of the mechanism for muscle contraction. To conclude, an entity can be a component of a mechanism only if it is involved in an activity, at least once while the mechanism proceeds from its beginning to its end.

Component entities, however, can be involved in activities in different ways. They can have an *active role* and initiate or maintain the activity (i.e., be "actors"; Glennan forthcoming), such as the ATP molecule that binds to the calcium pump. On the other hand, I think entities can also be *passively* involved in an activity and allow for or undergo the activity, such as the calcium pump to which the ATP molecule binds. This difference has not been recognized in the debate so far. Furthermore, there can be activities in which two or more entities with an active role are involved, such as the sliding of actin and myosin filaments past each other. According to the strict reading of the word, such cases would be examples of inter*actions*. In the debate, however, the term "interaction" is used in a wider sense as referring to any activity in which more than one entity is involved—let it be passively or actively (e.g., Tabery 2004; Glennan forthcoming).

Finally, it has been argued that, as components of mechanisms, entities must be "stable clusters of properties" (Craver 2007a, 131) and must "have a kind of robustness and reality apart from their place within that mechanism" (Glennan 1996, 53). For example, ATPases (i.e., transmembrane proteins that transport ions to produce ATP) can be found in a variety of different mechanisms in different organisms and they retain their properties also if studied in isolation; that is, if studied in different contexts than *in situ* (Kaiser 2015, 221–35). Even though *stability* and *robustness* may not succeed as criteria for identifying the components of mechanisms (see section 3), they might still be important features that many component entities share.

Main features of entities

(1) Entities are *material objects* (i.e., continuants).

(2) Entities are *bearers of properties*, which allow them to engage in specific activities.

(3) As components of mechanisms, entities necessarily *engage in activities* (at least once during the mechanism).

(4) Entities can be *actively or passively involved* in activities.

(5) Entities are relatively *stable* and *robust*.

I move now to considering what activities are (as components of mechanisms). These are the second kind of components that make up the mechanism for muscle contraction: activities such as sliding, releasing, and binding. "Activities are the things that the entities do" (Craver and Darden 2013, 16). As such, activities are necessarily *temporally extended* and possess characteristic durations, rates, and phases. For example, the binding of ATP to the calcium pump starts with the collision of the ATP molecule with a certain region of the pump, then certain chemical bonds are formed. Because activities do not only exist in time—as entities do as well—but are also extended in time and have temporal parts (i.e., stages), they belong to the metaphysical category of occurrents, which encompasses processes, events, and states (Kaiser and Krickel 2016).

Activities are thought to be "active rather than passive" and to be the "happenings" (Machamer 2004, 29) in a mechanism. It is this active nature of activities that seems to make them indispensable if one wants to account for the alleged activeness of mechanisms (MDC 2000, 5), and it is their active nature that is said to distinguish activities from, for instance, property instantiations or property changes (Glennan forthcoming rejects this; more on this in section 2). As I show elsewhere, it follows from the active nature of activities and from their ontological status as occurrents that activities must be *actualized* or manifest rather than merely potential or dispositional (Kaiser and Krickel 2016). Only manifestations of dispositions can be components of mechanisms; unmanifested dispositions cannot. It is the release of calcium ions that is a component of the mechanism for muscle contraction, not the unmanifested disposition of the sarcoplasmic reticulum to release calcium ions if a depolarization occurs.

Dualists claim that activities account not only for the active nature of mechanisms but also for their productivity (MDC 2000; Tabery 2004; Bogen 2008; Darden 2008). In their view, the order of activities in a mechanism exhibits a "productive continuity" (MDC 2000, 3): One activity *productively causes* the next, which ensures that a mechanism runs in its typical way from beginning to end (see Chapter 10). For example, the release of calcium ions into the cytosol causes the binding of calcium ions to troponin molecules, which causes the tropomyosin complex to move off the actin binding site, which causes the myosin head to bind to the actin filament, and so on. Hence, activities are characterized as the "producers of change" (MDC 2000, 3) and as the "causal components in mechanisms" (Craver 2007a, 5).[3]

Another central feature of activities is that they *require entities* that engage in them. As MDC put it, "there are no activities . . . that are not activities *of* entities" (2000, 5). Any activity requires at least one entity that engages in it. I think we can add to this that for each activity there must be at least one entity that is *actively* involved in it. That is, activities seem to require actors. For example, a gene can be passively engaged in the activity being transcribed but there must be another entity that is actively involved in the same activity, such as the DNA polymerase.

Different types of activities involve different numbers of entities. In other words, activities have "*unrestricted arity*" (Illari and Williamson 2013, 72; my emphasis). Binary activities, such as the binding of ATP to the calcium pump or the sliding of actin and myosin filaments past each other, are widespread; activities that involve many entities, such as transcription or osmosis, are common in the living world. In my view, there are also clear examples of activities that involve only one entity (so-called unary activities or "un-interactive activities"; Torres 2009, 243), such as the changing of the spatial conformation of a protein, the breaking of a chemical bond, and the closing of a stoma cell in a plant leaf. Other authors, however, question the existence of un-interactive activities (Tabery 2004; Fagan 2012). They claim that for an activity to be productive, it must involve no fewer than two entities and thus be an *inter*action (or a "jointly acting complex"; Fagan 2012, 464). Tabery (2004) argues that seemingly un-interactive activities, such as shifting conformation of a protein, still involve interactions on a lower level. This argument is not convincing because the claim was that there is no other entity (besides the protein) involved in the conformation shift. This is compatible with there being interactions among the parts of the protein. Furthermore, Fagan (2012) fails to provide an argument for why all component entities must form complexes that jointly act. I conclude that there are cases of unary activities and that interactions thus constitute a mere subset of the set of all activities.

One might want to add the sixth feature that activities do not only produce changes (i.e., act as causes) but also, themselves, *involve changes* in the properties of the engaged entities (i.e., are processes). An example that supports this claim is the release of calcium ions from the sarcoplasmic reticulum into the cytosol. This activity involves several changes, such as the calcium ions changing their location and the sarcoplasmic reticulum changing its concentration of stored calcium ions. Despite its initial plausibility, I think we should reject the claim that activities *must* involve changes because it overlooks the variety of mechanistic components. For the working of some mechanisms, it is crucial that certain properties are not changed but maintained (i.e., continuously instantiated) during a certain time span. For instance, it is essential for the mechanism of the action potential that some ion channels remain open for a certain period of time. Moreover, the mechanism of natural selection seems to consist also of "passive properties" (Skipper and Millstein 2005, 341), such as being camouflaged or being present (see Chapter 23). Examples like these show that not all component activities involve changes, which is why this feature is not listed below.[4]

Main features of activities

(1) Activities are *temporally extended* (i.e., occurrents).

(2) Activities are *actualized* (rather than merely potential).

(3) Activities *produce change* (i.e., are types of causes).

(4) Activities require at least one *actively involved entity*.

(5) Activities have *unrestricted arity* (i.e., involve one to many entities).

---

2. The relation between mechanistic components

This section examines the relation between entities and activities as components of mechanisms. Is the relation one of reduction? Can entities, for instance, be said to be more fundamental than activities? Or are both kinds of components ontologically on a par, for instance because they are mutually dependent? These questions trace back to the already introduced dispute between dualists (MDC 2000; Machamer 2004) and monists (Glennan 1996, 2002) that has been prominent in the new mechanical philosophy. In this section, I first clarify in what respects dualists and monists have different views about the relation between mechanistic components. Then I critically discuss the arguments that have been provided in favor of dualism.

To begin with clarifying the dualism versus monism debate, MDC argue that "[m]echanisms are composed of both entities (with their properties) and activities" (2000, 3). I call the claim that mechanisms consist of components of two kinds *Duality thesis*. This is quite a weak thesis because monists can accept it as well. Glennan argues that mechanisms are "complex systems" that consist of "parts" (1996, 52; 2002, 344) having relatively stable properties and interacting with each other in certain ways. Hence, even though Glennan (1996, 2002) denies that mechanisms are composed of entities and of activities, he also recognizes a second kind of components, namely interactions. In his recent work, Glennan (forthcoming) agrees with even MDC's version of the *Duality thesis* because he refers to the components of mechanisms as entities (or parts) and activities, and characterizes interactions as a special, important class of activities.

Contrary to dualists, however, monists reject the *Irreducibility thesis*. This is the claim that the two kinds of mechanistic components belong to two distinct ontological kinds in the sense that one cannot be reduced to the other. Glennan characterizes an interaction between parts as an "occasion on which a change in a property of one part brings about a change in a property of another part" (2002, 344). In other words, Glennan (1996, 2002) assumes that interactions are reducible to the properties of entities and rejects the *Irreducibility thesis*. He is thus only committed to the existence of entities (i.e., material objects) and their properties, including dispositional properties. By contrast, MDC (2000) postulate the existence of activities on top of the usual commitments to entities and properties. Because of this additional ontological category, MDC's approach is referred to as dualistic and contrasted with Glennan's monistic account.[5]

Dualism is characterized by two further assumptions. The first is the *Interdependency thesis* which says that entities and activities necessarily exist together and determine each other. MDC state that "no activities without entities, and entities do not do anything without activities" (2000, 8). This is why the new mechanists use terms such as "acting entities" (Craver 2007a, 189) or "working entities" (Darden 2008, 961). The *Interdependency thesis* does not reinforce the dualistic character of MDC's approach—one might even argue that it runs contrary to it—but it is central to their approach. Second, MDC (2000) defend what I call the *Parity thesis*. It is the claim that there is no priority of, for instance, entities over activities and that entities and activities are ontologically on a par (Illari and Williamson 2013, 70).[6] Strictly speaking, the *Parity thesis* is no independent thesis because it follows from the dualists' assumptions of irreducibility and interdependency.

In sum, MDC's (2000) dualism can be distinguished from Glennan's (1996, 2002) monism by the following three claims:

Core theses of dualism

(1) *Irreducibility thesis*: Entities and activities belong to two distinct ontological kinds because activities cannot be reduced to the properties of entities.

(2) *Interdependency thesis*: In mechanisms, entities and activities necessarily exist together and determine each other.

(3) *Parity thesis*: Entities and activities are ontologically on a par.

In his recent work, Glennan (forthcoming) seems to have partly converged to MDC's (2000) dualistic position. He speaks of entities (or parts) and activities as the components of mechanisms and he agrees with MDC in that there cannot be activities without entities, or entities without activities (which is the *Interdependency thesis*). Moreover, Glennan rejects the reproach that his account is guilty of an entity-bias. That is, he seems to want to accept the *Parity thesis*. This is, however, the point at which the remaining difference between MDC (2000) and Glennan (forthcoming) becomes apparent. Glennan rejects the label of dualism and the claim that entities and activities are two distinct ontological categories because he thinks that the ontological distinctness of entities and activities is incompatible with the interdependency of entities and activities. In other words, he sees an incompatibility between the *Irreducibility thesis* (in particular, the distinctness claim it contains) and the *Interdependency thesis*. On my view, this casts into doubt Glennan's approval of the *Parity thesis* because if entities and activities are not ontologically distinct and activities can be reduced to property changes of entities, it seems highly questionable that entities and activities are ontologically on a par. To conclude, Glennan (forthcoming) has converged to dualism by accepting the existence of activities and the *Interdependency thesis* but he still rejects the distinctness claim that the *Irreducibility thesis* contains and thus fails to convincingly defend the *Parity thesis*.

Having examined the nature of the disagreement, I will now go on to explore challenges to dualism. Even though entity-activity talk dominates the debate, dualism and the ontological category of activities are still disputed (e.g., Woodward 2002; Psillos 2004; Machamer 2004; Tabery 2004; Bogen 2008; Torres 2009; Illari and Williamson 2013; Glennan forthcoming). In this section, I introduce and critically discuss five major arguments that have been offered in defense of dualism.

The new mechanists agree with other philosophers that philosophy of science should pay close attention to actual scientific practice. They accept *"descriptive adequacy"* (MDC 2000, 8; Kaiser 2015, 9) as an important criterion of adequacy not only for epistemic but also for metaphysical claims (Kaiser and Krickel 2016). Dualists make use of this criterion and argue that only dualism is descriptively adequate because it accounts for the fact that scientists describe mechanisms in terms of entities and activities (i.e., by using nouns, such as "enzyme" and "repressor," and verbs, such as "inhibits") and because successful scientific practice is a practice of studying and manipulating activities (Illari and Williamson 2013, 73–5; see also Craver 2007a, 144–52). This might be convincing but it should be noted that the claim that activities play an important role in scientific practice does not represent a brute fact about science. It is already a philosophical interpretation or critical reconstruction of specific parts of scientific practice which can be contested (Kaiser 2015, chapter 2). In my view, a monist might have equally good reasons for viewing scientific practice as a practice of studying objects, their properties, and how these properties change through time. Descriptive adequacy can thus also be used as an argument *against* dualism.

A second argument stresses the *"epistemic adequacy"* (MDC 2000, 21) of dualism. Dualists argue that thinking about mechanisms in terms of entities and activities renders scientific phenomena intelligible because entity-activity language corresponds best to how we naturally describe the world (Machamer 2004, 31; Illari and Williamson 2013, 79). For example, we say that a neurotransmitter binds to a receptor (activity talk), not that a neurotransmitter changes from being unbound to being bound to a receptor (property talk), or that a neurotransmitter manifests its bonding capacity (disposition talk). Even if this claim raises questions about the naturalness of descriptions and even if it may have exceptions, I share the intuition that activity talk is more comprehensible than property or disposition talk and I agree that activity talk corresponds well to how scientists describe and explain phenomena.

When compared to an entity-disposition ontology, dualism is claimed to be epistemically more adequate because we can observe a disposition of an entity only through the disposition being manifested; that is, through the entity acting (MDC 2000, 4; Machamer 2004, 30; Illari and Williamson 2013, 78). This is true but proponents of an entity-disposition ontology can accept the conceptual dependency of dispositions on their manifestations without being committed to the existence of activities in particular. In general, one might object that all of these arguments refer to mere epistemic points that have no metaphysical relevance (Psillos 2004, 313). But I agree with the dualists that metaphysicians of science should pay attention to the epistemic practices in the sciences.

MDC also appeal to common *metaphysical intuitions* to support dualism. They claim that dualism is advantageous because it captures "healthy philosophical intuitions underlying both substantivalist and process ontologies" (2000, 4). The ontological category of entities accounts for the fact that activities or processes are not free-floating but that there is always an entity involved in an activity or process. The ontological category of activities, in turn, accounts for the fact that mechanisms are dynamic and "do things" (MDC 2000, 5) and thus cannot consist of entities having certain properties only (recall section 1; Kaiser and Krickel 2016). Dualism is able to account for these motivating intuitions because it conceives of entities and activities as ontologically distinct and on a par. The problem with this argument is that it remains unclear about to which kind of intuitions the dualists refer. The argument would be plausible if they referred to a sort of pre-theoretical intuitions that metaphysical theories must account for (even then, there might be disagreement about which intuitions are legitimate). However, MDC speak about philosophical intuitions and seem to have something more theoretical and metaphysically laden in mind.

Another argument in favor of dualism is that only activities sufficiently explain the activeness and the *productivity* of mechanisms (MDC 2000; Tabery 2004; Bogen 2008). Only activities specify *how* in a mechanism one property change brings about another property change (i.e., how one entity interacts with another). In Tabery's words, the "bringing about" in Glennan's interactions are "black boxes" (2004, 10) that can only be rendered intelligible by activities. For instance, the activity "releasing" is said to specify how the change of the sarcoplasmic reticulum's ion channel from closed to open brings about the shift in the location of some calcium ions. Dualists claim that if you reject the *Irreducibility thesis* and argue that activities are nothing but property changes, you will overlook the active and productive nature of mechanisms. This concern seems to be confirmed, for example, by Glennan's thesis that "mechanisms are things (or objects)" (2002, 345), such as watches or cells. As I argue elsewhere, from a dualistic perspective, it becomes clear that bare objects that are neither active nor productive (i.e., that do not involve any occurrent), such as a stopped watch, cannot be mechanisms (Kaiser and Krickel 2016). But the argument that only dualism accounts for the productive nature of mechanisms also encounters serious problems. First, dualists fail to provide a clear and metaphysically satisfying analysis of what productivity is (e.g., Woodward 2002; Psillos 2004). Second, the "black box problem" seems to be not a problem for monists but rather "is one for scientists to solve in the laboratory" (Torres 2009, 239) because scientists are the ones to discover how, exactly, one particular property change brings about another.

A final argument in favor of dualism pertains to *parsimony*. Illari and Williamson state that an entity-activity ontology is more parsimonious than an entity-disposition ontology (2013, 79–81). At first sight, this claim might be surprising because dualism accepts the existence of activities on top of the usual ontological commitments and thus entails a rather unparsimonious move. However, Illari and Williamson do not compare dualism with monism, but rather an entity-activity ontology with an entity-disposition ontology. They argue that, because of the restricted arity of dispositions (they attach to only one entity), an entity-disposition ontology results in a proliferation of indefinitely many dispositions, which is why an entity-activity ontology is more parsimonious. This argument raises the question of which kind of parsimony is most important. On which level of graininess should we evaluate parsimony? If the goal of metaphysics is to reveal the fundamental structure of reality, I doubt that it is worse to have fewer general ontological categories (entity, property) and more numerous ontological sub-categories (different entities, many different properties).

3. Individuating mechanistic components

The new mechanists agree that mechanisms consist of all and only those components that are relevant to the specific phenomenon that the mechanism is responsible for. The term "responsible for" (Bechtel and Abrahamsen 2005, 422; Illari and Williamson 2012, 123) can be specified in two ways: etiological mechanisms *cause* their phenomena, whereas constitutive mechanisms *constitute* their phenomena (Kaiser and Krickel 2016). Figure 9.1 illustrates the triangular relationship between mechanistic components, a mechanism, and a phenomenon.

Figure 9.1 raises the question, for instance, how we should spell out the relevance relation between components and phenomena. This question has an epistemic and a metaphysical side. Epistemically, we are asking for a criterion of explanatory relevance: What are the explanatorily relevant factors that must be represented in a mechanistic explanation of a given phenomenon? How do we distinguish adequate mechanistic explanations from mere descriptions?[7] Metaphysically, the question of relevance is a question about where the boundary of a particular mechanism runs: Which entities and activities belong to the mechanism for a specific phenomenon? Which criteria tell apart genuine components of a mechanism from mere pieces, correlates, redundant parts, or background conditions?

The two types of mechanisms allow us to distinguish two kinds of relevance relations. The components of etiological mechanisms are causally relevant to their phenomena, whereas the components of constitutive mechanisms are said to be constitutively relevant to their phenomena. Since the notion of causal relevance has been subject to extensive philosophical discussion, the new mechanists focus on the notion of "constitutive relevance" (Craver 2007a, 139–60, 2007b; see Chapter 14).[8] I conclude this section by reviewing different possible criteria for individuating the components of constitutive mechanisms.

One might first suggest identifying the boundaries of mechanisms with what I call "natural boundaries" (Kaiser 2015, 176) of biological objects (e.g., Darden seems to suggest this; 2008, 960). Paradigmatic examples of natural boundaries are the cell membrane, the exoskeleton of insects, the skin, the blood–brain barrier, and the chain of mountains that borders a particular ecosystem. I agree that the existence of natural boundaries is essential to the working of many mechanisms. For example, the mechanism for muscle contraction requires that the muscle fiber is surrounded by a membrane that functions as a selective barrier and receives extracellular

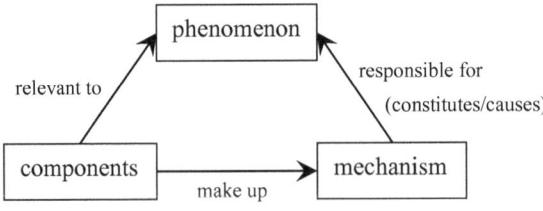

Figure 9.1 Relationships between components, mechanisms, and phenomena

signals. In most cases, however, these natural boundaries will not be the boundaries of the mechanism but of the corresponding biological object (e.g., of the muscle fiber).

According to my account, the boundary of a mechanism differs in two respects from the boundary of the corresponding object (what Craver refers to as "S" in his characterization of the phenomenon "S's ψ-ing"; Kaiser and Krickel 2016). First, mechanisms frequently transgress natural boundaries (Craver 2007b, 9). For instance, the mechanism for muscle contraction is composed also of entities, such as neurotransmitters, that are located outside the membrane of the muscle fiber. Also the gecko adhesion mechanism includes entities that belong to the gecko's environment (Kaplan 2012, 552). Hence, natural boundaries seem to be crucial to identifying the parts of biological objects in general (e.g., muscle fibers or geckos; Kaiser forthcoming) but not to individuating the components of mechanisms (e.g., the mechanism for muscle contraction or the gecko adhesion mechanism).

Second, biological objects, such as cells or geckos, "do many things at once" (Glennan 1996, 52) and all of these characteristic behaviors are relevant to individuating the parts of the biological object (Kaiser forthcoming). By contrast, mechanisms are responsible only for a single phenomenon (e.g., muscle contraction or gecko adhesion) and mechanistic components must be relevant to this phenomenon only. Hence, a mechanism typically includes many fewer entities than are parts of the corresponding system. To conclude, natural boundaries fail to provide us with adequate criteria for demarcating mechanisms because the condition of being located inside a natural boundary is neither necessary nor sufficient for an entity to be a component of a mechanism.

A second idea for how to identify the boundaries of mechanisms is using strength of interactions. The idea that the intensity of interactions determines part–whole relations can be traced back to Simon (1962). This claim is based on the observation that, at least in "nearly decomposable" (1962, 474) systems, interactions among the parts of an object are generally stronger and more frequent than the interactions between an object's parts and its environment. For example, the forces holding together a molecule are much weaker than those holding together atoms within the molecule. Other authors have picked up Simon's idea and applied it to the boundaries of mechanisms (e.g., Haugeland 1998; Wimsatt 1974, 2007).

This account, however, faces various problems. In general, it is quite difficult to assess and compare the strength of interactions. We seem to need to specify a threshold that will depend on pragmatic considerations of researchers. In addition, the strengths-of-interaction criterion seems unable to exclude background conditions and sterile effects as components of mechanisms (Craver 2007a, 143f.). Finally, when looking at actual biological practice we recognize that the strength of interactions may play a crucial role in individuating biological objects such as populations or ecosystems (e.g., Huneman 2014; Kaiser forthcoming) but not in individuating the components of biological mechanisms.

As a third possibility, we might turn to an account of constitutive relevance. Craver (2007a, 139–60, 2007b) has offered the most elaborated and widely discussed account of constitutive relevance. He identifies two conditions under which an acting entity (X's φ-ing) is constitutively relevant to a phenomenon (S's ψ-ing) and thus is a component of the mechanism for this phenomenon: the parthood condition and the mutual manipulability condition.

The parthood condition requires that "X is a part of S" (Craver 2007a, 153) or that X is "contained within S" (Craver and Tabery 2017). Hence, Craver seems to identify parthood with spatial inclusion. According to my view, the assumption that any component entity X must be spatially included in the corresponding object or system S encounters two major objections. First, it overlooks that temporal relations between component activities and the phenomenon of interest are important as well (Kaiser and Krickel 2016). Second, it conflicts with the fact that many mechanisms transgress natural boundaries (a fact that Craver, himself, recognizes; 2007a, 141) and include also entities and activities that are spatially located outside the corresponding object or system S (recall the discussion of natural boundaries in this section).

The second condition of Craver's account of constitutive relevance relies on Woodward's interventionist theory of causation (2003) and requires mechanistic components and phenomena to be mutually manipulable. This means that there must be an ideal intervention on the putative component X's φ-ing that changes the phenomenon S's ψ-ing and there must be an ideal intervention on S's ψ-ing that changes X's φ-ing. This part of Craver's account of constitutive relevance has recently attracted much philosophical attention (e.g., Harbecke 2010; Couch 2011; Leuridan 2012; Baumgartner and Gebharter 2016; Harinen forthcoming). It has been objected that Craver's account is merely epistemic and does not provide a satisfactory metaphysical analysis of what constitutive relevance or constitution really is (Harbecke 2010; Couch 2011). It seems to me that this criticism is too premature. Even though Craver's analysis of the explanatory and investigative practices of neuroscience (2007a) is not intended to be a metaphysical analysis, it can easily be used to draw metaphysical conclusions about the relation of constitution (Kaiser and Krickel 2016). Another criticism is that Craver's notion of constitutive relevance turns out to be a subtype of causal relevance and thus requires rethinking either Woodward's notion of an ideal intervention or the mutual manipulability account altogether (Leuridan 2012; Baumgartner and Gebharter 2016; Harinen forthcoming; see also Woodward 2014).

Finally, I will consider what are known as "INUS conditions." Regularity theories of constitution (Harbecke 2010; Couch 2011) develop criteria for individuating mechanistic components by making use of Mackie's idea of a cause being an INUS condition (i.e., an Insufficient but Non-redundant part of an Unnecessary but Sufficient condition for the effect; Mackie 1965). The general idea is that a component is an insufficient but necessary part of an unnecessary but sufficient mechanism for a phenomenon. This idea nicely captures the fact that a single component is insufficient for a phenomenon. A component must always be a non-redundant member of a "team" (Gillett 2013, 311) of components, which, together, are sufficient for the phenomenon. The criterion also accounts for the possibility that different particular mechanisms might be sufficient for a certain type of phenomenon.

It is problematic that regularity theories of constitution either require mechanistic phenomena to be capacities, which is an inadequate view of what constitutive mechanistic phenomena are (Kaiser and Krickel 2016), or they conceive of constitution as a relation between types only, which presupposes a realism about biological kinds that is heavily contested. Moreover, the claim that a mechanistic component must be a necessary member of a team of components raises the well-known problem of how to exclude mere correlates from being components (see also Craver and Tabery 2017).

4. Are boundaries of mechanisms real?

A scientific realist might begin with the intuition that mechanisms exist out there in nature independently of the scientists discovering and representing these mechanisms. Where one mechanism ends and another starts seems to be an objective feature of reality, which depends neither on our ability to individuate mechanistic components nor on our having an account of constitutive relevance. Mechanisms seem to be real things with real boundaries.

Although many mechanists share these realist intuitions, a realist interpretation of the boundaries of mechanisms encounters serious challenges. A commonly held assumption is that how one identifies the components of mechanisms depends upon the phenomenon one seeks to understand (recall section 3). Characterizing the phenomenon differently will yield different sets of relevant components and thus result in drawing the boundary of a mechanism differently. Since the choice of the phenomenon is relative to scientists' explanatory interests, some authors.have argued that the identification and decomposition of mechanisms is an inherently "perspectival" matter (Darden 2008, 960; see also Kauffman 1971; Wimsatt 1972, 2007). From here it seems to be only a small step to the antirealistic claim that mechanisms do not exist as well-delineated things in the world but that scientists impose boundaries on mechanisms relative to their explanatory purposes (Bechtel 2015). The aim of this section is to discuss whether such an antirealistic move is inevitable and which arguments a realist might invoke to defend the independent existence of boundaries of mechanisms.

Let us begin with the claim that characterizing a phenomenon guides and constrains the individuation of the components of a mechanism. According to the new mechanists, characterizing the phenomenon confines the space of possible mechanisms in so far as it uses the language of a given field and implicitly calls up a set of explanatory concepts and a "store of accepted entities, activities, and organizational structures that people in the field are licensed to use" (Craver and Darden 2013, 67). The relation between mechanisms and phenomena is thought to be reciprocal: not only does the nature of the phenomenon constrain what is to be included in the mechanism, but findings about the mechanism can also force one to "reconstitute the phenomenon" (Bechtel and Richardson 2010, 173).

But does this imply that the boundaries of mechanisms are not real in the sense of being no objective feature of reality? I think it does not. A realist can agree that for different token phenomena there exist different token mechanisms with different boundaries and that the choice of scientists of which phenomena to study may change over time and is dependent on pragmatic factors such as the scientist's perspective and explanatory interests. A realist would, however, emphasize that as soon as the phenomenon of interest is sufficiently specified and fixed, the boundaries of the mechanism for this phenomenon are fixed too, and do not depend on pragmatic factors. This way a realist can agree that the choice of the phenomenon is relative to the scientist's interests and perspective while holding on to the idea that for each token phenomenon there exists a token mechanism out there in the world that has a well-delineated, real boundary. Such a realistic position is taken up, for instance, by Glennan (forthcoming), and it seems to be the position that Craver is also committed to (not least because of his ontic conception of explanation; 2014).

Other authors reject the realistic assumption that boundaries of mechanisms exist in nature independently of the fact that scientists take specific perspectives and pursue certain explanatory goals when decomposing mechanisms. Wimsatt, for instance, argues that in the case of "interactionally complex" systems, different "theoretical perspectives" give rise to different non-isomorphic decompositions of a system into parts with "non-coincident boundaries" (Wimsatt 1974, 69–71). This sounds quite antirealistic, as if Wimsatt claimed that parts and boundaries would exist only relative to theoretical perspectives. But at other places in his work, Wimsatt explicitly defends realism—though he is keen to add that his realism is a realism for a "messy world" and is "piecemeal and usually satisfied with a local rather than a global order" (2007, 5f.; see also 1974, 672).

We can find antirealism about the boundaries of mechanisms more explicitly in the recent work by Bechtel. He draws on research in neurobiology to show that "mechanisms as bounded entities don't exist" (2015, 84). Marom's (2010) argument that the time-course of many phenomena in the life sciences is scale-free, in Bechtel's view, does not challenge the mechanistic account. It merely shows that the common view of mechanisms as temporally and spatially well-delineated entities is literally false—even though it is an idealization that facilitates developing scientific explanations. Bechtel argues that it is "the scientists who impose boundaries around entities and activities in nature and impose a time scale on which their functioning is characterized" (2015, 85). A realist may object that if boundaries of mechanisms do not exist in the actual world, mechanisms do not exist either, because it is indeterminate which entities and activities belong to which mechanism. Bechtel's antirealism seems to result in the implausible view that the world consists of loose entities and activities only. Alternatively, Bechtel's antirealism can be read as turning the mechanistic account into a purely epistemic account that has been detached from almost all ontological commitments.

Notes

1 I do not use the term "constituent" because some mechanisms are said to constitute their phenomena (Kaiser and Krickel 2016) and the relation between mechanisms and phenomena should be kept apart from that between components and mechanisms.
2 This claim presupposes endurantism (Lewis 1986).
3 In the debate, there is no agreement on what exactly a cause is. Some hold that causes are difference-makers (e.g., Craver 2007a), in the sense that if they had not occurred, the purported effect would not have occurred. Others claim that causation is production (e.g., Bogen 2008; Glennan 2010), where causes are actively producing their effects.
4 Alternatively, one might stick to the claim that activities must involve changes and argue that mechanisms consist of more than just entities and activities—namely property instantiations or maintenances (i.e., states).
5 Strictly speaking, these numbers are false because material objects and properties are already two ontological categories.
6 Only Machamer (2004) deviates from the *Parity thesis* insofar as he sometimes suggests the priority of activities.
7 According to the ontic conception of explanation, the question of explanatory relevance is not an epistemic but also a metaphysical question because explanations are regarded as things in the world (Craver 2014).
8 Whether Craver's account of constitutive relevance succeeds in keeping apart constitutive and causal relevance or whether it collapses into an account of causal relevance is, however, contested (e.g., Leuridan 2012; Baumgartner and Gebharter 2016; Harinen forthcoming).

References

Baumgartner, M., Gebharter, A. (2016): "Constitutive relevance, mutual manipulability, and fat-handedness", *British Journal for the Philosophy of Science* 67 (3), 731–756.
Bechtel, W. (2006): *Discovering Cell Mechanisms: The Creation of Modern Cell Biology*, Cambridge: Cambridge University Press.
——— (2008): *Mental Mechanisms: Philosophical Perspectives on Cognitive Neurosciences*, New York: Routledge.
——— (2015): "Can mechanistic explanation be reconciled with scale-free constitution and dynamics?", *Studies in History and Philosophy of Biological and Biomedical Sciences* 53, 84–93.
Bechtel, W., Abrahamsen, A. (2005): "Explanation: a mechanist alternative", *Studies in History and Philosophy of Biological and Biomedical Sciences* 36, 421–441.
Bechtel, W., Richardson, R.C. (2010): *Discovering Complexity: Decomposition and Localization as Strategies in Scientific Research*, Cambridge, MA: MIT Press.
Bogen, J. (2008): "Causally productive activities", *Studies in History and Philosophy of Science* 39, 112–123.
Couch, M.B. (2011): "Mechanisms and constitutive relevance", *Synthese* 183, 375–388.
Craver, C. (2007a): *Explaining the Brain*, Oxford: Clarendon Press.
——— (2007b): "Constitutive explanatory relevance", *Journal of Philosophical Research* 32, 3–20.

—— (2014): "The ontic conception of scientific explanation", in M.I. Kaiser, O. Scholz, D. Plenge, A. Hüttemann (eds.), *Explanation in the Special Sciences – The Case of Biology and History*, Dordrecht: Springer, 27–52.

Craver, C., Darden, L. (2013): *In Search of Mechanisms*, Chicago: University of Chicago Press.

Craver, C., Tabery, J. (2017): "Mechanisms in science", in E.N. Zalta (ed.) *The Stanford Encyclopedia of Philosophy* (Spring 2017 Edition). Available at: <https://plato.stanford.edu/archives/spr2017/entries/science-mechanisms/>

Darden, L. (2006): *Reasoning in Biological Discoveries: Essays on Mechanisms, Interfield Relations, and Anomaly Resolution*, Cambridge: Cambridge University Press.

—— (2008): "Thinking again about biological mechanisms", *Philosophy of Science* 75, 958–969.

Fagan, M.B. (2012): "The joint account of mechanistic explanation", *Philosophy of Science* 79, 448–472.

Gillett, C. (2013): "Constitution, and multiple constitution, in the sciences: using the neuron to construct a starting framework", *Minds and Machines* 23, 301–337.

Glennan, S. (1996): "Mechanisms and the nature of causation", *Erkenntnis* 44, 49–71.

—— (2002): "Rethinking mechanistic explanation", *Philosophy of Science* 69, 342–353.

—— (2010): "Mechanisms, causes, and the layered model of the world", *Philosophy and Phenomenological Research* 81, 362–381.

—— (forthcoming): *The New Mechanical Philosophy*.

Harbecke, J. (2010): "Mechanistic constitution in neurobiological explanations", *International Studies in the Philosophy of Science* 24, 267–285.

Harinen, T. (forthcoming): "Mutual manipulability and causal inbetweenness", *Synthese*.

Haugeland, J. (1998): *Having Thought*, Cambridge, MA: Harvard University Press.

Huneman, P. (2014): "Individuality as a theoretical scheme. II. About the weak individuality of organisms and ecosystems", *Biological Theory* 9, 374–381.

Illari, P., Williamson, J. (2012): "What is a mechanism? Thinking about mechanisms *across* the sciences", *European Journal for Philosophy of Science* 2, 119–135.

—— (2013): "In defense of activities", *Journal for General Philosophy of Science* 44, 69–83.

Kaiser, M.I. (2015): *Reductive Explanation in the Biological Sciences*, Cham: Springer.

—— (forthcoming): "Individuating part-whole relations in the biological world", in O. Bueno, R.-L. Chen, M.B. Fagan (eds.), *Individuation, Process and Scientific Practices*, Oxford: Oxford University Press.

Kaiser, M.I., Krickel, B. (2016): "The metaphysics of constitutive mechanistic phenomena", *The British Journal for the Philosophy of Science*, DOI: 10.1093/bjps/axv058.

Kaplan, D.M. (2012): "How to demarcate the boundaries of cognition", *Biology and Philosophy* 27, 545–570.

Kauffman, S.A. (1971): "Articulation of parts explanation in biology and the rational search for them", *PSA: Proceedings of the Biennial Meeting of the Philosophy of Science Association* 1970, 257–272.

Leuridan, B. (2012): "Three problems for the mutual manipulability account of constitutive relevance in mechanisms", *British Journal for the Philosophy of Science* 63, 399–427.

Lewis, D.K. (1986): *On the Plurality of Worlds*, Oxford: Blackwell.

Machamer, P. (2004): "Activities and causation: the metaphysics and epistemology of mechanisms", *International Studies in the Philosophy of Science* 18, 27–39.

Machamer, P., Darden, L., Craver, C.F. (2000): "Thinking about mechanisms", *Philosophy of Science* 67, 1–25.

Mackie, J.L. (1965): "Causes and conditions", *American Philosophical Quarterly* 24, 245–264.

Marom, S. (2010): "Neural timescales or lack of thereof", *Progress in Neurobiology* 90, 16–28.

Psillos, S. (2004): "A glimpse of the *secret connexion*: harmonizing mechanisms with counterfactuals", *Perspectives on Science* 12, 288–319.

Simon, H.A. (1962): "The architecture of complexity", *Proceedings of the American Philosophical Society* 106, 467–482.

Skipper, R.A., Millstein, R.L. (2005): "Thinking about evolutionary mechanisms: natural selection", *Studies in History and Philosophy of Biological and Biomedical Sciences* 36, 327–347.

Tabery, J.G. (2004): "Synthesizing activities and interactions in the concept of a mechanism", *Philosophy of Science* 71, 1–15.

Torres, P.J. (2009): "A modified conception of mechanisms", *Erkenntnis* 71, 233–251.

Wimsatt, W.C. (1972): "Complexity and organization", *Proceedings of the Philosophy of Science Association* 1972, 67–86.

—— (1974): "Reductive explanation: a functional account", *PSA: Proceedings of the Biennial Meeting of the Philosophy of Science Association* 1974, 671–710.

—— (1994): "The ontology of complex systems: levels of organization, perspectives, and causal thickets", *Canadian Journal of Philosophy* 24, 207–274.

—— (2007): *Re-Engineering Philosophy for Limited Beings: Piecewise Approximations to Reality*, Cambridge: Harvard University Press.

Woodward, J. (2002): "What is a mechanism? A counterfactual account", *Philosophy of Science* 69, 366–377.

—— (2003): *Making Things Happen: A Theory of Causal Explanation*, New York: Oxford University Press.

—— (2014): "A functional account of causation; or, a defense of the legitimacy of causal thinking by reference to the only standard that matters—usefulness (as opposed to metaphysics or agreement with intuitive judgment)", *Philosophy of Science* 81, 691–713.

10

MECHANISMS AND THE METAPHYSICS OF CAUSATION[1]

Lucas J. Matthews and James Tabery

1. Introduction

Whether this volume is your first exposure to the new mechanical philosophy or you have been contributing to the field for decades, one thing that quickly becomes obvious about the literature is the great diversity of causal concepts. There is talk of "activities," "interactions," "causal processes," and "counterfactual difference makers," to name but a few. These concepts are not synonymous. They have come to represent different approaches to thinking about causation in the philosophy of mechanisms, and debate has ensued as advocates of each point out their preferred approach's virtues alongside the limitations of the others.

This chapter has two, interrelated purposes. The first is to provide a brief history that traces these approaches to understanding causation back to their sources. That there are different approaches with different sources will not be a novel contribution (see, for example, Andersen 2014a, 2014b; Craver and Tabery 2015). However, the tracing will be. That is, we link the activity approach back to Elizabeth Anscombe, the counterfactual difference-making approach back to David Lewis, the causal process approach back to Wesley Salmon, and the mechanical causation approach back to Stuart Glennan, and most philosophers will recognize the influence of these figures on these approaches. What has not received attention is the way in which Anscombe, Lewis, Salmon, and Glennan all introduced their ideas about causation explicitly in response to Hume's problem of causation and the regularity view of causation that dominated philosophy after Hume. Anscombe, Lewis, Salmon, and Glennan were all responding to Hume, but they did so in quite different ways with quite different resulting desiderata. What we will show is how the differences in those responses and desiderata shaped the subsequent approaches to understanding causation that entered the new mechanical philosophy.

Our second purpose is to draw on this history to disentangle some of the debates that are now playing out in the philosophy of mechanisms literature concerning the metaphysics of causation. It has become common for advocates of these approaches to point out problems with the others—to complain, for example, that Salmon's approach doesn't apply outside fundamental physics, or to criticize proponents of activities for not providing a general account of causation. By tracing the various approaches back to their roles in Anscombe, Lewis, Salmon, and Glennan's original responses to Hume, we'll diagnose how such criticisms often overlook the different desiderata that shaped those approaches from their very beginning and so hold the approaches to standards that they were never designed to meet.

2. Hume and the problem of causation

Hume, in the Abstract to *A Treatise of Human Nature*, famously asked his readers to consider a billiard ball at rest on a table with another ball rolling toward it.

> They strike; and the ball, which was formerly at rest, now acquires a motion. This is as perfect an instance of the relation of cause and effect as any which we know, either by sensation or reflection. Let us therefore examine it. 'Tis evident, that the two balls touched one another before the motion was communicated, and that there was no interval betwixt the shock and the motion. Contiguity in time and place is therefore a requisite circumstance to the operation of all causes. 'Tis evident likewise, that the motion, which was the cause, is prior to the motion, which was the effect. Priority in time is therefore another requisite circumstance in every cause. But this is not all. Let us try any other balls of the same kind in a like situation, and we shall always find, that the impulse of the one produces motion in the other. Here therefore is a third circumstance, viz. that of a constant conjunction betwixt the cause and effect. Every object like the cause, produces always some object like the effect.
>
> *(Hume 2000 [1738–40], p. 409)*

Hume was an empiricist, requiring that knowledge be based on experience. When it came to causation of the sort "A caused B," all experience told Hume was that A came before B (priority), that A led immediately to B (contiguity), and that A was always joined by B following it (constant conjunction). And that's it. "In whatever shape I turn this matter, and however I examine it, I can find nothing farther" (ibid). Hume readily acknowledged that his readers would be inclined to think there was something more to causation—that there must be some *necessary connection* to the causal relationship, that B *must* follow A. Hume countered that such a necessary connection could not be observed by the empiricist; rather, the apparent sense of necessity just habitually built up by experiencing the priority, contiguity, and constant conjunction over and over and over again. This was Hume's problem of causation—the challenge to find a necessary connection to causation above and beyond priority, contiguity, and constant conjunction. Hume thus defined a cause in two ways in the *Treatise*, one definition pertaining to a relationship between objects in the world: "An object precedent and contiguous to another, where all the objects resembling the former are plac'd in like relations of precedency and contiguity to those objects that resemble the latter," and one definition pertaining to a relationship between objects in the world and human minds:

> A cause is an object precedent and contiguous to another, and so united with it, that the idea of the one determines the mind to form the idea of the other, and the impression of the one to form a more lively idea of the other.
>
> *(Hume 2000 [1738–40], p. 114)*

Hume scholars have debated what Hume himself meant with his definitions of cause—about the relationship between his two definitions in the *Treatise*, as well as the relationship between those definitions and the one he provided in his subsequent *An Enquiry Concerning Human Understanding*[2] (see, for example, Read and Richman 2000). Hume exegesis, however, is not the purpose of this chapter. It is sufficient simply to state that a particular view of causation inspired by Hume dominated philosophy for the next 250 years—the regularity view of causation, wherein causation was deemed to be patterns of regular succession (patterns of priority, contiguity, and constant conjunction). By the mid-twentieth century, this regularity view of causation faced substantial challenges. It struggled, for example, with cases of common cause and cases where there was constant conjunction but no causal relation, phenomena recognized to be common enough in nature that any legitimate view had to accommodate them. Now, to be sure, philosophers worked hard to bolster the view to surmount the identified weaknesses (for a review, see Psillos 2009). Still, by the 1970s, the regularity view was sufficiently compromised to encourage philosophers to abandon the project entirely and look elsewhere for a philosophical analysis of causation.

3. Four responses to Hume

Elizabeth Anscombe admitted that Hume upended long-held assumptions about causation; however, at the same time she thought he reinforced the idea that causation was equated with necessity by way of his focus on constant conjunction (Anscombe 1993 [1971]). Indeed, on Anscombe's reading, the causation-necessitation link only grew post-Hume as subsequent philosophers (ranging from Kant to Russell) took the Humean challenge to be finding the necessary connection that was not observable. Anscombe challenged this causation-necessitation link. Necessity assumed an outcome was predetermined, but the outcomes of complex systems (such as the presence or absence of infection following exposure to a disease) cannot be specified in advance. In part, Anscombe explained, this was a feature of humans' cognitive limitations; we simply cannot keep track of the many contributors to a complex system's outcome. Additionally, it was a feature of a world governed in part by indeterministic physics; if the world simply was not predetermined, then no account of causation that assumed it could be accurate. Anscombe admitted that philosophers and scientists alike pointed to apparent predetermined necessity and made universal generalizations regarding regularity. But to get necessity in a complex system, Anscombe argued, measurements of the system had to be made at an abstract enough level to wash out the underlying indeterminism. And to make universal generalizations work, Anscombe countered, all sorts of qualifications had to be inserted concerning the system in question behaving under "normal conditions" (ibid., p. 94).

Importantly, Anscombe's abandonment of necessity (and, in turn, the search for Hume's necessary connection) was not simultaneously an abandonment of a philosophical analysis of causation. It was just that she didn't think that the analysis benefited from being tied up with artificially constructed generalizations and abstracted necessitation. Rather, she advised taking the "messy and mixed up conditions of life" at face value and building an analysis of causation off that reality (ibid., p. 95). This involved shifting attention from the development of a general theory of causation to an understanding of and appreciation for how humans obtained knowledge of specific causal concepts with specific effects in specific instances, like "*scrape, push, wet, carry, eat, burn, knock over, keep off, squash, make* (e.g. noises, paper boats), *hurt*" (ibid., p. 93, emphasis in original). All that could be generally said about causation, Anscombe summarized, was that "causality consists in the derivativeness of an effect from its causes. This is the core, the common feature, of causality in its various forms. Effects derive from, arise out of, come of, their causes" (ibid., pp. 91–2). Even this derivativeness, however, was not to be interpreted as a general conceptualization of causation; rather, it was a placeholder for specific causings, like pushes and scrapes. Necessity and universality—the traditional target of philosophical focus—were just add-ons to this derivativeness according to Anscombe.

David Lewis, just two years after Anscombe, also looked for an alternative to the Humean regularity view of causation. But he suggested the substitute could actually be found with Hume himself (Lewis 1973). As mentioned above, Hume defined cause twice in his *Treatise* and then twice again in his *Enquiry* (see note 2). In the *Enquiry*, Hume started by defining the concept similar to how he defined it in the *Treatise*: "we may define a cause to be an object, followed by another, and where all the objects similar to the first, are followed by objects similar to the second." But then he continued with a new addition: "Or in other words, where, if the first object had not been, the second never had existed." It was this idea of counterfactually understanding the difference-making quality of causation that Lewis worked to develop. "We think of a cause as something that makes a difference, and the difference it makes must be a difference from what would have happened without it" (Lewis 1973, p. 557).

Lewis drew on possible world semantics to ground a theory of counterfactual difference-making, where the truth conditions of counterfactuals were based on similarity relations between possible worlds. The essential idea was that causation could be understood via counterfactual dependence. To say that "A caused B" was to say that "If A happens, then B happens" *and* "Had A not happened, B would not have happened." Importantly, Lewis thought the counterfactual dependence understanding of causation could handle the cases that posed problems for the regularity view. For example, if a common cause (*c*) was responsible for first one effect (*e*) and then another effect (*f*), the regularity view would interpret the first effect (*e*) as the cause of the second (*f*) because it was prior to, contiguous with, and in constant conjunction with that effect, even though *e* and *f* were in fact both effects of *c*. Using Lewis' counterfactual approach, however, offered a solution to the problem. The common cause *c* was the cause of both the first effect *e* and the second effect *f* because, had *c* not occurred, *e* and *f* would not have followed; however, *e* was not the cause of *f* because it was possible for *e* to not have occurred (because of some breakdown in the relation between *c* and *e*) but *f* to still have occurred.

Whereas Anscombe and Lewis looked for ways around Hume's challenge to identify the observable necessity of causation (either by abandoning necessity itself or by turning to the counterfactual alternative that depended on non-observable possible worlds), Wesley Salmon instead attempted to take Hume's empiricist challenge head-on. Starting with a series of papers in the 1970s and then culminating with his *Scientific Explanation and the Causal Structure of the World*, Salmon developed a theory of causation that based causal processes on their ability to transmit marks (Salmon 1975, 1977, 1984; on Salmon's indebtedness to Reichenbach for this focus on transmission, see Williamson 2011). Consider Salmon's favorite example: a rotating spotlight in the center of a large, circular room. The light from the spotlight to the wall was a causal process, but the light circling the room was a pseudo-process; we know this because a red filter experimentally inserted in the beam itself would alter or mark the transmission from the spotlight to the wall everywhere, but a red filter experimentally added anywhere on the wall would not alter or mark the signal except where the light passed over it. Salmon saw this mark-transmission aspect of causal processes as the key to answering Hume's challenge: "Ability to transmit a mark can be viewed as a particularly important species of constant conjunction—the sort of thing Hume recognized as observable and admissible. It is a matter of performing certain kinds of experiments" (Salmon 1977, p. 220).

Salmon's claim to have answered Hume's challenge made it an exciting development for philosophers but also a target for criticism. The most severe charge against the mark-transmission solution was that it was simply a counterfactual theory in disguise because the only way to distinguish the causal processes from the pseudo-processes was to counterfactually assess what would have happened under experimental intervention (Kitcher 1984; Dowe 1992). Salmon was repulsed by this idea; unobservable counterfactuals were not empiricist in nature, and so Salmon could not rightly claim to have met Hume's challenge on Hume's terms if he relied on counterfactuals. Fortunately, one of his critics also offered a solution. Phil Dowe suggested switching from the transmission of marks to the transmission of conserved quantities (Dowe 1992). Salmon was receptive to the shift. His mark-transmission reliance on experiments revealed how to identify causal processes, but it could not explicate the very concept of a causal process that he sought (Salmon 1994, p. 303). The updated Salmon-Dowe account of causation linked causal processes with the transmission of conserved quantities, such as momentum, mass-energy, and electrical charge. Basing the account on actual, physical things—the conserved quantities—led Salmon to believe that he had avoided counterfactuals and ultimately met Hume on his empiricist terms:

> the idea was to present a "process theory" of causality that could resolve the fundamental problem raised by Hume regarding causal connections. The main point is that causal processes, as characterized by this theory, constitute precisely the objective physical causal connections which Hume sought in vain.
>
> *(ibid., p. 297; see also Salmon 1997, 1998)*

Stuart Glennan's "Mechanisms and the Nature of Causation" (1996) is rightly recognized as one of the earliest contributions to the new mechanical philosophy; indeed, it is in that paper that we receive the first definition of a mechanism by a new mechanical philosopher. What has not been emphasized, however, is the fact that Glennan introduced that definition and his discussion of mechanisms more generally with a very specific purpose in mind—answering Hume's problem. Hume, recall, challenged philosophers to find the necessary connection between cause and effect that was above and beyond repeated exposures to priority, contiguity, and constant conjunction. In response, Glennan argued that two events are causally connected when there is a mechanism that connects them (Glennan 1996, p. 64). To use his example: we do not think turning the key in an automobile's ignition causes the engine to turn over because we have been exposed to that process over and over again; rather, we think turning the key causes the engine to turn over because we believe that there is a *mechanism* in the automobile that brings about the ignition of the engine.

It was because a mechanism was at the heart of causal connection that Glennan introduced his definition of the concept: "A mechanism underlying a behavior is a complex system which produces that behavior by of [sic] the interaction of a number of parts according to direct causal laws" (ibid., p. 52). Phenomena ranging from blood flow in the human body to the regulation of water level in a toilet, Glennan pointed out, can be causally understood as products of complex systems whose interacting parts produce the phenomena in question. Now, Glennan admitted that this process of understanding phenomena at one level by appeal to the mechanism(s) responsible for that phenomena at a lower level eventually bottomed out at the fundamental laws of physics, which seemed to defy further mechanical decomposition. But, Glennan countered, the vast majority of causal reasoning (both scientific and in everyday life) took place above this fundamental level. So while Glennan was willing to grant that, "At the level of fundamental physics, Hume's problem still remains," he also claimed that, with his mechanical theory of causation, "Hume's problem is not a universal one" (ibid., p. 68). At higher levels, where regularities could be mechanically explicated, mechanisms offered the necessary connection.

Anscombe, Lewis, Salmon, and Glennan all shared dissatisfaction with the Humean regularity view of causation and all offered up alternatives to that view. Those alternatives, however, were quite unique from one another, and they resulted in different desiderata that judged the success of an account of causation. Salmon and Glennan both demanded that an account of causation address Hume's problem of causation by identifying the necessary connection. Salmon offered this by going all the way down to the level of fundamental physics with the transmission of conserved quantities. Glennan, in contrast, stopped just short of fundamental physics, arguing that mechanisms connected causes and effects at higher levels. Anscombe and Lewis, each in their own way, dismissed Hume's challenge. Anscombe replaced the search for a general account of causation's necessary connection with a singularist search for the way that specific effects are brought about by specific causings. Lewis instead switched the focus to Hume's remark about the difference-making quality of causation, demanding of an account of causation that it capture this feature (which he did with counterfactual dependence).

At the most basic level, the differences between the four approaches can be seen by returning to Hume's famous billiard balls. For Anscombe, the causal relation was one of derivativeness in that instance; the first ball struck the second, and the second ball's motion derived from or arose out of the first ball's striking. It was about counterfactual difference-making for Lewis; had the first ball not struck the second, the second ball would not have moved. For Salmon it was about the transmission of a conserved quantity; the first ball transmitted an actual, physical thing to the second ball. And for Glennan the entire set-up could be understood as a complex system with parts (e.g. billiard balls) and interactions (e.g. a collision) that are mechanically explicable, and where the causal connection is a function of that mechanism.

4. Derivativeness, counterfactual difference-making, causal processes, and mechanical causation in the philosophy of mechanisms

The new mechanical philosophy emerged around the turn of the twenty-first century, as philosophers of science looked for an alternative framework to logical empiricism. Logical empiricists addressed a range of philosophical issues by formally characterizing the logical and mathematical structures constitutive of science, and by focusing on the abstract and epistemic features of science. New mechanical philosophers instead focused on detailed case studies of actual scientific practice, and drew attention to the material elements of the world as the object of scientific inquiry. What emerged was a new framework for thinking about the classic issues in philosophy of science (Craver and Tabery 2015). Fundamental to this shift was the idea that science is in the business of elucidating the mechanisms that are causally responsible for phenomena under investigation. As a result, considerations of causation have been embedded in the new mechanical philosophy from its beginning; however, different contributors to the framework have made sense of that causation in different ways.

Anscombe's approach to causation first appeared in the new mechanical philosophy with the publication of Peter Machamer, Lindley Darden, and Carl Craver's "Thinking about Mechanisms" (2000). "Mechanisms," they said, "are entities and activities organized such that they are productive of regular changes from start or set-up to finish or termination conditions" (ibid., p. 3). Anscombe, recall, dismissed the search for a general account of causation and instead encouraged turning to specific instances where effects derive from their causes—scrapes, pushes, burns. These are Machamer, Darden, and Craver's activities; they are types of causings (ibid., p. 6).

In the ensuing years, a number of philosophers worked to develop this Anscombian-inspired concept of an activity. Machamer (2004) picked up on Anscombe's reference to child development of verb-acquisition, pointing to research that explored how children procure concepts of things by way of interacting with and manipulating what those things do (i.e. the entities' activities). Bogen (2005, 2008) alternatively emphasized Anscombe's dismissal of necessity, highlighting examples of research where the elucidation of mechanisms provided explanations even if those mechanisms did not operate with generalizable regularity. Darden (2006) documented ways in which the study of activities figured into scientific discoveries, especially in molecular biology and genetics.

Despite these differences in foci, what Machamer, Bogen, and Darden all shared was an Anscombian abandonment of the search for a general theory of causation, in favor of attention to how the understanding of specific activities figured into specific causal mechanisms.[3] As Machamer put it:

> The problem of causes is not to find a general and adequate ontological or stipulative definition, but a problem of finding out, in any given case, what are the possible, plausible, and actual causes at work in any given mechanism.
>
> *(Machamer 2004, pp. 27–8)*

Or take Bogen: "Instead of suggesting a general, uniformly applicable, answer to the question of what differentiates causes from non-causes, what I take to be Anscombe's view calls for piecemeal treatments of questions about how specific effects are produced" (Bogen 2008, p. 114). And Darden, referencing the productive nature of activities, wrote, "Rather than seeking a general definition of production, it is more insightful to consider specific kinds of activities and the means for discovering them" (Darden 2013, p. 25).

A difference-making understanding of causation made its way into the philosophy of mechanisms by way of James Woodward's manipulationist understanding of causal relevance (Woodward 2003). The basic idea for Woodward is that scientists causally explain when they know how to manipulate. The manipulations are understood counterfactually and comparatively. If some particular variable is a cause of some outcome, then manipulating the value of the variable would be a way of manipulating the outcome. These counterfactual experiments formulate and then answer what-if-things-had-been-different questions; and, in so doing, they establish a pattern of counterfactual dependence that determines what difference a cause makes. Counterfactual dependence is understood with the closely related concepts of intervention and invariance. An intervention consists of an idealized experimental manipulation of the value of some variable, thereby determining if it results in a change in the value of the outcome. So the counterfactuals are formulated in such a way that they show how the value of the outcome would change under the interventions that change the value of a variable. Invariance, then, is a characterization of the relationship between variables (or a variable and an outcome) under interventions on Woodward's account. When there is an invariant relationship between a variable and an outcome, then that relationship is potentially exploitable for manipulation, and because of this it is a causal relationship.

Woodward himself applied his manipulationist framework to the new mechanical philosophy, building his invariance-under-intervention formulation into a characterization of mechanistic causation where "the behavior of each component is described by a generalization that is invariant under interventions" (Woodward 2002, S375). The move brought an account of how the difference-making quality of causal claims could be understood in sciences that elucidate mechanisms; moreover, the concept of invariance and degrees in it was attractive to philosophers because it offered a resource for characterizing mechanistic regularities that were robust but not law-like universal generalizations. Woodward's manipulationist framework has influenced a number of philosophers of mechanism; for example, Craver took Woodward's approach to understanding causal relevance in an etiological system and updated it to account for constitutive relevance in a multi-level system, where the idea was that coordinated top-down and bottom-up interventions could establish constitutive relevance by way of mutual manipulability (Craver 2007).

Woodward's manipulationist framework certainly introduced a counterfactual difference-making understanding of causation into the philosophy of mechanisms. But it is not obviously a *metaphysical* understanding. That is, Lewis' development of possible worlds and modal realism about those possible worlds was an explicit attempt to build a metaphysics of counterfactual causation. But Woodward's manipulationist framework is not so easily categorized as such. His is an account of causal relevance and the role of causal relevance in causal claims. As such, it attends more to the semantics and epistemology of causation than it does to a deeper metaphysics. Indeed, even Craver, in his embracing of Woodward's manipulationist approach, was careful to say that he was using Woodward's understanding of the difference-making quality of causation because of its epistemological implications for the role of causal relevance in scientific explanations, not for its metaphysical solution to Hume's problem (Craver 2007, pp. 105–6, 225–6).[4]

Salmon's development of the concept of a causal process was part of his more general revolution in philosophy of science aimed at reorienting philosophers toward the causal structure of the world. When it came to questions of causation and answering Hume's problem, as we saw above, he appealed to the transmission of conserved quantities. When it came to questions of explanation, he encouraged abandoning the epistemic conception that appealed to laws of nature and derivation of phenomena from those laws of nature, and replacing it with the ontic conception of explanation that situated phenomena in the physical structure of the world (Salmon 1984, 1998).

Although Salmon never identified with the new mechanical philosophy, his push for a philosophy of science that thought about causation and explanation in terms of being physically situated in the causal structure of the world certainly shaped the subsequent philosophy of mechanisms (Campaner 2013). For example, Salmon's focus on causal processes placed a priority on being able to trace the causal chain from the causal structure of the world through to the phenomenon of interest; this attention to tracing a causal chain can be seen in Machamer, Darden, and Craver's (2000) emphasis on productive continuity which renders intelligible the various stages of a causal mechanism. Salmon's ally—Phil Dowe—most explicitly applied their causal process approach to mechanisms, generating an account of "causal process mechanisms" (Dowe 2011). On this account, mechanistic causation is the transmission of conserved quantities within and by a mechanism, and so an ontic explanation is provided by the elucidation of a causal process mechanism when the propagation of the conserved quantities through the mechanism can be traced.

Glennan, in contrast to Anscombe, Lewis, and Salmon, is unique in that he both replied to Hume and then personally took that reply and incorporated it into the philosophy of mechanisms. Indeed, Glennan has wrestled with the metaphysics of causation perhaps more than any other new mechanical philosopher (see, for example, Glennan 2002, 2009, 2010a, 2010b, forthcoming; Kuhlmann and Glennan 2014). Certain aspects of his account have evolved over the years. For example, as described earlier, in his 1996 definition of a mechanism he based the interaction of parts on "direct causal laws." But then, in his 2002, those interactions were based on "direct, invariant, change-relating generalizations." This, however, was more of a terminological switch than a deep philosophical concession; Glennan used "laws" loosely in his 1996 to describe a range of generalizations, both the mechanically inexplicable ones of fundamental physics and the mechanically explicable ones of higher levels. And so, while Glennan dropped the philosophically loaded term "laws" in 2002, he was still committed to the basic idea that many generalizations above the level of fundamental physics were mechanically explicable ones. And, most important, he stayed true to the idea that causes and effects were causally connected when a mechanism links the cause to the production of the effect.

In more recent years, Glennan turned to considering the relationship between accounts of causation that focused on causal production (like his, Anscombe's, and Salmon's) and accounts of causation that focused on causal relevance (like Lewis' and Woodward's). Mechanisms remained at the heart of the account; causal production was brought about by way of a mechanism that linked a cause with an effect. But the mechanism also provided the foundation for claims about causal relevance; relevant properties embedded in the mechanism determined in what sense the causal production was difference-making (Glennan 2009, 2010a, forthcoming). "On the account I am offering," Glennan summarized, "productivity and relevance are not competing approaches to analyzing the nature of causal relationships, but rather are complementary concepts which refer to different features of the causal structure of the world" (Glennan forthcoming; see also Chapter 7 of this volume).

5. Assessing the contemporary debates

Tracing the current approaches to understanding causation in the mechanistic literature back to the unique responses to Hume by Anscombe, Lewis, Salmon, and Glennan helps evaluate recent criticisms of those approaches. It becomes evident that mechanistic accounts of causation are often held to desiderata that were never intended to be met. For example, some new mechanical philosophers have demanded that an account of causation be *non-reductionistic*. Not to be confused with *non-reductive* (which we'll discuss later in this section), a *non-reductionistic* account of causation explicitly avoids reducing talk of "causes" across the sciences to the lowest levels of analysis. Causal process approaches to causation in the mechanisms literature, that is, have been criticized on the grounds that they fail to account for how causation is understood across *all* the sciences—especially those beyond the level of fundamental physics. This *non-reductionistic* desideratum becomes most evident in light of how the new mechanists have responded to Salmon's causal process approach. A number of new mechanists, for example, have complained that the focus on the transmission of conserved quantities cannot explain causal language outside the domain of fundamental physics. Machamer, Darden, and Craver (2000), for example, argued that, although Salmon's analysis captures causal interactions in fundamental physics, it "is silent as to the character of the productivity in the activities investigated by many other sciences" (p. 7). On a similar point, Glennan (2002) argued that, while Salmon's account "has the advantage of characterizing interactions in terms of physical theory rather than the semantics of counterfactuals, it has the disadvantage that it obscures similarities between kinds of interactions among higher-level entities" (S346). Glennan challenged readers to explain commonplace causal interactions by appeal to the exchange of conserved quantities alone. Informed by investigative and explanatory practices in modern neuroscience, Craver (2007) argued that when "one begins to talk about causal relations that arise when parts are organized into mechanisms, the transmission view loses traction; its austere descriptive vocabulary no longer applies" (p. 76).

On this criticism of Salmon's approach that reveals the *non-reductionistic* desideratum, perhaps Jon Williamson (2011) said it best:

> conserved quantities are too low-level, far removed from most of our causal claims. Consider a claim like eliminating the 10% tax band caused an increase in inequality: not only does such a claim not explicitly mention conserved quantities, it is very hard to see how it could be about conserved quantities at all. Most of our causal claims are high-level claims like this, and it appears that the process theory has some work to do before it can provide a convincing interpretation of such claims.
>
> *(Williamson 2011, p. 427)*

Fair enough. But remember that Salmon's goal was to develop a notion of causal process that provided the necessary connection Hume said he could not observe. The goal wasn't to develop a notion of causal process that accounts for causal mechanical explanations across the sciences.

So this criticism holds Salmon's account of causation to a specific desideratum—*non-reductionistic*—that he never intended to meet. One could certainly say that Dowe's causal process mechanism restricts talk of mechanistic causation to cases where conserved quantities can be traced, thus leaving out mechanistic causation in a higher-level science like macro-economics. But then Dowe could simply respond that macro-economics doesn't provide any understanding of mechanistic causation.

Alternatively, consider criticisms of the activity-based approach to causation in the mechanisms literature. A number of new mechanists have criticized the activity-based approach on the grounds that all it does is list a series of examples such as "diffuse," "push," and "bond," or vaguely say only that activities are the "producers of change." Tabery (2004), for example, worried that introduction of the concept of activity fails to inform a general view of how activities actually produce changes in the world (beyond listing examples of change-production). Psillos (2004) made a similar point, highlighting that because production itself is an activity, then "we are not given an illuminating account of that which some things share in common, in virtue of which they are activities" (pp. 311–12). These criticisms of activity-based accounts of causation hint at a specific desideratum regarding what one might expect out of a promising metaphysical account of causation: *conceptual generality*. The thought is that any account of causation should be expected to say what causation as a general concept is, to say what distinguishes all causings from all non-causings.

Remember, however, that the activity-based approach to understanding causation is a direct descendant of Anscombe's response to Hume, and fundamental to that response was the abandonment of any attempt to generate a general account of causation beyond specific instantiations of it (like diffusion, pushing, and bonding). In effect, abandonment of Hume's call to provide a general account of causation is precisely the Anscombian move. So, again, holding activity-advocates to the *conceptual generality* desideratum that they've never acknowledged charges them with breaking a rule in a game they've explicitly refused to play.

Counterfactual difference-making approaches to causation in the mechanisms literature have also been held to an unacknowledged desideratum. From Salmon to contemporary new mechanical philosophers, a number of authors have argued that a fault with the counterfactual approach to causation is its understanding of causation in terms of what could have been, rather than in terms of what actually is or was (Salmon 1984, 1998). For example Machamer (2004) argued that "only ontological (or even ontic) principles or descriptions belong in a 'real' definition of causality" (p. 28). In other words, a satisfactory understanding of causation should be grounded in terms of what actually exists in the world, as opposed to what could, would, or might pertain to the actual world. Bogen (2004, 2005, 2008) criticized counterfactual dependence on similar grounds, arguing that while it may be the case that counterfactual reasoning influences experimental theorizing in scientific practice, it does not follow that causal claims are actually grounded in counterfactuals. Ultimately, the truth of causal claims in scientific practice is grounded in what actually happens—not what would have happened:

> But it does not follow from the fact that regularities are of great epistemic, theoretical and practical interest that actual or counterfactual regularities are constitutive of the ontological or conceptual difference between the causes of an effect and the non-causal items that accompany its production.
>
> *(Bogen 2008, p. 122)*

Further, and in a very similar vein, Waskan (2011) defended an explicitly non-counterfactualist mechanistic (what he called an "actualist-mechanist") account of causation in mechanisms. What is actual, the thought went, is more fundamental than what counterfactually could have been, and so an approach to understanding causation that settles for counterfactual dependence is missing something essential to causation.

We will follow Waskan in naming this the *actualism* desideratum of causation. Machamer, Bogen, and Waskan (as well as Salmon before them) all demanded that a metaphysical account of causation be based on actual causal connections between actual causes and actual effects. Here the truth-makers for counterfactual causes are actual features of the world. The problem with this criticism, however, is that it charges counterfactualists with missing something essential to causation, when in fact counterfactualists simply have a different idea about *what* is essential to causation. Ever since Lewis, employers of the counterfactual approach take causation to be essentially about difference-making, and so the various formulations of it (from possible world semantics to manipulationist interventions) have been designed to make sense of this difference-making quality of causation. On this particular disagreement it is important to remember that the actualism desideratum only works against counterfactual approaches to causation that hold the truth-makers of counterfactual claims to be features of other/possible/non-actual worlds; alternatively, if the truth-makers are based on features of the actual world, then the actualism desideratum still holds (see the discussion of causal pluralism in Glennan forthcoming for an example).

Lastly, consider a common criticism of Glennan's mechanical theory of causation. Most notably, his view that causal connections exist in virtue of mechanisms has been criticized on the grounds that it is uninformative because it fails to offer an analysis of causation that does not ultimately rely on the concept of "cause" itself. Kistler (2009), for example, argues that,

> On closer inspection, it appears that the concept of mechanism presupposes that of causation, far from being reducible to it. . . . The crucial point is that each step of the analysis of a mechanism makes essential use of the notion of cause, and thus presupposes it. . . . It follows that the concept of mechanism cannot be used to analyse the concept of causation and that, quite on the contrary, the concept of causation is among the irreducible conceptual instruments of mechanistic analysis.
>
> *(pp. 599–600)*

This criticism of Glennan's mechanistic analysis of causation appeals to the *reductive* desideratum (not to be confused with *reductionistic*). The concern, that is, is that the mechanistic account fails to reduce our understanding of causation to something other than causation.

True, the mechanistic account is *non-reductive*—but it is not clear that Glennan is moved by this desideratum. Glennan (forthcoming, chapter 6) responds to objections of this ilk by openly endorsing a non-reductive analysis of causation, arguing that more important than being reductive is being *informative*. While it may be the case that the concept of causation is not reduced by Glennan's mechanistic approach, that is, it does not follow that it fails to further develop our philosophical understanding of the causal structure of the world. On this view, the intricate mechanical story that connects causes and effects—the orchestration of organized parts, engaged in causally productive activities via mechanisms—is explanatory, despite the fact that it is non-reductive. Thus there is a sense in which even Glennan's mechanistic account of causation is criticized for failing to meet a desideratum that he never intended to meet.

6. Conclusion

The philosophy of mechanisms presents a new framework for thinking about classic issues in the philosophy of science. Under that general framework, however, there are a number of debates.

One of those debates pertains to how mechanistic causation should be understood, and because causation is fundamental to the philosophy of mechanisms, this is a debate that has repercussions throughout the framework.

We have showed how different approaches to understanding causation in the philosophy of mechanisms trace back to four separate responses to Hume's problem of causation: Anscombe's derivativeness approach, Lewis' counterfactual difference-making approach, Salmon's causal process approach, and Glennan's mechanical causation approach. Recognizing the relationship between current approaches in the philosophy of mechanisms and their historical ancestors best reveals the differences between those approaches and where they came from. It also allows for diagnosing when criticisms of a certain approach mistakenly hold that approach to a standard it was never designed to meet. As the debate over the metaphysics of mechanistic causation continues, be it in the direction of advocating for one approach in the face of the others or in the direction of finding common ground among the approaches, it is essential to keep this history in mind, as it offers a guide to where the fundamental differences between the approaches lay.

Notes

1 We are grateful to Jim Bogen, Stuart Glennan, and Phyllis Illari for providing us with valuable feedback on an earlier draft of this chapter.
2 In the *Enquiry*, Hume defines cause as such: "Similar objects are always conjoined with similar. Of this we have experience. Suitably to this experience, therefore, we may define a cause to be an object, followed by another, and where all the objects similar to the first, are followed by objects similar to the second. Or in other words, where, if the first object had not been, the second never had existed. The appearance of a cause always conveys the mind, by a customary transition, to the idea of the effect. Of this also we have experience. We may, therefore, suitably to this experience, form another definition of cause; and call it, an object followed by another, and whose appearance always conveys the thought to that other" (Hume 1993 [1748], p. 51).
3 Anscombe's influence reached out beyond the philosophical debates about causation. Nancy Cartwright's well-known book *How the Laws of Physics Lie* was very much a development of Anscombe's point about generalizations working only on artificial, idealized systems (Cartwright 1983). Because this chapter is about causation and not about laws of nature, we leave these other legacies of Anscombe's influence aside.
4 We are particularly grateful to Stuart Glennan for driving this point home.

References

Andersen, Holly (2014a), "A Field Guide to Mechanisms: Part I", *Philosophy Compass* 9: 274–283.
—— (2014b), "A Field Guide to Mechanisms: Part II", *Philosophy Compass* 9: 284–293.
Anscombe, G. E. M. (1993 [1971]), "Causality and Determination", in Ernest Sosa and Michael Tooley (eds.), *Causation*. Oxford: Oxford University Press, pp. 88–104.
Bogen, Jim (2004), "Analysing Causality: The Opposite of Counterfactual is Factual", *International Studies in the Philosophy of Science* 18: 1–26.
—— (2005), "Regularities and Causality; Generalizations and Causal Explanations", *Studies in the History and Philosophy of the Biological and Biomedical Sciences* 36: 397–420.
—— (2008), "Causally Productive Activities", *Studies in the History and Philosophy of Science* 39: 112–123.
Campaner, Raffaella (2013), "Mechanistic and Neo-mechanistic Accounts of Causation: How Salmon Already Got (Much of) It Right", *Metatheoria* 3: 81–98.
Cartwright, Nancy (1983), *How the Laws of Physics Lie*. Oxford: Oxford University Press.
Craver, Carl F. (2007), *Explaining the Brain: Mechanisms and the Mosaic Unity of Neuroscience*. Oxford: Oxford University Press.
Craver, Carl and James Tabery, (2015) "Mechanisms in Science", *The Stanford Encyclopedia of Philosophy* (Winter Edition), Edward N. Zalta (ed.), URL = <http://plato.stanford.edu/archives/win2015/entries/science-mechanisms/>.

Darden, Lindley (2006), *Reasoning in Biological Discoveries: Essays on Mechanisms, Interfield Relations, and Anomaly Resolution*. Cambridge: Cambridge University Press.

—— (2013), "Mechanisms Versus Causes in Biology and Medicine", in Hsiang-Ke Chao, Szu-Ting Chen, and Roberta Millstein (eds.), *Mechanisms and Causality in Biology and Economics*. Dordrecht: Springer, pp. 19–34.

Dowe, Phil (1992), "Wesley Salmon's Process Theory of Causality and the Conserved Quantity Theory", *Philosophy of Science* 59: 195–216.

—— (2011), "The Causal-Process-Model Theory of Mechanisms", in Phyllis McKay Illari, Federica Russo, and Jon Williams (eds.), *Causality in the Sciences*. Oxford: Oxford University Press, pp. 865–879.

Glennan, Stuart (1996), "Mechanisms and the Nature of Causation", *Erkenntnis* 44: 49–71.

—— (2002), "Rethinking Mechanistic Explanation", *Philosophy of Science* 69: S342–S353.

—— (2009), "Productivity, Relevance, and Natural Selection", *Biology & Philosophy* 24: 325–339.

—— (2010a), "Mechanisms, Causes, and the Layered Model of the World", *Philosophy and Phenomenological Research* 81: 362–381.

—— (2010b), "Ephemeral Mechanisms and Historical Explanation", *Erkenntnis* 72: 251–266.

—— (Forthcoming), *The New Mechanical Philosophy*. Oxford: Oxford University Press.

Hall, Ned (2004), "Two Concepts of Causation", in John Collins, Ned Hall, and L. A. Paul (eds.), *Causation and Counterfactuals*. Cambridge: The MIT Press, pp. 225–276.

Hume, David (2000 [1738–40]), *A Treatise of Human Nature: Being an Attempt to Introduce the Experimental Method of Reasoning into Moral Subjects*. Oxford: Oxford University Press.

—— (1993 [1748]), *An Enquiry Concerning Human Understanding*, second edition. Indianapolis: Hackett Publishing Company.

Kistler, M. (2009), "Mechanisms and Downward Causation", *Philosophical Psychology* 22: 595–609.

Kitcher, Philip (1984), "Explanatory Unification and the Causal Structure of the World", in Philip Kitcher and Wesley C. Salmon (eds.), *Scientific Explanation*. Minneapolis: University of Minnesota Press, pp. 410–505.

Kuhlmann, Meinard and Stuart Glennan (2014), "On the Relation between Quantum Mechanical and Neo-Mechanistic Ontologies and Explanatory Strategies", *European Journal for the Philosophy of Science* 4: 337–359.

Lewis, David (1973), "Causation", *The Journal of Philosophy* 70: 556–567.

Machamer, Peter (2004), "Activities and Causation: The Metaphysics and Epistemology of Mechanisms", *International Studies in the Philosophy of Science* 18: 27–39.

Machamer, Peter, Lindley Darden, and Carl Craver (2000), "Thinking about Mechanisms", *Philosophy of Science* 67: 1–25.

Psillos, Stathis (2004), "A Glimpse of the Secret Connexion: Harmonizing Mechanisms with Counterfactuals", *Perspectives in Science* 12: 288–319.

—— (2009), "Regularity Theories", in Helen Beebee, Christopher Hitchcock, and Peter Menzies (eds.), *The Oxford Handbook of Causation*. Oxford: Oxford University Press, pp. 131–157.

Read, Rupert and Kenneth A. Richman, eds. (2000), *The New Hume Debate*. London: Routledge.

Salmon, Wesley C. (1975), "Theoretical Explanation", in Stephan Körner (ed.), *Explanation*. Oxford: Basil Blackwell, pp. 118–145.

—— (1977), "An 'At-At' Theory of Causal Influence", *Philosophy of Science* 44: 215–224.

—— (1984), *Scientific Explanation and the Causal Structure of the World*. Princeton, NJ: Princeton University Press.

—— (1994), "Causality without Counterfactuals", *Philosophy of Science* 61: 297–312.

—— (1997), "Causality and Explanation: A Reply to Two Critics", *Philosophy of Science* 64: 461–477.

—— (1998), *Causality and Explanation*. Oxford: Oxford University Press.

Tabery, James (2004), "Synthesizing Activities and Interactions in the Concept of a Mechanism", *Philosophy of Science* 71: 1–15.

Waskan, Jonathan (2011), "Mechanistic Explanation at the Limit", *Synthese* 183: 389–408.

Williamson, Jon (2011), "Mechanistic Theories of Causality: Part I", *Philosophy Compass* 6: 421–432.

Woodward, James (2002), "What Is a Mechanism? A Counterfactual Account", *Philosophy of Science* 69: S366–S377.

—— (2003), *Making Things Happen: A Theory of Causal Explanation*. Oxford: Oxford University Press.

—— (2011), "Mechanisms Revisited", *Synthese* 183: 409–427.

11

MECHANISMS, COUNTERFACTUALS, AND LAWS[1]

Stavros Ioannidis and Stathis Psillos

1. Introduction

There have been two traditions concerning how the "link" between cause and effect is best understood (Hall 2004; Psillos 2004). According to the first tradition, which goes back to Aristotle, there is a *productive relation* between cause and effect: the cause produces, generates, or brings about the effect. This productive relation between cause and effect has been typically understood in terms of powers, which in some sense ground the bringing about of the effect by the cause. According to the second tradition, which goes back to Hume, the link is some kind of robust relation of dependence between what are taken to be distinct events. On this account, the chief characteristic of causes is that they are difference-makers: the occurrence of the cause makes a difference to the occurrence of the effect (for Hume's theory of causation, see Chapter 10). There are various ways to understand the notion of difference-making (e.g. in terms of laws or probabilities); but arguably the core notion of difference-making is counterfactual, i.e. based on contrary-to-fact hypotheticals. That is, a causal claim of the form "A caused B" would be understood as implying: if A hadn't happened, B wouldn't have happened either. It is in this sense that A *actually* makes a difference for B.[2]

The currently most popular version of the production approach cashes out the link between cause and effect by reference to mechanisms. The central thought behind a mechanistic account of causal production is that two events are causally connected, if and only if there is a mechanism that connects them. Hence, where there is causation, there is mechanism. As we shall see, however, various mechanists tie together a power-based account of production and a mechanistic account. In this chapter, we shall focus our attention on mechanisms and aim to compare mechanistic accounts with counterfactual accounts of causation.

As we will see, some mechanists tend to refrain from using counterfactuals. For others, counterfactuals are needed to ground the laws that characterize the interactions between the components of a mechanism; counterfactuals may in turn be grounded in lower-level mechanisms. Yet other mechanists try to dispense with both counterfactuals and laws, in favor of activities. Hence, understanding the relation between mechanisms and counterfactuals requires clarifying the relation between mechanisms and laws (on the relationship between mechanisms and laws, see also Chapter 12). Laws will thus be central in the argument of this chapter. The key question then, for our purposes, is: can there be a conception of mechanism which does not ineliminably rely on some non-mechanistic account of counterfactual dependence?

To be exact, we want to investigate whether a mechanistic theory of causation ultimately relies on a counterfactual theory (and hence, whether it turns out to be a version of the dependence approach); or whether it constitutes a genuine version of the production approach (either because it altogether dispenses with the need to rely on counterfactuals, or because it grounds counterfactuals in mechanisms). To be clear on these issues presupposes, as we will argue in section 2, a more careful analysis of the notion of mechanism. Here is the central line of argument we want to investigate: since a mechanism is composed of *interacting* components, the notion of a mechanism should include a characterization of these interactions; but if (i) these interactions are understood in terms of difference-making relations and (ii) these difference-making relations are not in turn grounded in mechanisms, then there is a fundamental asymmetry between mechanistic causation and causation as difference-making; for to offer an adequate account of the former presupposes an account of the latter.

As we will see, not all philosophers that stress the importance of mechanisms for thinking about science are after an account of causation. For some of them, mechanisms are important in understanding scientific explanation and theorizing, but it is not the case that causation *itself* is mechanistic (see, for example, Craver 2007, 86). Yet even if these philosophers do not have to provide a full-blown account of the ontology of mechanisms, they have to explain the modal force of mechanisms; hence the issue of the relation between mechanisms and counterfactuals is crucial. So, we can formulate our central question in a more comprehensive way, as follows: given that a mechanism consists of components that interact in some manner, and thus cause changes to one another, does an account of these *interactions* require a commitment to counterfactuals? As we shall see, ultimately, the issue turns not around the need or not to posit relations of counterfactual dependence but around what the suitable truth-makers for counterfactuals are.

2. Mechanisms-for vs mechanisms-of

Two traditions have tried to reclaim the notion of mechanism in the philosophical literature of the twentieth century. In exploring the relation between mechanisms and laws/counterfactuals, it is important to distinguish between these two very different senses of "mechanism."

In the fairly recent literature, mechanisms are always understood as mechanisms *for certain behaviors* (see Chapter 8). In other words, mechanisms are individuated in terms of what they do. For example, there are mechanisms *for* DNA replication, or *for* mitosis. What the mechanism does, what the mechanism is a mechanism *for*, determines the boundaries of the mechanism and the identification of its components and operations (see Chapter 9 for more on components and boundaries of mechanisms). Such mechanisms, which we call *mechanisms-for* (i.e. mechanisms *for* certain behaviors/functions), are the mechanisms that, according to many authors, figure in explanations in biomedical sciences and elsewhere, and are what many scientists aim to discover. Mechanisms-for we find, among others, in Machamer, Darden, and Craver (2000), Bechtel and Abrahamsen (2005), Craver (2007), and Bechtel (2008). This kind of conception of mechanisms is, arguably, the dominant one in various philosophical studies of mechanisms and their role in the various sciences.[3]

The second sense of mechanism is typically found in the context of mechanistic theories of *causation*. These theories aim to characterize the causal link between two events (to fathom Hume's "secret connexion") in terms of a "mechanism." For this second sense, what the mechanism *does* is not important; what is important is that it is actually there underlying or constituting a certain kind of process. More precisely, what makes a process *causal* is the presence of a mechanism which mediates between cause and effect (or whose parts or moments are the "cause" and the "effect"). We call such mechanisms *mechanisms-of*. Mechanisms-of are the mechanisms discussed in, for example, theories that view causation as mark-transmission (Salmon 1984), persistence, transference, or possession of a conserved quantity (Mackie 1974; Salmon 1997; Dowe 2000).

Talk of "mechanisms" in relation to causation goes back to John Mackie (1974), who took it that causation consists in a "causal mechanism"; that is, "some continuous process connecting the antecedent in an observed . . . regularity with the consequent" (1974: 82). His preferred account of a causal mechanism in terms of qualitative or structural continuity, or *persistence*, exhibited by certain processes, faced significant problems which led Wesley Salmon (1984) to argue for an account of causal mechanism that is based on the notion of structure-transference (see Psillos 2002: section 4.1; also Chapter 10 of this book). Salmon kept the basic idea that "[c]ausal processes, causal interactions, and causal laws provide the mechanisms by which the world works; to understand *why* certain things happen, we need to see *how* they are produced by these mechanisms" (1984: 132). But he claimed that the distinguishing characteristic of a causal process (and hence of a mechanism) is that it is capable of transmitting its own structure or modifications of its own structure. Generalizing Hans Reichenbach's (1956) idea that causal processes are those that are capable of transmitting a mark, Salmon noted that any process, be it causal or not, exhibits "a certain structure."

A causal process is then a process capable of *transmitting* its own structure. But, Salmon added, "if a process—a causal process—is transmitting its own structure, then it will be capable of transmitting certain modifications in the structure" (1984: 144). But as many critics noted, the very idea of structure-transference (*aka* mark-transmission) cannot differentiate causal processes from non-causal ones, since *any* process whatever can be such that *some* modification of *some* feature of it gets transmitted after a single local interaction. A typical example was the shadow of a car with a dent—this is a "dented" shadow, and the mark is transmitted with the shadow for as long as the shadow is there. In response to this Salmon strengthened his account of mark-transmission by requiring that for a process P to be causal, it is necessary that "the process P would have continued to manifest the characteristic Q if the specific marking interaction had not occurred" (1984: 148). It should be clear, though, that this kind of modification takes us back to persistence! In effect, the idea is that a process is causal if (i) a mark made on it (a modification of some feature) gets transmitted after the point of interaction and (ii) in the absence of this interaction, the relevant feature would have *persisted*, where the required persistence is counterfactual.

Salmon did modify this view further by adopting Phil Dowe's (2000) conserved quantity theory, according to which "it is the possession of a conserved quantity, rather than the ability to transmit a mark, that makes a process a causal process" (2000: 89). On what has come to be known as the Salmon-Dowe theory, a *causal process* is a world line of an object that possesses a conserved quantity. And a *causal interaction* is an intersection of world lines that involves exchange of a conserved quantity.

Dowe fixes the characteristic that renders a process causal and, consequently, the characteristic that renders something a mechanism. A conserved quantity is "any quantity that is governed by a conservation law" (2000: 91), e.g. mass-energy, linear momentum, charge, and the like. Apart from various issues that have to do with the question of whether this theory can avoid counterfactuals (see Psillos 2002: chapter 4), the main practical concern is that this account of mechanism is too narrow. For even if *physical* causation—and hence physical mechanism—was a matter of the possession of a conserved quantity, it's hard to see how this account of mechanism can even start shedding any light on causal processes in domains outside of physics (biological, geological, medical, social). These will have to be understood either in a reductive way or in non-mechanistic (Dowe-Salmon) terms.

A rather liberating conception of causal mechanism was offered by Rom Harré in the early 1970s. Harré connected the traditional idea of power-based causation with the traditional idea that causation involves a mechanism. What he called "generative mechanism" can be put thus:

generative mechanism = powers + mechanisms

As he (1972: 121) put it: "The generative view sees materials and individual things as having causal powers which can be evoked in suitable circumstances." And he added: "The causal powers of a thing or material are related to what causal mechanisms it contains. These determine how it will react to stimuli" (1972: 137). For example, an explosion is caused both by the detonation and the power of the explosive material, which it has in virtue of its chemical nature.

On this view of causation, the ascription of a power to a particular form has the following form:

X has the power to A = if X is subject to stimuli or conditions of an appropriate kind, then X will do A, *in virtue of its intrinsic nature.*

But this is not a simple conditional analysis of powers, since as Harré stressed, power-ascriptions involve two *analysans*:

a *specific conditional* (which says what X will or can do under certain circumstances and in the presence of a certain stimulus); and
an *unspecific categorical* claim about the nature of X.

The claim about the nature of X is unspecific, because the exact specification of the nature or constitution of X in virtue of which it has the power to A is left open. (Discovering this is supposed to be a matter of empirical investigation.)

It is a fair complaint that, as stated above, the ascription of powers is explanatorily incomplete unless something specific is (or can be) said about the *nature* of the particular that has the power. Otherwise, power-ascription merely states what needs to be explained, viz. that causes produce their effects. This is where mechanisms come in. Specifying the generative mechanism is cashing the promissory note. As Harré put it: "Giving a mechanism . . . is . . . partly to describe the nature and constitution of the things involved which makes clear to us what mechanisms have been brought into operation" (1970: 124). So the key idea in this mechanistic view of causation is this: causes produce their effects because they have the power to do so, where this power is grounded in the mechanism that connects the cause and the effect and the mechanism is grounded in the nature of the thing that does the causing.

This, as one of us has noted elsewhere (Psillos 2011), is a broad and liberal conception of causal mechanism. Generative mechanisms are taken to be the bearers of causal connections. It is in virtue of them that the causes are supposed to produce the effects. But there is no specific description of a mechanism (let alone one that is couched in terms of physical quantities). A generative mechanism is virtually *any* relatively stable arrangement of entities such that, by engaging in certain interactions, a function is performed, or an effect is brought about. As Harré explained, he did not "intend anything specifically mechanical by the word 'mechanisms'.

Clockwork is a mechanism, Faraday's strained space is a mechanism, electron quantum jumps [are] . . . a mechanism, and so on" (Harré 1970: 36).

Though this was not quite perceived and acknowledged when Harré was putting forward this conception, this liberal conception of mechanism pointed to a shift from thinking of mechanisms exclusively as the vehicle of causation (mechanisms-of) to thinking of mechanisms as whatever implements a certain behavior or performs a certain function (mechanisms-for). On this broader view, a mechanism is a complex system that consists of some parts (its building blocks) and a certain *organization* of these parts, which determines how the parts interact with each other to produce a certain output. The parts of the mechanism should be stable and robust; that is, their properties must remain stable, in the absence of interventions. The organization should also be stable; that is, the complex system as a whole should have stable dispositions, which produce the behavior of the mechanism. Thanks to the organization of the parts, a mechanism is more than the sum of its parts: each of the parts contributes to the overall behavior of the mechanism more than it would have achieved if it had acted on its own.

One natural question may arise at this point. Can a mechanism be *both* what we called a mechanism-for and what we called a mechanism-of? That is, can it be the case that a mechanism *both* underlies or constitutes a causal process *and* is a mechanism for a specific behavior? Though Harré adopted this view, this position acquired new strength in the early 1990s when Stuart Glennan developed his own mechanistic theory of causation. For him, mechanisms are both what underlie or constitute causal connections between events and thus provide the missing link between cause and effect (mechanisms-of) and at the same time complex systems that are responsible for certain behaviors (mechanisms-for) and are thus individuated in terms of them.

But, mechanisms-for are *not* necessarily mechanisms-of. Conceptually this is obvious if we think of a mechanism as a causal process with various characteristics (such as those discussed above—e.g., they possess a conserved quantity or some kind of persisting structure). There is no reason to think that this kind of mechanism (e.g., the process by means of which the sum of kinetic and potential energy is conserved in some interaction) is a mechanism *for* any particular behavior. Conversely, if we think of a mechanism as a complex system such that the interactions of its parts bring about a certain behavior, there is no *ipso facto* reason to adopt a mechanistic account of causation. In light of this, we arrive at a tripartite categorization of "mechanistic" accounts present in the literature (or at *three* independent notions of what a mechanism is): mechanisms can be mechanisms-for, or mechanisms-of, or both.[4]

With this map of the conceptual landscape of the philosophical literature on mechanisms in mind, our task now is to examine each case in turn and investigate the relations between each sense of "mechanism" and laws/counterfactuals.

3. Mechanisms-of

Let us first focus on mechanisms-of that are not at the same time mechanisms-for. As noted already, the best known such causal mechanisms are those discussed by Salmon and Dowe. Though these accounts of causation are presented as being compatible with singular causation, it should be quite clear that they rely on counterfactuals. We noted already that in Salmon's account counterfactuals loom large. In fact, counterfactuals play a *double role* in his theory. On the one hand, they secure that a process is causal by making it the case that the process does not just possess an actual uniformity of structure, but also a counterfactual one. On the other hand, they secure the conditions under which an interaction (the marking of a process) is causal: if the marking would have occurred even in the absence of the supposed interaction between two processes, then the interaction is not causal.

On Dowe's account, the very idea of a possession of a conserved quantity for a process to be causal implies that both laws and counterfactuals are in the vicinity. Conserved quantities are individuated by reference to conservation laws and it is hard to think of a process being causal without the conserved quantity that makes it causal being governed by a conservation law. Counterfactuals are also necessary for Dowe's account of causation. Not just because laws imply counterfactuals, but also because an appeal to counterfactuals is necessary for claiming that the process is causal. That is, it seems that without counterfactuals there is no way to ground the difference between objects to which conserved quantities may be applied and objects to which they may not (e.g. a single particle with zero momentum vs. a shadow with zero quantity of charge; the particle, but not the shadow, is a causal process precisely because it could enter into interactions, which could make its momentum non-zero (see Psillos 2002: 126)).

4. Mechanisms-for and mechanisms-of

Let us now turn to an account such as Glennan's, i.e. to an account that takes mechanisms to be both mechanisms-for *and* mechanisms-of. There are two parts in Glennan's definition of mechanisms. First, a mechanism consists of components that interact—in this, it is similar to Salmon's account of a mechanism-of as causal process. However, for Glennan, the mechanism itself is a complex system with a stable arrangement of components (see his 1996, though in more recent work he drops the stability requirement for some kinds of mechanisms—see section 5). So, in contrast to the view of mechanisms-of as processes (which can in principle be singular causal chains of events), such mechanisms are "types of systems that exhibit regular and repeatable behavior" (Glennan 2010: 259).

How should we understand the interactions among the components of such mechanisms? Should they be understood in terms of counterfactuals or not? To answer this question, let us briefly review various possible options.

The first general case we will consider is interactions with laws. Interactions can be governed by laws, where laws are understood in some robust metaphysical sense. For example, according to Dretske (1977), Tooley (1977), and Armstrong (1983), laws are necessitating relations between universals. So, if there is a necessitating relation between universals A and B, there will be a law between them and as a result of this law when A is instantiated, so will be B. Suppose we transfer that to the components of a mechanism: when component X instantiates A at some time t_1, some other component Y will instantiate universal B, perhaps at a later time. Or take the rival view (but metaphysically robust too) that laws are embodied in relations between powers. If this is the preferred account of laws, interactions will be understood in terms of powers. Powers are properties possessed by components of a mechanism, and produce specific manifestations under specific stimuli. Whereas for Dretske, Tooley, and Armstrong the interactions within the mechanism are grounded in the external relation of nomic necessitation, in the powers view, interactions are grounded in the internal relations between the powers of the components of the mechanism. Alternatively, interactions between the components of the mechanism may be viewed as being governed by metaphysically thin laws; e.g., by (Humean) regularities. Here, component A can be said to interact with component B, in virtue of the fact that this interaction is an instance of a regularity.

If, for the time being, we bracket laws, can we understand the interactions among the components of the mechanism differently? Perhaps counterfactuals can be of direct help here. So Glennan (2002), following Woodward (2000, 2002, 2003), understands interactions in terms of change-relating generalizations that are invariant under interventions. Such generalizations are change-relating in the sense that they relate changes in component A to changes in component B. They involve counterfactual situations in that they concern what would have happened to component B regarding the value of quantity Y possessed by it, if the value of quality X possessed by component A had changed. These generalizations are invariant under interventions, in that they are about relations between variables that remain invariant under (actual or counterfactual) interventions. These change-relating generalizations, then, are grounded in counterfactuals (called interventionist counterfactuals by Woodward—on Woodward's theory of causation, see also Chapter 10).

But if we are to understand interactions between components in terms of counterfactuals, the next question is: *what grounds these counterfactuals?* In particular, in virtue of what are interventionist counterfactuals true? The answers here are well known (see Psillos 2004, 2007). Counterfactuals can be grounded in laws or not. If they are grounded in laws, following what we said in the previous paragraph, these laws can be either metaphysically robust laws of the sort adopted either by Armstrong or power-based accounts of lawhood, or thin Humean regularities, instances of which are particular token-interactions between components. If the counterfactuals are not grounded in laws, then it's likely that there are counterfactuals "all the way down"; that is, that there are primitive modal facts in the world (see Lange 2009).

In *any* of these accounts of law-governed within-mechanism interactions, counterfactuals have a central place: either by directly accounting for interactions (as in Woodward's theory), or by being part of an account of the nomological dependences that ground the interactions, or as a primitive modal signature of the world.[5] So, if laws regulate the interactions between the components of the mechanism, we cannot do away with counterfactuals in grounding within-mechanism interactions. Before, for completeness, we consider the prospects for a non-law-governed account of interactions, let us discuss an attempt to have *mechanisms themselves* ground counterfactuals. This suggestion is put forward by Glennan (1996). For him, although interactions are understood in terms of interventionist counterfactuals, these counterfactuals are in turn grounded in (lower-level) mechanisms.

Here is Glennan's suggestion in more detail. Interactions among components of a mechanism are governed by laws, which are understood in terms of interventionist counterfactuals; these laws are "mechanically explicable," i.e. there are other mechanisms that ground them; but these (lower-level) mechanisms themselves contain parts, the interactions among which are understood in terms of counterfactuals, and which are in turn grounded in yet other mechanisms, until we finally reach a level where we run out of mechanisms to explain the laws that govern the interactions among components, and thus to ground the relevant counterfactuals. At this fundamental level, interactions among components are *directly grounded in counterfactuals*. But notwithstanding these not mechanically explicable laws, Glennan insists that at all other levels mechanisms can ground interactions. So, even if we need to introduce counterfactuals to account for interactions, mechanisms have priority over counterfactuals, and thus the account is supposed not to be a version of a difference-making theory of causation, but a genuinely mechanical account.

However, given the existence of not mechanically explicable laws, it is not clear how mechanisms can ground counterfactuals *at any level*. That is, given that the mechanisms at the lowest level depend on counterfactuals, the mechanisms at a level exactly above the fundamental must be equally dependent (albeit *derivatively*) on the fundamental counterfactuals, and so on for every higher level. In other words, to ground counterfactuals at any level, we need the whole lower hierarchy of mechanisms *and* counterfactuals, and since we ultimately arrive at a level where there are either only counterfactuals, or only laws (or both), it seems that there is a fundamental asymmetry between mechanisms and laws/counterfactuals. The only way to block the asymmetry would be to argue that the whole hierarchy is not needed to ground the counterfactuals at higher levels. Even if this were to be granted for purposes of explanation—that is, even if explanation in terms of mechanisms at level *n* does not require *citing* lower-level mechanisms—metaphysically the whole hierarchy constitutes the grounds for the mechanism.[6]

In sum, if laws are admitted in our notion of mechanism, a reliance on counterfactuals is inevitable. But can we perhaps avoid counterfactuals if we account for within-mechanisms interactions in some other way?

We move now to a second approach. We have reviewed various options to understand interactions of components of mechanisms, where these interactions are viewed as law-governed. The question now is: can we have interactions without some notion of law in the background (either in terms of regularities, or in some more metaphysically robust sense)? If yes, then this could be a way to have mechanisms-of as complex systems, without the need to put laws and counterfactuals in the picture.

For some mechanists, the interactions of components have to be understood in terms of *activities*. Activities are a new ontological category that, together with entities, are said to be needed for an adequate ontological account of mechanisms (Machamer, Darden, and Craver 2000; Machamer 2004). Activities are meant to embody the causally productive relations between components. Causation in terms of activities is viewed as a type of singular causality, where the causal relation is a local matter, i.e. it concerns what happens between the two events that are causally connected, and not what happens at other places and at other times in the universe (as is the case for the regularity theorist). Activities, thus, have been taken to obviate the need for laws.

A key argument in favor of activities turns on the claim that causation is *singular*. But, does singular causation imply that there are no laws? It would be too quick to infer from singular causal claims that laws are not part of causation. By singular causation we may simply mean that there exist genuine singular causal connections, i.e. causal connections between particular event-tokens. But this is not enough to prove that there are no laws in the background. For it is consistent with the existence of singular causal sequences that there are laws under which these causal sequences fall. To use a quick example, on Armstrong's account of laws, singular causation is *ipso facto* nomological causation since the nomic necessitating relation that relates two universals relates the instances of the two universals too (Armstrong 1997). Interestingly, the same is true if we take singular causation to be grounded in the powers possessed by objects; powers are again *wholly* present in the complex event that constitutes the singular causal sequence. And though there is no nomic relation that relates the two powers, the regular instantiation of the two powers implies the presence of a regularity. So, what both these cases show is that even singular causation can be nomological, i.e. subsumed under laws.[7]

Thus, singular causation does not, on its own, constitute an argument in favor of viewing interactions among components of mechanisms as not being law-governed, or more generally, as not depending on difference-making relations. So, friends of activities need to (i) give more reasons to justify the introduction of this new ontological category, and (ii) explain why activities qua producers of change are themselves counterfactual-free. Although it's conceivable that singular causation just amounts to the local activation of powers which in turn ground activities, powers being universals, it's upon the friends of powers to show that we can understand this co-instantiation without also assuming that there is a law present.[8]

5. Mechanisms-for

So far we have argued that mechanisms-of (mechanisms considered as underlying or constituting causal processes) require laws, and thus difference-making relations must be included in the notion of a mechanism-of. But what about mechanisms-for, mechanisms as complex systems responsible for certain behaviors? What is the relation between mechanisms-for and laws/counterfactuals? Recall that a mechanism-for is a mechanism as a complex system such that the interactions of its parts bring about a certain behavior-function. A mechanism-for need not commit us to a mechanistic (e.g. à la Salmon-Dowe) account of the causal interactions between its parts.

In light of what was said in the previous section, there is an argument as to why mechanisms-for have to incorporate laws and/or counterfactuals: a mechanism-for involves components that interact with one another; but laws and/or counterfactuals are needed to account for these interactions; hence, mechanisms-for need to incorporate laws and/or counterfactuals. However, Jim Bogen (2005) has taken the existence of mechanisms that function *irregularly* as an argument against the view that laws and regular behavior have to characterize the function of mechanisms. In this section we will deal with this argument from irregular mechanisms.

The first point that we want to stress is that irregular and unrepeatable mechanisms are not as ubiquitous as some philosophers want us to believe. So, consider a claim made by Leuridan (2010). He thinks that mechanisms as complex systems ontologically depend on stable regularities, since there can be no such mechanisms (i) without macrolevel regularities (i.e. the behavior produced by the mechanism), and (ii) without microlevel ones (i.e. the behaviors of the mechanism's parts). Kaiser and Craver (2013) have replied to this that Leuridan's first claim is "clearly false" since "[o]ne-off mechanisms are mechanisms *without* a macrolevel regularity," where "one-off mechanisms" are the causal processes discussed by Salmon and others (*mechanisms-of* in our terminology). Moreover, they point to examples where scientists seem to be interested in exactly this kind of mechanism, i.e. when they try to explain how a *particular* event occurred (for example, a particular speciation event).

This kind of reply confuses the two different senses of mechanism we have tried to disentangle. It is not the case that "singular, unrepeated causal chains . . . are a special, limiting case of [complex system] mechanisms, not something altogether different," as Kaiser and Craver insist. For it is not at all clear that such mechanisms-of are at the same time mechanisms-for, i.e. mechanisms *for* a certain behavior. Similarly, we remain unpersuaded by Glennan's (2010) claim that the mechanisms that produced various historical outcomes are mechanisms-for (he calls them "ephemeral mechanisms"). In any case, it would be very implausible to insist that any arbitrary causal chain is *for* a certain behavior (which is identified with the outcome of the causal chain or, alternatively, with the (higher-level) event constituted by the causal chain). For instance, the reflection of a light-ray on a surface is a clear case of a mechanism-of (since it constitutes a causal process), but it is not clear at all that it is a mechanism *for* a certain behavior (unless of course we follow Glennan (forthcoming) and equate "behavior" with "phenomenon"; that is, with causal effect).[9]

So, it is not enough to point to singular causal chains to argue that there can be irregular mechanisms, or one-off mechanisms (mechanisms that function only once) (see also DesAutels 2011; Andersen 2012). Still, one may wonder: can there be mechanisms-for *without* a corresponding macrolevel regularity? Thus, the issue that must be clarified is: what are the conditions for being a mechanism *for* a behavior? Is it merely to have a function (which is the mechanism's behavior), or should we, in addition, require that the behavior be regular?

This is not the place to discuss at any length the concept of function (on the relation between functions and mechanisms, see Chapter 8); for the purposes of the current argument, let us interpret this requirement in a wide sense, i.e. as not requiring that for something to have a function it has to be the product of conscious design or the result of natural selection. In other words, we are going to take a function in the sense of Cummins (1975), for whom functions are certain kinds of dispositions (see Craver 2001 for such an approach to the functions of mechanisms). In particular, what it is for a mechanism M to have a function F is to have a disposition to F, which contributes to a disposition of a larger system that contains M. Such functions need not be restricted to living systems or artifacts.[10] Yet, not anything whatsoever can be ascribed a Cummins function. In particular, unrepeated causal chains of events, which might well be Salmon's and Dowe's mechanisms-of, need not have a function. We can follow Cummins and say that talk of functions only makes sense when we can apply what Cummins calls the analytical strategy, i.e. explain the disposition of a containing system in terms of the contributions made by the simpler dispositions of its parts.

There can be systems with Cummins functions that exhibit the corresponding behavior only once; so, there are many biological functions, the realization of which requires that the biological entity that has the function cease to exist. An example is the mechanism for apoptosis, i.e. programmed cell death. Here, the relevant mechanism has a Cummins function, i.e. it causally contributes to the death of the cell; however, this is a function that, when successfully carried out, can occur only once. But even in such cases, the behavior of a particular mechanism of apoptosis is a token of *a type of behavior* that occurs countless times every second.

Can there be genuinely *irregular* mechanisms-for; that is, mechanisms-for without a corresponding macrolevel regularity? Bogen (2005) has offered the case of the mechanism of neurotransmitter release as an example of a mechanism-for that behaves irregularly. As this mechanism more often than not fails to carry out its function, there exists no corresponding macrolevel regularity; but moreover, and more importantly, Bogen thinks that within-mechanism interactions must themselves be irregular, and thus we must abandon the regularity account of causation in favor of activities.

We do not think that this last conclusion follows from Bogen's example. To see why this is the case, it is useful to distinguish between three cases of what we may call "irregular" mechanisms-for. The irregularity of mechanisms-for may be only contingent (irregular$_1$), stochastic (irregular$_2$), or (let us assume) more radical (irregular$_3$).

Irregular$_1$ mechanisms-for are mechanisms that could function regularly, but they in fact do not. A defective machine that only functions once in a while is a case in point. Such a machine (i) is a mechanism for a behavior and (ii) functions irregularly. However, it is certainly not the case that a successful operation of the machine is not subject to laws (e.g. laws of electromagnetism, gravity, or friction). (Nor is it the case that defective machines falsify the regularity account of causation.)

Irregular$_2$ mechanisms are like irregular$_1$ mechanisms in that they operate in accordance with laws, but in this case the laws are probabilistic. So, the existence of such mechanisms does not show that within-mechanism interactions need not be law-governed (or even that the regularity account of causation is false—regularities can be stochastic). What if the operation of a mechanism is completely chancy (e.g. because it involves the radioactive decay of a single atom)? Even if we do not have a law here (perhaps because the relevant law concerns a population of atoms rather than a single one), it is not at all clear to us that such a chancy "mechanism" could be an example of a mechanism-for.[11]

Finally, we can imagine an irregular$_3$ mechanism-for; such a *sui generis* mechanism only operates once, and its unrepeatability is supposed not to be a contingent matter, but this is because the interactions among its components cannot *in principle* be repeated. We are not sure that the notion of an irregular$_3$ mechanism-for actually makes sense. But this is the only kind of example we can imagine, where the irregularity or unrepeatability of a mechanism would be a reason to think that its operation is not law-governed. If that's where the friends of genuinely irregularly operating mechanisms can pin their hopes for a non-law-governed account of mechanism, then so be it!

6. Conclusions

In this chapter, we have examined the relation between mechanisms and laws/counterfactuals by revisiting the main notions of mechanism found in the literature. We distinguished between two different conceptions of "mechanism." What we have called *mechanisms-of* tally with the general conceptions of mechanisms offered in discussions of causation. A "mechanism" in these views is what underlies or constitutes a causal process or connects the cause with the effect. What we have called *mechanisms-for*, on the other hand, are complex systems that function so as to produce a certain behavior. According to some mechanists, a mechanism fulfils both of these roles simultaneously.

We have argued that for both mechanisms-of and mechanisms-for, counterfactuals and laws are central for understanding within-mechanism interactions. We have examined two main arguments in more detail. Concerning mechanisms-of, we have seen that singular causation is compatible with several quite different ways of understanding within-mechanism interactions, in all of which laws and counterfactuals are central. Concerning mechanisms-for, we have argued that the existence of irregular mechanisms is compatible with the view that mechanisms operate according to laws. Both of these arguments point to an asymmetrical dependence between mechanisms and laws/counterfactuals: while some laws and counterfactuals must be taken as primitive (non-mechanistic) facts of the world, all mechanisms depend on laws/counterfactuals.

Notes

1 We wish to thank Phyllis Illari and Stuart Glennan for valuable comments on an earlier draft of this chapter.
2 On a nomological account of causal dependence (i.e. B depends on A if there is a law that connects the two), counterfactuals are required to account for the modal strength of laws (for more on this see Psillos 2002). So, even if it were to be admitted that the alternative notions of dependence are distinct, counterfactuals play a key role in all versions of the dependence approach to causation.
3 We take it that to be a mechanism-for is tantamount to being a system performing a function; this is not always made explicit in general accounts of mechanisms-for (but see Bechtel and Abrahamsen (2005), Garson (2013), and Chapter 8). The minimal mechanism of Glennan and Illari (see Chapter 1), which is defined as a mechanism for a phenomenon, can be understood as either mechanism-of or mechanism-for in our terminology, according to how we understand "phenomenon," as a function or as causal process.
4 See Levy (2013) and Andersen (2014a, 2014b) for similar distinctions.
5 In Lange's (2009) theory of laws it is a counterfactual notion of stability that determines which facts are lawful and which are accidental. In other theories of lawhood, counterfactuals come "for free," so to speak, as they must be part of any metaphysically robust theory of laws (such as that of Dretske, Tooley, and Armstrong): any such theory must show why laws support counterfactuals. It is not obvious how exactly counterfactuals are part of a regularity view of laws. But note that this is a problem (if at all) for the regularity theorist, and not for the view that interactions have to be understood in terms of laws/counterfactuals. For an attempt to reconcile regularity theory with counterfactuals see Psillos (2014).
6 See Glennan (2011) for an attempt to respond to this argument, and Casini (2015) for a detailed criticism; see also Campaner (2006).
7 There is debate among friends of powers whether such a powers-ontology yields an account of laws in terms of powers (Bird 2007), or a lawless ontology (Mumford 2004). But this need not concern us here.
8 For more on activities see Chapters 9 and 10; see also Waskan (2011) for a mechanist account of the contents of causal claims that is not based on counterfactuals and Woodward's (2011) answer that causation as difference-making is fundamental in understanding mechanisms; Menzies (2012) provides an illuminating account of mechanisms in terms of the interventionist approach to causation within a structural equations framework; lastly, Glennan (forthcoming: chapters 5 and 6) offers a detailed treatment of mechanistic causation as a productive account of causation not reducible to difference-making relations.

9 In many examples where scientists refer to "mechanisms," it may not be clear whether it is the notion of mechanism-of or the notion of mechanism-for (or both, or neither) that they have in mind. For example, such uses as "mechanism of chemical reaction," "mechanism of speciation," "mechanism of action (of a drug)" can be construed in various ways; to insist on a widening of the concept of mechanism-for based on scientific practice (without further argument) seems too quick.

10 However, note that for some (e.g. Kitcher 1993) we cannot ascribe even Cummins functions to entities that are not products of (either conscious or nonconscious) design, i.e. that are neither artefacts nor living systems.

11 For a notion of "stochastic" mechanism see DesAutels (2011), as well as Andersen (2012).

References

Andersen, H. (2012) "The Case for Regularity in Mechanistic Causal Explanation," *Synthese* 189: 415–432.

—— (2014a) "A Field Guide to Mechanisms : Part I," *Philosophy Compass* 9: 274–83.

—— (2014b) "A Field Guide to Mechanisms : Part II," *Philosophy Compass* 9: 284–93.

Armstrong, D. M. (1983) *What Is a Law of Nature?*, Cambridge: Cambridge University Press.

—— (1997) "Singular Causation and Laws of Nature," in *The Cosmos of Science*, J. Earman and J. Norton (eds), Pittsburgh, PA: Pittsburgh University Press, pp. 498–511.

Bechtel, W. (2008) *Mental Mechanisms: Philosophical Perspectives on Cognitive Neuroscience*, New York: Routledge.

Bechtel, W. and A. Abrahamsen (2005) "Explanation: A Mechanistic Alternative," *Studies in History and Philosophy of Biological and Biomedical Sciences* 36: 421–41.

Bird, A. (2007) *Nature's Metaphysics: Laws and Properties*, Oxford: Oxford University Press.

Bogen, J. (2005) "Regularities and Causality; Generalizations and Causal Explanations," *Studies in History and Philosophy of Biological and Biomedical Sciences* 36: 397–420.

Campaner, R. (2006) "Mechanisms and Counterfactuals: A Different Glimpse of the (Secret?) Connexion," *Philosophica* 77: 15–44.

Casini, L. (2015) "Can Interventions Rescue Glennan's Mechanistic Account of Causality?," *British Journal for the Philosophy of Science*, doi:10.1093/bjps/axv014.

Craver, C.F. (2001) "Role Functions, Mechanisms and Hierarchy," *Philosophy of Science* 68: 31– 55.

—— (2007) *Explaining the Brain: Mechanisms and the Mosaic Unity of Neuroscience*, Oxford: Oxford University Press.

Cummins, R. (1975) "Functional Analysis," *Journal of Philosophy* 72: 741–64.

Davidson, D. (1967) "Causal Relations," *Journal of Philosophy* 64: 691–703.

DesAutels, L. (2011) "Against Regular and Irregular Characterizations of Mechanisms," *Philosophy of Science* 78: 914–25.

Dowe, P. (2000) *Physical Causation*, Cambridge: Cambridge University Press.

Dretske, F. I. (1977) "Laws of Nature," *Philosophy of Science* 44: 248–68.

Garson, J. (2013) "The Functional Sense of Mechanism," *Philosophy of Science* 80: 317–33.

Glennan, S. (1996) "Mechanisms and The Nature of Causation," *Erkenntnis*, 44: 49–71.

—— (2002) "Rethinking Mechanistic Explanation," *Philosophy of Science*, 69: S342–S353.

—— (2010) "Ephemeral Mechanisms and Historical Explanation," *Erkenntnis* 72: 251–66.

—— (2011) "Singular and General Causal Relations: A Mechanist Perspective," in P.M.K. Illari, F. Russo and J. Williamson (eds) *Causality in the Sciences*, Oxford: Oxford University Press, pp. 789–817.

—— (forthcoming) *The New Mechanical Philosophy*, Oxford: Oxford University Press.

Hall, N. (2004) "Two Concepts of Causation" in J. Collins, N. Hall & L. Paul (eds) *Causation and Counterfactuals*, Cambridge, MA: MIT Press, pp. 225–76.

Harré, R. (1970) *The Principles of Scientific Thinking*, London: Macmillan.

—— (1972) *The Philosophies of Science: An Introductory Survey*, Oxford: Oxford University Press.

Kaiser, M. and C.F. Craver (2013) "Mechanisms and Laws: Clarifying the Debate," in H.K. Chao, S.T. Chen and R.L. Millstein (eds), *Mechanism and Causality in Biology and Economics*, Dordrecht: Springer, pp. 125–45.

Kitcher, P. (1993) "Function and Design," *Midwest Studies in Philosophy* 18: 379–97.

Lange, M. (2009) *Laws and Lawmakers: Science, Metaphysics, and the Laws of Nature*, New York: Oxford University Press.

Leuridan, B. (2010) "Can Mechanisms Really Replace Laws of Nature?," *Philosophy of Science* 77: 317–40.

Levy, A. (2013) "Three Kinds of New Mechanism," *Biology & Philosophy* 28: 99–114.

Machamer, P. (2004) "Activities and Causation: The Metaphysics and Epistemology of Mechanisms," *International Studies in the Philosophy of Science* 18: 27–39.

Machamer, P., L. Darden and C. F. Craver (2000) "Thinking about Mechanisms," *Philosophy of Science* 67: 1–25.

Mackie, J.L. (1974) *The Cement of the Universe*, Oxford: Clarendon Press.

Menzies, P. (2012) "The Causal Structure of Mechanisms," *Studies in History and Philosophy of Biological and Biomedical Sciences* 43: 796–805.

Mumford, S. (2004) *Laws in Nature*, London: Routledge.

Psillos, S. (2002) *Causation and Explanation*, Montreal: Acumen & McGill-Queens University Press.

—— (2004) "A Glimpse of the Secret Connexion: Harmonising Mechanisms with Counterfactuals," *Perspectives on Science* 12: 288–319.

—— (2007) "Causal Explanation and Manipulation," in J. Person and P. Ylikoski (eds) *Rethinking Explanation*, Boston Studies in the Philosophy of Science, vol. 252, Dordrecht, the Netherlands: Springer, pp. 97–112.

——(2011) "The Idea of Mechanism," in P.M.K. Illari, F. Russo and J. Williamson (eds) *Causality in the Sciences*, Oxford: Oxford University Press, pp. 771–88.

——(2014) "Regularities, Natural Patterns and Laws of Nature," *Theoria* 79: 9–27.

Reichenbach, H. (1956) *The Direction of Time*, Berkeley: University of California Press.

Salmon, W. (1984) *Scientific Explanation and the Causal Structure of the World,* Princeton, NJ: Princeton University Press.

——(1997) "Causality and Explanation: A Reply to Two Critiques," *Philosophy of Science* 64: 461–77.

Tooley, M. (1977) "The Nature of Laws", *Canadian Journal of Philosophy* 7: 667–98.

Waskan, J. (2011) "Mechanistic Explanation at the Limit," *Synthese* 183: 389–408.

Woodward, J. (2000) "Explanation and Invariance in the Special Sciences," *British Journal for the Philosophy of Science* 51: 197–254.

—— (2002) "What Is a Mechanism? A Counterfactual Account," *Philosophy of Science* 69: S366–S377.

—— (2003) *Making Things Happen: A Theory of Causal Explanation*, New York: Oxford University Press.

—— (2011) "Mechanisms Revisited," *Synthese* 183: 409–27.

12

WHAT WOULD HUME SAY?

Regularities, laws, and mechanisms

Holly Andersen

1. Introduction

Contemporary discussion of mechanisms sprang from many failures of laws to do what philosophers of science wanted them to. Yet the expectations at that time for what laws should be able to do were impossible to achieve. Is there still room for a better construal of laws in understanding explanation and the nature of the world as studied by science, or do mechanisms provide everything that is required for scientific explanations? How do genuine laws relate to mechanisms in a mechanistic worldview?

After a survey of the discussion around laws and mechanisms, I offer an original argument for two main ways in which laws and mechanisms can relate without either conceptually or ontologically reducing to the other, using Hume to illustrate. The first connection between laws and mechanisms involves the recognition of the "brute" character of laws, such that they can explain but are not themselves further explicable. This is illustrated by Hume's skeptical realism, where the world's hidden springs and secret principles are forever covered over from our epistemic access. The second connection between laws and mechanisms involves a unique role for laws via distinctively mathematical explanations, in which laws can provide a further degree of necessity, via mathematics, than can any collection of mechanisms, no matter how comprehensive. This is illustrated in terms of the evidence for such laws, which may include the sort Hume labeled "relations of ideas," rather than solely the type of evidence we have for mechanisms that Hume labeled "matters of fact."

2. A short recent history of laws and mechanisms

Mechanisms as a characteristic form of explanation were originally developed as an alternative to a then-dominant way of understanding explanation as necessarily involving laws. The deductive-nomological model of explanation stated that explanations involve subsuming the explanandum (that which is to be explained) under a general law. The explanatory work was done by the law: event A led to event B because A was an F, B was a G, and all Fs are Gs (e.g. Hempel 1962). The explanans derived its power from the nomologicity of the law that figured in it. As such, laws needed to be general, or ideally, universal in that they apply everywhere at all times. They also needed to be more than merely true descriptions of what did or will happen. They had to involve a stronger degree of necessity, and convey what must happen, given those laws. Nomologicity is thus a kind of necessity that is stronger than mere actuality—it is more than what does happen; and it is weaker than what is sometimes called metaphysical or logical necessity, which involves what must or could not happen regardless of the laws.

In this potted mini-history, it is incredibly important to emphasize that the notion of laws that served in deductive-nomological accounts of explanation were of a peculiar logical sort. Traditionally, laws have been taken to be part of the world itself, and descriptions of those parts of the world yielded statements that had a special status as descriptions. The term "law" would be equally applied to both the features of the world being described, and to the description itself. For instance, in the law F = ma, force, mass, and acceleration, as well as the necessary relationship between them, are all part of the natural world, and are a special part such that learning that such a relationship holds among those quantities allows us to make far more predictions than we otherwise could with individual cases of acceleration, mass, and force. The written mathematical form describing the relationship among the quantities, "F = ma," is also called a law. As such, the term law applied to both the nomological features of the world and equally to the linguistic devices such as mathematical formulas used to describe those features of the world.

Laws were taken to bear the mark of their necessity in their very syntactic structure, so that only the schematic structure of the law, rather than its actual content or meaning, was required to identify a genuine law from a merely true generalization. Much work in the twentieth century went into trying to find ways to represent laws as sentences in logical notation such that lawlike and non-lawlike statements were clearly differentiable by syntactic structure alone, without reference to actual content of the laws (e.g. Ayer 1956). Their idea was that laws in the world could be identified by their description. Genuine laws, and only laws, could be written out logically such that only the form of the sentence matter (all Fs are Gs) was required to identify such sentences as lawlike, without having to say anything about what the variables (the F and the G) stood for. This project failed, gruesomely (Goodman 1983). There is no unique logical structure of laws such that all and only laws, but no accidentally true generalizations such as "all the coins in my pocket are silver," have that structure.

This failure to find some logical structure unique to laws was fruitful in generating new ideas about what makes laws lawlike, and about other forms of explanation that don't require such peculiarly structured laws or deductive relationships. Laws in physics, even under best-case scenarios, do not resemble the logical creatures required for deductive-nomological accounts (e.g. Cartwright 1983). Unification of apparently disparate phenomena, for instance, is a historically well-substantiated form of explanation that does not rely on the logical notion of laws (Kitcher 1981). Explanations in the so-called special sciences were noted to rarely, if ever, use laws of the form presumed to be found in physics (Fodor 1974). Many causal explanations need not ever involve such laws (e.g. Woodward 2005; Strevens 2006).

This opened the way for mechanisms to gain attention as a key form that explanations take in certain sciences (e.g. Bechtel and Richardson 1993). Mechanisms differ from both the logically distinctive but mythical creatures sought for the D-N model of explanation, and from the best examples of actual laws found in physics and related disciplines. Mechanisms can be indefinitely local, particular, and contingent. They can proliferate in number, rather than be eliminated via reduction. They can serve explanatory roles for which laws, even if they were available, would be somewhat unsatisfactory and brute (Andersen 2011). Some authors, such as Glennan (1996), even argued that mechanisms could replace the characteristic kind of nomological necessity associated with laws.

It is important to recognize the heterogeneity of "mechanism" in current discussions for reasons of clarity and philosophical precision. Some have argued that there is a growing consensus on the character of mechanisms (see Chapter 1). Others have argued that the situation is more of a muddle than a consensus, where very distinct senses of the term mechanism are used without sufficient care as to the differences between them (Andersen 2014a, 2014b). Minimally, there are relevant differences in emphasis between approaches to mechanisms. Some (e.g. Craver and Darden 2013) concentrate on the details of scientific practice and the role that mechanisms play in structuring discovery and investigation. Others (notably Glennan 2010) generalize the notion to offer a mechanistic worldview strikingly similar to those of the early modern era, which is more of an alternative to atomism or process ontology than it is an alternative to specific explanatory practices in contemporary science. Some take mechanisms to *involve* the notion of causation (such as Bogen 2005 and Machamer 2004) while others take mechanisms to *be* causation (such as Glennan 1996, or for a different notion of mechanism, Salmon 1998). Each of these conceptions of mechanism involves a different relationship to modal notions like nomologicity, which in turn changes the details of how such mechanisms are related to laws.

To leave as little metaphysical space, as it were, for laws, I will take the broadest possible construal of mechanisms (Glennan 2010) to push the question of what if any role for laws remains in such a mechanistic worldview. In such a construal, explaining the often surprisingly simple regularities in the world involves describing them with laws and then fleshing out laws with supporting mechanisms that explain *why* such lawfully describable regularities hold. There are two views about the relationship between laws and mechanisms that might follow from this strongest construal of mechanism. On the first, laws are a useful step in describing regularities, but cannot be the last step, since laws themselves require explanation in terms of the mechanisms that support or sustain them. On the second view, mechanisms are a useful way to represent higher-level structures in the natural world, but they cannot themselves be fundamental, since if one were to go "all the way down" with mechanisms, there would still be lawful connections between the elements or stages of a mechanism, such that the fundamental explanatory role would require laws and their nomologicity, which would remain brute in the sense of providing explanations but being themselves not further explicable.

If we focus on physics as providing the more fundamental account of the natural world, laws are often taken to be required for mechanisms. If we focus on biological sciences, laws and mechanisms appear more like distinctive stages in the process of discovery, investigation, and explanation of phenomena of interest. It is a difficult and sometimes underappreciated task to even locate genuine regularities and describe them in a sufficiently precise fashion that they might qualify as a law in the first place (see also Mitchell 2000). Once a given regularity is identified and then described in terms of a law, though, that very law is itself a target for explanation. Why does this law hold? What conditions give rise to or sustain the regularity? Why do the relevant parameters have these value ranges rather than some other value ranges? Thus, a difference in initial emphasis between physics versus biology can yield a difference in how prominent laws versus mechanisms appear in terms of explanation and fundamentality.

3. Contemporary discussions of regularities, mechanisms, and laws

The term "regularity" is often used to highlight a feature of the world, so that it picks out regularly recurring patterns of behavior. Such regularities provide the grounds for individuating phenomena for description and investigation. Using a historical example, there is a regularity in the way in which heavy objects' velocities change after being dropped, such that Galileo labels it the Law of Odd Numbers (Cohen 1985). He found a surprising regularity that depended only on how long the object had been falling, and not, for instance, on the weight of the object, how large or small it was, and more. The regularity in question was formulated in terms of the amount by which the velocity increased between each time interval—it went up by 1 unit, then 3, then 5, and so on. This simple progression in the rate at which the velocity of dropped objects increased was surprising, generally observable for any object (for the right sorts of objects, anyway), and not merely a coincidence. The Law of Odd Numbers describes a particular regularity that we now know how to derive from the uniform acceleration of massive objects in a gravitational field. Galileo's version of the law is subsumed by Newton's version as a particular case of more general laws about mass, force, and acceleration. Newton's laws thus describe a different, more abstractly picked out, regularity, of which the regularity that Galileo described is a proper subset.

This example highlights how some degree of generalizability is required to qualify as a regularity. Regularities must be at least minimally regular, holding over some changes in a variety of background circumstances and potentially relevant parameters. When scientists move from identifying an intriguing regularity and calling the resulting description a law, there is a commitment to the belief that the observed regularity does not hold merely for the circumstances already observed, but *will* hold for unobserved and future circumstances. Importantly for the connection with mechanisms, we can be very mathematically precise about how an object described by the Law of Odd Numbers or Newton's laws will accelerate by knowing only a few parameters about the strength of the gravitational field, and without knowing anything about the way in which a gravitational field brings about acceleration. Thus, laws describe regularities in a way that commits to a certain scope of generalizability. Laws are regularities that *must* hold, rather than ones that happen to hold. It is this necessity that makes them especially useful in explanations and in understanding the underlying structure of the natural world.

However, generalizability does not necessarily require a law. There are many regularities in fields such as biology, sometimes expressed in mathematical terms, that are often taken to be generalizable and describable in mathematically precise ways. There are mathematical relationships regarding the distribution of traits across generations that hold across species and are taken to hold into the future as well. They must hold, with the necessity characteristic of laws rather than of accidental generalizations. Yet these regularities don't seem to be part of the fundamental structure of the world; if we take laws to reveal the fabric of the world, then these generalizations fail to be part of it. Beatty (1995), for instance, has argued that if we ran the tape of evolution again, we would end up with a radically different outcome. Such incredible contingency seems at odds with the idea of lawfulness. If Beatty were right, then on the then-current understanding of laws, nothing in biology (or at least, no products of evolution) could be explainable by recourse to nomological necessity. This raised a pointed question for philosophers of biology about the status of generalizations in biology versus physics.

Are there such regularities as would require laws in biology? And would the resulting laws be of the same type as what goes under the name of law in physics? Cartwright (1983) has argued persuasively that even physics doesn't have the kinds of laws for which philosophers of science went looking in the twentieth century. Rather than try to get biological regularities to look like laws in physics, Mitchell (1997, 2000) argues for laws that apply to both physics and biology. This involves rejecting the binary dichotomy between full laws and mere accidental generalizations. Pragmatic laws are generalizations that are stronger than mere accidental regularities, but not as strong as universal, exceptionless laws.

The strength of pragmatic laws can be measured along multiple distinguishable dimensions. The first dimension is stability: across what range of background conditions and parameter values does the generalization hold? A generalization that holds under a wider range of conditions is more stable, with the limiting ideal of the universal law that holds everywhere at all times. The stability of the conditions on which a law depends can vary, while nevertheless yielding generalizations that bear the right degree of necessity under the right conditions. The second dimension is strength: how strong is the necessity associated with the generalization, for the conditions under which it holds? At one extreme is the exceptionless law, which is never violated in its domain. But there may be generalizations that hold almost always that are used in a lawlike way, even though there are exceptions to them. A third dimension is degree of abstraction. Generalizations that are more abstract are those that ignore more of the details, to pull out some broader pattern across a wider range of concrete examples. Woodward (2010) makes a very similar point, but leaves laws behind in favor of causal generalizations.

This brings us to mechanisms. Mechanisms are often defined explicitly as providing explanations for regularities in nature: why those regularities occur in the conditions that they do, why they don't occur under other conditions, how the underlying entities and activities are organized and causally connected such that they produce as a final stage or give rise to and sustain the observed regularity for which an explanation is sought (e.g. Bechtel and Richardson 1993; Machamer, Darden, and Craver 2000). A model of a mechanism for a given regularity may provide the ability to make predictions about what would happen under new conditions.

This highlights one complementary role that laws and mechanisms can play in science. While laws describe regularities, mechanisms may go further and explain them. It is a difficult, and often underappreciated, task to precisely describe a regularity. Finding the right variables, the right way to relate them, and the right way to delineate the conditions of application, in the form of a law that is both accurate and can be used to make precise predictions, is itself a major scientific achievement. To say that laws describe regularities is not to diminish the work that laws can do. The task of describing via a law is not trivial; it is a huge breakthrough that is often itself a prerequisite for further work on that phenomenon. Investigations into the mechanisms responsible for a lawful regularity can yield the mechanisms that sustain, produce, or give rise to the regularity. Mechanisms explain why that regularity holds as it does, and how it is that instances of that regularity occur when and how they do.

It is vital to note that there need not be one mechanism = one law equivalence. Multiple distinct kinds of mechanisms may give rise to one lawful regularity. It might be that a particular law, as the precise description of a recurrent regularity that involves some degree of necessity, holds because of several distinct mechanisms operating under different circumstances. Conversely, there may be many different laws that turn out to involve the same underlying mechanism in different contexts. Maxwell's laws illustrate both directions of this. According to Morrison (2007), the formalism Maxwell used allowed him to provide quite general laws without specifying the variety of physical causes that might be involved in any given instance to which the laws apply, allowing for multiple mechanisms underlying one law. At the same time, Maxwell's laws unified apparently disparate phenomena of electricity and magnetism, such that several older laws could be subsumed in terms of unified mechanisms of electromagnetism (e.g. p. 64).

Regularity can mean recurrent patterns of behavior that can be identified as phenomena of interest. There are also further notions of regularity that appear in discussions of mechanisms, focused on how tightly bound the stages of a mechanism are to count as a mechanism. Bogen (2005) has argued that regularity should not be a requirement on counting as a mechanism. Machamer, Darden, and Craver define mechanisms specifically in terms of regularity. "Mechanisms are regular in that they work always or for the most part in the same way under the same conditions" (2000, p. 3). Bogen (2005) argues that there are mechanisms (or what ought to count as mechanisms) that do not meet the requirement of "almost or for the most part."

Bogen invokes the example of the release of neurotransmitter vesicles given the stimulation of the neuron. Only about 10 percent of vesicles, in the right start-up conditions, actually release neurotransmitter. According to the MDC definition, the case should not count as a mechanism, since the "always or for the most part" condition is not met. Yet this release is still a key part in the mechanism for transmitting a signal across the synapse. How tightly bound together must a series of causal processes be to count as a mechanism, rather than a mere coincidental collection of nearby interactions? Bogen's aim is to push us toward a lower degree of regularity, such that the 10 percent counts as regular enough under the circumstances.

It is helpful to distinguish two possible loci of regularity at issue here: first, there is the degree to which a regularity exists in the world, as a phenomenon that could stand as the target of explanation; second, there is the degree of regularity within the mechanisms that explains such a phenomenon, where different organizational stages might have different probabilistic degrees of connection (see Chapter 13). I have responded (2012) to Bogen's challenge by arguing that mechanisms must be considered as at least minimally regular, in the sense of not being one-off causal chains, to be the target of *scientific* (rather than historical, for instance) explanation. If mechanisms are to do explanatory work for individual occurrences of a mechanism, it must be by dint of situating that individual instance as a member of a type of occurrence. A taxonomy of different degrees of regularity, and locations of different organizational stages of the mechanism, can convey a great deal of information about mechanism(s), and provide additional phenomena requiring further explanation. For example, the 10 percent figure for vesicle neurotransmitter release is part of a larger, embedding mechanism that essentially uses the 10 percent rate to calibrate neurotransmitter levels to reduce noisy synapse firings. Thus, the purportedly irregular 10 percent release rate is a consistent regularity in the other sense, that of a recurrent pattern of behavior requiring explanation for which a mechanism can be sought.

Notice how important the issue of generalization has been for construing the relationship between mechanisms and regularities. The very idea of a regularity contains within it the notion of recurrence: a singular event, that will only happen once, cannot be a regularity, and thus cannot be explained as a regularity. Insofar as it is situated within a group of other possible instances, even if they are only merely possible instances, it is already being treated as a kind of regularity.

Thus far, the relationship between laws and mechanisms has been more complementary than competitive. There is one contemporary debate that does pitch mechanisms and laws as competitors in the explanation business. A mechanism can explain why a law holds, and a law can connect the stages within a mechanism. Which is more explanatorily fundamental, mechanisms or laws? Suppose some regularity is identified and described as lawful. Further investigation turns up a mechanism that explains how that lawful regularity holds. But this mechanism is composed of organized entities and processes that are themselves lower-level lawful regularities. The mechanism that explained the original law requires further laws for its own operation. And each such law might be further explained by a mechanism, and the stages of *that* mechanism must involve laws to connect them, and so on downwards.

Where does this end? There are two main options. It could terminate with laws of physics at the most fundamental levels, such that there are no further mechanisms that could be posited to underlie the laws. They would be brute, in that they could explain but not be themselves explained; they would simply hold of the world. On the other option, even the laws of physics could themselves be mechanistically explained, such that mechanisms would be ontologically and explanatorily fundamental.

This framing of the question has several notable features, regardless of the eventual answer. The first is that it emphasizes the ontological or even metaphysical aspects of the question. It is not merely a question of what explains what. The explanatory consequences follow from ontological priority. A second feature is that the framing presupposes that either laws, or mechanisms, but not both, are ontologically primitive. This puts laws and mechanisms in a kind of explanatory competition, where it must be the same explananda that both endeavor to explain, but where there is only room for one genuine explanation, and only these two options on the table. A third feature is that it lacks criteria by which this question would be adjudicated. If string theory, for instance, turns out to provide the material for a grand unified theory, are the core features of the world that it postulates mechanisms or laws? How much change to the notion of mechanism would be required to accommodate such a heavily mathematical theory? At some point in stretching the meaning of mechanism to fit mathematical models, it simply can't be the same kind of mechanisms as are posited to explain the firing of a synapse (although a different conclusion is reached by Kuhlmann and Glennan 2014).

Finally, raising the question in terms of *laws* versus mechanisms elides the issue of *causation* versus mechanisms, or even in terms of nonmechanistic causation versus mechanistic causation. Mechanisms and counterfactuals are also taken to be opposing potentially ultimate categories (see especially Psillos 2004; Bogen 2005; Machamer 2004; Woodward 2005; Glennan 2010; Chapters 10 and 11 in this volume). The opposition is strikingly similar: counterfactuals govern the relationships between elements or stages in a mechanism, but those counterfactuals can be cashed out with yet lower-level mechanisms, and so on.

In these discussions of regularities, there is a kind of modal bump in the rug. This bump is the "oomph" that has been associated with causation, or the nomologicity that has been associated with laws, or the intricate architecture of a mechanism. The bump in the rug can be shifted, depending on other philosophical considerations, to be located at laws, causation, or mechanisms, but it has proven remarkably hard to just stomp it flat. When some regularity holds and we have reason to think that it would hold if poked and prodded in various ways, we need a way to capture the extra content that goes above and beyond merely describing what actually happened. This is not a merely semantic point: if some particular event really did have to happen a certain way, rather than merely happening to happen, leaving that out would be an incomplete description. Insofar as science is in the business of working toward not merely accurate but also complete descriptions of various parts of the natural world, it must be able to note these modal characteristics in a way that accurately portrays the degree of connection.

Where Mitchell offered pragmatic laws to do this work in fields such as biology, others such as Woodward (2010) offer very similar considerations, including scope, specificity, and stability, for causal generalizations in biology. Cartwright (2002) notes how discussions of explanation and scientific theories tried to eschew talk of causation by using talk of laws, and how that pendulum has now swung back to causation. Lewis' account of causation is based on regularities; Salmon's is based on mechanisms. Glennan (1996) also turns to mechanisms for causation.

There is an incredibly close link between laws and causation as the two leading candidates to account for the degree of connection or necessity beyond mere accident that we find in many of the most interesting generalizations in the sciences. We can attribute this necessity to causation, and cash it out one way, or to laws, and cash it out a different way. In this regard, either laws or causation, but not both, are required. This is not a tension per se, in that it needn't mean that *only* one of laws or causation are required (for instance, the idea of causal capacities or powers involve both; see Chapter 10). But it tends to go along with relying on one or the other to do the work of accounting for necessity and connection.

4. Two Humean roles for laws

We've now explored several subtle connections between laws and mechanisms as ways of explaining regularities in the world that bear some degree of necessity. In this section, I will lay out a schematic argument for two roles unique to laws that cannot be assimilated to mechanisms. Taking the broadest possible construal of mechanisms, and a weak construal of laws, consider: can mechanisms, if construed maximally broadly, do all the work that we wanted from laws, or is there still a role left for laws no matter how broadly one construes mechanism? This question is both perennial, in that it has arisen in a number of forms in philosophy since the early modern period, and Humean, in that it often arises anew in discussions of well-loved passages from Hume. I rely on Hume as a springboard for making the case for two new ways in which laws and mechanisms could relate, since both appear in his work. This is not Hume exegesis: I take it that no unambiguous answer can be given to the question of what Hume himself would actually say. Instead, this is a riff on Humean-style answers as a way of illustrating the point.

As Beebee (2006) has persuasively argued, there are at least two viable ways of understanding what Hume says about causation, and no further definitive answers about which is what he "really" meant (if we are even willing to assume Hume had a single consistent view across all his writings). One interpretation developed by Beebee is that Hume was a skeptical realist. The secret connection, which might bind cause and effect or primary and secondary quality, the connection that is tracked by the idea of force and purportedly conveys necessity along its chain—to a skeptical realist, such a secret connection exists, but the world is such that we are in principle barred from ever gaining genuine epistemic access to it. It *exists*, but we can't *know* it.

> It is confessed, that the utmost effort of human reason is to reduce the principles, productive of natural phenomena, to a greater simplicity, and to resolve the many particular effects into a few general causes, by means of general reasonings from analogy, experience, and observation. But as to the causes of these general causes, we should in vain attempt their discovery; nor shall we ever be able to satisfy ourselves by any particular explication of them. These ultimate springs and principles are totally shut up from human curiousity and enquiry.
>
> *(Hume [1748] 2007, p. 17)*

To the skeptical realist, laws may serve as the "the few general causes," the most general explanation that science may reach. These laws describe the phenomena that are produced by the "ultimate springs and principles." Yet the springs and ultimate principles themselves, the mechanisms producing those laws, are "totally shut up" from our investigations, off limits to knowledge.

In other passages, Hume denies that any amount of knowledge of what we would now call mechanisms is sufficient to discern what might happen with further instances prior to actual observation.

> Our senses inform us of the colour, weight, and consistence of bread; but neither sense nor reason can ever inform us of those qualities which fit it for the nourishment and support of a human body . . . . The bread, which I formerly eat, nourished me; that is, a body of such sensible qualities was, at that time, endued with such secret powers: but does it follow, that other bread must also nourish me at another time, and that like sensible qualities must always be attended with like secret powers? The consequence seems nowise necessary.
>
> *(Hume [1748] 2007, p. 33)*

Students encountering this passage for the first time often wonder what Hume would say in the face of modern science. Once we know the mechanisms by which chemicals interact with the microscopic processes in our digestive tracts, wouldn't we know whether a new piece of bread, or even an entirely new foodstuff given straight to scientists before eating, would harm us or nourish us? There is an intuitive appeal to the idea that this is a problem on which actual progress has been made by science—the secret powers of food are not so secret anymore.

But Hume himself considers this question. Experience has led us to understand some mechanisms about nourishment, certainly; but even those rest on further sensible qualities that ultimately must be supported by the hidden powers and secret connections that Hume is challenging. Food science merely defers the problem. It does not and cannot solve it.

One needn't even commit to the realist part of skeptical realism, leaving laws as sheerly brute. If there are no such hidden springs, the result is still skepticism that precludes any possible mechanistic explanation of laws. Thus, the first Humean role for laws is an especially poignant one. It allows for the possibility that there *are* mechanisms underneath the fabric of the universe, and it is these mechanisms that give the modal oomph to those laws by providing the secret connection. Yet they remain out of reach, if they even exist. The laws that are the last description of regularity before those hidden springs will remain brute, in that they can explain, but cannot be themselves explained. Even in the face of massive amounts of contemporary human knowledge of mechanisms, these ultimate regularities cannot be assimilated to them.

The second Humean role for laws that cannot be assimilated to mechanisms stems from the evidence for relations of ideas versus for matters of fact. The first role for laws, just discussed, is squarely within the realm of matters of fact: facts about regularities that are described lawfully, resolved into mechanisms, but ultimately come up against the opacity of the hidden springs. The second role for laws involves their status as mathematical relationships that fall at least partially under the category of relations of ideas and thus have additional evidentiary support compared to matters of fact.

Hume very famously distinguishes between knowledge in terms of the target of inquiry.

> All the objects of human reason or enquiry may be naturally divided into two kinds, to wit, *Relations of Ideas*, and *Matters of Fact*. Of the first kind are the sciences of Geometry, Algebra, and Arithmetic; and in short, every affirmation which is either intuitively or demonstratively certain. . . . Matters of fact, which are the second objects of human reason, are not ascertained in the same way; nor is our evidence of their truth, however great, of a like nature with the foregoing. The contrary of every matter of fact is still possible; because it can never imply a contradiction.
>
> *(Hume [1748] 2007, p. 25)*

He considers the kind of evidence we could have for the truth or falsity of claims about matters of fact versus relations of ideas. Since the contrary of any factual claim is not contradictory, we cannot use contradiction as a guide to truth and falsity for these claims, but must rely on evidence of the senses. "All reasonings concerning matter of fact seem to be founded on the relation of *Cause and Effect*. By means of that relation alone we can go beyond the evidence of our memory and senses" (ibid., p. 70). And of course, reasoning based on cause and effect is ultimately founded on sheer habit. It is in this gap between what we infer about matters of fact, and that from which we infer it, that Humean skepticism arises.

In contrast, relations of ideas have an entirely different evidentiary status. They do not ultimately rest on mere habit. They do not admit of the skepticism about knowledge to which matters of fact are subject. Claims about relations of ideas can be known with certainty and assurance, because their contraries are contradictions and therefore impossible. The claim I am offering here is that there is at least the possibility for mathematical laws to have some evidential support that is of the form of relations of ideas, and thus not reducible to mechanisms (do note this is an original argument about how to apply this distinction to laws, not an existing view in Humean scholarship).

Consider laws that are formulated mathematically (setting aside non-mathematical laws for now). Laws that are supported by evidence that is at least partially mathematical in character will have a different status than those based purely on matters of fact, even if the other part of the evidential support is drawn from matters of fact. Laws can be used in ways that rely on their mathematical features to draw conclusions that involve a markedly higher degree of necessity than mere nomologicity can convey, even though those same laws, used some other way, behave in a traditional way, conveying nomological but not mathematical necessity (for instance, see Lange 2013 and Andersen 2016).

Consider what "relations of ideas" evidence might be available for a candidate mathematical law. Insofar as a law is derived from other mathematical laws, plus additional mathematical machinery, some of the evidence for the new law taking the form that it does is drawn from those mathematical relationships. Not all of the evidence for such a law is. Any law with genuine physical content will require at least some evidence of matters of fact. The point I want to make is that some of those laws may *also* have additional evidentiary support from relations of ideas, and that such additional support is not even potentially available for mechanisms.

An intriguing potential example of this is leading versions of string theory, where this second Humean role for laws can account for why string theory is even being pursued as a viable physical theory despite the well-known lack of empirical confirmation. There is widespread agreement that string theory is not supported by empirical evidence, since it is extraordinarily difficult to even find ways to derive empirical predictions from it. In other words, string theory lacks evidence of matters of fact. Why is it even being considered as a potential theory, much less a fundamental one? The mathematical structures themselves, and the ways in which some mathematical relationships emerge as lawfully governing any such structure in the world, provide the kind of evidence that physicists find sufficiently compelling to continue working on it. This evidence is in the category of relations of ideas.

There is thus a philosophically pessimistic and a philosophically optimistic role for laws to play that cannot be assimilated to mechanisms, no matter how broadly construed. Each of these two roles have a distinctively Humean flavor. Pessimistically, laws might be brute and not further explicable; we can never know if mechanisms behind those laws even exist, much less what they are. Optimistically, laws can play a unique role by dint of mathematical relationships: this yields mathematical necessity as part of the nomologicity of laws, and an additional potential source of evidence, relations of ideas, that is not susceptible to inductive skepticism.

5. Conclusion

It is helpful in many contemporary discussions involving laws, mechanisms, regularities, and even causation to consider the recent trajectory of these ideas since the mid-twentieth century. The idea of a law was supposed to do an enormous amount of work in the development and use of scientific theories. Peculiarities in views about the nature of language painted philosophers into a corner. Laws were expected to shoulder the burden of explanation and to clarify nomologicity in logical terms. This impossible task failed, in interesting ways, not least of which was that it cleared the ground for mechanisms to surge as a locus of research for philosophers of science trying to capture the investigatory and explanatory practices of sciences like biology.

Yet alternative construals of laws, including but not limited to Mitchell's pragmatic laws, continue to offer something unique in terms of capturing the right degree of necessity associated with many scientific claims. I have argued here for two new ways to think about the relationship between mechanisms and laws. One is pessimistically Humean, where there may or may not be ultimate mechanisms under the very fabric of the universe which give rise to the laws, but which, if they even exist, are in principle locked away from us. The second is optimistically Humean, where mathematically formulated laws can offer a way to evade inductive skepticism at least partially, by relying on evidence drawn from mathematics, or from the relations of ideas rather than matters of fact.

References

Andersen, H., 2011. Mechanisms, laws, and regularities. *Philosophy of Science*, 78(2), pp. 325–331.

Andersen, H., 2012. The case for regularity in mechanistic causal explanation. *Synthese*, 189(2), pp. 415–432.

Andersen, H., 2014a. A field guide to mechanisms: Part I. *Philosophy Compass*, 9(4), pp. 274–283.

Andersen, H., 2014b. A field guide to mechanisms: Part II. *Philosophy Compass*, 9(4), pp. 284–293.

Andersen, H., 2016. Complements, not competitors: Causal and mathematical explanations. *The British Journal for the Philosophy of Science*, doi:10.1093/bjps/axw023.

Ayer, A.J., 1956. What is a law of nature?. *Revue Internationale de Philosophie*, 7, pp. 144–165.

Beatty, J., 1995. The evolutionary contingency thesis. In G. Wolters and J.G. Lennox (eds.), *Concepts, Theories, and Rationality in the Biological Sciences*. Pittsburgh: University of Pittsburgh Press. pp. 45–81.

Bechtel, W. and Richardson, R.C., 1993. *Discovering Complexity*. Princeton, NJ: Princeton University Press.

Beebee, H., 2006. *Hume on Causation*. London: Routledge.

Bogen, J., 2005. Regularities and causality: Generalizations and causal explanations. *Studies in History and Philosophy of Science Part C: Studies in History and Philosophy of Biological and Biomedical Sciences*, 36(2), pp. 397–420.

Cartwright, N., 1983. *How the Laws of Physics Lie*. Oxford: Oxford University Press.

Cartwright, N., 2002. In favor of laws that are not ceteris paribus after all. In *Ceterus Paribus Laws*. Dordrecht, the Netherlands: Springer. pp. 149–163.

Cohen, I.B., 1985. *The Birth of a New Physics*. WW Norton & Company.

Craver, C.F. and Darden, L., 2013. *In Search of Mechanisms: Discoveries across the Life Sciences*. Chicago: University of Chicago Press.

Fodor, J.A., 1974. Special sciences (or: the disunity of science as a working hypothesis). *Synthese*, 28(2), pp. 97–115.

Glennan, S., 1996. Mechanisms and the nature of causation. *Erkenntnis*, 44(1), pp. 49–71.

Glennan, S., 2010. Mechanisms, causes, and the layered model of the world. *Philosophy and Phenomenological Research*, 81(2), pp. 362–381.

Goodman, N., 1983. *Fact, Fiction, and Forecast*. Cambridge, MA: Harvard University Press.

Hempel, C.G., 1962. Deductive-nomological vs. statistical explanation. *Minnesota Studies in the Philosophy of Science*, 3, pp. 98–169.

Hume, D. [1748] 2007. *An Enquiry Concerning Human Understanding*. Fq Classics.

Kitcher, P., 1981. Explanatory unification. *Philosophy of Science*, 48(4), pp. 507–531.

Kuhlmann, M. and Glennan, S., 2014. On the relation between quantum mechanical and neo-mechanistic ontologies and explanatory strategies. *European Journal for Philosophy of Science*, 4(3), pp. 337–359.

Lange, M., 2013. What makes a scientific explanation distinctively mathematical?. *The British Journal for the Philosophy of Science*, 64(3), pp. 485–511.

Machamer, P., 2004. Activities and causation: The metaphysics and epistemology of mechanisms. *International Studies in the Philosophy of Science*, 18(1), pp. 27–39.

Machamer, P., Darden, L. and Craver, C.F., 2000. Thinking about mechanisms. *Philosophy of Science*, 67, pp. 1–25.

Mitchell, S.D., 1997. Pragmatic laws. *Philosophy of Science*, 64, pp. S468–S479.

Mitchell, S.D., 2000. Dimensions of scientific law. *Philosophy of Science*, 67(2), pp. 242–265.

Morrison, M. 2007. *Unifying Scientific Theories: Physical Concepts and Mathematical Structures*. Cambridge: Cambridge University Press.

Psillos, S., 2004. A glimpse of the secret connexion: Harmonizing mechanisms with counterfactuals. *Perspectives on Science*, 12(3), pp. 288–319.

Salmon, W.C., 1998. *Causality and Explanation*. Oxford: Oxford University Press.

Strevens, M., 2006. *Depth: An Account of Scientific Explanation*. Cambridge, MA: Harvard University Press.

Woodward, J., 2005. *Making Things Happen: A Theory of Causal Explanation*. Oxford: Oxford University Press.

Woodward, J., 2010. Causation in biology: Stability, specificity, and the choice of levels of explanation. *Biology & Philosophy*, 25(3), pp. 287–318.

13

PROBABILITY AND CHANCE IN MECHANISMS[1]

Marshall Abrams

Introduction

Though many authors recognize that mechanisms can involve stochasticity, there's been little discussion about the nature of the stochastic relationships in mechanisms. I try to elucidate some issues that arise with stochastic mechanisms and propose new ways of thinking about them. I'll focus mainly but not exclusively on what I call *recurrent mechanisms*—token mechanisms that operate in a similar manner at different times, or mechanism types that have many instances that operate in a similar manner. Scientists' uses of "mechanism" seem to refer primarily to these mechanisms, which have also been the main focus of recent discussions of mechanistic explanation by philosophers.

Probability-related terminology varies greatly. I'll use "stochastic" for anything that seems at least vaguely probabilistic, or chaotic, or erratic, or merely not guaranteeing specific outcomes, etc. "Probability," "probabilistic," "chance," and "random" will refer to things that involve a particular probability distribution, whether known or not. Thus, random behavior is always stochastic, as is erratic behavior (defined below); the converse doesn't hold.

I begin in section 1 with a description of part of the mechanism of bacterial chemotaxis, which provides an illustration I use throughout the chapter, and then make some general

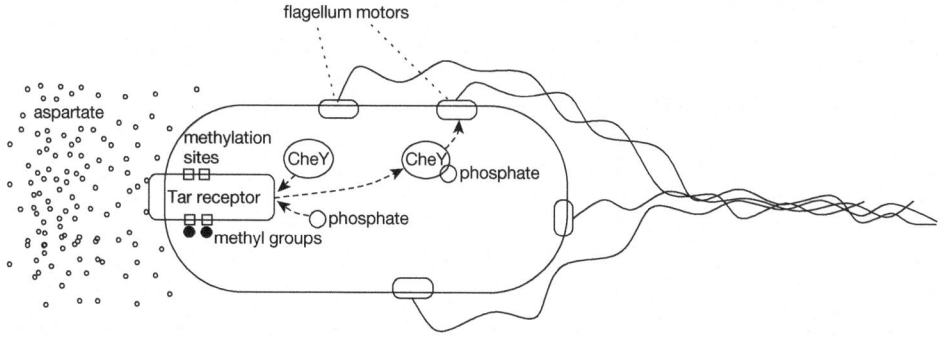

Figure 13.1   Schematic diagram showing some elements (entities and activities) involved in *E. coli* chemotaxis. Flagella are in their "pushy" state. See text

169

remarks about roles for probability in mechanisms. The rest of the chapter is somewhat arbitrarily divided into two sections (sections 2 and 3), focusing on metaphysical and episte-mological issues. Under metaphysics I include a brief general discussion of interpretations of probability, and a subsection in which I argue that those interpretations of probability relevant to the functioning of mechanisms are usually what I call *causal probability interpretations*. I also suggest that some stochasticity in mechanisms might not involve probabilities per se, but could instead involve what are known as imprecise probabilities. The epistemology section discusses various strategies for modeling probabilistic mechanisms, as well as evidence for probabilities in mechanisms. I conclude in section 4 with a discussion of non-recurrent mechanisms.

1. Chemotaxis in *E. coli*

The following simplified sketch of *Escherichia coli*'s chemotaxis mechanism[2] will help to clarify ideas below and suggest their connection to scientific practice (see Figure 13.1). Unfamiliar terms in this section can be treated as names for roles I'll describe.

Each *E. coli* bacterium has several flagella that allow it to swim toward beneficial substances and away from detrimental substances by sensing chemical gradients in water. In what I'll call *pushy* motion, flagella push in toward the body and soon wrap together into a bundle at the rear of the bacterium, pushing it forward. In *tumbly* motion, flagella turn in the other direction, pulling away from the cell membrane and causing the bacterium to tumble stochasti-cally. These two states alternate roughly once per second. By swimming for longer periods when pointed in the direction of increasing aspartate, for example, and swimming for shorter periods otherwise, the bacterium makes overall progress toward greater concentrations of aspartate. This is a good strategy because bacteria are so small that Brownian motion repeat-edly knocks them off course. The duration of pushy motion is controlled by several receptors embedded in an *E. coli*'s cell membrane.

For example, the Tar receptor responds to gradients in aspartate concentrations. Tar recep-tors in what's called the "active" state facilitate the phosphorylation of CheY molecules. Phosphorylated CheY molecules can bind to structures on a flagellar motor, increasing the chance that the flagellum will switch from its default pushy rotation to tumbly rotation. An active Tar receptor thus reduces the probability that directional movement will persist. The probability that a Tar receptor is in the active state is in turn a function of (a) the number of aspartate molecules bound to external sites on the receptor, and (b) the number of methyl groups bound to methylation sites on the internal side of the receptor. When there is a balance between the effects of bound aspartate molecules and effects of filled methylation sites, the receptor is more likely to shift to its active, CheY-phosphorylating state. An excess of bound aspartate molecules makes the receptor more likely to shift to its inactive state. Because bound methyl groups accumulate somewhat slowly, a balance between the two kinds of bindings means that there hasn't been a recent increase in the number of bound aspartate molecules on the outer side of the receptor. (I won't discuss the process by which methyl groups are removed from the receptor.) In sum, when there is an increase in the number of aspartate bind-ings, this initially prevents the active state, allowing pushy rotation to persist. However, unless there is a continuing increase in the number of aspartate bindings, methylation will gradually counteract effects of the unchanging aspartate bindings, tending to cause the receptor to switch to its active state, and increasing the chance of the flagella switching to tumbly rotation. All of these processes seem to be stochastic: given a certain number of aspartate bindings, bound methyl groups, etc., whether and when these processes occur varies from instance to instance, even within a single bacterium.[3]

Stochastic mechanisms

Machamer, Darden, and Craver (2000) (MDC) defined "mechanism" in this way:

> Mechanisms are entities and activities organized such that they are productive of regular changes from start or setup to finish or termination conditions.
>
> *(Machamer et al. 2000, p. 3)*

Roughly, entities are things—Tar receptors, CheY molecules, etc.—and activities are what they do—phosphorylation, movement of CheY molecules, etc. The activities associated with an entity can involve changes in that entity, as in the case of flagellar rotation, or in other entities, as in the effect of a methyl group on the state of the Tar receptor to which it is bound.

MDC also wrote that:

> Mechanisms are regular in that they work *always or for the most part* in the same way under the same conditions.
>
> *(Machamer et al. 2000, p. 3, emphasis added)*

MDC's discussion implies that activities whose setup conditions occur have (at least) a very high probability of occurring. In the case of the *E. coli* chemotaxis mechanism, we might think of the setup condition for the mechanism as an increase in the number of aspartate molecules binding to it, resulting in the Tar receptor remaining in its "inactive" state for longer.

It's now pretty clear that MDC-like views must be modified to allow for activities with probabilities far from 1.[4] The question of how probable activities must be was partially motivated by questions about whether natural selection counted as a mechanism (see Chapter 22), but some of the examples that have been central to philosophical discussion of mechanisms in recent years turned out to be probabilistic (or at least stochastic). For example, Craver (2007, p. 68), discussing a mechanism that leads to a certain kind of increase in neuronal efficiency, notes that it seems to be successful no more than about 50 percent of the time. My discussion of *E. coli* below provides another illustration. (I'll emphasize causation as relevance, i.e. difference-making, rather than production simply because probability is not only about what actually happens, but about what might happen, and because, as we'll see, scientific investigation of probabilistic mechanisms routinely involves relevance.)

I assume that scientists have some flexibility in what they treat as entities and activities of particular token mechanisms: Scientists model mechanisms, and different models can characterize different components of the same token mechanisms, or do so with different degrees of approximation (see Chapter 17). Claims about probabilities in a mechanism are then relative to a model that picks out certain ways of decomposing it. Since my focus is primarily on probabilities in the mechanism itself—i.e. in the world—I'll treat models as specifying aspects of the world about whose probabilities we then inquire.

Activity probabilities

Given a model of a mechanism that specifies its entities and activities, we can ask about the probability of an activity producing certain changes in entities; I'll call such changes *outcomes*, and I'll call this probability an *activity outcome probability*, or more simply an *activity probability*. A prima facie different question concerns the probability of one activity

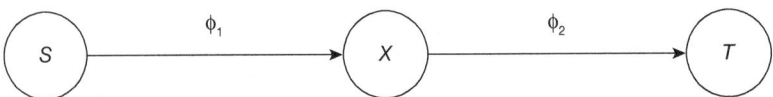

Figure 13.2 Three entities *S*, *X*, and *T*, and two possible activities ϕ_1 and ϕ_2 that might produce changes in *X* and *T* when caused to occur as a result of changes in *S* and *X*, respectively

occurring rather than another. This probability of activity occurrence may be equivalent to a probability of a change in an entity, or the probability of multiple entities interacting in a particular way. For the sake of a simpler, more unified discussion below, if there are probabilities for alternative activities ϕ_i to occur, I'll treat the process that determines which ϕ_i it is that occurs as itself an activity. Then the occurrence of one of the different ϕ_i's is this activity-determining activity's outcome, and we can talk about its own activity probability. (Where there is only a single activity that may or may not occur, we can treat its non-occurrence as a kind of null activity.)

I'll restrict my focus to activity probabilities in the preceding sense, but some authors discuss the probability of a mechanism as a whole exhibiting a particular behavior or phenomenon (e.g. DesAutels 2015; Andersen 2012; Krickel forthcoming). This is the probability of exhibiting particular termination conditions given particular setup conditions. However, the probability that a mechanism produces termination conditions given setup conditions reduces to activity probabilities and facts about the mechanism's structure. Consider, for example, a simple mechanism with (a) startup conditions represented by variable *S* taking the value *s*, (b) an entity *X* in one of two states x_1 or x_2, and (c) two possible termination conditions t_1 and t_2, represented as values of a variable *T*. Suppose there are two activities ϕ_1 and ϕ_2 that may or may not occur (see Figure 13.2). If we simplify part of the *E. coli* chemotaxis mechanism described above, the result would be a mechanism of this form: Let $S = s$ represent the presence of an additional methyl group at one of a Tar receptor's methylation sites, $X = x_1$ represent the Tar receptor remaining in its "active" conformation, and $T = t_1$ represent a flagellar motor with a bound CheY molecule. ϕ_1 could then represent the processes by which the Tar receptor changes from its active shape to inactive shape, and ϕ_2 could represent the complex activity consisting of a CheY molecule becoming phosphorylated, moving from the Tar receptor to a flagellar motor, and binding to the motor.

If we assume that both activities will occur, the probabilities in this model are as follows:

1) For ϕ_1: *S* being in state *s* results in *X* being in either state x_1 or state x_2 with probabilities $P(X = x_1 \mid S = s)$ and $1 - P(X = x_1 \mid S = s)$, respectively.
2) For ϕ_2: *X* being in one of its states results in one of two termination states t_1 or t_2. The probability of t_1 is either $P(T = t_1 \mid X = x_1)$ or $P(T = t_1 \mid X = x_2)$, depending on whether *X* is in state x_1 or x_2 (independent of *S*'s state). The corresponding probabilities of t_2 are $1 - P(T = t_1 \mid X = x_1)$ and $1 - P(T = t_1 \mid X = x_2)$.

Then the probability $P(T = t_1 \mid S = s)$ that T is in termination state t_1, rather than t_2, given that the mechanism begins operation (i.e. $S = s$), is the sum of the probabilities of the internal pathways that can lead to T being t_1. For example, the probability that T ends up in t_1 via *X* being in state x_1 is the product of the probability that *X* comes to be in x_1 and the probability of t_1 given x_1: $P(T = t_1 \mid X = x_1)P(X = x_1 \mid S = s)$. The probability of the mechanism as a whole exhibiting end state $T = t_1$ given startup condition $S = s$ is:

$$P(T = t_1 \mid S = s) =$$
$$P(T = t_1 \mid X = x_1)P(X = x_1 \mid S = s) + P(T = t_1 \mid X = x_2)P(X = x_2 \mid S = s) \qquad (13.1)$$

This simply follows mathematically from the above assumptions. Analogous computations for more complex mechanisms will, of course, be different, but the fact remains that *probabilities of termination conditions given setup conditions are a function of activity probabilities*. Thus, focusing solely on activity probabilities below will allow us to understand setup-to-termination probabilities as well. Exceptions, mentioned below, concern cases in which an interpretation of probability defines higher-level probabilities in ways that allow some variation in lower-level probabilities (Abrams 2015), or when there are activities affecting the mechanism that are not considered to be part of that mechanism.

I also discuss probabilities involved in constitutive relationships below.

2. Metaphysics

If it's not necessary that mechanisms "are regular in that they work always or for the most part in the same way under the same conditions" (Machamer et al. 2000, p. 3), then what is the difference between what are usually called mechanisms in science, and mere happenstance? I won't be able to do justice to several very valuable proposals. For example, Andersen (2012) proposes that mechanism-hood may be a matter of degree, depending, for example, on the frequency or probability of certain activity outcomes. Baetu (2013) argues that to be understood as a mechanism, a system must at least be regular enough—in the sense that activities occur and succeed with a certain relative frequency—that repeated interventions are possible. Krickel (forthcoming) argues that the regularity of an overall mechanism must be such that the phenomenon it explains occurs as a result of its operation more often than other outcomes, or that the phenomenon is brought about more often by that kind of mechanism than by others. I suspect that the degree and kind of regularity needed for mechanism-hood may also depend on other aspects of research contexts (cf. Abrams 2012a), even if it turns out that as a general rule of thumb, larger probabilities explain better than smaller probabilities (Strevens 2000).

Nevertheless, I think that we can get some insight about the nature of stochastic mechanisms by asking questions about the nature of activity probabilities. Are there constraints on the character of activity probabilities in order for a system to count as a mechanism—even if some probabilities represented in models of mechanisms need not correspond to reality?

Determinism and interpretations of probability

It will be helpful to begin with a few points about interpretations of probability. Mathematical probability is defined by a set of axioms (e.g. Grimmett and Stirzacker 1992), and anything that satisfies these axioms counts as probability. At a minimum, we need a set Ω of basic elements and a set of subsets ("outcomes") of Ω closed under unions and intersections. A probability measure P is any function that assigns 1 to Ω, 0 to the empty set, and such that $P(A \cup B) = P(A) + P(B)$ when A and B are disjoint subsets of Ω. An interpretation of probability is (supposed to be) a specification of a way that a set of properties realized in a part of the world (a "chance setup") can satisfy probability axioms. Some of these properties must also determine the numerical values of probabilities according to the interpretation. I'll assume familiarity with several well-known interpretations of probability: Bayesian credence, finite frequency, single-case propensity, long-run propensity, and Best System analysis probabilities (e.g. Hájek 2012; Earman 1992; Eagle 2004; Berkovitz 2015; Lewis 1980; Hoefer 2007).

Figure 13.3   A croupier's initial velocity distribution for a wheel of fortune. x-axis: angular velocity. y-axis: frequency or probability of spins at that velocity. Black regions: overall frequency or probability of spins resulting in the *black* outcome; gray regions represent the same thing for the *red* outcome. This distribution is for a wheel with black wedges that are twice as large as red wedges, which is why black regions are twice as wide as gray regions

It's worth mentioning recently developed "complex causal structure" (CCS) interpretations (e.g. Rosenthal 2010, 2012; Strevens 2011; Abrams 2012b; Beisbart 2016). While single-case propensities are often viewed as requiring fundamental indeterminism, as is thought to be present when quantum mechanical effects dominate a process, long-run propensities and CCS probabilities are usually viewed as consistent with either determinism or indeterminism. Indeed, one of the motivations for these interpretations is to develop a conception of objective probability that can underpin references to probability in higher-level processes that supervene on (nearly-)deterministic lower-level processes.[5] CCS probabilities seem particularly relevant to mechanisms such as those studied in molecular biology, because it's plausible that molecules interacting in fluids often realize the kind of causal structure on which CCS interpretations are based. Ignoring important subtleties of different CCS interpretations, the key idea is that some complex activities map input entity states to outcome states in such a way that small changes in input states easily lead to large differences in outcomes.[6] For example, a wheel of fortune[7] can be viewed as part of a mechanism in which a croupier's activity initiates an activity by which initial angular velocities are mapped to outcomes (*red*, *black*) in such a way that a small difference in velocity would cause a *red* rather than a *black* outcome. Figure 13.3 shows that unless a croupier were able to impart velocities significantly more often in very narrow regions of the input space, the shallow slope of the input distribution curve over any small contiguous region containing velocities leading to *red* and *black* means that the relative frequencies of spins leading to *red* and *black* within that region will be roughly equal to the relative widths of the intervals along the velocity (*x*) axis. Since this is true throughout the input space, frequencies will be close to ratios between wedge sizes. CCS interpretations use this insensitivity of outcome frequencies to variation in input frequencies to define interpretations of probability. Complex interactions between numerous swiftly moving molecules in fluid probably give rise to an analogous causal structure for many outcomes (cf. Strevens 2003, 2013). (CCS interpretations are not without serious challenges, but neither are the well-known interpretations mentioned above.)

Activity probabilities as causal probabilities

Despite the diversity of mechanisms in the world, their character and the way in which they are studied generally places constraints on what interpretations of probability are suitable as analyses of activity probabilities. Investigation of mechanisms typically involves interventions, manipulating either entities or activities and measuring resulting changes (Craver 2007; DesAutels 2011; Baetu 2013). When activities are stochastic, what must be measured are the effects of the intervention on frequencies of outcomes (Baetu 2013). This is true whether we are concerned with a token mechanism examined over time or a class of similar mechanisms. If we view these manipulable frequencies as depending on probabilities realized in the mechanism, then any account of what activity probabilities are must allow that manipulating probabilities—by manipulating properties that realize them—can also manipulate frequencies. In particular, manipulating the probability of outcome A should usually produce frequencies of A that are close to its probability. That manipulating probabilities manipulates frequencies in this way is true only for some interpretations of probability; I call such interpretations *causal probability interpretations*, and the probabilities they define *causal probabilities* (Abrams 2015). (This is a difference-making sense of "causal.")

Causal probability is not an interpretation of probability, but designates a property that applies to probabilities defined by some interpretations but not others. Causal probability supervenes on such interpretations. Causal probabilities can at least partially explain frequencies, in that the probabilities, as realized in the chance setup, partially control the frequencies. However, it's not required that a causal probability of an outcome always be roughly equal to its frequency in a large number of trials. That's just what usually happens. Despite the need for philosophical work on such claims, the vague idea that frequencies are usually roughly equal to probabilities, and manipulable by the means for manipulating probabilities, is implicit in assumptions that are widespread in successful science (Abrams 2015).

Which interpretations are causal probability interpretations? Notice, for example, that manipulating a person's credence that tosses of a coin will produce heads needn't affect frequencies of heads in tosses of that coin. Since manipulation of credences needn't affect frequencies, credences are not causal probabilities. Finite frequencies aren't causal probabilities either; in this case, the probabilities *are* the frequencies, so there is nothing further that can be affected by manipulating probabilities. Single-case and long-run propensities are plausibly causal probabilities, however. Consider the chance setup that consists of checking whether a particular Geiger counter clicks during any given minute in a particular day, with the possible outcomes *click, no click*. Suppose that the probabilities of these outcomes are single-case propensities. We can manipulate both this propensity and the usual frequencies of *click* by changing the radioactive substances that are near or in the Geiger counter, so such propensities are causal probabilities. CCS probabilities are also plausibly causal probabilities. For example, by manipulating the sizes of wedges on a roulette wheel, we can manipulate the probability of *red* as well as (usually) its relative frequency. On the other hand, Best System Analysis probabilities may not be causal probabilities if manipulating probabilities in a Best System just is manipulating frequencies in certain ways.

Causal probability helps to resolve a potential problem arising when different interpretations of probability apply to distinct activities in the same mechanism. Consider the activity probabilities involved in *E. coli* chemotaxis. These probabilities would usually be thought to be due to the statistical mechanical process of Brownian motion, which causes molecules to move rapidly through the cytoplasm (intracellular liquid) in a stochastic fashion. This is true of transitions between the active and inactive conformations of the Tar receptor as well (Hoffmann 2012, ch. 4). Quantum mechanical effects in this transition are small enough that they are difficult to calculate

(Kuriyan et al. 2013, ch. 6). Nevertheless, suppose it turned out that probabilities for active/inactive state shifts were quantum mechanical (cf. Jeknić-Dugić 2009; Luo 2014), and that they should be understood as propensities. Suppose that other chemotaxis activity probabilities, such as the probability of a CheY molecule becoming phosphorylated and subsequently binding to a flagellar motor, were CCS probabilities. We can capture this distinction by treating the probabilities on the right-hand side of equation (13.1) as being of different kinds. What is the nature of the overall probability $P(T = t_1 | S = s)$ of a CheY molecule binding to the flagellar motor given the Tar receptor's new aspartate binding? It appears to be some kind of weird mixture of propensity and CSS probabilities. Note that it won't do to say that $P(T = t_1 | S = s)$ is a single-case propensity simply because single-case propensities are more fundamental if, as is often assumed, single-case propensities depend on all factors that might affect what happens (Berkovitz 2015). This would seem to restrict the probability of a CheY-bound flagellar motor to values near 0 or 1, since for any particular token bacterium at a particular moment, the future movements of molecules would be largely determined by their current spatial relationships and states of interaction. Yet our current best understandings of chemotaxis treat these probabilities as non-extremal.

The concept of causal probability allows us to sidestep the preceding puzzle in a way that captures what is crucial about probability in mechanisms. A *complex chance setup* consists of several chance setups connected in such a way that whether a trial occurs on one or more setups depends on the outcome of an earlier trial on another setup (Abrams 2015). This is what a mechanism involving probabilistic activities is. For example, we can view the active and inactive states of the Tar complex as defining two setups, so that $P(T = t_1 | X = x_1)$ and $P(T = t_1 | X = x_2)$ are probabilities relative to two different setups, chosen by an outcome $X = x_i$ with probability $P(X = x_i | S = s)$. Then the overall probability $P(T = t_1 | S = s)$ of CheY-flagellar binding given aspartate binding is the probability for an outcome of this complex chance setup.

I argued in (Abrams 2015) that probabilities of outcomes from complex chance setups are causal probabilities if all of the probabilities from the component setups are causal probabilities. Since each component chance setup allows one to manipulate probabilities and frequencies in a generally coordinated way, and since frequencies of trials on later chance setups are controlled by frequencies of outcomes of trials on certain earlier chance setups, manipulating any of these component probabilities manipulates both probabilities and frequencies for the complex chance setup. So if activity probabilities are typically causal probabilities, the probability of a mechanism producing a phenomenon, for example, would typically be a causal probability as well, regardless of what interpretations of probability apply to the individual component activity probabilities.

Baetu (2013) argues that frequencies of outcomes of activities must be high enough that interventions can make an observable difference in frequencies, and Andersen (2012) and Krickel (forthcoming) also suggest that whether something counts as a mechanism depends on rough limits to the values of probabilities involved in its operation. I suggest that as long as activity probabilities are causal probabilities—so that they can make a difference in repeated tokenings of activities—a lower limit to probability values would depend on the nature of the mechanism and the phenomenon to be understood. Low probabilities can be investigated using larger samples, more money, advanced statistical methods, or convergent evidence.

Erraticity and imprecise probability

Objective probabilities, including causal probabilities, are only well-defined relative to (a) a chance setup and (b) a set of outcomes. Must every chance setup and set of outcomes define a set of causal probabilities? This seems unwarranted. Suppose I devise a way of classifying pieces of paper into three mutually exclusive categories, *low, medium, high,* according to the percentage of the paper's mass that consists of ink. Is there some causal probability that the average ink level among pieces of paper carried by the next ten people seen after reading this paragraph will be *high*? I don't see why there must be. (The question is not whether the probability of *high* is 0 or 1 for the nearly deterministic token process that you and your surroundings realize for the next day or two. That process isn't repeatedly realizable. Instead, think of the chance setup here as analogous to a coin toss, realized every time someone reads this paragraph.)

This example suggests that not all chance setups must give outcomes causal probabilities of some kind. Perhaps for some setups, outcomes are *erratic* (Hájek and Smithson 2012): Their frequencies have none of the systematicity that probabilities usually produce. The processes that produce them exhibit *erraticity*. For example, for all I know it may be that for a given strain of *E. coli*, there is a determinate probability that the number of Tar receptors will lie within a certain range, but the precise number of receptors within that range is erratic. This is not to say that there would be no reason at all for erratically produced outcomes *in particular tokenings* of the chance setup. In every single trial, these outcomes might be the result of a deterministic course of events, or might have single-case propensities close to 0 or 1, for example. However, where outcomes are erratically produced, the setup type and the outcomes are not such that realizations of it would usually exhibit the kind of systematicity that's typical of outcomes that have causal probabilities.

We can also consider a kind of systematicity intermediate between causal probability and pure erraticity. *Causal imprecise probability* exists, for example, when an erratically determined outcome $X = x$ in turn determines which of several different chance setups, with different probability distributions, will give rise to a subsequent outcome $Y = y$. For this complex chance setup, the imprecise probability of Y can be defined by the set of precise probabilities for Y that the value X might determine. In (Abrams MS) I provide details of this proposal.[8] Suppose, for example, that the number of Tar receptors in bacteria of a strain of *E. coli* is erratic as above, but that each number of Tar receptors defines a different probability distribution over frequencies of phosphorylated CheY molecules for a given aspartate gradient. Then it would turn out that for this *E. coli* strain, some of the activities that lead to the production of phosphorylated CheY would involve imprecise probabilities, neither fully probabilistic nor fully erratic.

For a mechanism to generate characteristic phenomena, not all of its activities need to involve (precise) probabilities. Biological and other mechanisms often involve entities and activities that are robust to variation in those entities and activities that produce them (Wagner 2005; Hermisson and Wagner 2004; cf. Wimsatt 1980, 2007). For example, average values of certain activities in the *E. coli* chemotaxis mechanism seem to be robust to some variations in the chemical composition of a bacterium (Barkai and Leibler 1997; Alon et al. 1999; Yi et al. 2000). Where this kind of robustness exists, a mechanism can function in a normal manner whether or not causally prior activities involve probabilities, imprecise probabilities, or full-fledged erraticity, as long as these activities' outcome frequencies tend to remain within certain ranges. Note, however, that imprecisely probabilistic activities can be modeled using probabilities when, for each outcome of an activity, all of the precise probabilities in the set that makes up the outcome's imprecise probability have similar values.

3. Epistemology

Modeling probabilistic mechanisms

Once we start thinking of mechanistic activities as probabilistic, it's natural to wonder whether Bayesian network models can provide a general way of modeling the causal structure of mechanisms. These models are defined in terms of mathematical assumptions about probabilistic relationships represented by arrows between nodes or variables (Pearl 2009; Spirtes et al. 2000). For example, Figure 13.2 and the assumptions about probabilities that I specified for it made it a representation of a Bayesian network. However, many mechanisms involve causal cycles (Bechtel and Abrahamsen 2005; Bechtel 2011). Bayesian networks can't directly represent cycles, but the framework can be extended to model causal cycles, though in slightly unnatural ways (Clarke et al. 2014; Gebharter and Kaiser 2014; Gebharter and Schurz 2016). Further, cyclic processes (such as those involved in the *E. coli* chemotaxis mechanism) can involve activities for which timing is important (Bechtel and Abrahamsen 2005; Bechtel 2011; Weber 2016), and current extensions of Bayesian network representations seem inadequate for modeling timing (Weber 2016; Gebharter and Kaiser 2014).

Bechtel (2011) notes that systems of differential equations are often appropriate for modeling such cyclic systems, and suggests that other ideas from dynamical systems theory are also useful. Differential equations can be used to model probabilistic systems, as Barkai and Leibler (1997) did for chemotaxis, but Firth and Bray (Morton-Firth and Bray 1998; Firth and Bray 2001) argued that differential equation models such as Barkai and Leibler's can make inaccurate predictions for aspects of intracellular processes that involve small numbers of molecules. Firth and Bray developed StochSim, an agent-based model in which "agents" are Tar receptors and molecules of several kinds with multiple states (Morton-Firth and Bray 1998; Morton-Firth et al. 1999; Firth and Bray 2001). Timing and outcomes of interactions between molecules in StochSim are probabilistic, with probabilities represented explicitly in the model. This model was able to reproduce several aspects of the empirical data on chemotaxis better than other models such as Barkai and Leibler's, suggesting further investigations.

Markov process theory also provides a wealth of mathematical tools for representing probabilistic processes such as those found in some mechanisms (e.g. Grimmett and Stirzacker 1992; Bharucha-Reid 1996 [1960]). For example, diffusion process models were originally developed to model Brownian motion, which plays a significant role in *E. coli* chemotaxis. Perhaps some Markov process models or computer simulations can capture some of the subtle complexity that comes from the role of spatial relationships in some mechanisms, which is difficult to capture with Bayesian networks (Kaiser 2016). In any event, we needn't think that there is some one privileged or standard way of modeling mechanisms. Just as there can be different sorts of diagrams that are useful for modeling mechanisms—even for modeling the same mechanism—there are many different kinds of models that may be useful for modeling mechanisms.

Modeling constitutive relationships

Discussion of mechanistic explanation sometimes focuses on ways in which some entities and activities constitute higher-level entities and activities (see Chapters 9 and 14). Craver (2007) proposes that constitutive relationships be understood in terms of manipulating entities and activities at one level and producing changes in entities at other levels. Since this view derives from Woodward's (Woodward 2003; Woodward and Hitchcock 2003) account of manipulation, in which interventions can affect probabilities, we can ask whether the state of an entity at one level could depend probabilistically on the state of entities that compose it, or vice versa. There has in fact been recent research on ways to extend Bayesian networks to represent certain probabilistic constitutive relationships in mechanisms (e.g. Casini et al. 2011; Gebharter 2014; Gebharter and Kaiser 2014).

There is a constraint on such models, unrecognized by those investigating them as far as I can see, which seems illuminating in its own right. The properties of an entity X at a higher level must supervene on the lower-level entities Y_i that constitute it and on the activities ψ_k involving these lower-level entities. Thus, *given* a specific set of lower-level entities Y_i in specific states y_j involved in specific activities ψ_k, the properties of the higher-level entity X that the Y_i constitute are determined. This means that the probability of X having the state constituted by those lower-level states $Y_i = y_j$ is equal to the probability of the conjunction of those states. Further, that probability is a mathematical function of activity probabilities whose outcomes determine whether $Y_1 = y_5$ and $Y_2 = y_7$, etc. Similarly, a probability that some lower-level entities Y_i have particular states given that a higher-level entity has a particular state $X = x_3$ is just the probability that lower-level activities cause the Y_i to have certain states, where the lower-level processes are restricted to those combinations of activities that can cause X to have state x_3. These stated identities between probabilities are not merely mathematical; they are identities between objective probabilities realized in the mechanism. Thus, *probabilities concerning constitutive relationships are really activity probabilities*. Models that postulate probabilities concerning constitutive relationships should be able to address this point.

Learning about activity probabilities

Background theory and similarity to related mechanisms may determine or suggest that certain activities are likely to be probabilistic, and that manipulation of conditions alters probabilities and frequencies in a coordinated fashion. For example, a transmembrane receptor in *E. coli*, the Tsr receptor, is known to have many structural and functional similarities to the Tar receptor (Parkinson et al. 2015), so it could be reasonable to assume that certain activity probabilities in the Tsr mechanism are similar to corresponding probabilities in *E. coli*. However, background theory or comparison to other mechanisms ultimately rests on earlier investigations, and therefore on other methods for learning probabilities. A few further details about *E. coli* chemotaxis research provide illustrations.

A natural way to measure causal probabilities is to measure how frequencies are manipulated by interventions, but this process is not always simple, and background theory can play a role. Consider Firth and Bray's (Morton-Firth and Bray 1998) use of Borkovich and colleagues' results (Borkovich et al. 1989, 1992). The Borkovich teams measured relationships between aspartate concentrations, numbers of methyl site bindings, and frequencies of CheY phosphorylation. These researchers took the last two frequencies to be manipulable both by varying aspartate concentrations and by varying methylation states of Tar receptors. The researchers didn't directly measure frequencies of active/inactive shifts in the Tar receptor, but background theory on the relationship between the Tar receptor's active and inactive conformations and CheY phosphorylation allowed Firth and Bray to use the Borkovich results to estimate probabilities of Tar receptor state changes in their computer simulation. Firth and Bray (apparently) assumed that the manipulated frequencies reflected probabilities, but used background theory to justify attributing probabilities to outcomes whose frequencies had not been measured. The same example also illustrates a role that computer simulations can play in learning about probabilities.

First note that in a computer simulation, probabilities are modeled using software implementing a deterministic random number-generating algorithm, but it's reasonable to assume in such cases that there is some interpretation of probability (perhaps yet unknown) in terms of which this software can be viewed as realizing a part of chance setup. The simulation doesn't just model probabilities; it realizes them as well (Glennan 1997; Abrams 2015).

When running their simulations Firth and Bray manipulated parameters so as to manipulate probabilities that were derived from the active/inactive switch probabilities. They then analyzed patterns of frequencies in results of simulation experiments. Because of the way in which frequencies were manipulated by manipulating probabilities, it appears that probabilities in the model are causal probabilities. Since these causal probabilities model probabilities in real *E. coli*, the probabilities that Firth and Bray assumed to exist in *E. coli* appear to be causal probabilities as well, with values like those in the simulations (cf. Abrams 2015).[9] In this way causal probabilities in real *E. coli* that had not been mathematically estimated from data were estimated through inference from a combination of observed frequencies, background theory, and computer simulation results.

Not all probabilities involved in a mechanism or its models must be causal probabilities. Sometimes a frequency is just a frequency. For example, one causal factor in bacterial chemo-taxis is the frequency of sites on a Tar receptor that are methylated at a particular time. This is simply a frequency—nothing more. Further, some models incorporate probabilities assumed for the sake of convenience or simplicity that are not thought to correspond to any prob-abilities in the world. Morton-Firth and Bray's (1998) chemotaxis simulation incorporated "pseudo-molecule" software objects that could be used to model spontaneous changes in a single molecule. The number of pseudo-molecules was chosen so that the probabilities of single-molecule changes would not result in these changes happening at a significantly slower rate than interactions between molecules. The probabilities of single-molecule changes in the model were not motivated entirely by theoretical or empirical factors; they were adjusted to accommodate the fact that the simulation modeled parallel, mostly independent activities as a sequence of activities that took place one at a time.

4. Non-recurrent mechanisms

So far I've focused on recurrent mechanisms, in which the same processes are repeated and realized either in a token mechanism or in many similar mechanisms. Some authors (Glennan 2009, forthcoming; Illari and Williamson 2012) advocate a broad conception of mechanism that need not involve the likelihood of recurrence. So defined, entities and activities might be organized in an entirely transient manner, and recurrence of such a mechanism could be unlikely. For example, Glennan (2009) argues that there is a mechanistic explanation of the literary critic Roland Barthes' accidental death due to being hit by a truck on the way home in Paris. The non-recurrent mechanism that produced Barthes' death is what Glennan (2009) calls an "ephemeral mechanism," where "the configuration of parts may be the product of chance or exogenous factors" and is "short-lived and non-stable, and is not an instance of a multiply-realized type" (Glennan 2009, p. 260).

As Glennan (2009, forthcoming) suggests, even where a non-recurrent mechanism consists of entities and activities in a short-lived configuration, *each individual activity* might neverthe-less be of a kind that is quite systematic. The coming together of these entities and activities might even be due to erraticity. Thus even if a mechanism as a whole is non-recurrent, it could consist of activities such as certain ways of trucks hitting pedestrians that regularly par-ticipate in mechanisms. For such activities, at least, we can apply arguments above that activity probabilities are causal probabilities discoverable through manipulation and modeling. Even if an activity in fact occurs only once, it may involve a causal probability, since that concept is defined in terms of counterfactuals about what frequencies usually would be seen in repeated trials. However, activities in non-recurrent mechanisms might also involve causal imprecise probabilities in some cases. Or some activities in non-recurrent mechanisms could simply be erratic—even so erratic that what happens just happened as it did for no reason other than the idiosyncrasies of lower-level processes.

5. Conclusion

This chapter looked at stochastic mechanisms through a discussion of the nature of probabilistic and other stochastic activities, primarily in recurrent mechanisms. I argued that activity probabilities in such mechanisms are nearly always what I call "causal probabilities," except, occasionally, when they are bare relative frequencies. I also argued that some activities might not be probabilistic, but instead involve imprecise probabilities or constrained erraticity. I discussed methods for modeling probabilistic mechanisms, and described several ways of learning about probabilities in mechanisms. Among other things, I illustrated a role that computer simulations can play in this process. I also suggested ways that the preceding ideas can be extended to ephemeral mechanisms that are unlikely to recur.

Notes

1 I'm grateful for Stuart and Phyllis's invitation to contribute this chapter, which drew me into fruitful and rewarding investigations I wouldn't have pursued otherwise, and for their feedback and patience during its writing. I got helpful feedback on ideas presented here from Holly Andersen, Murat Aydede, Paul Bartha, Sylvia Berryman, Scott Dixon, Kenny Easwaran, Chris French, Bruce Glymour, Alan Hajek, Chris Hitchcock, Daniel Malinsky, Ron Mallon, David McElhoes, Scott Peck, Alirio Rosales, Beckett Sterner, Johanna Thoma, and Jerry Tsui. Don Muccio and Scott Brande helped me understand the (nearly nonexistent) role of quantum mechanics in cellular processes.
2 My sketch is based on Eisenbach 1996; Barkai and Leibler 1997; Morton-Firth and Bray 1998; Morton-Firth et al. 1999; Firth and Bray 2001; Bray 2009; Parkinson Lab 2015.
3 *E. coli*'s so-called "exact" or "perfect" adaptation to chemical concentrations (e.g. Yi et al. 2000) refers to robustness in its ability to return to roughly the same state, on average, after persistent exposure to a concentration of a chemical such as aspartate. The states returned to exhibit significant stochastic variation (Alon et al. 1999).
4 See e.g. Bogen 2005; Craver 2007; Illari and Williamson 2012; Andersen 2012; Barros 2008; Baetu 2013; DesAutels 2011, 2015.
5 Rosenthal (2012) and I (2012b) have argued that our CCS interpretations don't define single-case probabilities, but Strevens (2011) argues that his does.
6 When I called my own CCS interpretation "mechanistic probability," I meant "mechanistic" in a very loose sense. CCS interpretations do always depend on processes that can be considered mechanisms according to some conception of mechanism or other, but these mechanisms would generally operate at a lower level than the activities that involve CCS probabilities. Because of this and because setup-to-termination probabilities of a mechanism reduce to activity probabilities (see above), I'm skeptical of DesAutels' (2015) apparent suggestion that such setup-to-termination probabilities might be mechanistic probabilities, unless component activity probabilities are CCS probabilities.
7 A wheel of fortune is like a roulette wheel, but with a fixed pointer rather than a ball indicating the wedge representing the outcome.
8 Related ideas and relevant mathematical details can be found e.g. in Troffaes and de Cooman 2014; Fierens et al. 2009; Bradley 2015, 2016; Dardashti et al. 2014.
9 I believe that analogous arguments can be given using mathematical models rather than simulations.

References

Abrams, M. (2012a), Measured, modeled, and causal conceptions of fitness, *Frontiers in Genetics* 3(196), 1–12.
Abrams, M. (2012b), Mechanistic probability, *Synthese* 187(2), 343–375.
Abrams, M. (2015), Probability and manipulation: Evolution and simulation in applied population genetics, *Erkenntnis* 80(S3), 519–549.
Abrams, M. (MS), Imprecise probability and biological fitness. draft paper.
Alon, U., Surette, M. G., Barkai, N. and Leibler, S. (1999), Robustness in bacterial chemotaxis, *Nature* 397(6715), 168–171.

Andersen, H. (2012), The case for regularity in mechanistic explanation, *Synthese* 189, 415–432.

Baetu, T. M. (2013), Chance, experimental reproducibility, and mechanistic regularity, *International Studies in the Philosophy of Science* 27(3), 253–271.

Barkai, N. and Leibler, S. (1997), Robustness in simple biochemical networks, *Nature* 387(6636), 913.

Barros, D. B. (2008), Natural selection as a mechanism, *Philosophy of Science* 75(3), 306–322.

Bechtel, W. (2011), Mechanism and biological explanation, *Philosophy of Science* 78, 533–557.

Bechtel, W. and Abrahamsen, A. (2005), Explanation: A mechanist alternative, *Studies in History and Philosophy of Science Part C: Studies in History and Philosophy of Biological and Biomedical Sciences* 36(2), 421–441.

Beisbart, C. (2016), A Humean guide to Spielraum probabilities, *Journal of General Philosophy of Science* 47(1), 189–216.

Berkovitz, J. (2015), The propensity interpretation of probability: A re-evaluation, *Erkenntnis* 80(S3), 629–711.

Bharucha-Reid, A. T. (1996 [1960]), *Elements of the Theory of Markov Processes and their Applications*, Dover.

Bogen, J. (2005), Regularities and causality: Generalizations and causal explanations, *Studies in History and Philosophy of Science Part C: Studies in History and Philosophy of Biological and Biomedical Sciences* 36, 397–420.

Borkovich, K. A., Kaplan, N., Hess, J. F. and Simon, M. I. (1989), Transmembrane signal transduction in bacterial chemotaxis involves ligand-dependent activation of phosphate group transfer, *Proceedings of the National Academy of Sciences* 86(4), 1208–1212.

Borkovich, K. A., Alex, L. A. and Simon, M. I. (1992), Attenuation of sensory receptor signaling by covalent modification, *Proceedings of the National Academy of Sciences* 89(15), 6756–6760.

Bradley, S. (2015), Imprecise probabilities, *in* E. N. Zalta (ed.), *The Stanford Encyclopedia of Philosophy*, Summer 2015 edition.

Bradley, S. (2016), Vague chance?, *Ergo* 3(18–21), 524–538.

Bray, D. (2009), *Wetware: A Computer in Every Living Cell*, Yale University Press.

Casini, L., Illari, P. M., Russo, F. and Williamson, J. (2011), Models for prediction, explanation and control: Recursive Bayesian networks, *THEORIA. An International Journal for Theory, History and Foundations of Science* 26(1), 5–33.

Clarke, B., Leuridan, B. and Williamson, J. (2014), Modelling mechanisms with causal cycles, *Synthese* 191(8), 1651–1681.

Craver, C. F. (2007), *Explaining the Brain: Mechanisms and the Mosaic Unity of Neuroscience*, Oxford University Press.

Dardashti, R., Glynn, L., Thebault, K. and Frisch, M. (2014), Unsharp Humean chances in statistical physics: A reply to Beisbart, *in* M. C. Galavotti, D. Dieks, W. J. Gonzalez, S. Hartmann, T. Uebel and M. Weber (eds), *New Directions in the Philosophy of Science*, Springer, pp. 531–542.

DesAutels, L. (2011), Against regular and irregular characterizations of mechanisms, *Philosophy of Science* 78(5), 914–925.

DesAutels, L. (2015), Toward a propensity interpretation of stochastic mechanism for the life sciences, *Synthese*, 192(9), 2921–2953.

Eagle, A. (2004), Twenty-one arguments against propensity analyses of probability, *Erkenntnis* 60(3), 371–416.

Earman, J. (1992), *Bayes or Bust*, MIT Press.

Eisenbach, M. (1996), Control of bacterial chemotaxis, *Molecular Microbiology* 20(5), 903–910.

Fierens, P. I., Rêgo, L. C. and Fine, T. L. (2009), A frequentist understanding of sets of measures, *Journal of Statistical Planning and Inference* 139, 1879–1892.

Firth, C. A. J. M. and Bray, D. (2001), Stochastic simulation of cell signaling pathways, *in* J. M. Bower and H. Bolouri (eds), *Computational Modeling of Genetic and Biochemical Networks*, MIT Press, chapter 9, pp. 263–286.

Gebharter, A. (2014), A formal framework for representing mechanisms?, *Philosophy of Science* 81(1), 138–153.

Gebharter, A. and Kaiser, M. I. (2014), Causal graphs and biological mechanisms, *in* M. I. Kaiser, O. Scholz, D. Plenge and A. Hüttemann (eds), *Explanation in the Special Sciences: The Case of Biology and History*, Springer, pp. 55–86.

Gebharter, A. and Schurz, G. (2016), A modeling approach for mechanisms featuring causal cycles, *Philosophy of Science* 83(5), 934–945.

Glennan, S. (1997), Probable causes and the distinction between subjective and objective chance, *Noûs* 31(4), 496–519.

Glennan, S. (2009), Ephemeral mechanisms and historical explanation, *Erkenntnis* 72(2), 251–266.

Glennan, S. (forthcoming), *The New Mechanical Philosophy*. Oxford University Press.

Grimmett, G. R. and Stirzacker, D. R. (1992), *Probability and Random Processes*, 2nd edition, Oxford University Press.

Hájek, A. (2012), Interpretations of probability, *in* E. N. Zalta (ed.), *The Stanford Encyclopedia of Philosophy*, Winter 2012 edition. http://plato.stanford.edu/archives/win2012/entries/probability-interpret/.

Hájek, A. and Smithson, M. (2012), Rationality and indeterminate probabilities, *Synthese* 187(1), 33–48.

Hermisson, J. and Wagner, G. P. (2004), The population genetic theory of hidden variation and genetic robustness, *Genetics* 168(4), 2271–2284.

Hoefer, C. (2007), The third way on objective probability: A sceptic's guide to objective chance, *Mind* 116(463), 449–596.

Hoffmann, P. M. (2012), *Life's Ratchet: How Molecular Machines Extract Order From Chaos*, Basic Books.

Illari, P. M. and Williamson, J. (2012), What is a mechanism? Thinking about mechanisms across the sciences, *European Journal for Philosophy of Science* 2(1), 119–135.

Jeknić-Dugić, J. (2009), The environment-induced-superselection model of the large-molecules conformational stability and transitions, *The European Physical Journal D* 51(2), 193–204.

Kaiser, M. I. (2016), On the limits of causal modeling: Spatially-structurally complex phenomena, *Philosophy of Science* 83(5), 921–933.

Krickel, B. (forthcoming), A regularist approach to mechanistic type-level explanation, *British Journal for the Philosophy of Science*.

Kuriyan, J., Konforti, B. and Wemmer, D. (2013), *The Molecules of Life: Physical and Chemical Principles*, 1st edition, Garland Science.

Lewis, D. (1980), A subjectivist's guide to objective chance, *in* R. C. Jeffrey (ed.), *Studies in Inductive Logic and Probability*, Vol. II, University of California Press. Reprinted in (Lewis 1986).

Lewis, D. (1986), *Philosophical Papers*, volume II, Oxford University Press.

Luo, LiaoFu (2014), Quantum theory on protein folding, *Science China Physics, Mechanics and Astronomy* 57(3), 458–468.

Machamer, P., Darden, L. and Craver, C. F. (2000), Thinking about mechanisms, *Philosophy of Science* 67(I), 25.

Morton-Firth, C. J. and Bray, D. (1998), Predicting temporal fluctuations in an intracellular signalling pathway, *Journal of Theoretical Biology* 192(1), 117–128.

Morton-Firth, C. J., Shimizu, T. S. and Bray, D. (1999), A free-energy-based stochastic simulation of the Tar receptor complex, *Journal of Molecular Biology* 286(4), 1059–1074.

Parkinson, J. S., Hazelbauer, G. L. and Falke, J. J. (2015), Signaling and sensory adaptation in *Escherichia coli* chemoreceptors 2015 update, *Trends in Microbiology* 23(5), 257–266.

Parkinson Lab (2015), An overview of *E. coli* chemotaxis, http://www.pdn.cam.ac.uk/groups/comp-cell/StochSim.html. University of Utah Department of Biology, John Parkinson Lab. Downloaded October 17, 2015.

Pearl, J. (2009), *Causality: Models, Reasoning, and Inference*, 2nd edition, Cambridge University Press.

Rosenthal, J. (2010), The natural-range conception of probability, *in* G. Ernst and A. Hüttemann (eds), *Time, Chance, and Reduction: Philosophical Aspects of Statistical Mechanics*, Cambridge University Press, pp. 71–90.

Rosenthal, J. (2012), Probabilities as ratios of ranges in initial-state spaces, *Journal of Logic, Language, and Inference* 21, 217–236.

Spirtes, P., Glymour, C. N. and Scheines, R. (2000), *Causation, Prediction, and Search*, 2nd edition, MIT Press.

Strevens, M. (2000), Do large probabilities explain better?, *Philosophy of Science* 67(3), 366–390.

Strevens, M. (2003), *Bigger Than Chaos: Understanding Complexity through Probability*, Harvard University Press.

Strevens, M. (2011), Probability out of determinism, *in* C. Beisbart and S. Hartmann (eds), *Probabilities in Physics*, Oxford University Press, chapter 13, pp. 339–364.

Strevens, M. (2013), *Tychomancy: Inferring Probability from Causal Structure*, Harvard University Press.

Troffaes, M. C. M. and de Cooman, G. (2014), *Lower Previsions*, Wiley.

Wagner, A. (2005), *Robustness and Evolvability in Living Systems*, Princeton University Press.

Weber, M. (2016), On the incompatibility of dynamical biological mechanisms and causal graphs, *Philosophy of Science* 83(5), 959–971.

Wimsatt, W. C. (1980), Randomness and perceived randomness in evolutionary biology, *Synthese* 43, 287–329.

Wimsatt, W. C. (2007), *Re-Engineering Philosophy for Limited Beings: Piecewise Approximations to Reality*, Harvard University Press.

Woodward, J. (2003), *Making Things Happen*, Oxford University Press.

Woodward, J. and Hitchcock, C. (2003), Explanatory generalizations, Part I: A counterfactual account, *Noûs* 37(1), 1–24.

Yi, T.-M., Huang, Y., Simon, M. I. and Doyle, J. (2000), Robust perfect adaptation in bacterial chemotaxis through integral feedback control, *Proceedings of the National Academy of Sciences* 97(9), 4649–4653.

14

MECHANISTIC LEVELS, REDUCTION, AND EMERGENCE[1]

Mark Povich and Carl F. Craver

1. Why levels?

In *The Sciences of the Artificial*, Herbert Simon (1969) offers the parable of the watchmakers, Tempus and Hora. Tempus builds watches holistically, holding each part in place until it forms a stable whole. Hora builds watches modularly, first assembling stable components and then organizing them into watches. Each is interrupted now and then. When Tempus is interrupted, she loses all her work on the watch and has to start again from scratch. When Hora is interrupted, she loses only her work on a single component. Hora thrives; Tempus struggles. The Moral: Evolved systems are likely to have (nearly) decomposable architectures. Their working parts are likely to be organizations of working parts, which are themselves organizations of parts, and so on. In other words, evolved systems likely exhibit regular mechanistic organization at multiple levels.

One might object to this argument on the ground that watches do not reproduce or that biological evolution does not assemble components sequentially (Bechtel and Stufflebeam 2001). As Simon points out, this would miss his point, which has now been repeated across many domains of science and philosophy: viz., that nearly decomposable (or modular) systems are more stable than holistic systems; they are to be expected when we find improbable stability in the face of random challenges. Contemporary work in network analysis, for example, shows that modular networks (those with high intra-modular connectivity and sparse inter-modular connectivity) are more "robust" than networks that are less modular; e.g., their mean minimum path length remains low in the face of random attacks. This is important, for example, in telecommunication, air traffic control, and neural networks (Albert et al. 2000). Such "modularity" is a widely accepted principle of both development and evolution (see Schlosser and Wagner 2004). Systems with causally independent parts are expected, in part, because nearly decomposable components can be independently assembled, regulated, damaged, and repaired without disturbing the other components. In short, everywhere we look from abstract theoretical considerations to concrete details of biology, we find reasons to expect dynamic but stable systems to be arranged in nearly decomposable hierarchies. Where we find embedded decompositions of working parts within working parts, we find nested mechanistic explanations.

Indeed, the assumption of near decomposability underlies the strategy of reverse engineering, of discovering how something works by learning how its parts interact. Kauffman (1970) calls this practice "articulation of parts explanation"; Haugeland (1998) calls it "explanation by

185

system decomposition"; Cummins (1975) calls it "functional analysis"; Fodor (1968), Craver (2007), and others (Machamer 2004; Glennan 2002; Menzies 2012) call it "mechanistic explanation." Mechanistic explanations explain a phenomenon by situating it in the causal structure of the world. In constitutive mechanistic explanations (as opposed to etiological and contextual mechanistic explanations; see Craver 2001), one looks to a lower level, within the phenomenon, to reveal its internal causal structure. In many such systems, one can explain the behavior of the whole in terms of the organized behaviors of its parts, and one can explain the behaviors of the parts in terms of the organized behaviors of their parts. Levels, on this view, are simply embedded mechanistic explanations.

In this chapter, we explicate this mechanistic view of levels (section 2), contrast it with other senses of "level" (section 3), and sketch its implications for emergence (section 4) and reduction (section 5), for the ontological status of higher-level phenomena (section 6), and for thinking about the lowest level(s) in such hierarchies (section 7).

2. Mechanistic levels

We use the term "mechanism" permissively to describe causal systems in which parts are organized such that they collectively give rise to the behavior or property of the whole in context (cf. the notion of minimal mechanism in Chapter 1). As mentioned above, not all mechanisms are modular at intermediate levels. In regular networks of simple nodes, for example, there is no nearly decomposable structure between the behavior of the mechanism as a whole and the activities of the parts. All mechanisms involve the organization of interacting parts, but not all mechanisms have intermediate levels of organization.

Mechanisms so construed contrast with aggregates (see section 3) in that mechanisms are organized. The parts of mechanisms have spatial (e.g., location, size, shape, and motion), temporal (e.g., order, rate, and duration), and active or otherwise causal (e.g., feedback or other motifs of organization; see Levy and Bechtel 2013) relations to one another such that they work together. This is why mechanistic hierarchies are often called "levels of organization." Aggregates, in contrast, are the lower limit of mechanistic organization: No between-component interactions are relevant to the true aggregate.

Figure 14.1 depicts three mechanistic levels. At the top is the activity of some mechanism as a whole (S's ψ-ing). S's ψ-ing includes the mechanism as a whole (e.g., the protein synthesis mechanism) doing what it does (e.g., synthesizing proteins). In this diagram, ψ is the topping-off activity of the mechanism; it is the activity to which all of the lower-level components are relevant. Beneath S's ψ-ing are the activities (ϕ_i) and components (X_j) organized such that together they ψ. Beneath that is an iteration of the levels relation, in which one of the X's ϕ-ings is decomposed into the organized ρ-ings (rho-ings) of Ps. As Simon suggested, this process of decomposition might continue until we run out of either relevant or practically salient decompositions. Individually, the X_j and ϕ_i are organized to compose S's ψ-ing. Collectively, the X_j and ϕ_i exhaustively constitute (or, as is often said, "realize") S's ψ-ing in context.

X's ϕ-ing is at a lower mechanistic level than S's ψ-ing if and only if X and its ϕ-ing are component parts and activities in S's ψ-ing. A component is a part whose activities contribute to the behavior of the mechanism as a whole. The components (X_j) are spatially contained within S (because components are necessarily parts of the things they compose). Furthermore, each component is relevant to S's ψ-ing: X's ϕ-ing should contribute to (or be operative in) S's ψ-ing (see Martin and Deutscher 1966).

There is some debate about how this interlevel relevance relationship should be understood. Craver (2007) defines it in terms of the mutual manipulability of the whole and the part: X

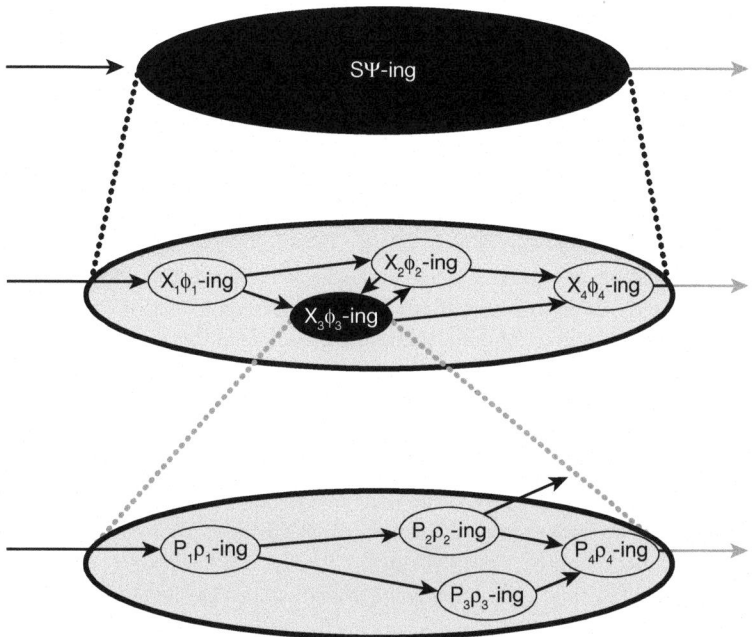

Figure 14.1 A schematic of mechanistic levels

and its ϕ-ing are component parts and activities of S's ψ-ing if one can manipulate S's ψ-ing by intervening on X's ϕ-ing, and one can manipulate X's ϕ-ing by manipulating S's ψ-ing. This is a sufficient, but not necessary, condition for componency, as there might be parts of a mechanism that do not change during its operation (consider the walls surrounding pistons in a car engine) or redundant parts that can be independently manipulated with no effect on the behavior of the mechanism as a whole. Couch (2011) argues that componency should be understood in terms of Mackie's notion of an INUS condition. Components are <u>I</u>nsufficient and <u>N</u>on-redundant parts of an <u>U</u>nnecessary but <u>S</u>ufficient set of contributors to S's ψ-ing. Still others (such as Harinen 2014 and Romero 2015; see also Craver 2007) understand constitutive relevance as causal betweenness: a component of a mechanism is causally intermediate between the mechanism's start conditions and its termination conditions (see Chapter 9 on parts and boundaries of mechanisms).

Furthermore, Figure 14.1 is ambiguous: Two distinct relations are represented in both the top and bottom (corresponding roughly to what Gillett (2002) calls dimensioned and flat realiza-tion). The first is a part–whole relation between S's ψ-ing and each of the X's individual ϕ-ings. We use the term "mechanistic level" exclusively for this relation. The second is a whole–whole relation between two ways of describing the same thing. We can talk about S's ψ-ing or we can talk about the organized collection of ϕ-ing Xs. These are two ways of describing one and the same object. Likewise, at the bottom of the figure, we can talk about a given X's ϕ-ing or we can talk about the organized collection of ρ-ing Ps in virtue of which X ϕs. Again, these are two levels of description that apply to one and the same object. Marr's levels of realization are like this: the computational level, the algorithmic level, and the implementation level are three ways of describing one and the same whole (Marr 1982; see also Chapter 29). We return to this issue in section 6.

The above explication of mechanistic levels has several conceptual benefits. First, it accurately describes the multilevel explanatory structures one finds in biology and other special sciences, and makes explicit the kinds of evidence required to evaluate such explanations. Second, it satisfies many of our pre-analytic intuitions about the levels of explanation. Things at higher mechanistic levels are typically larger (and necessarily no smaller) than things at lower levels because the latter are parts of the former. Things at lower levels often take less time than things at higher levels because activities at lower levels compose activities at higher levels. And the idea of mechanistic levels captures the common idea of "levels of organization" because higher-level mechanisms are made up of organized parts and activities. Third, this understanding of levels clarifies why interlevel causation is at least *prima facie* mysterious: causation between mechanistic levels must involve parts interacting with their wholes (see section 4).

Finally, this understanding of levels helps to undermine the thought that nature can be usefully divided into monolithic levels of, e.g., atoms, molecules, cells, organs, organisms, and societies (Oppenheim and Putnam 1958). For mechanistic levels, there is no unique answer to the question of when two items are at the same mechanistic level. There is only a necessary condition: X's ϕ-ing and S's ψ-ing are at the same level of mechanisms only if (i) X's ϕ-ing and S's ψ-ing are components in the same mechanism, (ii) X's ϕ-ing is not a component in S's ψ-ing, and (iii) S's ψ-ing is not a component in X's ϕ-ing. Unlike size levels or levels defined in terms of the types of objects found at a given level, mechanistic levels are defined by the componency relationship between things at higher and lower levels. If two things are not related as part to a whole, they are not at different levels; if they are in the same mechanism, then they are in this limited sense "at the same level." But one might just as easily say on this basis that sameness of level has no deep conceptual significance for mechanistic levels.

Note, for example, that the levels of nearly decomposable structure in stars do not correspond to the levels in starfish. Each has parts and wholes, but the parts and wholes are of different kinds. It is an empirical question, answered case by case, how many levels there are in a system. It makes no sense to ask whether hippocampi are at a higher or lower mechanistic level than horseshoes. They are not components in the same mechanism; they are not related as parts to wholes. For some, this is tantamount to abandoning the idea of levels (e.g., Eronen 2015). We see our account rather as a distillate of the ordinary scientific concept, an extraction of an explanatorily and metaphysically central idea, leaving behind as residue the problematic commitments inherent in our inchoate, folk talk of "levels."

3. Comparisons and contrasts

The ordinary concept of "level" is inchoate, in part, because the term is used promiscuously to describe many distinct kinds of relata and relations. Here we distinguish mechanistic levels from aggregates, size levels, causal levels, and Oppenheim and Putnam's levels.

As noted above, mechanisms contrast with aggregates. In aggregates, the property of the whole is literally a sum of the properties of its parts. The concentration of a fluid is an aggregation of particles; allelic frequency is a sum of individual alleles. Aggregate properties change linearly with the addition and removal of parts, they don't change when their parts are rearranged, and they can be taken apart and reassembled without any special difficulty. This is because in true aggregates, spatial, temporal, and causal organization are irrelevant (Wimsatt 1997). Mechanisms, in contrast, are literally more than the sums of their parts: they change nonlinearly with the addition and removal of parts, their behavior is disrupted if parts are switched out, and this is because their spatial, temporal, and causal organization make a difference to how the whole behaves.

Though distinct, mechanisms and aggregates are nonetheless species of a genus: each involves a relationship between the properties or activities of wholes and the relevant properties or activities of their parts. Many mechanisms (from steam engines to action potentials) rely on both mechanistic and aggregative relations. As noted above, aggregation is the limit as mechanistic organization goes to zero.

Mechanistic levels also contrast with size levels. The relata in size levels are space-involving entities (like cells), and they are higher- or lower-level than one another because they are bigger or smaller, respectively. They are "at" a level when they have the same (or similar) sizes. Mechanistic levels are also orderable by size, but the size relationship between mechanistic levels follows from the more fundamental, compositional relationship. Like Wimsatt's (1976) classic image of levels as peaks of regularity and predictability, nearly decomposable mechanisms will tend to carve most naturally at interfaces between components, thereby identifying isolable pockets of regularity and predictability. Wimsatt describes but does not explain why these levels of regularity and predictability cluster at different size scales. In mechanistic levels, the clustering is explained by patterns of near decomposability in the mechanism's causal organization.

Sometimes the term "level" is used to describe relations among the stages in a causal pathway, as when one distinguishes "higher-level" and "lower-level" visual areas (e.g., area MT and V1, respectively) or, arguably, when one speaks of "higher-level executive functions." These causal relations are clearly distinct from the compositional relation involved in mechanistic levels. MT is not part of V1, but downstream in a causal process from V1; lower-level cognitive functions are not parts of higher-level cognitive functions, but asymmetrically controlled or dominated by higher-level cognitive functions (see Churchland and Sejnowski 1992).

The term "levels" is also associated with Oppenheim and Putnam (1958), who structure their view of the unity of science around a monolithic conception of levels. They carve the world into roughly six ontological strata (societies, organisms, cells, molecules, atoms, and elementary particles). Each stratum corresponds to a distinct field of science, from economics at the top to particle physics at the bottom. Each level has its distinctive theory. The unity of science consists in the explanation of higher-level theories in terms of lower-level theories.

Oppenheim and Putnam could easily have embraced mechanistic levels as the ontic component of their picture; things at higher levels are wholes made up of things at lower levels. As Wimsatt (1976) notes, however, this minor amendment does violence to the simplicity of Oppenheim and Putnam's vision of scientific unity. If levels are compositional, and different mechanisms decompose into different kinds of parts, then we should not expect a uniform decomposition of all things into the same kinds of parts (compare stars to starfish and hippocampi to horseshoes).

Oppenheim and Putnam's vision of scientific unity is descriptively inadequate, as they would likely acknowledge. Both sciences and theories run rough-shod over levels. Models and theories often span levels of organization, linking phenomena studied by different fields of science (Darden and Maull 1977; Craver 2007). Scientific fields are also increasingly transdisciplinary (e.g. neuroscience, ecology). Things that are the same size can usefully be investigated by altogether distinct fields of science: Cytologists, anatomists, and electrophysiologists all study cells with different tools. The relation between fields of science, levels of theory, and ontological levels is thus many to many to many.

The notion of mechanistic levels provides a compelling alternative to the Oppenheim and Putnam model. Mechanisms span multiple levels of organization. Scientists approach these structures and processes with diverse tools and from diverse theoretical vantage points. The unity that results from this collaboration is more like a mosaic than a layer cake: it is achieved when different scientists with different instruments and modeling tools use their diverse expertise to understand the same mechanism (Craver 2007).

4. Levels, emergence, and interlevel causation

Do things at higher levels "emerge" from things at lower levels? And can things at different levels causally interact? If we think of levels as mechanistic levels, these questions have clear content, and it is clear what is at stake in answering them.

Mechanisms are not aggregates. A property or activity at a higher level of mechanistic organization (e.g., S's ψ-ing) is literally more than the sum of the properties of its parts. So, if emergence is defined as the failure of aggregativity (call this *organizational emergence*; Wimsatt 1997), then things at higher mechanistic levels emerge organizationally from things at lower levels. No ontological extravagance is required: two stacked toothpicks have the organizationally emergent capacity to catapult raisins. Neither toothpick can catapult raisins alone: so the whole has capacities the parts alone do not possess.

Often the term "emergence" is used in an *epistemic* sense to refer to the inability to predict the properties or behaviors of wholes from properties and behaviors of the parts. Epistemic emergence can result from our ignorance, such as failing to recognize a relevant variable, or from failing to know how different variables interact in complex systems. It might also result from limitations in human cognitive abilities or current-generation representational tools (Boogerd et al. 2005; Richardson and Stephan 2007). The practical necessity of studying mechanisms by decomposing them into component parts raises the epistemic challenge of conceptually putting the parts back together so they work (Bechtel 2013). Epistemic emergence results from the limits of our knowledge or of our representational capacities, not from discontinuity in the world's causal structure (see Chapter 27).

Ontic emergence is suspect or promising (depending on one's perspective) precisely because it involves such discontinuity: there are higher-level properties and capacities that have no sufficient (ontic) explanation in terms of the parts, activities, and organizational features of the system in the relevant conditions. Some say life, consciousness, or intentionality are emergent in this sense. Mechanistic levels, in contrast, are levels of mechanistic dependence: they are defined in terms of componency (Craver 2014). If that ontic relationship is severed, then the sense in which emergent properties are at a "higher level" must be different than the sense in mechanistic levels. Indeed, it is unclear why properties that ontically emerge should be thought of as higher-level at all (rather than, e.g., effects of a cause). Ontic emergence, whatever its virtues, is mysterious precisely because it is distinct from, and so gains no plausibility from verbal association with, organizational or epistemic emergence. Organizational and epistemic emergence are unmysterious both in scientific common sense and common sense proper (Van Gulick 1993; Kim 1998). Appeal to ontic emergence, on the other hand, arouses suspicion because it is committed to the existence of phenomena that have no sufficient (ontic) mechanistic explanation.

Can things at different mechanistic levels properly be said to causally interact with one another? Many common assumptions about causation appear to block this thought. For Hume and Lewis, the relata in a causal relationship must be distinct,

> and distinct not only in the sense of nonidentity but also in the sense of nonoverlap and nonimplication. It won't do to say that my speaking this sentence causes my speaking this sentence; or that my speaking the whole of it causes my speaking the first half of it; or that my speaking causes my speaking it loudly, or vice versa.
>
> *(Lewis 2000, p. 78)*

Wholes and parts overlap; a token S's ψ-ing as a whole includes every part of S's ψ-ing, and that includes all the X_j, ϕ_i, and organizing relations. On Salmon's process view of causation, the relata in causal relations must intersect in space-time, exchange conserved quantities, and maintain those changes after the intersection ends. But parts and wholes are always everywhere together; the whole has no additional conserved quantity to pass to its parts. The relationship between LTP and the opening of NMDA receptors during LTP induction is directly analogous to the relationship between speaking the whole of a sentence and speaking its first half. The induction of LTP is partly constituted by the opening of the NMDA receptor. The would-be cause in this top-down causal claim already contains the would-be effect; talk of causation across levels of mechanism thus appears inappropriate (Craver and Bechtel 2007).

What about the bottom-up case? Here we must guard against an ambiguity. Sometimes mechanists describe the "phenomenon" as an activity or process that starts with the mechanism's setup conditions and ends with its termination conditions (Machamer et al. 2000). For example, Long-Term Potentiation (LTP) is described as a process beginning with rapid and repeated stimulation of the presynaptic neuron and ending with enhanced synaptic transmission. Other times, mechanists describe the phenomenon as the product of that process (as one of its termination conditions). For example, the mechanism of LTP produces a potentiated synapse. This second phrasing leads us to seek the antecedent causes: the tetanus and the subsequent changes in the NMDA receptor. But if we think about the phenomenon as a process, an input–output relation starting with the tetanus and ending with a potentiated synapse, then we should not say the NMDA receptor is a cause of that, if we wish to speak accurately about the ontic structure of the situation at hand. The NMDA receptor is a part of that causal process; it causes neither itself nor its antecedents in the mechanism.

5. Reduction

Like "level," "reduction" is used many ways. Sometimes, it expresses a thesis about *explanation*, about how one theory or law is explained by other theories or laws. Other times, it is a thesis about *scientific integration* and unity, about the relationships among the diverse branches of science. And still other times, it is a *metaphysical thesis* about the structure of the world (section 6). The idea of mechanism offers considerable insight into explanation and scientific integration. Although the mechanistic view of levels alone does not resolve the deepest ontological puzzles, it nonetheless offers a descriptively adequate image of the kind of ontological structure by which the manifest phenomena of our world might be (and arguably are) related to its most fundamental parts and activities.

According to the covering law (CL) model, reductive explanations involve deriving higher-level theories from lower-level theories with the aid of bridge laws connecting their distinct theoretical vocabularies (Nagel 1961; Schaffner 1993). This model of explanatory reduction serves as a valuable ideal in the search of mechanisms. A description of the behavior of the mechanism should be shown to *follow from* a description of the parts and their interaction, as one understands them. For example, Hodgkin and Huxley showed that one could recover the shape of the action potential from measured values of ionic concentrations and experimentally determined ion-gating functions coupled with laws governing electrical circuitry. Yet this epistemic test of the sufficiency of an explanation should not be confused with the explanation itself.

Not all predictively adequate models are explanatory. We can, as Hempel (1965) argued, derive Boyle's law from the conjunction of Boyle's law and Kepler's laws, but not in a way that explains it. We can predict the behavior of a part from the description of the behavior of the whole and the behaviors of a few other parts, despite the fact that many expect such micro-reductive explanations to work only from the bottom up. We can predict the behavior of a system by tracking the right correlations in the system irrespective of their causal relations to one another (as in diagnosis and prognosis) and would not thereby explain the system's behavior. Finally, the classical model of reduction is not equipped with an account of explanatory relevance, a way to determine which predictively relevant features are explanatorily relevant and which are not. If one thinks of reduction not as primarily an epistemic, deductive relationship between descriptions of the behavior of a mechanism and of the organized activities of its component parts but as a matter of learning and revealing the causal structure of a mechanism, these problems for the classic theory of micro-reductive explanation are directly addressed (Craver 2007).

Given the multilevel structures found in mechanistic explanations, it's clear that scientific integration will not conform to the layer cake pattern Oppenheim and Putnam describe across science as a whole. Interlevel linkages will be of more local significance, relativized to particular explanandum phenomena. Different fields of science collaborate with one another both to bridge across mechanistic levels and to bring their unique perspectives to bear on one and the same thing. A kind of mosaic unity results from having the same mechanism as the target of one's scientific investigation; integration occurs when the findings from different fields and perspectives mutually constrain our understanding of how the mechanism works.

This mechanistic perspective provides an attractive successor to empiricist views of explanation as a model of interfield integration. First, it is based on a more accurate and informative view of levels. Second, it provides significantly more insight into the diverse evidential constraints by which interlevel bridges are evaluated and into the forces driving co-evolution across different levels (Craver and Darden 2013). Constraints on the parts, their causal interactions, and their spatial, temporal, and hierarchical organization all help to flesh out an interfield theory. Finally, many mechanists argue that classical reduction models are myopically focused on downward-looking explanations at the expense of a fuller account of the details by which interlevel and interfield theories more generally are constructed and evaluated. This is especially true when one must look up to higher levels to see how a part contributes to a mechanism of which it is a component and in which two fields combine findings "at" rather than across mechanistic levels.

6. The ontological status of higher-level phenomena

We turn now to the metaphysical questions central to many philosophical discussions of reduction: what is the relationship between higher-level phenomena and things at lower levels? We disambiguated two questions implicit in Figure 14.1. The first is: Is the behavior of the mechanism as a whole greater than the sum of the behaviors of its parts? We defend an affirmative answer above. The second question is: Is the behavior of the mechanism as a whole greater than the organized interaction of its parts in context (commonly called the *realizer* of the mechanism's behavior)?

Simple examples suggest that the relationship between the phenomenon and its realizer is more intimate than correlational, causal, and nomologial relationships are. In a standard mousetrap, if the trigger is depressed, the catch slides, and the impact bar completes its arc (collectively, M), then the trap necessarily fires (F). We are precluded from imagining that all the stages of M have occurred but that the trap has not fired precisely because there is nothing more to be done, no activity or property to add, to transform M in context, into an instance of F. We are similarly precluded from imagining in a population of mousetraps composed of identical parts organized and interacting identically with one another in the same conditions that only 30 percent of these mechanisms should fire and 70 percent should not. Correlational, causal, and nomological relationships are not necessarily like this. The intimacy between a phenomenon and its realizing mechanism thus appears to be stronger, more metaphysically necessary, than a simple matter of regularity, causation, or law. The behavior of the whole contains the behaviors of the parts, and the behaviors of the parts collectively and exhaustively constitute the behavior of the whole. Dualism, parallelism, and emergentism do not share this commitment and allow that there are properties and activities of wholes that are not exhaustively constituted by lower-level mechanisms.

This metaphysical commitment functions as a constitutive ideal in the search for mechanistic explanations (Haugeland 1998). If one knows all of the relevant entities, activities, and organizational features, and knows all the relevant features of the mechanism's context of operation, and can in principle put it all together, then one must know how the mechanism will behave. This was Hempel's important insight about the epistemology of explanation. It is an epistemic warning sign if features of the mechanism's behavior cannot be accounted for in terms of our understanding of its parts, activities, organization, and context. Mechanists thus operate with a background assumption that the phenomenon is exhaustively explained (in an ontic sense) by the organized activities of parts in context.

One is tempted to say that higher-level phenomena (in a given context) *are identical to* the organized activities of the mechanism's parts (in that context). By this one might assert a type identity, according to which ψ-ing in general is identical to the organized φ-ing of X's, or a token identity, according to which each instance of ψ-ing is identical to some organized φ-ing of X's. Classical reductionists favor type identity, though this thesis is largely out of fashion (though see Polger 2004). As Putnam (1967) and Fodor (1974) argued, such type-identity statements are false when the phenomenon is or can be multiply realized in distinct mechanisms. There are many ways to trap a mouse and many ways to produce an action potential. If so, the same type of phenomenon might be instantiated on different occasions in different mechanisms; the realized type is not the same as the realizer type.

Bechtel and Mundale (1999) counter that this argument rests on an illusory matching of grains between the characterization of the phenomenon and the characterization of the mechanism. If one characterizes the phenomenon abstractly, then many mechanisms can or could give rise to it. If one characterizes the phenomenon in fine detail (including, for example, all its excitatory, inhibitory, and modulatory conditions), then the class of actual and potential realizers might shrink to one. If the phenomenon and the mechanism are characterized at comparable degrees of abstraction, then the thesis of type identity begins to look more promising.

But perhaps a deeper problem for type identity is lurking in the background and not so easily addressed: the question of whether there is, in general, an "appropriate" grain of analysis that adequately captures the phenomenon and the mechanism at once. Even some token things are multiply realizable: One and the same rook in a chess game might be instantiated at different times by different figurines (imagine replacing a lost plastic castle midgame with a Fig Newton; Haugeland 1998). One and the same kidney is made up of different parts over a person's lifetime. To answer whether ψ and the organized collection of φ-ers are the same thing, we need to be able to answer the question "The same what?" with a non-dummy sortal (Marcus 2006). The conditions under which something is the same ψ-er might be different from the conditions under which it is the same organized collection of φ-ers, in which case there are no individuation conditions that cover the thing both as a ψ-er and as an organized collection of φ-ers, and there is no sense to the question of whether they are identical. Items at different mechanistic levels are no doubt intimately related, but our ordinary notions of token and type identity appear poorly equipped to describe that relation. For these reasons, we prefer to speak of a particular organized φ-ing of X's as exhaustively constituting a particular ψ-ing at or over a time.

Instead of asking about identity, one might ask rather about whether phenomena have causal powers in a given context that the organized collection of interacting parts do not have. As noted above, it is relatively uncontroversial that wholes can do things their parts cannot do (see Kim 1998, p. 85). Do realized phenomena have causal powers their realizing mechanisms do not have in the same contexts? If one thinks of the phenomenon as identical to the behaving mechanism, then the answer is no: the two have all and only the same powers. Non-reductive physicalists have traditionally attempted to argue that although higher-level phenomena are all ontically explainable in terms of more fundamental mechanisms, there remains a sense in which the higher-level phenomena have causal powers of their own.

Again, issues of multiple realization arise. Suppose a neuron generates an action potential, causing neurotransmitters to be released from the cell. Whenever it does, we suppose there is a determinate set of entities (ions, channels, etc.) organized and interacting in determinate ways that fully constitutes it at that time. Now suppose we ask: which feature of the world is relevant to the release of neurotransmitters: the action potential or the determinate organization and interaction of ions and channels? One standard test of causal relevance is to ask what would happen if the cause had been different. Experimental investigation will reveal that the action potential makes a substantial difference to the probability of neurotransmitter release by raising membrane voltage. The precise arrangement of ions and ion channels makes no difference except insofar as it affects membrane voltage. Many differences among the realizers make no difference to the effect in question. If causal relevance is a matter of making a difference (Woodward 2003), the fine details about the realizing mechanism are not causally relevant. Each action potential is exhaustively constituted by a mechanism, but the causally relevant difference is the occurrence of the action potential, not the precise mechanism that produced it (see Craver 2007, chapter 6).

Just how this mechanistic perspective best fits with metaphysical views about, e.g., identity and causal powers remains to be decided. A determinate view about multiple realization and its implications requires metaphysical commitments that go beyond the skeletal framework of levels presented here. Views on multiple realization differ depending, *inter alia*, on how one thinks about properties (Heil 1999), the realization relation (Shapiro 2004), and theories (Klein 2013). Similarly, the prospects for token physicalism depend, *inter alia*, on one's account of events (Kim 2012) and identity (Marcus 2006). One goal of future work is to integrate the mechanistic perspective with a metaphysics that produces a coherent and plausible view of the structure of the mechanistic world.

7. The lowest level

Questions also arise concerning the bottom of a hierarchy of mechanisms. Two options seem to be available: either (1) the hierarchy has a bottom or (2) it does not, i.e. it continues ad infinitum. Though both options are metaphysically live, some facts about quantum mechanics seem to favor the first.

It has been argued, for example, that the non-separability of ontological units in the theories of quantum mechanics entails that there are no distinct objects to be identified as components beneath a certain level of organization. Kuhlmann and Glennan (2014) argue that for any local, classical phenomenon, there is a lowest classical level at which mechanistic explanation bottoms out. Precisely where it bottoms out varies across phenomena because decoherence—the physical process by which the quantum realm becomes classical—is itself local. Below this level, where properties like spatial location are indeterminate, prospects for mechanistic explanation are murky. Mechanistic explanation may still be possible if quantum phenomena can be explained by causal organization among non-local "objects" (for example, see the discussion of laser light (Kuhlmann and Glennan 2014, p. 354)), but in quantum holistic cases where the properties of subsystems don't appear to determine the properties of the whole system (for example, EPR experiments), the constitutive ideal may be unreachable.

8. Conclusion

Our goal has been to sketch the mechanistic approach to levels, to contrast it with near neighbors, and to explore some of its metaphysical implications. This perspective provides some important content missing from standard metaphysical discussions of levels. Most notably, it offers a clear sense of what it means for things to be at different levels. It also allows us to distinguish mechanistic levels from realization relations. And finally, it provides a clearer vision of the structure of multilevel explanations, the evidence by which such explanations are evaluated, and the form of scientific unity that results from building them. The mechanistic approach does not, by itself, solve all the metaphysical questions one might have about physicalism. Yet it does provide a skeletal framework for thinking about how the macroscopic phenomena of our world are or might be related to its most fundamental parts and activities. Simon's parable and its many contemporary analogs give us reason to think that this skeletal picture captures a central and important pattern in the causal structure of our world.

Note

1 Thanks to Mike Dacey, James Gulledge, Ben Henke, Eric Hochstein, Emily Prychitko, and Felipe Romero for invaluable comments, and to Pamela Speh for the design of Figure 14.1.

References

Albert, R., H. Jeong, and A.-L. Barabási. (2000) "Error and Attack Tolerance of Complex Networks," *Nature*, 406, 378–382.

Bechtel, W. (2013) "Addressing the Vitalist's Challenge to Mechanistic Science: Dynamic Mechanistic Explanation,": in S. Normandin and C. T. Wolfe (eds), *Vitalism and the Scientific Image in Post-Enlightenment Life Science 1800–2010*, Dordrecht: Springer, 345–370.

Bechtel, W. and J. Mundale. (1999) "Multiple Realizability Revisited: Linking Cognitive and Neural States," *Philosophy of Science*, 66, 175–207.

Bechtel, W. and R.S. Stufflebeam. (2001) "Epistemic Issues in Procuring Evidence about the Brain: The Importance of Research Instruments and Techniques," in W. Bechtel, P. Mandik, J. Mundale and R.S. Stufflebeam (eds), *Philosophy and the Neurosciences: A Reader*, Blackwell: Malden, MA, 55–81.

Boogerd, F.C., F.J. Bruggeman, R.C. Richardson, A. Stephan and H.V. Westerhoff. (2005) "Emergence and Its Place in Nature: A Case Study of Biochemical Networks," *Synthese*, 145, 131–164.

Churchland, P.S. and T. Sejnowski. (1992) *The Computational Brain*, Cambridge, MA: MIT Press.

Couch, M. (2011) "Mechanisms and Constitutive Relevance," *Synthese*, 183, 375–388.

Craver, C.F. (2001) "Role Functions, Mechanisms, and Hierarchy," *Philosophy of Science*, 68, 53–74.

Craver, C.F. (2007) *Explaining the Brain*, Oxford: Oxford University Press.

Craver, C.F. (2014) "The Ontic Account of Scientific Explanation," in M.I. Kaiser, O.R. Scholz, D. Plenge, and A. Hüttemann (eds), *Explanation in the Special Sciences: The Case of Biology and History*, Dordrecht: Springer, 27–52.

Craver, C.F. and W. Bechtel. (2007) "Top-down Causation without Top-down Causes," *Biology and Philosophy*, 22, 547–563.

Craver, C.F. and L. Darden. (2013) *In Search of Mechanisms: Discoveries across the Life Sciences*, Chicago: University of Chicago Press.

Cummins, R. (1975) "Functional Analysis," *Journal of Philosophy*, 72, 741–765.

Darden, L. and N. Maull. (1977) "Interfield Theories," *Philosophy of Science*, 44, 43–64.

Eronen, M.I. (2015) "Levels of Organization: A Deflationary Account," *Biology and Philosophy*, 30, 39–58.

Fodor, J. (1968) *Psychological Explanation*, New York: Random House.

Fodor, J. (1974) "Special Sciences: Or the Disunity of Science as a Working Hypothesis," *Synthese*, 28, 97–115.

Gillett, C. (2002) "The Dimensions of Realization: A Critique of the Standard View," *Analysis*, 62, 316–323.

Glennan, S. (2002) "Rethinking Mechanistic Explanation," *Proceedings of the Philosophy of Science Association*, 3, S342–S353.

Harinen, T. (2014) "Mutual Manipulability and Causal Inbetweenness," *Synthese*, 1–20.

Haugeland, J. (1998) *Having Thought: Essays in the Metaphysics of Mind*, Cambridge, MA: Harvard University Press.

Heil, J. (1999) "Multiple Realizability," *American Philosophical Quarterly*, 36, 189–208.

Hempel, C. (1965) *Aspects of Scientific Explanation*, New York, NY: Free Press.

Kauffman, S.A. (1970) "Articulation of Parts Explanation in Biology and the Rational Search for Them," *PSA: Proceedings of the Biennial Meeting of the Philosophy of Science Association*, 1970, 257–272.

Kim, J. (1998) *Mind in a Physical World: An Essay on the Mind-Body Problem and Mental Causation*, Cambridge, MA: MIT Press.

Kim, J. (2012) "The Very Idea of Token Physicalism," in C. Hill and S. Gozzano (eds), *New Perspectives on Type Identity: The Mental and the Physical*, Cambridge: Cambridge University Press, 167–185.

Klein, C. (2013) "Multiple Realizability and the Semantic View of Theories," *Philosophical Studies*, 163, 683–695.

Kuhlmann, M. and S. Glennan. (2014) "On the Relation Between Quantum Mechanical and Neo-Mechanistic Ontologies and Explanatory Strategies," *European Journal for Philosophy of Science*, 4, 337–359.

Levy, A. and W. Bechtel. (2013) "Abstraction and the Organization of Mechanisms," *Philosophy of Science*, 80, 241–261.

Lewis, D. (2004) "Causation as Influence," reprinted in J. Collins, N. Hall and L.A. Paul (eds), *Causation and Counterfactuals*, Bradford: MIT Press, 75–106.

Machamer, P. (2004) "Activities and Causation: The Metaphysics and Epistemology of Mechanisms," *International Studies in the Philosophy of Science*, 18, 27–39.

Machamer, P., L. Darden and C.F. Craver. (2000) "Thinking about Mechanisms," *Philosophy of Science*, 67, 1–25.

Marcus, E. (2006) "Events, Sortals, and the Mind-Body Problem," *Synthese*, 150, 99–129.

Marr, D. (1982) *Vision*, San Francisco: Freeman Press.

Martin, C.B. and M. Deutscher. (1966) "Remembering," *Philosophical Review*, 75, 161–196.

Menzies, P. (2012) "The Causal Structure of Mechanisms," *Studies in History and Philosophy of Science Part C: Studies in History and Philosophy of Biological and Biomedical Sciences*, 43, 796–805.

Nagel, E. (1961) *The Structure of Science: Problems in the Logic of Scientific Explanation*, New York: Harcourt, Brace and World, Inc.

Oppenheim, P. and H. Putnam. (1958) "Unity of Science as a Working Hypothesis," in H. Feigl, M. Scriven, and G. Maxwell (eds), *Concepts, Theories, and the Mind-Body Problem,* Minnesota Studies in the Philosophy of Science II, Minneapolis: University of Minnesota Press, 3–36.

Polger, T. (2004) *Natural Minds*, Cambridge, MA: MIT Press.

Putnam, H. (1967) "Psychological Predicates," in W.H. Capitan and D.D. Merrill (eds), *Art, Mind, and Religion*, Pittsburgh: University of Pittsburgh Press, 37–48.

Richardson, R.C. and A. Stephan. (2007) "Emergence," *Biological Theory*, 2, 91–96.

Romero, F. (2015) "Why There Isn't Inter-level Causation in Mechanisms," *Synthese*, 192, 3731–3755.

Salmon, W. (1984) *Scientific Explanation and the Causal Structure of the World*, Princeton, NJ: Princeton University Press.

Schaffner, K. (1993) *Discovery and Explanation in Biology and Medicine*, Chicago: University of Chicago Press.

Schlosser, G. and G. Wagner (eds) (2004) *Modularity in Development and Evolution*, Chicago: University of Chicago Press.

Shapiro, L. (2004) *The Mind Incarnate*, Oxford: Oxford University Press.

Simon, H. (1969) *The Sciences of the Artificial*, Cambridge, MA: MIT Press.

Van Gulick, R. (1993) "Who's in Charge Here? And Who's Doing All the Work?" in J. Heil and A.R. Mele (eds), *Mental Causation*, Oxford: Oxford University Press, 233–256.

Wimsatt, W.C. (1976) "Reductionism, Levels of Organization, and the Mind-Body Problem," in G. Globus, G. Maxwell and I. Savodnik (eds), *Consciousness and the Brain: A Scientific and Philosophical Inquiry*, New York: Plenum Press, 202–267.

Wimsatt, W.C. (1997) "Aggregativity: Reductive Heuristics for Finding Emergence," *Philosophy of Science*, 64, 372–384.

Woodward, J. (2003) *Making Things Happen*, Oxford: Oxford University Press.

15

MECHANISMS AND NATURAL KINDS[1]

Emma Tobin

Scientists divide the particulars that they study into kinds and classes. Chemists classify particular pieces of gold, as members of the class gold. We cluster things together and find this useful for the purposes of prediction and explanation. For example, we think we can study one piece of gold and learn things about other pieces of gold. Debates about classification in philosophy have traditionally focused on the delineation of natural kinds of substances. More particularly, we ask whether these natural kinds reflect real natural divisions in nature or whether alternatively they merely reflect the theoretical interests of the scientists that use them. The discussion has more recently invoked a more generous notion of natural kind, going broader than substances to consider whether we can have natural kinds of processes, relations, and/or property clusters (Ellis, 2001).

As we will see, some views of natural kinds (Boyd, 1991) have even invoked the notion of a mechanism to provide a robust realist account of natural kinds. In the meantime, the mechanisms literature has occasionally discussed the reality of mechanisms, but has seldom addressed the question of whether mechanisms actually form natural kinds. A closer analysis of the relationship between natural kinds and mechanisms is therefore required. The analysis of this chapter reframes the question as to what a more liberal account of classification might look like, one which includes mechanisms. Moreover, I will show that thinking seriously about how we classify mechanisms in scientific practice significantly alters our understanding of what classification is supposed to do.

Boyd's (1991) homeostatic property cluster (HPC) account is one much-discussed view of natural kinds, and it invokes the notion of a mechanism as a factor sustaining the clusters of properties that are considered to be natural kinds. On Boyd's view, natural kinds are reliable for the purposes of explanation and inductive inference, precisely because clusters of properties are reliably sustained by homeostatic mechanisms. However, an ontological account of mechanisms is not sufficiently developed by those holding this view, nor is the relationship between kinds and mechanisms precisely cashed out. Kinds, we are told, are understood as property clusters sustained by homeostatic mechanisms. Nevertheless, the relationship between the two, namely what counts as a mechanism and how homeostasis is achieved, is not sufficiently developed.

To develop this view, a closer analysis of how mechanisms are involved in classification is required, which directly addresses the relationship between kinds and mechanisms. An additional question is raised from considering this relationship, which is how we might classify mechanisms themselves. Can mechanisms be sorted into more and less similar mechanisms, also allowing us to study one token mechanism and think we can learn something about similar mechanisms? If so, perhaps we need to broaden further our notion of natural kinds to accommodate mechanisms as well as substances, processes, relations, and/or property clusters.

One ambition of addressing these questions would be to provide a more robust realist HPC view of natural kinds, which precisely specifies the role of mechanisms in maintaining natural property clusters, and my primary aim in this chapter will be to examine whether this is feasible. Craver (2009) is skeptical because he claims that conventional elements are involved in deciding when two mechanisms are of the same kind and also in identifying when one mechanism ends and the other begins. Kind classifications are affected by facts about how humans find it useful to cluster properties. This is a key conventionalist challenge to the HPC account, which undermines a realist interpretation of mechanisms and their natural property clusters. My secondary aim in this chapter will be to examine how thinking seriously about classifying mechanisms can illuminate both the mechanisms literature and the classification literature.

I am going to do two things. In section 1, I will bring together the classification literature with the mechanisms literature in an attempt to show how the two debates impact on one another. In section 2, I will address two pragmatic challenges facing mechanism classification. In the mechanisms literature there has been a worry about drawing the boundaries of mechanisms. This boundary problem, as I will call it, is about token mechanisms. Theorists are concerned about how we can know where one mechanism begins and another ends. Related to this is the concern about how we delimit token mechanisms from the environment. In contrast to the mechanisms literature, boundary problems in the kinds literature have been about how we sort things into kinds, or, in other words, how we sort tokens into types. Given that the HPC view is an attempt at combining both mechanisms and natural kinds, a positive view will need to get clear on these various kinds of boundary issues. I examine these issues looking at two case studies from scientific practice in biochemistry and chemistry, showing how there are ways in which classifying mechanisms runs into familiar problems from the classification debate. In particular, I argue that there is an overlapping problem for mechanisms similar to the problem that has been discussed in the natural kinds literature. I will also show that the dynamic and complex picture of kinds and mechanisms that emerges from practice transforms how we should think about classification.

In section 3, I return to views of classification, and explore an apparent circularity that infects the HPC view. The difficulty is that mechanisms were introduced to help with the delineation of the property clusters on the HPC account of natural kinds, but addressing the issues that emerge when identifying mechanisms in scientific practice seems to require kinds of property clusters. This is because the property clusters will be the phenomena that the mechanisms are invoked to explain, and the phenomena partially individuate the mechanisms. In the end, I argue that the HPC view must answer both of the boundary problems as well as the issues of circularity if it is to succeed as an account of classification informed by scientific practice.

1. Philosophical approaches to natural kinds

The two traditional theoretical accounts of natural kinds are essentialism and conventionalism. Essentialists hold that there are essential properties that all and only members of a natural kind share, which makes them categorically distinct (Ellis, 2001; Lowe, 2006). These properties provide necessary and sufficient conditions for kind membership. For example, chemical elements are solely individuated by their atomic number, namely the number of protons in their atomic nuclei. Silver is the element with 47 protons in its atomic nucleus. A substance with an additional proton, 48 in total, will be the chemical element cadmium and not silver. The boundaries between chemical elements are such that an atomic number is a necessary and sufficient condition for kind membership.

Conventionalists hold that there is no objective distinction between natural kinds and any other cluster of properties (Hacking, 1991a, 1999). Instead, the kinds that are deemed natural are those that have theoretical and practical salience for the scientific practitioner. For example, LaPorte (2004) uses the following example to make clear how the practical commitments of a group of scientific practitioners can impact the decisions about natural kind membership. Jade is microstructurally composed of two distinct chemical structures; jadeite and nephrite. For many centuries the jade used in China was just nephrite. Toward the end of the eighteenth century jadeite from Burma was introduced to China. Practitioners knew that it was different micro-structurally, but nevertheless, decided to regard it as jade.[2] A conventionalist would argue that in this case contingent and conventional elements come into play in deciding what counts as a member of the natural kind jade.

In Bird and Tobin (2008), we distinguish two forms of realism about natural kinds, which sit on the spectrum between essentialism and conventionalism. *Strong realism* about natural kinds is the view that there exist real entities that correspond to the natural kinds. Alternatively, *weak realism* is just naturalism about natural kind classes. In terms of our earlier example, the difference between silver and cadmium is not just a difference between two natural classes of thing, but is a difference between two distinct and real entities on this view.

Boyd's homeostatic property cluster account of natural kinds is often seen as a way between the traditional essentialist and conventionalist account of natural kinds. Boyd originally proposed this view of natural kinds as a way of articulating the kinds of the special sciences (e.g. biology, psychology, the social sciences, etc.). In particular, Boyd's view was introduced to try to refute the view in the philosophy of biology that species should be understood as individuals rather than as natural kinds (Ghiselin, 1974, 1987, 1997; Hull, 1976, 1978). He acknowledged that essentialism could not accommodate special science kinds, and wanted to formulate a weaker version of realism that would show how the kinds of the special sciences are objective and sufficiently robust to support inductive inference and explanation, without involving a commitment to essentialism. The kinds in the special sciences appear difficult to accommodate on traditional accounts because they are changing and dynamic. For example, take a particular species of "rabbit," like the Cashmere Lop found in Britain. This is not static, but has come into being at a particular time and place and continues to evolve. So an account of natural kinds that is static and uncompromising does not appear fit for purpose for all kinds.

According to Boyd's homeostatic property cluster view of natural kinds, natural kinds are clusters of co-occurring properties. Boyd's account is an extension of the traditional account of natural kinds. He claims that this extension is appropriate just to the extent that the kinds in question are employed for induction and explanation. Reference to a particular natural kind is appropriate for certain sorts of inductive generalization, because it is explained by *property correlations* that obtain independently of the understanding. These property correlations are sustained by homeostatic mechanisms. Studying gold allows us to learn something about other samples of gold, and studying a particular species of rabbit or a population of rabbits tells us something about other rabbits or populations; this is why these are candidate natural kinds. Boyd's view then might be regarded as a strong form of realism, rather than a merely naturalist view, because there are real entities, the property clusters that correspond to the natural kinds. These clusters are complex and dynamic in nature, which is why it is so difficult to draw their boundaries.

The key question that arises is: why do these property correlations obtain? Boyd's own answer is that clusters of properties are found together in virtue of homeostatic mechanisms.

This view portrays a dualist ontic picture including (a) real property clusters and (b) real homeostatic mechanisms. The HPC view argues that natural kinds are real property clusters in the world because they are sustained by underlying mechanisms. For Boyd, a homeostatic mechanism is defined loosely as a mechanism, which maintains a cluster of properties. For example, in the case of a population (or species) of rabbits (e.g. the Cashmere Lop), we can explain the clustering together of properties of members of the species by the homeostatic mechanisms in virtue of which members all form the same population or species, such as common ecological niche or gene exchange within this population.

Or take another example at the sub-organismic level: consider the heart as the means by which the circulatory system pumps the blood around the body. In Boyd's terms, the heart is a clustering together of properties, which is maintained by the homeostatic mechanisms underlying it. The property of having a regular heart rhythm is maintained by the heart's natural pacemaker in the sinoatrial node. The pacemaker sends out regular electrical impulses from the top chamber, causing it to contract and pump blood into the lower ventricle. The pacemaking mechanism allows the heart to contract in a continuous fashion, which allows it to pump blood to the body and the lungs. These might be considered to be the underlying homeostatic mechanism which allows the heart to be considered as a stable grouping of properties, namely as a HPC kind.

The attractiveness of this view is that it moves away from the traditional static view of classification and allows for change within the clustering of properties over time. It is therefore consistent with the view that kinds like those in biology might evolve over time. The HPC view allows a more dynamic version of classification, which sidesteps some of the issues discussed in philosophy of biology surrounding the so-called species problem.

The ontological addition of mechanisms might appear promising to those interested in providing a realist account of natural kinds, in so far as the mechanistic structure of the world could serve as an objective foundation for an adequate taxonomy of kinds. In other words, kind concepts cut nature at its joints, and to find nature's joints, we must analyze how the homeostatic mechanisms can set the boundaries of the property clusters. In cases where property clusters are dynamic and may even change over time, the aspiration is that the mechanisms can help us to sort token property clusters into types. This is not just of theoretical interest, but also provides a practical rationale for how scientists might approach taxonomy in practice, in that it provides hope of knowing how to go about identifying kinds in practice. There are also additional benefits for classification, primarily because this view is less stringent than its essentialist alternative. It allows for natural kinds to change some of their properties over time within any given cluster, while remaining relatively stable over time. This is a very attractive view for taxonomists, particularly in disciplines like biology, where natural kinds evolve over time: it seems promising for rabbits. In fact, it is the inability of the old essentialist view to straightforwardly accommodate these changes, which has been the chief objection to it.

The HPC view, then, presents us with property clusters sustained—particularly sustained together—by underlying mechanisms. Parallel to the classification literature, a significant literature about the nature of mechanisms has developed (Bechtel and Abrahamsen, 2005; Craver, 2007; Glennan, 1996, Machamer, Darden, & Craver, 2000). Glennan and Illari argue that these new mechanists have a set of core commitments about what mechanisms are that are embodied in a characterization of mechanisms they call minimal mechanism:

> A mechanism for a phenomenon consists of entities (or parts) whose activities and interactions are organized so as to be responsible for the phenomenon.
>
> *(Chapter 1 of this volume; see also Illari and Williamson,*
> *2012; Glennan, forthcoming)*

Here we have a mechanism responsible for a phenomenon, which aligns in many ways with the idea of an underlying mechanism sustaining a property cluster.

In Chapter 7 of this volume, Glennan and Illari suggest that a scheme for classifying things begins with a specification of the set of things to be classified. In the case of mechanism, the things to be classified are captured by minimal mechanism, which they use as the beginning of a classification of varieties of mechanisms. Minimal mechanism is an intentionally broad and permissive characterization of mechanism, which does not necessarily reject the more restrictive sense of mechanisms in earlier accounts, but rather treats these as types of mechanisms construed more broadly. Taking this broader version of mechanism would seem to offer the best path for providing a view of the complex and dynamic kinds that have proved difficult to characterize in realist accounts of natural kind classification.

Craver (2009) is the first to directly bring together the recent mechanisms literature and the HPC approach to kinds, and I will examine and develop his view in this through the rest of this chapter. According to the HPC view, the taxonomies of natural kinds are adequate for the purposes of explanation and induction only because the kinds in that taxonomy are sustained by mechanisms. This view provides a justification for realism about natural kinds, because the properties cluster together in virtue of these mechanisms. Natural kind concepts cut nature at its joints and Craver suggests this fact can provide a normative constraint, if mechanisms have real boundaries and so sustain real boundaries between the property clusters they sustain. In Craver's own term's, "nature's joints are located at the boundaries of mechanisms" (2009: 575).

An interesting question then arises of what exactly is meant by the boundaries of mechanisms. In the mechanisms literature, the question of the boundaries of mechanisms is a pragmatic challenge about how a practitioner would know which entities, activities, and organizational features are part of a mechanism (or kind of mechanism) and which are not (see Chapter 9). Now interestingly, the boundary question in the kinds literature (as we saw earlier) is a question about how to sort token kind members into types. These are two separate questions. What makes the HPC view so interesting is that it attempts to use mechanisms to solve the kinds boundary problem; namely, the reason why we can sort property clusters into types is because they are sustained by homeostatic mechanisms. However, this will require additionally an answer to the first mechanisms boundary question. So, to say which homeostatic mechanisms are responsible for property clusters, we will need to be able to clearly draw the boundaries of token mechanisms, and perhaps also sort those mechanisms into kinds of mechanisms. This serves to make clear that there is a serious challenge in conflating the boundary issues, and a better understanding of the relationship between property clusters and mechanisms will require a much deeper analysis than has been provided in the literature thus far.

The addition of mechanisms successfully provides an important pragmatic strategy for allowing the practitioner to divide the world into kinds. This strategy, elaborated by Craver (2009), is called the lump and split methodology: if you find that a single cluster of properties is sustained by more than one kind of mechanism, split the cluster into subset clusters, each of which is explained by a single kind of mechanism. Alternatively, if you find that two or more distinct kinds are explained by the same kind of mechanism, lump the kinds into one. For example, take the traditional yet controversial example of species delineation in biological taxonomy. In the case of a speciation event where reproductive isolation occurs, reproductive isolation changes the mechanism for the subset of the species population that becomes isolated. In this case, the practitioner is forced to split the species into two separate species given the operation of this new mechanism.

Despite the initial attractiveness of the view, Craver (2009) raises two concerns, which suggest a huge impact of conventional elements. In the end, he argues that conventional elements come into play in deciding (a) when two mechanisms are mechanisms of the same type, and (b) where one particular mechanism ends and another begins. In the next section, I will look at how the issues of overlapping have been dealt with in the classification literature so far. Additionally, I will address how issues of overlap impact some real cases from scientific practice. An analysis of these cases will prove helpful in looking at the relationship between mechanisms and property clusters in practice.

2. Overlapping: mechanism and kind classification

In the classification literature, there has been much discussion of the no-overlap principle, which is intended to allow a categorical distinction between natural kinds from a realist perspective. As discussed earlier, chemical elements have clear boundaries and cannot overlap. The atomic number of an element makes it categorically distinct from the elements that come either side of it on the periodic table. The hierarchy thesis for natural kinds is one way for natural kinds realists to accommodate apparent cases of the overlapping of kind members. Theorists claim that natural kinds form a hierarchy; if any two kinds overlap, then one must be subsumed under the other like a species under a genus (Kuhn, 2000: 228–52, Ellis, 2001: 67–76, 97–100). In Tobin (2010a), I discussed the following case, which would appear to support both the categorical distinctness of natural kinds and the hierarchy thesis. The elements magnesium (Mg) and promethium (Pm) are classified together as metals. The elements lanthanum (La) and promethium (Pm) are classified together as lanthanides. Therefore, in accordance with the hierarchy thesis, magnesium and lanthanum must also be classified together. This is the case because all lanthanides are classified as metals.

However, *ipso facto* cases of crosscutting natural kinds in scientific practice provide a serious challenge to the no-overlap principle for classification (Khalidi, 1998; Dupré, 1993; Tobin, 2010a). For example, in Tobin (2010a), I examine how crosscutting occurs in the case of proteins and enzymes, which appears to contradict both categorical distinctness and the hierarchy thesis. For example, albumin and renin can be classified together as proteins. Renin and the hairpin ribozyme can be classified together as enzymes. But the hairpin ribozyme and albumin cannot be classified together at all, because albumin is not an enzyme. So, enzymes and proteins are not categorically distinct. As we have seen with the problem of boundaries in the last section, the problem of boundaries recurs with mechanisms, and is essentially the issue pointed out by Craver when he claims that there is a difficulty in deciding where one particular mechanism begins and another ends. As stated earlier, there are two different issues in that the boundary problem for natural kinds would appear to be one of how we group tokens into types, whereas the boundary problem for mechanisms appears to be one about how we draw the boundaries between the token mechanisms themselves. Given that the HPC view involves both kinds and mechanisms, the HPC view sits awkwardly at the juncture of these two boundary problems.

Lumping and splitting provides a practical methodology for coping with the boundary issue for mechanisms. But it is worth pointing out that the strategy of lumping and splitting also adheres to the same kind of no-overlap principle as we saw in the natural kinds literature; namely, mechanisms that overlap should be lumped together, and those that do not overlap should be split. For example, in cognitive science, spatial selective attention and spatial working memory have largely been studied in isolation and so traditionally they have been split into two separate mechanisms, because they were thought not to overlap (Jonides and Awh, 2001).

The HPC view introduces mechanisms to provide an ontological entity, which allows us to draw the boundaries between property clusters. However, it would seem that this kicks the pebble down the path somewhat, since the boundary problem for kinds is replaced by an additional boundary problem for mechanisms. Moreover, even if the boundary problem for token mechanisms can be answered or at least dealt with pragmatically, then an additional sorting problem of how we group token mechanisms into types also arises. The mechanisms literature has rather loosely conflated these issues and separating them out will allow for a much more nuanced discussion of the relationship between mechanisms and natural kinds.

Despite these theoretical worries, the lump and split strategy is one used and in fact introduced by scientific practitioners themselves. The strategy was originally described by the geneticist Victor McKusick in 1969. Additionally, it is worth noting that a number of arguments in favor of the lump and split strategy have been provided in the mechanisms literature. The lump and split strategy is successful to the extent that mechanisms are categorically distinct, they develop independently of one another, they have evolved separately, and they can be manipulated or suffer loss of function independently of one another (Craver, 2009: 581).

However, on closer analysis it looks as though scientific practice also suggests a problem when mechanisms are added. Putting it in the terms of the classification debate, clearly mechanisms overlap, but in some cases they are not lumped together or subsumed one under the other, and in other cases they are not split. For example, take the case above from cognitive science. A more recent study has discovered that actually the mechanisms for spatial attention and spatial working memory interact and overlap in performing early sensory processing. In fact, spatial memory recruits top-down processes from spatial attention and these modulate the earliest stages of visual analysis. Despite this clear overlap, these mechanisms continue to be treated as distinct. Craver's argument essentially is that sometimes the decision to lump or split involves conventional and contextual elements.

The problem with the strategy of lumping and splitting is that there is an assumption that once a mechanism is found to be responsible for a property cluster, this alone is sufficient for delineating the boundaries of that cluster. However, in fact it is more likely that there are a number of mechanisms that are responsible for a given clustering of properties. The boundary problem in the mechanisms debate has been concerned with drawing the boundaries between these related mechanisms. In other words, it might be possible in some cases to lump or split the same property cluster differently depending on which mechanism the practitioner uses or is focusing on. This is made worse by the fact that, in scientific practice, it is in fact normal that multiple mechanisms might sustain the same property cluster. For example, Craver notes that there might be multiple etiological pathways, which could result in the same clustering of properties. This is certainly the case for psychiatric disorders, which often have multiple distinct etiological mechanisms. For example, prenatal smoking exposure, low birth weight and exposure to high amounts of lead are distinct risk factors for attention-deficit/hyperactivity disorder (ADHD). In other words, there might be multiple causal routes that could result in a similar functional output. Moreover, there might be different kinds of mechanisms, which produce the same property cluster, some etiological and some constitutive, and depending on which one we are using, the outcome of whether to lump or split may be different.

The problem of overlapping will be made clearer by considering some cases from scientific practice. By considering two cases in detail, I hope to be able to show how the different boundary problems play out, but also give some clarity on whether these problems are at the level of tokens or at the level of types. Protein synthesis is a much-discussed case study in the mechanisms literature, since the complexity of protein folding is illustrative of mechanistic behavior. I would like to focus here on the case of protein classification more particularly. The protein data bank is the primary repository of protein structures. Scientists use experimental techniques such as X-ray crystallography, NMR spectroscopy and cryo-electron microscopy to determine the location of each atom relative to the others in the molecule. Then, this information is deposited in the protein data bank. However, the sheer volume and complexity of this data mean that the scientific community has had to design *hundreds* of secondary databases that categorize the data according to different criteria, to serve their different purposes.

Before using any one of these different databases to classify (e.g. SCOP, CATH, or DDD), scientists have to split the proteins into domains. Domains are distinct globular parts, which are considered to function independently. These globular parts are tertiary structures which are the result of a simple mechanism where the molecule's apolar amino acids are bound to the molecule's interior and the polar amino acids are bound outwards, allowing dipole-to-dipole interactions. In effect, they are the simple mechanisms of protein folding. Once a newly solved structure is discovered, the information is placed, with its unique PDB identifier, in the protein data bank.

In incremental classification, given a newly solved structure, the new protein structure is partitioned into domains, and each new domain is compared to domains of the existing classification to identify the most similar folds. If no such fold exists (dependent on the parameters of the classification), then a new fold is added to the classification and the particular domain is added to it. In other words, if the domain cannot be lumped then it is split. In practice, at least superficially, this looks like a case of support for the lumping and splitting strategy, in so far as there appears to be a clear methodology for when we should lump and split protein domains.

However, this makes the case look more straightforward than it is. In practice, scientists use many different algorithms for domain identification. Splitting proteins into domains makes the boundary problem for mechanisms quite clear. Domains are not distinct parts of proteins, which can be clearly delineated and are categorically distinct; there are different ways to draw the boundaries, because of the different mechanisms of protein folding.

In other words, the concept of an independent functioning unit is not enough to allow us to delineate the boundaries of protein domains absolutely. Scientists end up having to design algorithms to interrogate the primary data so as to artificially agree upon a boundary for the protein domains. This is made more problematic by the fact that the algorithms used do not always agree on what the boundaries are and often "domain choice," as scientists call it, is relatively subjective.

So, before the practitioner can even make the decision to lump and split, the scientific practitioner must make a domain choice. The domain is an artificial freeze frame of the part of a protein that is sustained by the underlying mechanisms of protein folding. Depending on the algorithm or category used (e.g. family, superfamily) plus which database is used (there are over 100), then it looks as though quite a few conventional elements come into the picture in practice. This case makes clear the complexity of the boundaries of the protein domains and the practical tools scientists need to use before the work of classification can even begin.

There is a further problem that we might consider when drawing the boundaries for token mechanisms found in different contextual settings. In Tobin (2010b), I discussed the case of moonlighting proteins. Moonlighting proteins are proteins that have different functional roles in different parts of the organism. Another example is crystallins, which have a structural function in the lens of the eye, but also aid catalytic activity in cells. Crystallins are examples of what scientists call intrinsically unstructured proteins, which means that the mechanism for protein folding differs slightly depending on the context. One and the same primary structure can assemble different kinds of macromolecules (quaternary structures) with different functional roles in different parts of the organism. This is because of the structural malleability of the region involved in binding, which allows for different protein folds because of the interactions with other biochemicals found in different environments.

The question raised by this case is whether the two-tokens mechanism can be grouped under the same mechanism type—namely mechanisms for crystallins—when they are found to be distinct in different parts of the organism. However, it would appear that the decision to lump together the two mechanisms under the same type—as mechanisms for crystallins—is taken because they are considered to be two tokens of the same type, despite their different functional roles in different contexts.

To sum up, drawing the boundaries between mechanisms for proteins is extremely difficult. The first issue is the difficulty of creating a type of protein domain (which is an idealized model of the dynamic part of a protein sustained by the background mechanisms of protein folding). The second issue is that protein folding itself can differ in different contexts, to the extent that we sometimes have to split proteins because of their great functional difference in different contexts. Importantly, the biochemical milieu is intimately connected to the mechanisms of protein folding, so there is a real issue of drawing the boundaries between the mechanisms and their environment in these complex biochemical cases.

A similar and interesting case is the use of computational methods in the classification of chemical reactions. Ratcliffe (2015) examines the following case in chemical classification. She discusses the use of Quantitative Structure Activity Relations (QSAR) and neural networks in the classification of chemical reactions. For classification by QSAR and neural networking to take place, a preliminary classification of reactions into general types of changes in entities takes place. Scientists seek to identify the structural properties that correlate with the instantiation of certain types of chemical reactions. Scientists have to use this methodology because of the infinite number of chemical reactions that are possible. They seek to find a mechanism that is responsible for the general change perceived in a chemical reaction of a certain type. This is then further refined to identify the subtypes of reactions within that type. This case allows us to see the complexity involved in the mechanisms for chemical reactions. As a result of this complexity, scientists in practice have designed and used pragmatic techniques to classify dynamic entities and their activities. Scientists need to apply QSAR methods to identify the relevant data at the level of token mechanisms for chemical reactions before classification into types can even commence.

Ratcliffe also (2015: 74) discusses the ambiguity in deciding the exact physicochemical properties that best determine the route of any kind of chemical reaction. These will in fact vary on a case-by-case basis because identifying the correct properties is a matter of determining the structure-activity relation governing a particular reaction. In some cases, the difference in partial atomic charges, which describe the polarity of the bond, will be responsible, while in other cases it will be effective bond polarizability. In other words, the chemical reaction can occur via different causal routes, which amount to distinct mechanisms even though they determine one and the same kind of chemical reaction. Again, which causal route operates will be highly context sensitive.

Both of these cases make clear how fuzzy the boundaries of mechanisms can be. The mechanisms that regulate kinds in practice are so dynamic and complex that scientists in both these cases have had to develop techniques to identify the relevant data before classification can begin. In both cases, conventional and contextual elements often dictate the choice of the boundaries of the mechanisms involved.

Recall that on the HPC view, mechanisms are introduced to give a rationale for the boundaries of property clusters. It is these sustaining mechanisms which make the natural kinds stable and robust for the purposes of induction and prediction. However, looking at the relationship between the two cases in scientific practice, it would appear that drawing the boundaries of the underlying mechanism is itself often dictated by conventional elements and pragmatic techniques designed by the scientists to cope with the complexity and dynamism both of the mechanism themselves and of the clusters that are maintained by those mechanisms.

3. Circularity

Consideration of the cases in scientific practice suggests there is a problem of circularity that is faced by the HPC view of natural kinds. Given that the boundaries of mechanisms are vague and can be drawn in different nuanced ways, it is perfectly legitimate to ask: when can we say that token mechanisms are mechanisms of the same type? In particular, when distinct mechanisms are responsible for a property cluster, do we need to look at the resulting cluster to say whether the mechanisms are mechanisms of the same kind? This would be particularly true in cases where quite distinct mechanisms might both regulate a property cluster.

To make this clear, let's return to our population of rabbits. We saw earlier that multiple mechanisms might be involved in holding a property cluster together. So, for example, in the case of rabbits, shared ecological niche and gene exchange in the population are two mechanisms which make the property cluster a reliable one for drawing the boundaries of the population. But these two heterogeneous mechanisms are not mechanisms of the same type. They have in common only that they sustain the same property cluster. This evokes a problem of circularity given that the mechanisms were introduced to help us to delineate the property clusters on the HPC account. But now it looks as though one way of dealing with the boundary problem for mechanisms is to look at what mechanisms are responsible for sustaining; namely, the resulting property clusters.

We saw earlier that the mechanisms literature has traditionally been focused on solving boundary problems for token mechanisms, whereas the kinds literature has been more interested in how we group tokens into types or classes. An interesting question that emerges from considering these cases in chemical practice is whether this type-level organization of mechanisms should be considered to have any real ontological significance, because many heterogenous mechanisms are often responsible for sustaining a property cluster and these can also differ depending on the environment and context. Token-level differences between the mechanisms, which are responsible for kind clusters in practice, make the identification of the boundaries for types of mechanisms very difficult (perhaps even impossible). I have also shown that because of the complexity of mechanistic classification, practitioners are often forced to use pragmatic techniques to interrogate voluminous data about these complex causal routes. In fact, these considerations suggest that grouping mechanisms into types may be mostly a matter of epistemic parsimony and convenience.

If this is the case, then it now becomes much clearer that the "homeostatic mechanism" part of the HPC view is not a simple type-level component, but rather a catchall for a very complex set of mechanism tokens, which in scientific practice more often than not requires very complex technologies for the boundaries to be drawn. The impression from scientific practice, in fact, is that we often draw the boundaries in an artificial or idealized way, depending on which technology is used. In the case of protein domains, for example, different algorithms will result in distinct drawing of boundaries.

Originally, the aspiration of the HPC view was to allow a strong form of realism about dynamic property clusters, allowing a more liberal account of natural kinds. However, if my analysis of scientific practice is correct, then the addition of mechanisms brings with it some deep challenges about how we can draw the boundaries. If it is a form of realism, given all of the pragmatics that come into play, it is a very promiscuous form of realism indeed.

4. Conclusion

Thus far, I have discussed the HPC view of natural kinds by attempting a more in-depth discussion of the relationship between property clusters and mechanisms. Several problems have been raised for this relationship. I have argued for theoretical reasons that there is a problem of circularity in the introduction of mechanisms for solving the boundary problem for kinds. I have also shown that in scientific practice the boundary problems are much more problematic than has been acknowledged in the mechanisms literature and that a conflation of token and type-level classification has proved unhelpful. This means that the "homeostatic mechanism" part of Boyd's account is not nearly clearly enough defined.

One possible response from the new mechanists would be to argue that within the new mechanistic framework there is no need for HPC kinds at all. Some new mechanists (Bogen, 2005; Machamer, 2004; Glennan, forthcoming) claim that both causes and mechanisms are singular, not general or universal. However, if the circularity worries above are correct, then to solve the boundary problem for mechanisms, we need (at the very least for pragmatic reasons) to group those mechanisms into property clusters or mechanism kinds. In fact, this is precisely what scientists do when using methods like QSAR and neural networking in the case of chemical reactions. The new mechanists need to either provide an argument for why we do not need property clusters or mechanism types, or they need to show why these groupings are merely epistemic and not of ontological significance.

Another puzzle for the mechanists is what I would call the problem of generality for a singularist conception of mechanisms. A core consensus view of the mechanisms literature in general is that we should do without laws of nature in many of their traditional roles. Particularly in the early literature on mechanisms, mechanisms are contrasted explicitly with laws of nature (Bechtel, 1988; Bechtel and Abrahamsen, 2005; MDC, 2000) and the mechanisms literature can rightly be viewed as an alternative view to laws-based accounts of explanation. This is certainly the case in the philosophy of biology literature in the discussion of *ceteris paribus* biological generalizations. On the mechanisms view, the resilience and necessity of these generalizations are because of the resilience of the mechanisms. However, if we allow for promiscuity in terms of the overlapping of mechanisms and we accept that we have to rely either on type-level mechanisms or the resulting property clusters which the mechanisms sustain, then are we not back to formulating generalizations about these, which look very much like laws of nature understood as universal generalizations?

Future work will need to addresses the question of whether the boundary problems for mechanisms commit the new mechanist to laws of nature. There are already some movements in this direction in recent literature. Glennan (forthcoming, ch. 4) makes the interesting suggestion that models are the tools that allow practitioners to classify particulars into kinds. In Chapter 12 of this volume, Andersen also offers an argument for how laws and mechanisms relate without either conceptually or ontologically reducing to one another.

In conclusion, despite the initial ambition of having a more liberal view of classification, which would allow for the joint classification of mechanisms and kinds, considerations from scientific practice together with theoretical concerns appear to undermine the HPC account significantly. Cases in scientific practice show that there is a boundary problem for locating the vague boundaries of mechanisms. I have argued that the mechanisms, which regulate kinds in practice, are so dynamic and complex that scientists in practice are often forced to develop techniques to identify the relevant data before classification can even begin. Moreover, depending on the techniques used, the boundaries are often drawn differently. Secondly, I argue that there is a theoretical problem of circularity because the vagueness of mechanism boundaries requires classification into kinds to avoid ambiguity and overlap. It seems clear from a closer examination of the relationship between mechanisms and kinds that mechanisms alone do not have the ontic resources to allow us to identify nature's joints.

Notes

1 I would like to thank Professor Stuart Glennan and Dr Phyllis Illari for excellent guidance and feedback on earlier drafts of this chapter. Their editorial role has been crucial in producing the final article. I am also grateful to Professor Alexander Bird and to staff and students at the Department of History and Philosophy of Science in Cambridge who all commented on earlier versions of this chapter.
2 It is worth noting that LaPorte uses this case not as an argument for conventionalism about natural kinds, but to reject Kripke's view that the essences of natural kinds are discovered a posteriori. I am using it here as a case to show how a conventionalist might argue that pragmatic and conventional elements play a role in the delineation of natural kinds.

References

Bechtel, W., 1988, *Philosophy of Science: An Overview for Cognitive Science*, Hillsdale, NJ: Erlbaum. Italian translation: Filosofia della scienza e scienza cognitiva, Gius. Laterza & Figli, 1995.
Bechtel, W. and A. Abrahamsen, 2005, "Explanation: A Mechanistic Alternative", *Studies in History and Philosophy of the Biological and Biomedical Sciences*, 36: 421–441.
Bird, A. and E. Tobin, 2008, "Natural Kinds", *Stanford Encyclopedia of Philosophy*, https://plato.stanford.edu/entries/natural-kinds/.
Bogen, J., 2005, "Regularities and Causality: Generalizations and Causal Explanations", *Studies in History and Philosophy of Science Part C*, 36(2): 397–420.
Boyd, R., 1991, "Realism, Anti-Foundationalism and the Enthusiasm for Natural Kinds", *Philosophical Studies*, 61: 127–148.
Boyd, R., 1999a, "Homeostasis, Species, and Higher Taxa", in R. Wilson (ed.), *Species: New Interdisciplinary Essays*, Cambridge, MA: MIT Press: 141–186.
Boyd, R., 1999b, "Kinds, Complexity and Multiple Realization", *Philosophical Studies*, 95: 67–98.
Craver, C.F., 2001, "Role Functions, Mechanisms & Hierarchy", *Philosophy of Science*, 68(1): 53–74.
Craver, C.F., 2007, *Explaining the Brain: Mechanisms and the Mosaic Unity of Neuroscience*, Oxford: Clarendon Press.
Craver, C.F., 2009, "Mechanisms and Natural Kinds", *Philosophical Psychology*, 22: 575–594.
Craver, C.F. 2015, "Mechanisms and Emergence: A Reply to Denis, C. Martin", in T. Metzinger & J.M. Windt (eds), *Open MIND*, Frankfurt am Main: MIND Group. doi: 10.15502/9783958571099.
Dupré, J., 1993, *The Disorder of Things: Metaphysical Foundations of the Disunity of Science*, Cambridge, MA: Harvard University Press.
Ellis, B., 2001, *Scientific Essentialism: Cambridge Studies in Philosophy*, Cambridge: Cambridge University Press.
Ghiselin, M.T., 1974, "A Radical Solution to the Species Problem", *Systematic Zoology*, 23: 536–544.
Ghiselin, M.T., 1987, "Species Concepts, Individuality and Objectivity", *Biology and Philosophy*, 2: 127–144.
Ghiselin, M.T., 1997, *Metaphysics and the Origin of Species*, Albany: State University of New York Press.
Glennan, S., 1996, "Mechanism and the Nature of Causation", *Erkenntnis*, 44: 49–71.
Glennan, S., 2002, "Rethinking Mechanistic Explanation", *Philosophy of Science*, 69: S342–S353.
Glennan, S., Forthcoming, *The New Mechanical Philosophy*, Oxford: Oxford University Press.
Hacking, I., 1991a, "A Tradition of Natural Kinds", *Philosophical Studies*, 61: 109–126.
Hacking, I., 1991b, "On Boyd", *Philosophical Studies*, 61: 149–154.
Hacking, I., 1999, *The Social Construction of What?* Cambridge, MA: Harvard University Press.
Hull, D., 1976, "Are Species Really Individuals?", *Systematic Zoology*, 25: 174–191.
Hull, D., 1978, "A Matter of Individuality", *Philosophy of Science*, 45: 335–360.
Illari, P., and J. Williamson, 2012, "What Is a Mechanism?: Thinking about Mechanisms Across the Sciences", *European Journal for Philosophy of Science*, 2: 119–135.

Jonides, J. and Awh, E., 2001, "Overlapping Mechanisms of Attention and Spatial Working Memory", *Trends in Cognitive Science*, 5(3): 119–126.

Khalidi, M., 1998, "Natural Kinds and Crosscutting Categories", *Journal of Philosophy*, 95: 33–50.

Kuhn, T., 2000, *The Road Since Structure*, Chicago: University of Chicago Press.

LaPorte, J., 2004, *Natural Kinds and Conceptual Change*, Cambridge: Cambridge University Press.

Lowe, E.J., 2006, *The Four-Category Ontology: A Metaphysical Foundation for Natural Science*, Oxford: Clarendon Press.

Machamer, P., 2004, "Activities and Causation: The Metaphysics and Epistemology of Mechanisms", *International Studies in the Philosophy of Science*, 18: 27–39.

Machamer, P.K., L. Darden, and C.F. Craver, 2000, "Thinking about Mechanisms", *Philosophy of Science*, 67: 1–25.

Ratcliffe. S., 2015, *A Metaphysics for the Classification of Chemical Reactions in Practice*, PhD Thesis, UCL Discovery. Available at: http://discovery.ucl.ac.uk/1464574/1/Steph%20Ratcliffe%20PhD%20Final%20copy.pdf.

Tobin, E., 2010a, "Crosscutting Natural Kinds and the Hierarchy Thesis", in Helen Beebee and Nigel Sabbarton-Leary (eds.), *The Semantics and Metaphysics of Natural Kinds*, London: Routledge: 1–179.

Tobin, E., 2010b, "Microstructuralism and Macromolecules: The Case of Moonlighting Proteins", *Foundations of Chemistry*, 12(1): 41–54.

PART III

Mechanisms and the philosophy of science

16

MECHANISTIC EXPLANATION AND ITS LIMITS

Marta Halina

1. Introduction

Many attempts have been made by philosophers to provide a satisfying account of what constitutes an explanation in the sciences. Over the last few decades, particular attention has been given to the role of mechanisms in providing such an account in the life sciences. Mechanistic explanations have proved to provide a powerful account of both the practices of biologists and the normative constraints on explanation. Even so, many philosophers are wary of explanatory hegemony, holding instead that a plurality of accounts of explanation will be needed to capture the diversity of scientific practice (Godfrey-Smith 2003).

This chapter examines the strengths and limitations of mechanistic explanation. It does this by considering the advantages of the mechanistic account over previous models of scientific explanation, as well as its descriptive adequacy with respect to various aspects of scientific practice. Concerning the latter, I focus on how mechanistic explanation accounts for the following three aspects of science: the appeal to non-mechanistic explanations, the use of abstract and idealized models, and the generality of explanation (for more discussion along these lines, see Chapter 17). Recently, critics of mechanistic explanation have cited these areas as posing problems for the mechanistic account. Examining whether and why this is the case will allow us to probe the boundaries of mechanistic explanation—highlighting its strengths and potential weaknesses. In each of these three cases, I discuss tools that the mechanist might use to address the problems advanced. As these areas represent active domains of research, I do not attempt to adjudicate them here. This chapter instead aims to introduce the reader to the current state of the art on mechanistic explanation, as well as potential directions for moving forward.

I begin in section 2 by introducing the idea of a general philosophical account of explanation and the various considerations and constraints that factor into the construction and evaluation of such an account. In section 3, I briefly introduce two precursors to mechanistic explanation before turning to mechanistic explanation in section 4. Section 5 examines the potential limitations of this model of explanation as they manifest in the three areas of scientific practice noted above.

2. What is explanation?

To provide an account of mechanistic explanation, it is important to first understand what is meant by "explanation." Generally, an explanation can be understood as an answer to a why question. We might ask why the sky is blue or why human skin wrinkles when submerged in water. The thing in need of an explanation (the blue sky or wrinkled skin) is the *explanandum*, whereas the explanation for this phenomenon is the *explanans*. Scientists may seek explanation as an end in itself—that is, to better understand a given phenomenon—or as a means to other ends, such as prediction and intervention (for a general introduction to scientific explanation, see Woodward 2014).

In developing a philosophical account of explanation, one must be clear about the standards that will be used to evaluate such an account. Should a philosophical account of explanation, for example, capture the way in which the term "explanation" is used in ordinary life? Or should it instead focus on the way it is used in a particular science, such as theoretical physics or molecular biology? Alternatively, should philosophers base their account of explanation on accounts of evidence or truth or justification, or should their main concern be capturing the explanatory practices of scientists? To date, philosophers have taken a variety of approaches to developing and evaluating general accounts of explanation. This has led to some confusion, as one might develop an account using one standard, such as capturing the normative practices of a particular scientific field, which might then be critiqued using a different standard, such as capturing ordinary language use (Ruben 2012).

For the purposes of this chapter, I adopt explanatory demarcation and normativity as the main standards by which to evaluate an account of explanation (see Craver 2014). *Explanatory demarcation* is the practice of distinguishing explanation from other activities, such as description or prediction. For example, identifying a bird as belonging to a particular species involves description and categorization; however, scientists do not view this as explanatory. *Explanatory normativity*, on the other hand, is the practice of distinguishing good explanations from bad. Biologists reject the claim that wet skin wrinkles because water enters the epidermis through osmosis, causing it to swell. Instead, they hold that this phenomenon is due to vein constriction triggered by the sympathetic nervous system. At present, the former is considered a bad explanation and the latter good.

The above criteria are widely adopted within philosophy of science as evaluative standards for philosophical accounts of explanation (Lipton 2004). One reason for this is that they enable philosophers to construct accounts of explanation that remain true to the practices of scientists. Although an account of explanation might be constructed that accurately describes our ordinary language use (see Wright 2012), such practices may fail to conform to the explanatory standards used in science. Insofar as this is the case, and our goal is to develop an account of scientific explanation, common sense and ordinary language practices may lead us astray. Within the broad realm of scientific practice, however, this chapter endorses a pluralist stance toward explanation. Discussions of explanation have often presumed that there is one explanatory form or relation that will capture the explanatory practices of all scientific disciplines (Woodward 2014). I follow others in rejecting this assumption (Godfrey-Smith 2003; Kellert, Longino, and Waters 2006). Perhaps we will discover that explanation in physics and anthropology and molecular biology all take the same form, but we should not begin our inquiry assuming that this is the case. With this in mind, the focus of this chapter will be on biological explanation (although some general examples will be provided for illustrative purposes). Lastly, it is important to note that explanatory demarcation and normativity are not the only features of explanation discussed by philosophers. Peter Lipton (2004), for example, highlights that explanations commonly have a self-evidencing nature (the explanandum may serve as evidence for the explanans) and lack explanatory regress (something can serve as an explanation even when it itself is unexplained). Although these may be important features of scientific explanation, I do not discuss them here due to limited space and the fact that they do not feature highly in the debates under consideration.

Before introducing precursors to mechanistic explanation in the next section, it would be useful to highlight a feature of explanation that is receiving increasing attention in the literature. This is the question of whether explanation should be understood as an epistemic activity or ontic feature of the world. Broadly, epistemic explanation concerns the models, representations, and activities used to communicate and elicit understanding of a target system. When I communicate to a group of students how photosynthesis works, I explain in the epistemic sense. Ontic explanation, in contrast, refers to that aspect of the world that explains why some worldly event happened (Craver 2007; Strevens 2008; Craver 2014). When I say that the ice on the road explains the car accident, I am using explanation in this ontic sense. The term "explanation" is often used in both ways (Illari and Williamson 2011; Colombo, Hartmann, and van Iersel 2014). For the purposes of this chapter, I adopt an ecumenical account that accepts both senses of explanation as important for understanding the explanatory practices of science (Illari 2013; but see Sheredos 2015). Distinguishing them is important, however, as they impose different constraints on good explanation, as we will see in section 5 (see also Illari 2013).

3. Precursors to mechanistic explanation

Contemporary discussions of scientific explanation find their origins in the development of the deductive-nomological (DN) model of explanation developed in the mid-1900s by Carl Hempel and others (Hempel and Oppenheim 1948; Hempel 1965). One of the main rivals to the DN model of explanation is the causal-mechanical account of explanation (Salmon 1984). A descendant of causal accounts of explanation, mechanistic explanation emerged at the turn of the century and has since become the dominant philosophical account of scientific explanation (Bechtel and Richardson 1993; Glennan 1996; Machamer, Darden, and Craver 2000). This section introduces the DN and causal models of explanation, reviewing their strengths and weakness, before turning to mechanistic explanation in the following section.

The DN model was the dominant model of scientific explanation in the twentieth century. According to this model, to successfully explain a phenomenon, one must derive it from a set of premises—premises that tell us why the conclusion is true. Under this view, the premises constitute the explanans, the conclusion the explanandum, and they relate to each as parts of a deductive argument. Not every deductive argument counts as an explanation, however. To count as an explanation, Hempel (1965) maintained that the premises must include at least one generality or law of nature. Hence the "deductive-nomological" account, where "nomological" derives from the Greek word "nomos" for law. For example, if someone were to point to a pot of boiling water and ask, "why is that liquid boiling?" an explanation would take the form of noting that water heated to 100°C at 1 atmosphere of pressure boils (law of nature), that the liquid in the pot is water (particular fact), and that it has been heated to 100°C at 1 atmosphere of pressure (particular fact). This explains why the liquid is boiling.

The DN model has many virtues. One of the virtues is that it captures the epistemic advantages of both explanation and prediction. Under the DN model, there is a symmetry between explanation and prediction: I can explain why the liquid is boiling by citing the above laws of nature and particular facts, but these laws and facts can also be used to predict what will happen when I heat water to the conditions of 100°C at 1 atmosphere of pressure. As noted in section 2, explanation is often thought to be valuable as a means for prediction and intervention. The DN model provides a nice account of why this is the case: explanation simply is a form of prediction.

Although the DN model has many virtues, it is now widely rejected as an account of explanation because of its well-known shortcomings (see Woodward 2014 for an overview). The most commonly noted shortcomings are that it is both over-permissive and too restrictive in what counts as an explanation. The DN model is over-permissive in that it includes many things as explanatory, which we generally do not accept as explanations. We take a drop in atmospheric pressure as explaining a storm, but not a drop in the reading of a barometer, even when both of these variables reliably vary with the onset of storms. However, there is nothing in the DN model that excludes us from explaining a storm by appealing to the regularity that a low barometer reading tends to precede storms. Also, we can insert irrelevant information into the premises of a deductive argument without affecting its validity and soundness. For example, I could add to our water boiling argument the premise that it is Wednesday. However, we would not want to say that the fact that it is Wednesday explains why the water boils. Again, the DN model is too permissive in counting this as explanatory. On the other hand, the DN model is too restrictive in that it requires the explanans to include a regularity or law of nature. Scientific explanations do not always include such laws, however (Sober 1997). Biological explanations, for example, often include complex, idiosyncratic systems that do not exhibit law-like regularities (but see section 5).

The shortcomings of the DN model have played a key role in the development and acceptance of the causal-mechanical account of explanation. Broadly, this account holds that a phenomenon is explained by its causes (Salmon 1984). Not only does this account overcome the problems of over-permissiveness and over-restrictiveness of the DN model, but it also explains why explanatory information is delimited in the ways described above. Why does the barometer reading not explain the storm? Why does the fact that it is Wednesday not explain the boiling water? According to the causal-mechanical account of explanation, the reason is that these factors play no causal role in producing the phenomena to be explained. Low atmospheric pressure is causally involved in the production of a storm and heating water to 100°C is causally involved in its boiling. It is in virtue of these facts that the phenomenon is explained. The causal-mechanical account also avoids the over-restrictiveness of the DN model because it does not require laws of nature for explanation. The First World War might have been caused—and thus explained—by the fact that the driver of Archduke Franz Ferdinand took a wrong turn in Sarajevo regardless of the fact that this was a one-time and unlikely event.

Mechanistic explanation is a form of causal-mechanical explanation in that it maintains that mechanisms are explanatory and causes are one of the four features of mechanisms (Craver and Tabery 2015). It also extends this account, however, to include constitutive causal-mechanical explanations in addition to etiological ones (Craver 2007). An etiological causal explanation involves explaining an event by citing its antecedent causes, such as when we explain the car accident by citing the ice on the road. A constitutive explanation, in contrast, explains a phenomenon by citing the mechanism responsible for it, such as explaining cancer by citing the mechanisms responsible for abnormal cell growth. Explanations of both varieties are used widely in the biological sciences (Bechtel and Richardson 1993). The rest of this chapter explicates the notion of mechanistic explanation and considers its limitations.

4. Mechanistic explanation

Contemporary notions of mechanistic explanation began to emerge in the 1990s, and various accounts have been developed over the last few decades (for an overview, see Craver and Tabery 2015; Chapter 1 of this volume). For our purposes, we can adopt the ecumenical definition of mechanism advanced by Illari and Williamson (2011): "A mechanism for a phenomenon consists of entities and activities organized in such a way that they are responsible for the phenomenon" (120). Philosophers have discussed and debated many aspects of the nature of mechanisms, including how to determine its boundaries, define its parts, identify its activities and functions, and more (see Part II of this volume). For our purposes, it is sufficient to identify the way in which mechanisms are taken to be explanatory. Generally, mechanistic explanations are taken to "explain *why* by explaining *how*" (Bechtel and Abrahamsen 2005, 422). One can, for example, explain why bones become brittle in old age by specifying the biological factors and processes involved in the reduction of bone mineral density. By providing a description of the mechanisms responsible for a phenomenon, one provides an explanation for why that particular phenomenon occurs and why it has the properties that it does. The purpose of identifying operating parts and determining their organization is to go beyond describing the phenomenon to showing how the working entities cause and constitute the phenomenon.

Mechanistic explanations inherit many of the advantages of the causal-mechanical account of explanation. They capture the asymmetry of explanation: those causes or mechanisms responsible for a phenomenon explain that phenomenon, but not vice versa (the influenza virus explains swine flu, but swine flu does not explain the influenza virus). They also account for why irrelevant information is not explanatory. The fact that the water boiled on Wednesday might be true, but it is not explanatory because it is not part of the causal mechanism responsible for the water boiling. Like the causal-mechanical account, mechanistic explanation also does not require laws of nature, so it is not restrictive in this respect. Mechanistic explanation has many of the virtues of the DN model as well. It provides an account of the epistemic advantages of prediction and intervention: understanding how something works gives us knowledge about how to effectively intervene on that system and predict its future states or how it will behave in a new context.

5. The limits of mechanistic explanation

Discussions concerning the possible limitations of mechanistic explanation are usefully framed in terms of the two criteria for a successful account of explanation introduced in section 2. A philosophical account of explanation should capture the scientific practices of distinguishing explanation from other activities (explanatory demarcation) and good explanations from bad (explanatory normativity). In what follows, I consider whether mechanistic explanation is successful in these two respects. Rather than attempting an exhaustive survey, however, I focus on those aspects of mechanistic explanation that have been points of recent controversy. These concern the use of non-mechanistic explanation in the sciences, abstract and idealized explanatory models, and the generality of scientific explanation. Examining these areas will give us a good idea of the current perceived limitations of mechanistic explanation and how these potential limitations might be overcome.

One criticism of mechanistic explanation is that it excludes certain things as explanatory when those things are indeed explanatory. This is a failure of explanatory demarcation insofar as mechanists claim that something is not an explanation when it in fact is. One of the most prominent criticisms of this type comes from advocates of dynamical modeling (see Chapter 20). Dynamical models track and predict the behavior of complex systems using mathematical tools such as difference and differential equations. These models can be extremely powerful in both their predictive success and ability to unify disparate phenomena. This has led some theorists to maintain that they are explanatory (Chemero and Silberstein 2008; Stepp, Chemero, and Tuvey 2011).

Mechanists have rejected dynamical explanation, however, for many of the same reasons that have led to the rejection of the DN model. They maintain that dynamical models are either explanatory in virtue of conveying information about the mechanism responsible for the phenomenon or not explanatory at all (Kaplan and Bechtel 2011; Kaplan and Craver 2011; Chapter 20 of this volume). If dynamical models fall in the former category, then they are simply instances of mechanistic explanation; if they fall in the latter, then they do not present a counterexample to mechanistic explanation because they are not explanatory. Mechanists hold that models in the latter category—those that describe and predict a phenomenon without providing information about how it works—are problematic because descriptions and predictions are well known to be insufficient for explanation (for the reasons discussed in section 3). Although dynamical models may be insufficient for explanation, it is important to note that this does not mean they do not contribute to our explanatory practices at all. As Bechtel and Abrahamsen (2010, 2011) show, dynamical modeling is often used to understand the behavior of complex mechanisms in fields like chronobiology and cognitive science. Under this view, tools like quantitative computational modeling are essential for explanation because they enable researchers to investigate how the properties of the parts and operations of a mechanism dynamically change over time.

Batterman and Rice (2014) and Ross (2015) argue that dynamical models are explanatory not in virtue of their descriptive and predictive accuracy alone, but also in virtue of abstracting from irrelevant details (see also Batterman 2001). They call such explanations "minimal models" and argue that they constitute not only an alternative to mechanistic explanation, but also to all "common features accounts" of explanation or those that take the explanatory power of a model to derive from its representation of the processes responsible for the explanandum phenomenon. According to these authors, minimal models explain by showing that a subset of possible and actual systems (including a minimal model and a target system) fall within the same universality class. This is despite the systems differing in their microdetails. Although the minimal model and target system will have other features in common (besides the macrobehavior in question), the proponents of this view hold that it is not these common features that explain the macrobehavior of the target system, but rather that these features are a "by-product of the mathematical delimitation of the universality class" (Batterman and Rice 2014, 362).

Whether the minimal model is a genuine alternative approach to mechanistic explanation is unclear, however. As Lange (2015) argues, minimal model explanations appear to be another form of common features account. First, as mentioned above, explanations are generally asymmetrical: the explanans explains the explanandum, but not vice versa. Minimal models appear to lack this property. If a minimal model explains a target system in virtue of falling into the same universality class, it is not clear why the target system cannot also explain the minimal model. This is inconsistent with the explanatory practices of scientists, however, as they generally do not take target systems to be explanatory in this way. One could solve this problem by interpreting minimal model explanations as explaining in virtue of identifying those features that lead to the production of the macrobehavior in question. Under this view, the minimal model explains because it includes all and only those features necessary to produce the macrobehavior (as opposed to the target, which includes many other features). Insofar as the target system has these features too (i.e., insofar as it shares the features exhibited by the minimal model), the explanation is successful. This account preserves the idea that minimal models explain by abstracting and idealizing away from irrelevant detail, but it does so by interpreting these models in terms of a common features account.

Before closing this discussion of potential non-mechanistic explanations, it is important to note that mechanists do not maintain that all scientific explanations are mechanistic. That is, they hold that there are many structures and relations that we take to be explanatory, but which are not causal or mechanistic. For instance, Craver (2014) maintains that, "attractors, final causes, laws, norms, reasons, statistical relevance relations, symmetries, and transmission of marks" might all be considered explanatory, depending on the field of inquiry (29). Strevens (2008) also cites things like moral and aesthetic relations, holding that the explanatory relevance of these relations stems from something other than their causal import (see also Lipton 2004). The dispute about non-mechanistic explanation then is not whether they exist, but rather what form they take and what renders them explanatory. Mechanists deny that dynamical and minimal models are explanatory independently of the information they convey about the workings of the target system. Advocates of these models need to do more to show that this is indeed the case.

One of the great strengths of the mechanist account of explanation has been its ability to elucidate a wide range of scientific practices—practices that were difficult to account for under traditional views of explanation, such as the DN model (Bechtel and Abrahamsen 2005). Recently, however, this aspect of mechanistic explanation has come under criticism, particularly with respect to the use of abstract and idealized models in science. The criticism holds that mechanistic explanation is committed to a representational ideal of completeness— the ideal that the more features and causal relationships a model represents, and the more accurately it represents them, the better the model (see Weisberg 2013 and Chapter 17 of this volume for discussion). If this is the standard to which we should hold explanatory models, however, it is in conflict with the actual practices of scientists. It is well known that most scientific models are highly abstract and idealized. They are abstract in the sense of omitting detail about the target system and idealized in the sense of distorting elements of that system. Explanatory models are thus intentionally incomplete and inaccurate; further, this is not viewed as a deficit by practicing scientists. So the representational ideal of completeness is not something to which scientists appear to adhere.

This ideal of completeness is attributed specifically to the views of mechanists like Machamer, Darden, Craver, and Kaplan (see Levy and Bechtel 2013; Chirimuuta 2014; Batterman and Rice 2014; Levy 2014; Love and Nathan 2014). For example, Levy and Bechtel (2013) write, "Machamer, Darden, and Craver appear to treat abstractions as, at best, templates for explanation. They regard the filling in of concrete detail as a hallmark of explanatory progress" (258). Similarly, Chirimuuta (2014) attributes to Craver and Kaplan the claim that, "the hypothetical, maximally complete and detailed representation of the mechanism is the one best explanatory model onto which all others aim to converge" (132). Incomplete models are sketches that fall short of their explanatory goal; the more detailed and precise they get, the better. This is what Chirimuuta refers to as the "More Details the Better" assumption (132). Levy (2014) also attributes the representational ideal of completeness to Craver, writing, "sketches are typically steps along the way to a better explanation. If all goes well, the gaps are filled and the mechanism is described in full detail. Once that occurs, the sketch is transformed into a satisfactory explanation" (479; see also Batterman and Rice 2014, 352).

The same point is made against mechanistic explanation with respect to idealization. The representational ideal of completeness requires that explanatory models are not only complete, but also accurate. This, however, is at odds with the widespread use of idealized explanatory models. As Love and Nathan (2014) write: "The idealization of causal relations—the intentional misrepresentation of how the mechanism produces the phenomenon—means that these models do not show how the mechanism actually works. Mechanistic explanations thus appear to fail according to their own criteria" (13).

The critics hold then that mechanists such as those above are committed to a representational ideal of completeness and that this renders their account of explanation descriptively inadequate. Abstract and idealized models are taken to be explanatory by practicing scientists despite the fact that they are incomplete and inaccurate. If the critics are right, this represents a crucial limitation of the mechanistic account. Mechanistic explanation fails to capture the practices of the explanatory normativity of science: what scientists take to constitute a good explanation is at odds with what the mechanist account maintains.

Are the critics right about this shortcoming of mechanistic explanation? I argue that they are not for the reason that completeness is not the sole ideal that guides mechanistic explanation. To see this, recall that under the mechanistic account, it is the causal mechanism that produces, constitutes, or maintains a phenomenon of interest that explains that phenomenon—the process by which plants convert solar energy into usable carbohydrates explains photosynthesis, for example. For those who hold the ontic view of explanation (introduced in section 2), it is the actual entities and activities found in plants that explain photosynthesis. Descriptions of photosynthetic processes (in the form of diagrams, linguistic descriptions, physical models, etc.) are explanatory in a derivative sense: we take them to be explanatory because they convey information about the real-world explanation. For those who hold the epistemic view, it is the models and communicative acts that we use to convey information about the mechanisms responsible for photosynthesis that constitute the explanation. These models and acts might be explanatory in virtue of conveying information about the target mechanism, but the mechanism itself is not explanatory. Both ontic and epistemic views, however, hold that mechanisms are the targets of explanatory models and that the explanatory power of a model is in part a function of the information that model conveys about this target (Wright 2012; Illari 2013; Craver 2014). Accurately representing the target mechanism then is an important success condition for explanatory models.

Mechanisms constrain those communicative acts, texts, and representations that we take to be explanatory. If this were the only dimension along which explanatory models were assessed, then the practices of abstraction and idealization in science would indeed be puzzling. Why settle for incomplete and inaccurate models when we can do better? Crucially, however, this is not the only constraint on explanatory models. As Kaplan and Craver (2011) write,

> the idea of an ideally complete how-actually model, one that includes all of the relevant causes and components in a given mechanism, no matter how remote, negligible, or tiny, without abstraction or idealization, is a philosopher's fiction. Science would be strikingly inefficient and useless both for human understanding and for practical application if it dealt in such painstaking minutiae.
>
> *(609–10)*

An explanatory model serves many aims. In addition to conveying information about the target mechanism, such a model must often be intelligible, easily shared with a broader community, practical to work with, and more. Some of these desiderata pull in different directions. For example, making a model more intelligible will often require simplifying it by removing detail. This may decrease the amount of information that the model conveys about the target mechanism, but this does not mean that we should avoid such simplifications. If explanatory models were unintelligible, researchers would be unable to use them to communicate findings, design experiments, interpret results, train students, etc. All of these activities are important for advancing research, including the further discovery of mechanisms.

Various accounts of scientific practice support the claim that multiple desiderata constrain the construction and evaluation of explanatory models. For example, Weisberg (2013) discusses how different goals give rise to different forms of idealization (where an idealization is the intentional distortion of a target system as mentioned above). Researchers engage in Galilean idealization when their goal is to increase the tractability of a model. In this case, elements of the target system are distorted not because they are unimportant, but because they interfere with a researcher's ability to understand and work with the model. In contrast, minimal idealization involves introducing distortions for the sake of capturing the core causal factors of a target system. Here distortions are introduced not for the purpose of making a model more intelligible, but to remove information about those parts of the world that are taken to be explanatorily irrelevant. Weisberg notes that desiderata such as simplicity and accuracy will often trade off of each other, in which case multiple models may be used for attaining one's research goals. Especially when the target is a complex system, we should expect that multiple models will be deployed, each of which will capture some aspect of the core causal mechanism, while simplifying other aspects to increase intelligibility. Taken together, these models may advance us toward capturing the target mechanism in a way that can be understood and applied.

Even the critics' descriptions of explanatory practice suggest the operation of multiple desiderata. As Love and Nathan (2014) note, "These kinds of [highly abstract and idealized] diagrammatic representations are common in textbooks. However, in more advanced discussions, we find increasingly detailed representations of eukaryotic gene expression and more precise narrative descriptions of the mechanism" (8). This suggests that intelligibility may take priority in pedagogical contexts; while conveying information about the target mechanism may become more important in those contexts where advanced researchers are attempting to understand and intervene on a target system.

Are mechanistic explanations generalizable? This is an important question because good explanations seem to be those that can be applied to a variety of systems and contexts. This question is all the more pressing in biology, where most mechanistic models are developed by investigating only a small handful of organisms, those known as "model organisms" (Weber 2005; Ankeny and Leonelli 2011). The mechanistic explanations developed by studying model organisms are often done so under the assumption that they will apply to a wide variety of organisms, not simply to those being investigated. Insofar as explanations in the sciences are general, a philosophical account of explanation should be able to account for this.[1]

Glennan (2002) holds that mechanistic explanations are generalizable, writing, "although any particular mechanism will occupy a particular region of space-time, it is an important feature of our world that it often contains many tokens of a single type of mechanism" (S345). One might ask what "important feature of our world" is responsible for the fact that similar mechanisms tend to appear again and again (for more discussion on the regularity of mechanisms, see Chapter 12). When it comes to biological organisms, evolution seems to provide an answer. Mechanisms that have been inherited by descendants of a common ancestor will have features in common. Hence, if an explanation is provided for a mechanism that is evolutionarily conserved in this way, that explanation will likely apply to a variety of organisms (Bechtel 2009; Halina and Bechtel 2013). Does this mean that mechanistic explanations should be confined to those mechanisms that are highly evolutionarily conserved? This would be unreasonable. Scientists provide mechanistic explanations for a wide variety of phenomena, whether they know them to be evolutionarily conserved or not. Furthermore, even conserved mechanisms exhibit variation.

Bechtel and Abrahamsen (2005) suggest that mechanisms might be categorized by how similar they are, rather than by whether they are exactly the same. This approach is promising as an account of how scientists identify mechanisms as belonging to a particular type (see Glennan forthcoming). Also, it is important to note that recognizing the similarity between mechanisms is a useful investigatory and explanatory tool, even if the mechanisms differ in other respects. Consider, for example, the use of the model organism *Drosophila* to study human ovarian cancer. In *Drosophila*, border cells in the ovary migrate in preparation for egg fertilization. Studying the migratory behavior of these cells has proven useful for understanding particular aspects of ovarian cancer in humans—namely, the movement of cancer cells from one site to another. The spread of ovarian cancer is a complicated process involving many variables of which the migration of the malignant cancer cells is only one part. However, the study of normal-functioning border cells in *Drosophila* has the potential to shed valuable light on at least this aspect of metastasis (Naora and Montell 2005). Although cells in *Drosophila* and human ovarian cancer cells are neither completely similar nor evolutionarily conserved processes, a mechanistic explanation of the former can still be usefully applied to the latter.

In addition to similarities serving as heuristics for mechanism discovery, differences can also play such a role. For example, one strategy for discovery in circadian rhythms research is to search for the homolog of a protein found in one species in another species. If this homolog then plays a different role in the second species, researchers identify this new role. This in turn leads them to look for a protein that would perform this new role in the original species (see Halina and Bechtel 2013 for a specific example of such a case). Such back-and-forth research fueled by the discovery and pursuit of homologs is common in the biological sciences (see Bechtel 2009; for more on heuristics for mechanism discovery, see also Craver and Darden 2013 and Chapter 19 of this volume).

Thus, to the extent in which mechanisms are evolutionarily conserved, we can expect that mechanistic explanations can be generalized. However, formulating models of mechanisms that are not evolutionarily conserved, or that exhibit variation across individuals or species, is still useful as a heuristic for discovery. Such heuristics are crucial for mechanistic explanations, as such explanations are explanatory precisely in virtue of providing information about the target mechanism. Little work has been done on the generality of mechanistic explanation to date, but it would be premature to say that this account is unable to capture this aspect of scientific practice, at least in biology. Whether the same is true of other sciences, such as physics and chemistry, is an open question. Additional candidates for generating similarities across mechanisms might include convergent evolution, self organization, cultural transmission, and intentional design (Glennan forthcoming).

6. Conclusion

Mechanistic explanation has played a pivotal role in developing our understanding of explanatory practices in biology and other sciences. This chapter has highlighted the advantages of the mechanistic approach over other accounts of explanation, as well as its descriptive adequacy with respect to the use of idealized, abstract, and general models. Our understanding of explanatory practices in sciences like biology is still rudimentary. As work in this area continues, the successful application and limitations of mechanistic explanation will become clearer. Until then, mechanistic explanation might serve as its own heuristic for discovery—finding the ways in which it coincides and diverges from the explanatory practices found in science will no doubt improve our understanding of those practices.

Note

1 Although I will not employ this distinction here, see Sheredos 2015 for a discussion of the difference between generality and scope and its implications for mechanistic explanation.

References

Ankeny, R.A. and Leonelli, S. (2011) "What's So Special about Model Organisms?" *Studies in History and Philosophy of Science*, 42, 313–323.

Batterman, R.W. (2001) *The Devil in the Details: Asymptotic Reasoning in Explanation, Reduction, and Emergence*, New York: Oxford University Press.

Batterman, R.W. and Rice, C.C. (2014) "Minimal Model Explanations," *Philosophy of Science*, 81, 349–376.

Bechtel, W. (2009) "Generalization and Discovery by Assuming Conserved Mechanisms: Cross-Species Research on Circadian Oscillators," *Philosophy of Science*, 76, 762–773.

Bechtel, W. and Abrahamsen, A. (2005) "Explanation: A Mechanist Alternative," *Studies in History and Philosophy of Biol & Biomed Sci*, 36(2), 21–21.

Bechtel, W. and Abrahamsen, A. (2010) "Dynamic Mechanistic Explanation: Computational Modeling of Circadian Rhythms as an Exemplar for Cognitive Science," *Studies in History and Philosophy of Science Part A*, 41(3), 321–333.

Bechtel, W. and Abrahamsen, A. (2011) "Complex Biological Mechanisms: Cyclic, Oscillatory, and Autonomous," in C.A. Hooker (ed.) *Philosophy of Complex Systems*. Handbook of the Philosophy of Science, Volume 10, New York: Elsevier, 257–285.

Bechtel, W. and Richardson, R.C. (1993) *Discovering Complexity: Decomposition and Localization as Strategies in Scientific Research*, Princeton, NJ: Princeton University Press.

Chemero, A. and Silberstein, M. (2008) "After the Philosophy of Mind: Replacing Scholasticism with Science," *Philosophy of Science*, 75, 1–27.

Chirimuuta, M. (2014) "Minimal Models and Canonical Neural Computations: The Distinctness of Computational Explanation in Neuroscience," *Synthese*, 191, 127–153.

Colombo, M., Hartmann, S. and van Iersel, R. (2014) "Models, Mechanisms, and Coherence," *The British Journal for the Philosophy of Science*, 66, 181–212.

Craver, C. (2007) *Explaining the Brain*, Oxford: Oxford University Press.

Craver, C.F. (2014) "The Ontic Conception of Scientific Explanation," in M.I. Kaiser, O.R. Scholz, D. Plenge, and A. Hütteman (eds.) *Explanation in the Special Sciences: The Case of Biology and History*, Dordrecht: Springer, 27–52.

Craver, C.F. and Darden, L. (2013) *In Search of Mechanisms: Discoveries across the Life Sciences*, Chicago: The University of Chicago Press.

Craver, C. and Tabery, J. (2015) "Mechanisms in Science," in E.N. Zalta (ed.) *The Stanford Encyclopedia of Philosophy*, https://plato.stanford.edu/.

Glennan, S.S. (1996) "Mechanisms and the Nature of Causation," *Erkenntnis*, 44, 49–71.

Glennan, S. (2002) "Rethinking Mechanistic Explanation," *Philosophy of Science*, 69, S342–53.

Glennan, S. (forthcoming) *The New Mechanical Philosophy*.

Godfrey-Smith, P. (2003) *Theory and Reality: An Introduction to the Philosophy of Science*, Chicago: University of Chicago Press.

Halina, M. and Bechtel, W. (2013) "Mechanism, Conserved," in W. Dubitzky, O. Wolkenhauer, H. Yokota, and K.H. Cho (eds.) *Encyclopedia of Systems Biology*, Dordrecht: Springer, 1201–1204.

Hempel, C.G. (1965) *Aspects of Scientific Explanation and Other Essays in the Philosophy of Science*, New York: Free Press.

Hempel, C.G. and Oppenheim, P. (1948) "Studies in the Logic of Explanation," *Philosophy of Science*, 15, 135–175.

Illari, P. (2013) "Mechanistic Explanation: Integrating the Ontic and Epistemic," *Erkenntnis*, 78, 237–255.

Illari, P.M. and Williamson, J. (2011) "Mechanisms Are Real and Local," in P.M. Illari, F. Russo, and J. Williamson (eds.) *Causality in the Sciences*, New York: Oxford University Press, 818–844.

Kaplan, D.M. and Bechtel, W. (2011) "Dynamical Models: An Alternative or Complement to Mechanistic Explanations?" *Topics in Cognitive Science*, 3, 438–444.

Kaplan, D.M. and Craver, C.F. (2011) "The Explanatory Force of Dynamical and Mathematical Models in Neuroscience: A Mechanistic Perspective," *Philosophy of Science*, 78(4), 601–627.

Kellert, S.H., Longino, H.E. and Waters, C.K. (2006) "Introduction: The Pluralist Stance," in S.H. Kellert, H.E. Longino, and C.K. Waters (eds.) *Scientific Pluralism*, Minneapolis: University of Minnesota Press, vii–xxix.

Lange, M. (2015) "On 'Minimal Model Explanations': A Reply to Batterman and Rice," *Philosophy of Science*, 82, 292–305.

Levy, A. (2014) "What Was Hodgkin and Huxley's achievement?" *The British Journal for the Philosophy of Science*, 65(3), 469–492.

Levy, A. and Bechtel, W (2013) "Abstraction and the Organization of Mechanisms," *Philosophy of Science*, 80(2), 241–261.

Lipton, P. (2004) *Inference to the Best Explanation*, London: Routledge.

Love, A.C. and Nathan, M.J. (2014) "The Idealization of Causation in Mechanistic Explanation," paper presented at the Philosophy of Biology at Madison Workshop, University of Wisconsin-Madison, 5 May-1 June.

Machamer, P., Darden, L., and Craver, C.F. (2000) "Thinking about Mechanisms," *Philosophy of Science*, 67(1), 1–25.

Naora, H. and Montell, D.J. (2005) "Ovarian Cancer Metastasis: Integrating Insights from Disparate Model Organisms," *Nature Reviews Cancer*, 5(5), 355–366.

Ross, L. (2015) "Dynamical Models and Explanation in Neuroscience," *Philosophy of Science*, 82(1), 32–54.

Ruben, D. (2012) *Explaining Explanation: Second Edition*, London: Routledge.

Salmon, W. (1984) *Scientific Explanation and the Causal Structure of the World*, Princeton, NJ: Princeton University Press.

Sheredos, B. (2015) "Re-reconciling the Epistemic and Ontic Views of Explanation (Or, Why the Ontic View Cannot Support Norms of Generality)," *Erkenntnis*, 1–31. doi: 10.1007/s10670-015-9775-5.

Sober, E. (1997) "Two Outbreaks of Lawlessness in Recent Philosophy of Biology," *Philosophy of Science*, 64, S458–S467.

Stepp, N., Chemero, A. and Tuvey, M.T. (2011) "Philosophy for the Rest of Cognitive Science," *Topics in Cognitive Science*, 3, 425–437.

Strevens, M. (2008) "Approaches to Explanation," *Depth: An Account of Scientific Explanation*, Cambridge, MA: Harvard University Press.

Weber, M. (2005) "Model Organisms: Of Flies and Elephants," in his *Philosophy of Experimental Biology*, Cambridge: Cambridge University Press, 154–187.

Weisberg, M. (2013) *Simulation and Similarity: Using Models to Understand the World*, Oxford: Oxford University Press.

Woodward, J. (2014) "Scientific Explanation," in E.N. Zalta (ed.) *The Stanford Encyclopedia of Philosophy*, https://plato.stanford.edu/.

Wright, C.D. (2012) "Mechanistic Explanation without the Ontic Conception," *European Journal for Philosophy of Science*, 2(3), 375–394.

17

MODELS OF MECHANISMS[1]

John Matthewson

1. Introduction

Mechanisms and models are two of the most intensively studied areas in recent philosophy of science, and there are clear points of connection between these fields. However, apart from some notable examples (for instance Glennan 2005; Craver 2006; Darden 2007; Levy and Bechtel 2013), there has been surprisingly little work on modeling in mechanistic sciences (see also Kaplan 2011: 346). As noted by Stuart Glennan over a decade ago, contemporary mechanistic philosophy has not been overly concerned with issues regarding modeling:

> Perhaps because of the realist tendencies of the philosophers involved, most of the literature has focused on the properties of mechanisms themselves and has not said much about the relationships between mechanisms and their models or theoretical representations.
>
> *(2005: 443–4)*

In fact, even when theoretical representation in mechanistic science is directly addressed, it is still not obvious how this work relates to the literature regarding scientific modeling. For example, mechanistic philosophy often considers the representational roles of mechanism sketches, mechanism schemas, and their instantiations. While these clearly relate to scientific models as discussed in philosophy of science more broadly, exactly how they connect to one another is not entirely certain.[2]

Similarly, the modeling literature has tended to neglect the centrality of mechanistic models in many branches of science, usually focusing instead on highly theoretical mathematical models. These lacunae are unfortunate, and represent genuine missed opportunities. Thinking about models casts light on the practice of mechanistic science, and thinking about how mechanisms are modeled is an essential component of understanding scientific models in general.

The next two sections of this chapter will describe some common and important topics in the philosophical literature regarding scientific modeling. The final two sections illustrate how these topics arise in the study of mechanistic science specifically. These latter sections are not intended to present a definitive list of the relevant issues, or a definitive treatment of those issues I do discuss. Hopefully, however, they will give a feel for the landscape where models and mechanisms interact, and where further work is required.

2. Some clarifications

Scientific models have been subject to philosophical scrutiny for the last fifty years or so, and the extent of this scrutiny continues to grow. This attention has resulted in an increasing number of involved and subtle positions, each with their own distinctions and terminology. Additionally, although much of the work regarding models is closely interconnected, subtle differences in the use of the word "model" can sometimes obscure important differences regarding what is actually under discussion. It is therefore important to clarify that our focus in this chapter is not the role models play in the structure of scientific theories as discussed in the semantic account, but the way models are used to represent mechanisms.[3] So the primary sense of "scientific model" in this context refers to the representations employed by scientists to describe, predict, or explain some system of interest (Downes 1992; Godfrey-Smith 2014: 19).

However, it is also worth noting a related but different philosophical project involving models, more concerned with the *practice* of modeling in science. Here investigators adopt an intentionally indirect methodology, investigating the system or phenomenon of interest by creating and exploring a model of that system. These models will usually be much simpler than the system they stand in for. For example, biologists interested in population growth may consider a "model population," where immigration and emigration are ignored, and size increases continuously rather than discretely. Nevertheless, this model population can then be used to make inferences about the growth of real populations on the basis of similarities between the model and actual populations in the world (Godfrey-Smith 2006; Weisberg 2007, 2013). There is a great deal of overlap between the study of models as representations and this practice of indirect modeling, and I will use insights from both of these traditions in what follows.

3. Models and scientific representation

There is a vast range of models used in scientific theorizing. Philosophical discussion often focuses on mathematical models described by equations and often depicted as trajectories through a state-space, or groups of entities, each acting according to a set of simple rules and represented on computer screens as changing patterns on a lattice. However, science also employs models that are totally different to these theoretical representations: living fruit flies and particular species of flowering plants, for example, or physically constructed miniature canals and boats (Sterrett 2002). Still other models are simply imagined scenarios, sketched out in a way akin to philosophical thought experiments (Godfrey-Smith 2006).

As pointed out by Stephen Downes (1992, 2009), in the face of such diversity the challenge is to find a middle ground between the claim that models are one single specific type of thing (which is clearly false) and allowing anything at all to function as a model (which would tell us nothing useful or interesting about scientific models). One way to approach this is to consider the role these models all play. A quite standard place to turn to here is the work of Ronald Giere (1988, 2004). Giere points out that the descriptions used in science often do not appear to be directly about the subject of interest. Scientific discussion is often concerned with entities such as pendulums operating in perfectly uniform gravity, instantaneously growing populations, or perfectly informed consumers. These aren't descriptions of any actual pendulums, populations, or consumers. Rather, they are descriptions of *models* of these entities. Giere argues that these models are useful to scientists to the extent that they are relevantly similar to the phenomenon under consideration. An understanding of the model and the ways in which it is similar to this phenomenon allows scientists to articulate, predict, or explain that particular part of the world.

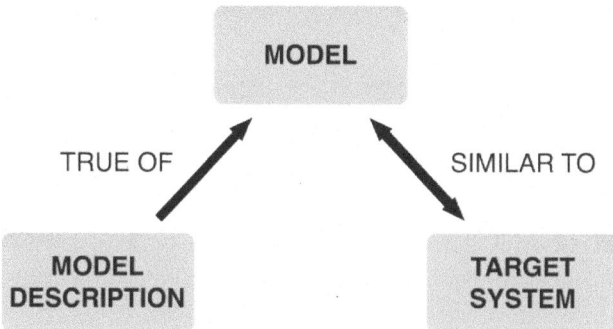

Figure 17.1 Diagram of the relations between description, model, and target, modified from Giere (1988: 83)

Note that according to this account, we distinguish between models and their descriptions. Models are not the equations or diagrams used to describe them any more than any other object can be equated with its description. Scientific modeling therefore involves three relata: the description of the model, the model itself, and the part of the world under investigation (the "target system"). The description is true of the model, while the model is connected to the target system via a relation of similarity (see Figure 17.1).

I will focus on three prominent issues regarding this framework: first, the ways in which models can be abstract and idealized—how they diverge from being complete and veridical representations of their targets; second, the relation of similarity that is meant to hold between model and target system; third, the ontology and individuation of different types of models, especially in the case of models that are mathematical or highly abstract. These three issues are key topics in the philosophy of modeling generally, but each of them will also prove important when we go on to consider mechanistic models in particular.

There are many different ways of understanding the terms "idealization" and "abstraction" (for example, see Cartwright 1989; Leonelli 2008). Here I will follow Peter Godfrey-Smith's framework, as it draws a distinction that will be important for some of the discussion that follows in later sections (Godfrey-Smith 2006; Love and Nathan forthcoming).

For Godfrey-Smith, idealization occurs when the model represents something known to be false of the target system, while abstraction involves the omission of certain aspects of that system. On this way of understanding the terms, only some representations are idealized, while essentially every representation must be abstract to some extent, given the impossibility of articulating every aspect of a target.

An example Godfrey-Smith uses to illustrate the difference is a model population where the investigators are only concerned with the density of its members, and so information regarding the population's actual size is not included. This is an abstract representation of some population, because it expresses *only* truths (let's say), but not *all* of the truths regarding that population. This differs from a model in population genetics that is stipulated to be infinitely large to ignore drift effects. Here we have a description that is false of any actual population, and is known by any scientist using it to be false. It is therefore a case of idealization.

Idealizations can be distinguished further, between what Michael Weisberg calls "Galilean" and "minimal" idealization (2007a, 2013: 98–103). In the former case, simplifications are introduced into the model for pragmatic reasons such as computational tractability or ease of use. In the latter, idealization is used in the service of explanatory generality. By removing some details, a model may gain generality, and thereby—on some accounts at least—gain explanatory power.

Weisberg notes that these two strategies will often generate the same models: minimizing the details in a model may make it both more general and easier to use. However, we can still differentiate these kinds of idealization according to their ultimate goals. Galilean idealization is usually something to be removed over time, as scientists and their technology become better able to manage complexity. In contrast, minimal idealizations may remain even as the science progresses, as this type of idealization is made for in-principle reasons. On this view, generality can be a desirable feature of the model, regardless of whether a more complex model would be tractable.[4]

Now we turn to consider similarity. As noted multiple times in the literature, it is not very illuminating to simply claim that representations or models must be similar to their targets (Goodman 1972; Godfrey-Smith 2006: 733). Rather, we have to understand the ways and extent to which a model must be similar to its target to successfully represent that part of the world. To complicate issues, these demands on similarity may vary depending on the phenomena studied and the purpose of the model.

To help clarify, I will again draw on concepts developed by Michael Weisberg (2007b, 2013), although my terminology is somewhat different. One way a model can be similar to its target is in terms of its behavior. In this sense, a model is similar to the phenomenon of interest to the extent that their outputs match when given the same inputs or initial conditions. A model that is similar to its target in this way will be useful for describing and predicting the behavior of the system it represents.

However, it is important to note that the underlying structure of such a model will not necessarily reflect any of the real structural elements or dependencies in the target system. We can see this in the setting of "phenomenological" models, where the model simply codifies correlations between the values of certain variables, with no commitment to the structure of the model matching anything particular in the world (Craver 2006). For example, we might construct a perfectly serviceable phenomenological model that connects the probability of a local storm to barometer readings. Such a model will allow us to make defeasible predictions about the weather, while making no claims that it resembles what actually underlies the probability of storm occurrence.

Alternatively, one might construct a model that includes cloud cover, air pressures, ocean temperature etc., and maps the interactions of these features to generate an output regarding the probability of a storm in a particular location. To the extent that this model's parts and their interactions match the actual processes involved in the production of local weather conditions, this model will thereby be similar to its target with respect to those underlying processes.

If a scientist is only interested in predicting the outputs of some target, then a purely behaviorally similar model may be exactly what they are after. There is no reason for the scientist to care about a match between the underlying structures of model and target, as long as the outputs are correct. However, if the modeler's objective is to understand what actually underlies that behavior, then the model will need to be structurally similar to its target in at least some important respects. As we will see, this type of similarity is central when we consider the adequacy conditions for mechanistic modeling.

It is also important to note that similarity to target might not be the modeler's sole objective. For example, a model may be intended to represent or unify a broad class of phenomena rather than a single system. Depending on the complexity and heterogeneity of these phenomena, a *trade-off* between generality and similarity may force scientists to be satisfied with a limited match between their model and any specific target (Levins 1966; Odenbaugh 2003; Matthewson and Weisberg 2009; Matthewson 2011).

The upshot of the foregoing is that the manner and extent to which a model should resemble its target will vary according to the task at hand. However, as pointed out by Susan Sterrett (2006: 72), this doesn't make the relevant similarities subjective, just contextual. Given a specific setting and goal, a scientist will have particular similarity requirements that the model needs to meet to be adequate for the task.

I now turn to the ontology of models. This is relatively clear-cut when the model is a physical object, such as a scale model of a boat or building. However, things are more problematic when we consider models that are not concretely instantiated in the world, such as frictionless projectiles or infinitely large populations. I will consider two prominent views regarding the ontology of these theoretical models: first that they are mathematical objects, and second that they should be treated as imaginary or hypothetical entities akin to literary fictions.

Many theoretical models are described by equations that relate the values of different variables and are depicted as (sets of) trajectories through a state-space. The values of these variables represent certain properties of the model system. Each variable can be ascribed a dimension in a space, so a full specification of all the model's variables can be described by a point in this space. One of the variables will often stand for time, and variable values will evolve over time to trace a trajectory through the space. According to some views regarding non-physical models, the model just is this mathematical object: a set of trajectories through an n-dimensional space, described by a collection of equations, and corresponding to a set of possible inputs (for more on this, see Weisberg 2004, 2007b, 2013; Odenbaugh 2008).

At least some modelers do appear to think of their models in this way at least some of the time. However, we might wonder how well this fits with what Martin Thomson-Jones (2006) calls the "face-value practice" of modeling: how scientists generally act and speak about these parts of their day-to-day work. Peter Godfrey-Smith (2006: 734–5) points out that the "folk ontology" of scientists (the term is from Deena Skolnick Weisberg) often tends to treat models more like concrete entities; bone fide populations or cell gates, for example, rather than a set of trajectories through an abstract multidimensional space.

Godfrey-Smith argues that for this and other reasons, such theoretical models are better thought of as "imagined concrete" objects (2006: 104, 734). These models might not physically exist (or at least are assumed to not exist), but if they did exist, they would be concrete entities.

There is a great deal that can be said about this, and it is certainly a controversial topic (see, for example, Giere 2009; Fine 2009; Frigg 2009). For example, the "imagined concrete" view arguably captures the richness and flexibility of the model-target similarity relation better than the "mathematical object" view. Unfortunately, it also appears to generate further problems of its own, such as how scientists could use a purely imaginary object to learn something new about the world (Godfrey-Smith 2009). However, rather than exploring the ontology of models further here, we will now turn to consider how this and the other issues discussed above manifest in the setting of mechanistic science, beginning with the question of what counts as a distinctively mechanistic, or mechanical model.

4. Models in mechanistic science

In the paper "Modeling Mechanisms," Stuart Glennan presents the view that "A mechanical model is (not surprisingly) a model of a mechanism" (Glennan 2005: 445). This is an intuitive starting point: what makes a model mechanical is that it represents a mechanism. However, at least two important issues immediately arise. First—as Glennan goes on to argue—for a model to truly be "about" a mechanism, it cannot merely mirror the mechanism's behavior, but must also represent the mechanism itself (i.e. its parts, activities, and their organization). This central idea is discussed further below, when we turn to mechanistic representation.

Second, although the ontology of models is controversial, views based on Giere's work generally agree that models are entities in their own right with their own properties. This raises the possibility that models might be mechanistic (or not) in and of themselves, regardless of their target (see also Craver and Tabery 2015: section 3.3).

An example used in (Matthewson and Calcott 2011) is the Meccano model of a VW gearbox designed and built by Alan Wenbourne (2006). The Wenbourne model is itself a mechanism, made of metal struts and fastenings, organized in such a way that they interact to produce a specific output. It would seem that the label "mechanistic model" is appropriate here in at least some sense, independent of what the model is ultimately used to represent. In this case, when we claim the model is mechanical or mechanistic, we might be saying something about the target of the model, but we might also be saying something about the model itself. So to avoid possible ambiguity or misunderstanding when discussing "mechanistic models," models *of* mechanisms should not be conflated with models that *are* mechanisms. (For further discussion regarding this way of understanding the ontology of mechanistic models, see Matthewson and Calcott 2011.)

Here the earlier discussion regarding model ontology becomes important. A model that is mechanistic "in its own right" must be able to exhibit the properties of a mechanism: it must be composed of identifiable, separable interacting parts that are organized in a way that brings about some output behavior. Although this is quite intuitive in the case of a physical model like the Wenbourne gearbox, it seems we should allow for non-physical models to be mechanistic in this sense also. After all, the mechanistic sciences aren't restricted to the use of concrete models.

Unfortunately, how a purely theoretical model can possess the properties of a mechanism is not entirely clear. One way to resolve this would be to adopt one of the views regarding theoretical models we considered in the previous section: that such models should be considered hypothetical concrete objects. On this understanding, *were* a mechanistic model to be realized concretely, it *would* be a mechanism in the same way actual concrete models can be mechanisms. For example, a gearbox does not need to be modeled with pieces of metal. One might instead describe the model with a series of equations articulating the various interactions involved, or draw a diagram of such a mechanism. In this case, the model so described isn't a concrete object, simply because it has not been realized concretely. However, if it were to be realized, it would be a mechanism.

In discussions of mechanistic modeling, however, authors are usually more concerned with the issue of how models can be mechanistic in Glennan's sense above: what does it take for a model to successfully represent a mechanism?

Recall that a model can be similar to its target with respect to behavior or underlying structure. In the case of mechanistic representation, the model must be capable of exhibiting both kinds of similarity, again as noted by Glennan:

> A mechanical model consists of (i) a description of the mechanism's behavior (the behavioral description); and (ii) a description of the mechanism that accounts for that behavior (the mechanical description).
>
> *(2005: 446)*

A purely phenomenological model reflects some dependencies that produce the target's behavior, but only those that involve external inputs or initial conditions. Whether such a model can be explanatory at all is controversial, but it is uncontroversial that such models cannot provide mechanistic explanations. To represent and explain mechanistically is to show how a mechanism produces the phenomenon of interest in virtue of its parts and their organized interactions. This requires a model that is not only similar to its target with respect to its behavior, but also with respect to its structure.

Carl Craver and David Kaplan have made similar points (Craver 2006; Kaplan 2011; Kaplan and Craver 2011). In recent work, they call this the "model-mechanism-mapping" constraint, and explicitly note how this might be understood in the setting of mathematical models:

> A model of a target phenomenon explains that phenomenon to the extent that (a) the variables in the model correspond to identifiable components, activities, and organizational features of the target mechanism that produces, maintains, or underlies the phenomenon, and (b) the (perhaps mathematical) dependencies posited among these (perhaps mathematical) variables in the model correspond to causal relations among the components of the target mechanism.
>
> *(Kaplan 2011: 347; see also Kaplan and Craver 2011)*

This all looks reasonably straightforward, details aside, and a great deal of mechanistic modeling can be understood in this way. However, there are of course complications. Up to now we have been talking as though the distinction between pure behavioral similarity and structural similarity is always clear-cut. This is not the case. We might ask when a model is "merely" phenomenological, and when it might be, or become, a truly mechanistic model. For example, debates regarding the famous Hodgkin–Huxley model of axon depolarization are at least in part concerned with when and how a model based on data might be confirmed in the right ways to represent the underlying mechanism (Weber 2008; Craver 2008; Levy 2013).

A further potential complication involves exactly what aspects of structural similarity must be met to explain mechanistically. For example, similarity of mechanistic structure might require an accurate spatiotemporal match between the parts of both model and target. However, structural similarity might also mean something less constrained, such as preserving just the causal mapping between parts and their interactions, or perhaps something even more abstract than this (see for example Illari and Williamson 2010).

A favorite example of mine here is the MONIAC: a model that uses fluid movements through valves and plastic tubing to represent a national economy. This is a mechanistic model in the sense that it is itself a mechanism, but furthermore, it is thought to explain the economy's behavior on the basis of its causal structure. The MONIAC's parts, such as idle balances, savings, and revenue, combine via their interconnections to show how certain economic outcomes might arise (Phillips 2000; Barr 2000).

Here is a case that has the hallmarks of mechanistic explanation—parts, interactions, and organization—and there are at least some abstract structural similarities between target and model. However, this mapping is not at all straightforward. For example, the MONIAC is made up of parts that are discrete and spatially separate, entirely unlike these "parts" in a real economy. Spatial separation in the MONIAC is used to represent causal independence in the model, but this comes at the cost of fidelity to the target in just these spatial respects. Indeed, it is not entirely clear what being spatially discrete would mean in economic terms. In turn, this means the structures of model and target are dissimilar in some ways that at least intuitively might be thought essential to mechanistic explanation. Nevertheless, the MONIAC does appear to explain its target by way of its internal structure (see also Matthewson and Calcott 2011).

This issue is potentially even more marked in the setting of mathematical models. Here, certain aspects of mechanistic structure may not even be applicable, where part-hood and organization may be expressed through the modularity of, and interactions between, the equations that express the model.

Such mismatches regarding spatial discreteness and location might appear to constitute a significant problem for representation in mechanistic sciences. However, it is at least not obvious that mechanistic representation and explanation require that the parts in a model are spatiotemporally similar to those of its target. For sure, spatiotemporal features will be essential in at least some instances of mechanistic explanation, but we have seen that similarity requirements can vary according to context, and mechanistic explanation of at least a kind seems possible when abstract causal structure is all that is similar between model and target. (For examples of this idea in the setting of systems biology, see Chapter 27.)

In summary, although it is correct that successful mechanistic representation requires structural similarity between model and target system, different instances of mechanistic explanation may have different requirements regarding what this comes to. This means a full picture of mechanistic modeling will be more complicated (and possibly more varied) than just stating that the model must be structurally similar to its target. With this in mind, we now turn to consider the role of abstraction and idealization in mechanistic science.

Abstraction is certainly present in mechanistic models, for the reasons discussed above, and at least some prominent mechanists are conscious of the presence and possible utility of abstraction in scientific work. For example, Lindley Darden has discussed abstraction in science in a number of settings, including the relationship between abstraction and generality (1996), and how abstract representations can function as parts of theories by subsuming various cases. These abstract representations are then to be instantiated with the relevant details to explain particular phenomena (2006: chapter 10).

However, given this, the general position in the philosophical literature appears to be that abstraction is not in itself a desirable aspect of mechanistic models (Machamer, Darden, and Craver 2000; Darden and Craver 2002), at least as far as explanation is concerned. Rather, it seems to be more a practical constraint or present for reasons of convenience; not something to be valued for its own sake.

This idea has recently been challenged by Arnon Levy and Bill Bechtel, who argue that abstraction allows for certain explanatory benefits, and so sometimes "less can be more" (Levy and Bechtel 2013: 241). For example, emphasizing causal structure at the expense of details regarding the specific parts of a mechanism may highlight features that truly make a difference to the phenomenon of interest. Levy and Bechtel's position is more in keeping with the literature regarding modeling and explanation more generally, where abstraction is often seen as a positive goal in at least some settings (e.g. Levins 1966; Wimsatt 1987; Strevens 2004, 2008).

This debate is only just beginning to be addressed in the context of mechanistic models, but clarifying this type of issue is exactly the kind of insight that philosophy of modeling may bring to the philosophy of mechanistic science. Might the inclusion of further detail sometimes actually reduce the representational adequacy of these models? Regardless of the final answer to this question, the outcome will be of interest to all parties. If mechanistic models are always improved by the inclusion of more detail when possible and practical, this is of relevance to arguments regarding the proper objectives of modeling (Levins 1966; Orzack and Sober 1993; Odenbaugh 2003; Potochnik 2007). On the other hand, if increased detail is sometimes a bad feature of mechanistic models, then this finding will substantially add to discussion regarding representation in the mechanistic sciences.

Idealization in mechanistic models has received even less attention than abstraction, but again the standard position appears to be that it is something generally to be avoided. On this view, the more that a mechanistic model diverges from the structure of its target, the worse that model will be. There may be pragmatic or pedagogical reasons to produce simplified models, but this is paying an epistemic price for an increase in ease of use, rather than a positive aspect of the model per se.

This position has also recently been questioned. In the paper "The idealization of causation in mechanistic explanation," Alan Love and Marco Nathan argue that mechanistic models are sometimes altered through judicious simplification to improve explanatory power. This can sometimes even include misrepresenting features known to be genuine difference-makers for the phenomenon of interest. For example, Love and Nathan note that many standard mechanistic explanations of gene expression represent the process in extremely simplified ways, omitting various necessary co-factors and converting multiple steps involving multiple molecules into a series of simple, unitary causal processes. Nevertheless, they maintain (and the scientific literature appears to agree) that these idealized representations are successfully explanatory (Love and Nathan forthcoming).

In fact, it is not difficult to find examples of quite markedly idealized mechanistic models. Furthermore, this idealization can occur in fields usually thought to lend themselves to a straightforward "machine-like" representation of the target system (Levy 2014). For example, many cellular phenomena are modeled as proceeding through a series of steps from start-up conditions to an end point, with each entity playing a specific role in the process. However, cellular behaviors such as protein synthesis and signaling pathways are often quite noisy "biased random walks" (Moore 2012: 8), and the relationships between cellular entities and their roles are often many–many, in ways not reflected in standard models (see, for example, Raser and O'Shea 2005; Viney and Reece 2013; Moore 2012; Zhao et al. 2009). This does not necessarily mean that a close match between model and target isn't usually desirable in mechanistic science. However, idealization is clearly present in many mechanistic models, and merits more attention than it has received so far.

5. Relevance for contemporary debates in the mechanisms literature

We have seen how the study of mechanistic models intersects with a number of concerns in the broader modeling literature. I will now outline a couple of examples to briefly illustrate how the consideration of modeling may speak to some core issues in mechanistic philosophy. First I will address the question of what counts as a mechanism in the setting of natural selection, and then the issue of whether mechanistic explanation is ontic or epistemic.

In section 4 of this chapter I argued that there can be successful instances of mechanistic modeling where the model and target are dissimilar in certain ways, including their spatiotemporal organization. One example of this occurs when scientists construct mechanistic models of populations. In such cases, although the parts and interactions which underlie certain population-level behaviors might arguably occur at the level of individual members, it may be possible to mechanistically model these behaviors using population-level properties. The earlier case of the MONIAC is one instance of this, but it is by no means the only one.

This type of scenario is interesting for a number of reasons, not least because it may have a bearing on the limits of mechanistic representation. For example, in the article "Thinking about Evolutionary Mechanisms: Natural Selection" (2005), Robert Skipper and Roberta Millstein address the question of whether natural selection can be represented as a mechanism in a way that fits standard philosophical accounts. They point out that scientists call natural selection a mechanism, and engage with natural selection as though it is a mechanism in much of their work. However, the then standard accounts of mechanisms (Machamer, Darden, and Craver 2000; Glennan 2002) seem unable to accommodate this, since—among other reasons—the entities involved and their interactions do not exhibit the right kinds of stability and regularity to qualify. This concern has since led to considerable discussion regarding whether natural selection can legitimately be considered a mechanism (e.g. Barros 2008; Havstad 2011; Levy 2012).

Our earlier points regarding how mechanistic models might misrepresent the discreteness and spatiotemporal structure of their target may be relevant to this issue. The "mechanism" of natural selection might be represented at the level of the population, where the effects of the environment, breeding, and so on can be modeled *as though* they occur discretely in space and time, washing out messy details of the actual ongoing individual events that underlie them. Even though such a model misrepresents certain aspects of the process of natural selection, it is certainly not a phenomenological model. Thinking of natural selection as a population-level mechanism still explains the process in a way that relies on organization and causal structure (see also Illari and Williamson 2010).

Here we see the importance of recognizing that similarity requirements can vary according to scientific purpose. For mechanistic explanation to be possible, there must be structural similarity between model and target. However, as argued above, at least sometimes this similarity may elide many of the spatiotemporal details. In this way, Skipper and Millstein can be right that natural selection is not strictly mechanistic as outlined in standard views (I do not make any claims either way here, but see Chapter 22 for more on this), while it is still the case that for some purposes, at least, scientists can legitimately think of and describe natural selection as a mechanism, and thereby gain some explanatory insights.

Another prominent internal disagreement within mechanistic philosophy is whether explanation is "ontic" or "epistemic." Proponents of the ontic view claim that "mechanisms in the world" explain phenomena, while their opponents instead see explanation as dependent on epistemic artifacts such as scientific representations (Machamer, Darden, and Craver 2000; Craver 2007; Bechtel and Abrahamsen 2005; Bechtel 2006).

It is not absolutely clear what this disagreement turns on (see Illari 2013 and Chapter 16, this volume, for further discussion), but one way to frame the issue is as an argument regarding priority. Everyone can agree that both mechanisms-in-the-world and their representations are required for mechanistic explanation, but there is still the question of how these are related. For example, if the mechanism itself is key to explanation, then the role of scientific representation will be merely to articulate that mechanism. On the other hand, if the epistemic artifact is central, then there may be cases where successful explanatory representations articulate something different to, or other than, the relevant mechanism.

Now the prior discussion regarding abstraction and idealization becomes important. Recall that mechanistic models are always at least somewhat abstract, and at least sometimes idealized. It might seem that these findings immediately militate in favor of the epistemic view. However, the issue is subtler than this, and the reasons that underlie such abstraction and idealization are relevant to the debate. For example, if the abstraction is only intended to omit non-essential features to make the relevant mechanism more transparent, this is certainly consistent with the ontic view. However, if abstraction occurs in the service of greater unification or generality, even at the expense of certain relevant mechanistic details, this would fit more neatly within an epistemic framework.

Even some kinds of idealization may be consistent with an ontic approach. If idealization in mechanistic science is generally Galilean (and so simply to serve pragmatic concerns such as tractability), this is perfectly in keeping with a mechanism-first account of explanation. Conversely, minimal idealization seems genuinely inconsistent with the ontic account. If mechanistic models can be improved through intentional misrepresentation of the mechanism for in-principle reasons, this finding would be in favor of the epistemic view (Weisberg 2013: 102).

For example, if Love and Nathan are correct, scientists consider at least some models of gene expression to be superior when they represent the relevant mechanism in ways known to be misleading. Intentionally including falsehoods for explanatory gain seems primarily an epistemic exercise, and it is not obvious how such an approach could be accommodated by the ontic' view. The more that mechanistic models are thought to be improved by omitting parts of the relevant mechanism, or by misrepresenting it, the less convincing it seems that the mechanism itself is the primary driver of a model's explanatory force.

Once again, these points are not intended to decide this extremely complex issue, and the arguments will turn on careful examination of particular cases of mechanistic representation. However, it is hopefully clear how consideration of concerns embedded in the philosophy of modeling can provide traction regarding key problems in the philosophy of mechanisms. These examples represent only part of the work that has been done here, let alone the work that remains to be done. The intersection of modeling and mechanisms will undoubtedly prove to be an area of increasingly important work in philosophy of science.

Notes

1 My sincere gratitude to Stuart Glennan and Phyllis Illari for their patience, advice, and support.
2 For example, in a recent article Dana Matthiessen treats schemas as at least akin to scientific models (Matthiessen 2015), while Lindley Darden has noted that any of these representations of mechanisms might be treated as models in a general sense (Darden 2007: 145). To be fair, difficulties in establishing a clear or consistent view here are likely to be at least partly due to the various ways the term "model" is used in the philosophical literature.
3 According to the semantic account, theories are sets of models, where these models are interpretations of, or structures that satisfy, the sentences of the prior "syntactic" account identified with scientific theory (Suppe 1977; van Fraassen 1980; Odenbaugh 2008). As a potential contrast to the semantic view, Lindley Darden has suggested that theories may be characterized as sets of mechanism schemas (2007: 142).
4 Weisberg also identifies a third type of idealization he calls "multiple model" idealization, which I won't consider here. However, this kind of idealization has been discussed in the specific setting of mechanistic models (Love and Nathan forthcoming).

References

Barr, Nicholas. 2000. "The History of the Phillips Machine." In *A. W. H. Phillips: Collected Works in Contemporary Perspective*, edited by R. Leeson, 89–114. Cambridge University Press.
Barros, D. Benjamin. 2008. "Natural Selection as a Mechanism." *Philosophy of Science* 75 (3): 306–22.
Bechtel, William. 2006. *Discovering Cell Mechanisms: The Creation of Modern Cell Biology*. Cambridge University Press.
Bechtel, William, and Adele Abrahamsen. 2005. "Explanation: A Mechanist Alternative." *Studies in History and Philosophy of Biological and Biomedical Sciences* 36 (2): 421–41.
Cartwright, Nancy. 1989. *Nature's Capacities and Their Measurement*. Oxford University Press.
Craver, Carl. 2006. "When Mechanistic Models Explain." *Synthese* 153: 355–76.
——. 2007. *Explaining the Brain: Mechanisms and the Mosaic Unity of Neuroscience*. Oxford University Press.
——. 2008. "Physical Law and Mechanistic Explanation in the Hodgkin and Huxley Model of the Action Potential." *Philosophy of Science* 75 (5): 1022–33.
Craver, Carl, and James Tabery. 2015. "Mechanisms in Science." In *The Stanford Encyclopedia of Philosophy*, edited by Edward N. Zalta, Winter 2015. http://plato.stanford.edu/archives/win2015/entries/science-mechanisms/.
Darden, Lindley. 1996. "Generalizations in Biology." *Studies in History and Philosophy of Science Part A* 27 (3): 409–19.
——. 2006. *Reasoning in Biological Discoveries: Essays on Mechanisms, Interfield Relations, and Anomaly Resolution*. Cambridge University Press.
——. 2007. "Mechanisms and Models." In *The Cambridge Companion to the Philosophy of Biology*, edited by David L. Hull and Michael Ruse, 139–59. Cambridge University Press.

Darden, Lindley, and Carl Craver. 2002. "Strategies in the Interfield Discovery of the Mechanism of Protein Synthesis." *Studies in History and Philosophy of Science Part C* 33 (1): 1–28.

Downes, Steven. 1992. "The Importance of Models in Theorizing: A Deflationary Semantic View." *Proceedings of the Philosophy of Science Association* 1: 142–53.

———. 2009. "Models, Pictures, and Unified Accounts of Representation: Lessons from Aesthetics for Philosophy of Science." *Perspectives on Science* 17: 417–28.

Fine, Arthur. 2009. "Science Fictions: Comment on Godfrey-Smith." *Philosophical Studies* 143: 117–25.

Frigg, Roman. 2009. "Models and Fiction." *Synthese* 172 (2): 251–68.

Giere, Ronald. 1988. *Explaining Science: A Cognitive Approach.* University of Chicago Press.

———. 2004. "How Models Are Used to Represent Reality." *Philosophy of Science* 71 (5): 742–52.

———. 2009. "Why Scientific Models Should Not Be Regarded as Works of Fiction." In *Fictions in Science: Philosophical Essays on Modeling and Idealization*, edited by Mauricio Suárez, 248–58. Routledge.

Glennan, Stuart. 2002. "Rethinking Mechanistic Explanation." *Proceedings of the Philosophy of Science Association* 3: 342–53.

———. 2005. "Modeling Mechanisms." *Studies in History and Philosophy of Biological and Biomedical Sciences* 36 (2): 443–64.

Godfrey-Smith, Peter. 2006. "The Strategy of Model Based Science." *Biology and Philosophy* 21: 725–40.

———. 2009. "Models and Fictions in Science." *Philosophical Studies* 143: 101–16.

———. 2014. *Philosophy of Biology.* Princeton University Press.

Goodman, Nelson. 1972. *Problems and Projects.* Bobbs-Merrill.

Havstad, Joyce C. 2011. "Problems for Natural Selection as a Mechanism." *Philosophy of Science* 78 (3): 512–23.

Illari, Phyllis. 2013. "Mechanistic Explanation: Integrating the Ontic and Epistemic." *Erkenntnis* 78 (2): 237–55.

Illari, Phyllis, and Jon Williamson. 2010. "Function and Organization: Comparing the Mechanisms of Protein Synthesis and Natural Selection." *Studies in History and Philosophy of Biological and Biomedical Sciences* 41 (3): 279–91.

Kaplan, David Michael. 2011. "Explanation and Description in Computational Neuroscience." *Synthese* 183 (3): 339–73.

Kaplan, David Michael, and Carl Craver. 2011. "The Explanatory Force of Dynamical and Mathematical Models in Neuroscience: A Mechanistic Perspective." *Philosophy of Science* 78 (4): 601–27.

Leonelli, Sabina. 2008. "Performing Abstraction: Two Ways of Modelling Arabidopsis Thaliana." *Biology and Philosophy* 23: 509–28.

Levins, Richard. 1966. "The Strategy of Model Building in Population Biology." *American Scientist* 54 (4): 421–31.

Levy, Arnon. 2012. "Three Kinds of New Mechanism." *Biology & Philosophy* 28 (1): 99–114.

———. 2013. "What Was Hodgkin and Huxley's Achievement?" *The British Journal for the Philosophy of Science* 65: 469–92.

———. 2014. "Machine-Likeness and Explanation by Decomposition." *Philosophers' Imprint* 14 (6): 1–15.

Levy, Arnon, and William Bechtel. 2013. "Abstraction and the Organization of Mechanisms." *Philosophy of Science* 80 (2): 241–61.

Love, Alan, and Marco Nathan. Forthcoming. "The Idealization of Causation in Mechanistic Explanation." *Philosophy of Science.*

Machamer, Peter, Lindley Darden, and Carl Craver. 2000. "Thinking about Mechanisms." *Philosophy of Science* 67 (1): 1–25.

Matthewson, John. 2011. "Trade-Offs in Model-Building: A More Target-Oriented Approach." *Studies in History and Philosophy of Science* 42: 324–33.

Matthewson, John, and Brett Calcott. 2011. "Mechanistic Models of Population-Level Phenomena." *Biology and Philosophy* 26: 737–56.

Matthewson, John, and Michael Weisberg. 2009. "The Structure of Tradeoffs in Model Building." *Synthese* 170: 169–90.

Matthiessen, Dana. 2015. "Mechanistic Explanation in Systems Biology: Cellular Networks." *The British Journal for the Philosophy of Science*, doi:10.1093/bjps/axv011.

Moore, Peter B. 2012. "How Should We Think about the Ribosome?" *Annual Review of Biophysics* 41 (1): 1–19.

Odenbaugh, Jay. 2003. "Complex Systems, Trade-Offs and Mathematical Modeling: Richard Levins' 'Strategy of Model Building in Population Biology' Revisited." *Philosophy of Science* 70: 1496–1507.

——. 2008. "Models." In *Blackwell Companion to the Philosophy of Biology*, edited by S. Sarkar and A. Plutynski, 506–24. Blackwell.

Orzack, Stephen, and Elliot Sober. 1993. "A Critical Assessment of Levins's 'the Strategy of Model Building in Population Biology' (1966)." *Quarterly Review of Biology* 68 (4): 533–46.

Phillips, A. W. 2000. "Mechanical Models in Economic Dynamics." In *A. W. H. Phillips: Collected Works in Contemporary Perspective*, edited by R. Leeson, 68–88. Cambridge University Press.

Potochnik, Angela. 2007. "Optimality Modeling and Explanatory Generality." *Philosophy of Science* 74 (5): 680–91.

Raser, Jonathan, and Erin O'Shea. 2005. "Noise in Gene Expression: Origins, Consequences, and Control." *Science* 309 (5743): 2010–13.

Skipper, Robert, and Roberta Millstein. 2005. "Thinking about Evolutionary Mechanisms: Natural Selection." *Studies in History and Philosophy of Biological and Biomedical Sciences* 36 (2): 327–47.

Sterrett, Susan. 2002. "Physical Models and Fundamental Laws: Using One Piece of the World to Tell about Another." *Mind and Society* 3: 51–66.

——. 2006. "Models of Machines and Models of Phenomena." *International Studies in the Philosophy of Science* 20: 69–80.

Strevens, Michael. 2004. "The Causal and Unification Approaches to Explanation Unified—Causally." *Nous* 38 (1): 154–76.

——. 2008. *Depth: An Account of Scientific Explanation*. Harvard University Press.

Suppe, Frederick. 1977. "The Search for Philosophic Understanding of Scientific Theories." In *The Structure of Scientific Theories*, edited by Frederick Suppe, 1–241. Urbana–Champaign, IL: University of Illinois Press.

Thomson-Jones, Martin. 2006. "Models and the Semantic View." *Philosophy of Science* 73: 524–35.

van Fraassen, Bas. 1980. *The Scientific Image*. Oxford University Press.

Viney, Mark, and Sarah E. Reece. 2013. "Adaptive Noise." *Proceedings of the Royal Society of London B: Biological Sciences* 280 (1767), doi:10.1098/rspb.2013.1104.

Weber, Marcel. 2008. "Causes without Mechanisms: Experimental Regularities, Physical Laws, and Neuroscientific Explanation." *Philosophy of Science* 75 (5): 995–1007.

Weisberg, Michael. 2004. "Qualitative Theory and Chemical Explanation." *Philosophy of Science* 71 (5): 1071–81.

Weisberg, Michael. 2007a. "Three Kinds of Idealization." *The Journal of Philosophy* 104 (12): 639–59.

——. 2007b. "Who is a Modeler?" *The British Journal for the Philosophy of Science* 58: 207–33.

——. 2013. *Simulation and Similarity: Using Models to Understand the World*. Oxford University Press.

Wenbourne, Alan. 2006. "Computer Controlled DSG Transmission—South East London Meccano Club." Accessed January 13, 2016. http://www.selmec.org.uk/article_0004_computer_controlled_dsg_transmission.aspx.

Wimsatt, William. 1987. "False Models as a Means to Truer Theories." In *Neutral Models in Biology*, edited by M. Nitecki and A. Hoffmann, 23–55. Oxford University Press.

Zhao, Ping, Lu Yang, Jamie A. Lopez, Junmei Fan, James G. Burchfield, Li Bai, Wanjin Hong, Tao Xu, and David E. James. 2009. "Variations in the Requirement for v-SNAREs in GLUT4 Trafficking in Adipocytes." *Journal of Cell Science* 122 (19): 3472–80.

18

EXPLAINING VISUALLY USING MECHANISM DIAGRAMS

Adele Abrahamsen, Benjamin Sheredos,
and William Bechtel

Scientists are prolific purveyors of diagrams and other visual representations. Their publications are replete with them; their lab meetings are organized around them. With the mechanistic turn in philosophy of science, many of the New Mechanists and Social Scientific Mechanists assembled in this handbook under the broad "minimal mechanism" umbrella (Chapter 1; Illari and Williamson 2012; Glennan in press) have incorporated some of these as figures within their otherwise text-heavy case studies. Several have devised their own diagrams as well, to help make sense of the science (see Chapters 9, 13, and 14). Our own chapter therefore could have aimed to explore the visualizations used in the physical sciences, social sciences, and philosophy of science in all of their diversity. Instead, we have chosen to focus on the one type of visual representation that most directly captures the machine metaphor at the heart of the researchers' endeavor, which we call *mechanism diagrams*. These have deep historical roots (which are unearthed as far back as Descartes' biology; see Chapter 3).

In constructing a mechanism diagram, a scientist lays out the parts of a proposed mechanism and the operations (activities) they perform so as to highlight spatiotemporal organization (e.g., Figure 13.1 in Chapter 13). Our focus on mechanism diagrams reflects an epistemic, rather than ontological, construal of mechanism, beginning with what we call *basic mechanistic explanation* (relying on strategies of decomposition and localization) and progressing to *dynamic mechanistic explanation* (which adds computational modeling of the mechanism's dynamics to its otherwise merely ordinally specified temporal organization; see Bechtel and Abrahamsen 2012). Recently we have dubbed these Mechanism 1.0 and 1.1 and argued for a Mechanism 2.0 in which stable parts and other strictures of the machine metaphor can be abandoned (Levy and Bechtel forthcoming). Since most scientists' diagrams reside comfortably within Mechanism 1.0, however, most of this chapter focuses on basic mechanistic diagrams (sections 1–5). It is bookended with consideration of a broader range of diagrams and other visual representations: first, a brief review of work on the functions diagrams serve for scientists and others (section 1); later, an annotated mechanism diagram linking a basic mechanism to a computational model of its dynamics (section 7); and finally, visual representations even further afield—such as those displaying data from experiments, or causal relations between variables—which potentially may contribute to mechanistic understanding despite not portraying parts and operations (section 8).

1. Functions served by diagrams

The key function of diagrams, shared by almost of their specific uses, is to provide an external aid that supports and expands the reach of mental work. (An exception is the archival function, but even that preserves the potential for diagrams to aid future mental work.) More specific functions include the use of a diagram by an individual (or interactively by a group) as an aid in discovering or modifying a scientific account, solving a problem, and understanding a particular mechanistic explanation or other sort of content. Mechanism diagrams in particular not only support conceptualizing and reasoning about a proposed mechanism but also help with relating the proposal to existing explanations, considering the evidence for new components, trying out a reorganization of existing components, and so forth (see Sheredos and Bechtel in press-b). For example, after adding an arrow for a recently discovered molecular interaction to a diagram, a scientist may stare and sketch for some time considering what further changes to the previous explanation this entails.

Diagrams can also be used for communication, either one-to-one or one-to-many; for example, an organizational chart may be handed to a new executive or emailed to all. For scientists, if a journal article includes a mechanism diagram it typically is the first figure (raising a question) or last figure (displaying a mechanism that fits the findings) or, increasingly, serves as a graphical abstract (capturing the crux of the contribution). Multiple mechanism diagrams may be included in an integrative review or oral presentation. When instead the communication is from teacher to student—pedagogy—the diagram may have a different design that takes into account students' limited knowledge of both content and diagram conventions in the field under study (Tversky 2005) and congruity with the desired mental representation.

A few philosophers of science have offered insightful inquiries into scientists' various uses of diagrams, especially for discovery. Construing discovery as an extended, iterative process, Darden and colleagues have introduced *mechanism schemas* and the related notions of *mechanism sketches* and *schema types*. Their case studies of the discovery of particular biological mechanisms, many of which are brought together in the book by Craver and Darden (2013), include some of the mechanism diagrams scientists have used to convey proposed schemas. In Chapter 19 of this handbook, Darden offers a very clear précis (sans diagrams). Another window into the discovery process is provided by unpublished diagrams produced over the course of a single investigation (Burnston, Sheredos, Abrahamsen, and Bechtel 2014). Incorporating both published and unpublished diagrams, and arguing that they form a coupled system with mental models, Nersessian (2008) contributed a penetrating account of Maxwell's development of his theory of electromagnetism. Moving forward from discovery, Perini (2005) emphasized the final stage in which scientists include diagrams and other visual representations in their publications to support their claims. She later compared three visual formats used by biologists (pictorial, schematic, and compositional) with respect to the different ways each related form to content. She pointed out that this relationship was least transparent for the compositional format, as illustrated by a basic mechanistic diagram of ATP synthesis in which wedge shapes represented the (not wedge-shaped) enzyme ATPase (Perini 2013). Finally, Woody (2014) explored yet another use of diagrams—as problem-solving aids—by focusing on the periodic table and other chemistry diagrams.

These philosophical inquiries tie into the much larger literature on diagrams in cognitive science in which experimenters present research participants with tasks involving diagrams. In a pioneering study, Larkin and Simon (1987) had participants perform problem-solving tasks in which the same information was provided in different formats. Inferences were made more effectively with diagrams than with sentential representations because, they argued, diagrams enabled faster search and the use of perceptual recognition. One of the most ambitious later studies of this type (Cheng 2011) demonstrated superior learning by students completing a two-session mini-curriculum in which they learned to use probability space diagrams to solve probability problems. In a review Hegarty (2011) noted that visual-spatial displays, such as the weather maps promoting inference in one of her own studies, support cognitive activities by organizing information spatially, storing it externally, and enabling it to be processed visually. Tversky is another cognitive scientist providing astute theoretical roadmaps in addition to empirical studies of diagrams (e.g., Tversky 2005, 2011).

2. A simple diagram of a proposed circadian mechanism

We employ as exemplars of mechanism diagrams those published in a single field, circadian biology, to show the variety of options available even when the potential content is the same. Daily oscillations in organisms' behaviors, such as plants folding and unfolding their leaves, have been observed since ancient times. It was not until the twentieth century, however, that researchers established that these rhythms are internally generated but entrainable to local day–night cycles.

The rise of molecular biology in the mid-twentieth century provided the tools to start uncovering the molecular mechanisms responsible for a wide range of phenomena, including behavioral rhythms. The first circadian gene, *per (period)*, was identified in the 1970s in fruit flies. Like most genes, it is transcribed into *mRNA*, which is transported into the cell's cytoplasm and translated there into a protein. But for *per*, concentrations of these products oscillate on a 24-hour cycle (with *per mRNA* about 8 hours ahead of the PER protein). To explain these oscillations, Hardin, Hall, and Rosbash (1990) proposed a negative feedback mechanism.

To make their explanation accessible to nonbiologists, we designed a mechanism diagram showing the parts, operations, and functional organization of the simplest version of this molecular clock. (Their own working diagram contains alternative loops and a number of question marks—a good example of Darden's mechanism sketches and useful for supporting the scientists' own reasoning, but too complicated and idiosyncratic for the purpose here.) As shown in Figure 18.1, after PER molecules have been produced, they are transported back into the nucleus and inhibit further expression of *per* (though less so as they degrade, and later found to require the help of additional components). The result of this negative feedback is a cycle in which increases in PER lead to increasing inhibition, which leads to decreases in PER and hence less inhibition, which leads to increases in PER once again—a new turn of the cycle.

If the specific molecular labels are removed, the remaining generic labels and arrows constitute a schema type (see Darden's discussion in Chapter 19)—the simplest version of the large family of *transcription-translation feedback loops* exemplified by the *per* loop. While Hardin, Hall, and Rosbash proposed this feedback loop to explain circadian rhythms in fruit flies (*Drosophila*), its basic organization is remarkably well conserved. There is a similar mechanism within the neurons of the suprachiasmatic nucleus (SCN) in mouse brains, as we will see in the mechanism diagrams in sections 4–6. These have more loops, but two of them involve homologs of *per* (labeled as m*Per1* and m*Per2* in Figure 18.2 and—omitting the optional "m" for mouse—as *Per1/2* in Figure 18.3). For all mammals, including humans, the SCN relies on these molecular loops to serve its function as a central clock that adapts its timing to input from the retina and sends output signals to synchronize the cycles of other cells throughout the body.

Figure 18.1 illustrates some common graphical conventions: regular arrows for most operations, a flat-edged arrow for inhibitory operations (negative feedback), and an irregular arrow for degradation. The additional interacting loops uncovered since 1990 inspired graphical innovations that provide a variety of ways to fit multiple feedback loops into a single mechanism diagram. We will encounter some of these in sections 4–6, but begin by sorting through the toolbox of components from which both simple and complex diagrams are constructed.

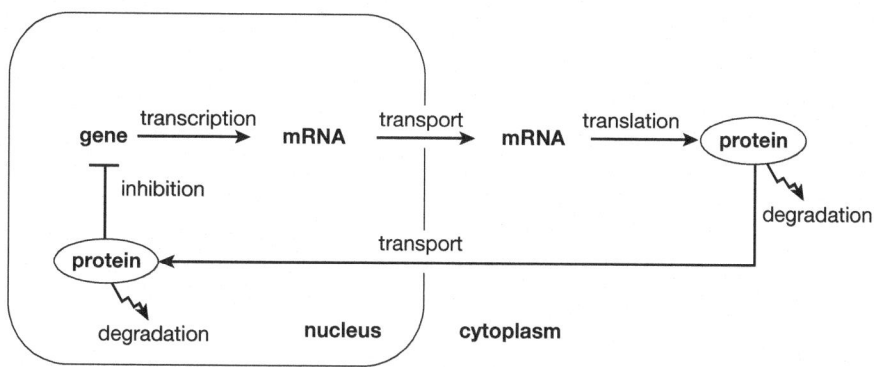

Figure 18.1   A simple mechanism diagram showing the key parts and operations in the transcription-translation feedback loop proposed by Hardin, Hall, and Rosbash (1990) to account for circadian rhythms. (The roles of additional interacting loops were discovered later.)

3. Components of mechanism diagrams

Most mechanism diagrams are composed of several different kinds of elements. Some of the most powerful are in a category that Tversky (2011) calls *glyphs*. These are simple elements that are "prevalent across a wide range of graphics" and are "readily understood in context";

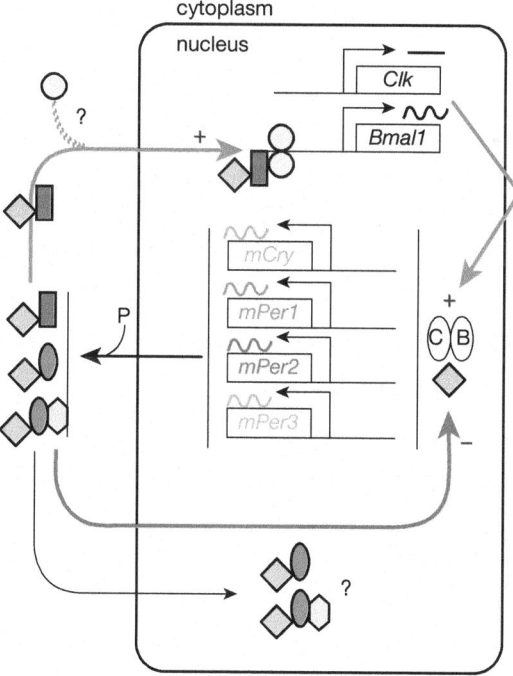

Figure 18.2   A diagram of the molecular clock mechanism for circadian rhythms in mammals, exemplifying several types of elements used in mechanism diagrams. From Reppert and Weaver (2001), figure 1. Reproduced with permission of *Annual Review of Physiology*, Volume 63 © 2001 by Annual Reviews, http://www.annualreviews.org. See plate 1

moreover, "Like words in language, [they] can be combined in various ways to create varying meanings. Like words in language, there are constraints on how they can be combined" (Tversky 2005, p. 141). In the following subsections we first describe the glyphs that are most prevalent in mechanism diagrams: shapes, lines, and arrows. We note how color can combine with these, and then describe three more specialized types of elements (text, iconic symbols, and graphic symbols). We introduce these by discussing the choices made by Reppert and Weaver (2001) in their diagram of the mammalian circadian clock mechanism (Figure 18.2; see plate 1). We also point out alternative choices made by others in Figures 18.3–18.5 (more fully discussed in sections 4–6) and note one additional type of element—embedded data graphs—within Figure 18.4. Finally, what we call a *conventional complex* is a multi-element building block seen in Figures 18.2–18.5.

Shapes

Different types of parts in a mechanism typically are represented by different geometric shapes, which sometimes are labeled or color-coded to further distinguish them. At the center and upper right of Figure 18.2 are several oblong boxes, each labeled in color with the abbreviated name of the gene it represents (*mCry*, *mPer1*, etc.). Elsewhere, shapes in matching colors represent proteins produced from these genes: gold-colored diamonds for the protein mCRY, blue ovals for mPER1, red rectangles for mPER2, and light-green hexagons for mPER3. The two slightly larger ovals with the letters C and B in them represent the proteins CLOCK and BMAL1. Thus, two different ways of using shapes were chosen here for the same basic task of representing proteins: distinctive shapes with redundant color-coding for the four central proteins vs. a plain shape with a distinctive initial letter for two other proteins. (An additional shape, the small gray circle, represents a hypothesized molecule.) This diagram also illustrates how the positioning of shapes can convey relationships between entities: the ovals labeled C and B touch each other to indicate that CLOCK and BMAL1 have dimerized (formed a two-part complex).

Lines and arrows

Lines and arrows can have a variety of interpretations, depending on context or stipulation. Among those for arrows are simple temporal sequencing, abstract causal relationships, and transformations without change of location. However, the arrows in this chapter have several different more specific interpretations, requiring viewers to keep track of which arrows have which interpretation—even within a single diagram. This is easy for the single feedback loop in Figure 18.1, because each arrow represents one operation and is labeled. However, most diagrams of circadian mechanisms include multiple loops, pushing designers toward more compact representations of each loop. Figure 18.2 makes demands on viewers by mixing different strategies in representing its six loops. All of them incorporate the convention of a small arrow bent at a 90° angle next to a gene to represent its transcription. For *Clk* and *Bmal1*, the subsequent parts and operations in their loops are combined into a single glyph—a large, angled green arrow—directed at the C and B ovals that represent the resulting CLK:BMAL1 dimer.

The green arrow simplifies Figure 18.2 visually, but makes demands on the viewer's background knowledge—a frequently encountered tradeoff. Here, the viewer needs to know not only the parts and operations involved but also that the CLK:BMAL1 dimer, having been transported to the nucleus, activates expression of the four central genes.

A different strategy was used to depict the central genes' loops. The four genes are bundled together by two vertical lines so they can share the thick black arrow representing the two operations following transcription—transport and translation of mRNA—as well as the curved line from **P** representing phosphorylation of the resulting proteins. A single vertical line bundles together the color-coded glyphs for the proteins, which are arranged to indicate the complexes they form (dimers and trimers). The fact that some tokens of each type of complex are transported back to the nucleus, where they inhibit the activation activity of the CLK:BMAL1 dimers, is separately represented by the long red arrow with a minus sign (an alternative to the flat-edged inhibitory arrows in Figures 18.1, 18.3, and 18.5). Other tokens return to the nucleus for more specialized operations only partly understood in 2001, as indicated by the two arrows accompanied by copies of the relevant complexes and question marks.

To make the use of lines and arrows in Figure 18.2 understandable, we have alluded to the use of color and such elements as labels, question marks, and minus signs. These are among the elements discussed in the rest of this section.

Colors

Color can interact with glyphs in a variety of ways that aid understanding. Often one shape represents a class of objects (e.g., ovals for proteins), and color is used to make distinctions within the class. For example, in Figure 18.3 (see plate 2), proteins and polypeptides are represented by circles that are colored gold for PER, green for CRY, and violet for VIP. Reppert and Weaver took a different approach: each protein in Figure 18.2 has its own particular shape as well as color. This visual redundancy facilitates tracking each protein and linking it to its gene (which is labeled in the same color). For example, focusing on gold takes a viewer from the gene *mCry* to the protein CRY and then to its various interactions with other proteins. Yet another way of color-coding glyphs is showcased in Figure 18.5, where color identifies functionally similar components of the circadian mechanism not only in the same organism but also across species (discussed in section 6; see plate 4). As for lines and arrows, they are usually black, but Figure 18.2 (as well as Figure 18.5) illustrates the convention that a green arrow indicates activation and a red arrow inhibition. This helps viewers see the overall organization of the mechanism as combining positive and negative feedback.

Text

Although diagrams are quintessentially visuospatial, rather than linguistic, most incorporate textual elements. At one extreme, Figure 18.1 makes extensive use of text and only limited use of space and glyphs. Abbreviations alone identify parts (except for the oval around "PER"), and arrows are labeled to identify operations. At the other extreme, Figure 18.2 has color-coded shapes depicting parts (adding text labels only where necessary) and unlabeled arrows. Both figures have an enclosed nucleus with word labels distinguishing nucleus from cytoplasm. A final use of text in Figure 18.1, prompted by our pedagogical aims in crafting it, is for classifying the parts as gene, mRNA, or protein. Various other styles of incorporating text can be seen in Figures 18.3–18.6, ranging from single characters to multiword phrases and (in the original articles) lengthy captions.

Iconic symbols (icons)

As an alternative to text, iconicity is a powerful tool for conveying meaning with visual immediacy. An iconic symbol (icon) resembles its referent in certain respects, and its meaning is specific and stable across contexts. Designers can draw upon a variety of conventional icons or invent their own.

Every transcription arrow in Figure 18.2 points toward a conventional icon: either a wavy line indicating that the gene's transcription activity exhibits circadian rhythmicity or (for *Clk* only) a straight line indicating continuous transcription. Figure 18.4 (discussed in section 5; see plate 3) shows one variation on another convention: the use of a large 24-hour clock face to mark timing for the rest of the diagram. (A common alternative involves two small icons: the sun marking daytime vs. the moon marking nighttime.) Figure 18.4 also has good examples of less conventional icons: line drawings of mice and humans asleep or awake, with the obvious meanings. More than purely aesthetic features, they enable quick interpretation of the diagram without having to consult textual labels or figure captions.

Graphic symbols

Certain text elements, by taking on a life of their own, have acquired the impact of iconic symbols despite their noniconicity. Figure 18.2 includes four examples of single characters that, at least for biologists, are visually compelling and immediately interpretable: + for activation, − for inhibition, **P** for the phosphate groups that bind with circadian proteins, and **?** for gaps in knowledge. The + and − symbols have stable core meanings (positive vs. negative), but their more specific meanings depend on context (for example, in arithmetic, rating scales, chemistry, or electricity). The phosphate symbol (**P**) is familiar on its own or inside a small circle (and in Figure 18.4, tiny circles are used without **P**). Finally, the question mark (**?**) occurs with surprising frequency in mechanism diagrams and is perhaps the most interesting of the graphic symbols. Those in Figure 18.2 indicate uncertainty about the next operations performed by known proteins (bottom) or an unidentified, hypothesized protein (top left). In Figure 18.5 they mark proteins and operations that are unspecified, hypothesized, or for which evidence is weak (with readers left to consult the text for the precise import of each question mark). Occasionally a diagram with question marks will be the first figure in the article, setting the questions to be addressed by the research that follows. Yet another function of question marks is to aid ongoing reasoning, which is especially evident in working diagrams produced throughout the research process and sometimes even published (e.g., that of Hardin, Hall, and Rosbash 1990, as mentioned in section 2).

Embedded data graphs

We have distinguished mechanism diagrams from data graphs, but occasionally data graphs are inserted into mechanism diagrams. One purpose is to display quantitative evidence about the operation of a given part. Burnston (2016) argues that such graphs play an explanatory role complementary to the basic mechanistic proposal.) Another purpose is to characterize the phenomenon explained by the mechanistic proposal. For example, in Figure 18.4 (see plate 3), the two thumbnail plots show that SCN neural firing is less intense in nighttime than daytime—one of the phenomena to be explained by the intracellular mechanism in the center of the diagram. When graphs are abstracted even further, they may give rise to iconic symbols; for example, line graphs of transcription activity across 24 hours inspired the wavy and straight-line icons in Figure 18.2.

Conventional complexes

Diagrams are created by flexibly combining different kinds of elements. Efficiency trumps flexibility, though, when certain combinations of elements become standardized and take on a life of their own. Consider the boxes for genes and the lines and bent arrows that accompany them in Figure 18.2. Molecular biology students learn to see these arrangements (and variations on them in other figures) as skeletal depictions of the spatial layout of different functional units along a DNA strand. For each gene in Figure 18.2, the box is the coding region and the horizontal line is noncoding DNA (including the promoter regions—not specifically depicted here, but distinguished by labeled glyphs in Figure 18.3 and thickened line segments in Figure 18.5). The bent arrow's placement on the line loosely represents the transcription start site between the promoter and coding regions. Since this whole arrangement has a specific, stable meaning in diagrams throughout molecular biology, we are inclined to count it as a conventional complex rather than as an arrangement of glyphs—and given its spatiality, one that is somewhat iconic. The complex does retain some degree of flexibility in which shapes and labels do or do not get added to the horizontal line (and for Figure 18.2, the addition of wavy vs. straight-line icons as well).

4. Space in mechanism diagrams: representing physical vs. functional organization

Since a diagram is laid out in space and a mechanism occupies space, it is natural to use the two-dimensional space of a diagram to situate parts of the proposed mechanism in a way that approximates its *spatial* organization. In Figure 18.2, for example, in addition to the one-dimensional use of space in complexes representing genes, a well-delineated region representing cytoplasm surrounds a smaller region representing the nucleus. Though simplified, the spatial relation of enclosure in the diagram corresponds to that in a cell. The various molecules are placed as appropriate in either the nucleus or cytoplasm, and their movements between these two regions are depicted by arrows. All of these design decisions use space as space. Within the two regions of the diagram space, however, the molecular parts and operations are laid out to satisfy nonspatial objectives, such as clearly conveying their *functional* organization into feedback loops, grouping together parts that behave similarly, and avoiding crossed arrows. In the nucleus, for example, two sets of genes play different roles: the two peripheral genes produce proteins that mediate the auto-inhibitory loop of the four central genes. Grouping their glyphs in different locations, and using shared arrows where possible for each group's operations, makes this functional organization clear. Likewise, the glyphs and arrows for the four types of proteins produced by the central genes are the minimum needed to show the major operation they have in common (red inhibitory arrow) as well as their distinctive operations. Finally, there is no attempt to show that in the physical cell, numerous proteins of each type are distributed through much of the physical space of the cytoplasm and nucleus. These last few design decisions show how space can be used to clearly convey the functional (rather than spatial) organization of the mechanism. Once again there is a tradeoff: the clarity gained by showing types rather than tokens vs. the risk of downplaying the stochasticity of molecular mechanisms (see Chapter 13—especially its mechanism diagram, Figure 13.1, which shows multiple tokens of the component types).

When a diagram is used to represent activities in multiple spaces, it is often necessary to simplify what is shown in each space. For example, in Figure 18.3 (see plate 2) DeWoskin, Myung, Belle, Piggins, Takumi, and Forger (2015) attempt to relate the familiar molecular mechanism within individual cells to molecular pathways that extend to the extracellular environment. The large ovals represent the cytoplasm (lilac) enclosing the nucleus (purple). It is notable that only a select few of the parts and operations of the intracellular clock mechanism are shown: the loops involving *Clk* and *Bmal1* are absent here, as are numerous additional components that had been discovered between 2001 and 2015. Only the *Per* and *Cry* homologs and their

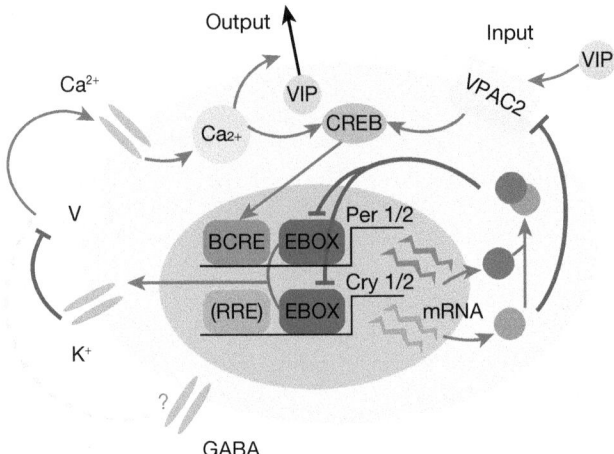

Figure 18.3 A mechanism diagram showing some of the key parts of the molecular clock mechanism and how, in SCN neurons, they interact with pathways extending into the extracellular environment. From DeWoskin, D., Myung, J., Belle, M. D., Piggins, H. D., Takumi, T., and Forger, D. B. (2015) "Distinct roles for GABA across multiple timescales in mammalian circadian timekeeping," *Proc Natl Acad Sci USA*, figure 1, 112(29): E3911-9. See plate 2

mRNA and proteins are shown, using different conventions than Figure 18.2. Each pair of homologs shares one minimalist representation of gene (the vertical and upper horizontal line) and adjacent DNA (lower horizontal line); the bent arrow indicating transcription is omitted. However, there are labels for the genes (*Per 1/2* for *Per1* and *Per2*; *Cry 1/2* for *Cry1* and *Cry2*) and for specific promoter regions indicated by colored ovals (EBOX, CRE, and RRE). The transcription-translation feedback loops are shown schematically: (1) The cyclic transcription of each pair of genes is denoted by two color-coded wavy lines terminating in the word "mRNA." (2) Solid arrows represent transport of the mRNA into cytoplasm and translation there into proteins (denoted by color-coded circles). (3) Two merging arrows indicate dimerization. (4) The negative feedback, in which PER1:CRY1 and PER2:CRY2 dimers inhibit their own genes' transcription, is represented by an inhibitory arrow that splits, with each branch terminating on one of the EBOX glyphs. (As partly shown in Figures 18.2, 18.4, and 18.5, but not here, the inhibition is mediated by removal of CLK:BMAL1 from the EBOX promoters.)

By omitting the operations through *Clk* and *Bmal1* from their diagram, DeWoskin et al. gained the space to show pathways for modulating *Per* expression that extend out to the extracellular environment. Potassium ions (K$^+$) play a role, but we will focus on two pathways that can be thought of as beginning with the entry of calcium ions into the neuron and ending with CREB proteins enhancing *Per* transcription by binding to the CRE promoter region. In the simplest case, the calcium ions directly induce phosphorylation of CREB. The more elaborate pathway involves cell-to-cell communication via VIP. Each neuron signals its activity by releasing VIP into the extracellular environment (with vesicle release aided by calcium). To the extent that such signals are received in turn from other neurons via the VPAC2 receptors, CREB is up-regulated (though the inhibitory arrow indicates it is also subject to cyclic down-regulation). Since DeWoskin et al. show just one cell, it is up to the viewer to imagine this mechanism replicated in numerous SCN cells, such that the level and timing of gene expression in each cell is regulated in part by that of other cells via intercellular VIP signaling. This intercellular level of organization is only hinted at here by the words "Output" (indicating where VIP is excreted to affect other cells) and "Input" (signifying where the VIP produced in other cells affects this one cell).

5. Representing time in static diagrams

Mechanisms function in time. This is conveyed minimally by arrows in a mechanism diagram. In Figure 18.3, for example, the arrows from calcium to CREB and from CREB to CRE inform us of two operations occurring in sequence. However, there is no indication of the duration of each operation or its onset and offset times. Do they overlap, occur in immediate sequence, or does one end hours before the other begins? Also, are there changes in their rate or intensity? For circadian rhythms, timing is critical: the oscillations generated by the transcription-translation feedback loop must have a period of approximately 24 hours. Given the speed with which most chemical reactions occur, delays must be built into the molecular clock. None of this is conveyed in the figures above. Animations provide one solution: beyond using space as space, they can also

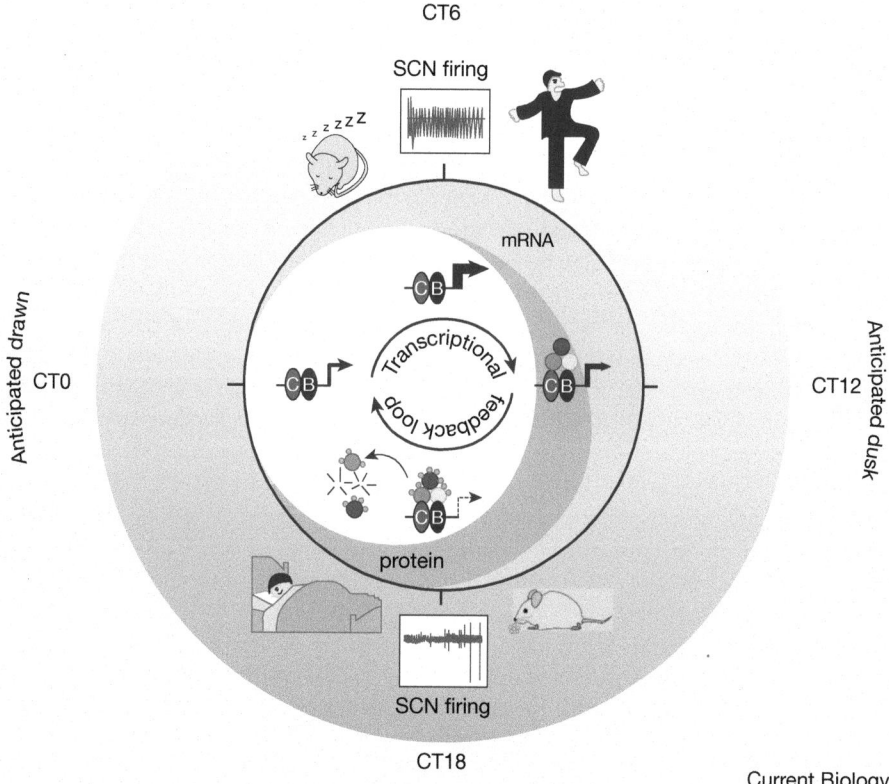

Figure 18.4 Using a 24-hour clock face to show the timing of behavioral and neural activity (gray region) and the underlying molecular clock (inner circle). Appropriately located iconic symbols contrast the activity of humans (diurnal) with that of mice (nocturnal). Reprinted from *Current Biology*, Volume 18, Hastings, M. H., Maywood, E. S., and O'Neill, J. S., Cellular circadian pacemaking and the role of cytosolic rhythms, figure 2, © (2008), with permission from Elsevier. See plate 3

use time as time—more or less veridically, and more or less sped up or slowed down. Circadian animations are well worth viewing, but for printed books and articles circadian researchers have introduced some apt strategies for representing time indirectly (and perhaps more effectively; see Tversky 2011, p. 526). For example, some figures lay out separate subdiagrams for the same mechanism in different phases, arranged around a circle representing a 24-hour cycle. (A similar strategy produces circular data graphs, which are among the specialized circadian data graphics discussed by Bechtel, Burnston, Sheredos, and Abrahamsen 2014.)

The circular displays in mechanism or data diagrams have their roots in familiar physical clocks with the numbers 1–12 placed in a circle, but instead use a 24-hour clock face. When researchers place an organism into constant conditions (for animals, typically by switching from light-dark cycles to constant darkness), time is conventionally reported in *circadian time* (CT). CT0 is the time at which the organism's internal clock should expect light ("anticipated dawn"). In Figure 18.4 (from Hastings, Maywood, and O'Neill 2008; see plate 3), CT0 is placed at the left side of the clock face. Labels for CT6, 12 ("anticipated dusk"), and 18 provide a timescale for the rest of the figure, in which a gray region depicting behavioral and neural activity surrounds an inner circle depicting molecular activity. The variable shading of the gray region provides visual support for the CT labels: lightest at anticipated midday (CT6) and darkest at anticipated midnight (CT18). Within it, diagram elements are appropriately indexed to CT6 and CT18: (a) icons show mice and humans as awake vs. asleep at opposite times, and (b) neural firing rate is indicated *via* thumbnail data graphs obtainable from single-cell recording in mouse SCN (~10 Hz top, ~1 Hz bottom).

In the center of Figure 18.4, multiple representational tools are used creatively and compactly to convey information about the state of the molecular mechanism at different phases of the circadian cycle. Two arrows, labeled "transcriptional feedback loop," indicate the type of molecular mechanism. The parts and operations are represented not by a single detailed diagram, but rather by four minimalist versions of the conventional complex used to depict genes and their transcription. These are indexed to circadian times 0, 6, 12, and 18 and emphasize changes in the state of the mechanism. Each depiction uses a shared, unlabeled horizontal line to represent all of the *Per* and *Cry* genes and adjacent DNA plus the thickness of the attached bent arrow to represent different rates of transcription. They also include (a) labeled oval glyphs for the proteins CLK and BMAL1, to be understood as a dimer binding to the promoter regions for *Per* and *Cry*, and (b) small color-coded circles for the proteins PER and CRY and Casein Kinase. The binding of PER and CRY to the dimer around CT12 (day)—indicated by contact—is followed by their phosphorylation and consequent degradation around CT18 (night). Yet another diagrammatic technique—the positioning of crescents—is used to show how changes in concentration of *Per* and *Cry* mRNA (tan crescent) precede those of PER and CRY proteins (blue crescent).

Overall, once this diagram is unpacked and its various representational strategies interpreted, it conveys considerable information about circadian phenomena (in the gray region) and the time-locked phases of the molecular mechanism responsible for them (in the center).

6. Comparing two or more mechanisms

Diagrams provide a unique and powerful tool for comparing mechanistic accounts. In circadian rhythm research, proposed molecular clock mechanisms are similar enough across different species and even different biological orders that insights from research conducted on one species has often guided research on distant species—quite a valuable shortcut for the discovery process. For diagrams to facilitate this kind of research guidance, as well as integrative thinking

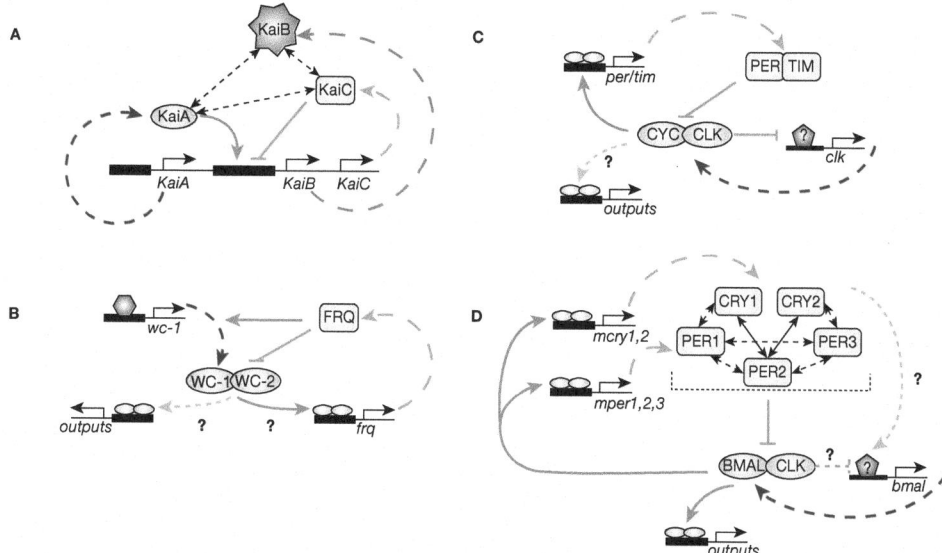

Figure 18.5 Comparison of the molecular clock mechanisms, as understood in 2001, in (A) cyanobacteria, (B) fungi, (C) fruit flies, and (D) mice. From Harmer, S. L., Panda, S., and Kay, S. A. (2001), figure 2. Reproduced with permission of *Annual Review of Cell and Developmental Biology*, Volume 17 © 2001 by Annual Reviews, http://www. annualreviews.org. See plate 4

across the phylogenetic spectrum, consistency is needed in the choice of elements and use of space. (To see how difficult comparisons would be otherwise, consider how differently the *Per* loop was diagrammed in Figures 18.1–18.4.) Figure 18.5 (see plate 4) shows how Harmer, Panda, and Kay (2001) used consistently designed diagrams to compare the clock mechanism in (A) cyanobacteria, (B) fungi, (C) fruit flies, and (D) mice. At the time all of these were thought to centrally involve transcription-translation feedback loops.

Each diagram uses the familiar horizontal line to represent each gene and its associated DNA regions, with the line thickened for the noncoding promoter region. (Homologs and other genes playing similar roles, such as *per* and *tim* in panel C, share a single line.) The gene(s) are denoted by a label under the line, and transcription by the usual bent arrow. The remaining gene expression operations are denoted by a color-coded dashed arrow terminating at the glyph(s) for the resulting protein(s). These glyphs use distinctive shapes and colors to highlight which proteins play comparable roles not only within a species (e.g., PER1,2,3 homologs in mice) but, crucially, in different species. Most noticeable is the use of light-blue rounded rectangles for the key proteins: KaiC in cyanobacteria, FRQ in fungi, the PER:TIM dimer in fruit flies, and dimers (indicated by bidirectional arrows) of PER1,2,3 and CRY1,2 in mice. Each of these proteins indirectly inhibits its own further production as its concentration increases, shown visually by the red inhibitory arrow directed at one or two other proteins (labeled yellow ovals). The role of these other proteins—KaiA, WC-2, CYC:CLK, and BMAL1:CLK in A–D respectively—is to activate the gene producing the key protein. This is depicted by a green arrow pointing from the yellow ovals to smaller yellow ovals, which indicate binding of the proteins to the gene's promoter region. That this binding is disrupted in the inhibitory phase must be inferred. Each panel also includes rose-colored shapes for other proteins (known or hypothesized) and black lines for protein–protein interactions.

The use of color-coding is particularly important in supporting the comparison across species, since it did not prove feasible to locate the corresponding parts of each mechanism in the same region of the spatial layout. This was especially helpful for seeing ways in which the cyanobacteria clock (panel A) is similar to those of the other species. (The diagrams were less successful in conveying how it differs.) For the other species (panels B–D), an important benefit can be seen in the question marks. Although some simply flag issues not yet settled by research within one species, others are motivated by exactly the kinds of inter-species comparisons Figure 18.5 aims to facilitate. In the mammalian case (D), the arrow to the gene labeled "outputs" has no question mark, since the researchers were confident that the genes generating the clock mechanism's output to other systems were directly activated by the BMAL1:CLK dimer. This motivates a corresponding arrow in the fruit fly diagram (C) from CYC:CLK to the output genes there. Harmer et al. (2001) made this arrow dotted (rather than solid or dashed) and appended a question mark, inviting researchers to seek evidence of this direct link or, failing that, to find unknown intermediaries. The text of the article must be consulted for many details and nuances, but these diagrams uniquely serve as a visual heuristic by which researchers can offload some of the cognitive burden of tracking similarities and differences, hence facilitating understanding as well as discovery. The diagrams also make a theoretical contribution by highlighting mechanism schema types held in common across disparate species but sometimes filled in differently—one way of moving toward "unification in biology . . . through abstraction from causal details for the purpose of identifying generic organizational patterns" (Chapter 27, p. 371).

7. Linking computational models to mechanism diagrams

Here we return to the simple *per* feedback loop, as understood by fruit fly researchers in the early 1990s, to illustrate the benefits gained when computational models (discussed as well in Chapters 17, 20, and 33) are linked to mechanism diagrams. Goldbeter (1995) proposed a computational model to capture the cyclic dynamics of the *per* loop, using what we call a *computationally annotated mechanism diagram* (Figure 18.6) to link variables and parameters in his model to the mechanistic account. It is similar to Figure 18.1 in its austere reliance on arrows and text labels, but has mathematical symbols appended. Next to each word denoting a part is a variable tracking its changing concentration (M, P_0, P_1, P_2, P_N). Next to each arrow denoting an operation is a parameter. Three of the parameters directly represent the rate of the corresponding operation: *vs* for the accumulation of *per* mRNA in cytoplasm as it is transported from the nucleus, *k1* for transport of PER2 into the nucleus, and k_2 for a new proposed operation, return of some of the PER2 back into cytoplasm. The other parameters represent enzyme actions influencing the rate of their operations.

The computational model comprises five differential equations, one for each variable. Running the model using parameter values that he regarded as biologically plausible, Goldbeter demonstrated that the values of the relevant variables exhibited sustained oscillations—a successful simulation of *per*'s rhythmicity. In the first equation, for example, oscillations in the value of M (mRNA concentration) are obtained by subtracting a term that includes *vm* from a term that includes *PN* and *vs*. (Each term also incorporates constants that would have cluttered the diagram if shown.)

Figure 18.6 emphasizes those components of the mechanism that figure most prominently in the five equations of the model. For example, it explicitly distinguishes the two partly reversible phosphorylation steps that result in fully phosphorylated PER$_2$ since these steps correspond to two equations in the model. In contrast, gene transcription and its inhibition are represented sparsely and somewhat idiosyncratically within the dashed rectangle, with *per* not even explicitly

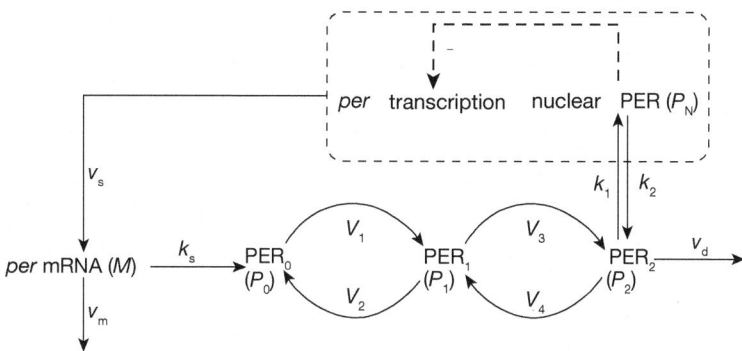

Figure 18.6 Goldbeter's computationally annotated mechanism diagram of the transcription-translation feedback loop for *per*. Goldbeter, A. (1995) A model for circadian oscillations in the *Drosophila* period protein (PER), *Proceedings of the Royal Society of London. B: Biological Sciences*, 261, 1362, figure 1, by permission of the Royal Society

shown as a part of the mechanism. This is because only products of *per* have varying concentrations, not the gene itself, and oscillations in its transcription are incorporated in the nonlinear term of the equation tracking *M*.

Pursuing mechanistic and computational modeling in tandem produces what we call *dynamic mechanistic explanation* (see Chapter 20 and Bechtel and Abrahamsen 2012, 2013). Rather than viewing the two approaches as unrelated or as competitors, researchers can reap the benefits of building a whole that is greater than the sum of its parts. If at least a sketch of a mechanistic account is in place first, it can suggest candidate variables or parameters for a computational model of the mechanism's dynamics. In the example here, these quantified the concentrations of parts and rates of operations; much work remained to select which to track, which to use in which equations, which to omit, which would be variables vs. parameters, what additional parameters to include and how, and so forth. For a given mechanism diagram, many different computational models can be developed—but the diagram provides a starting point. It can then be modified and annotated repeatedly, with the positioning of the mathematical symbols serving the modeler as a *locality aid* (Jones and Wolkenhauer 2012). On the other hand, when a computational model comes first or very early, it can help shape an initial mechanism sketch or schema and influence what research is undertaken to build on it.

For some purposes, such as assessing whether a mechanistic model can generate the right dynamics, this kind of coordination is not just advantageous but essential in moving discovery forward. Without computational modeling, mechanistic researchers rely on a strategy researched by Hegarty (1992): mentally animating their diagrams. This is satisfactory for sequential mechanisms with no feedback loops or nonlinearities, and can even yield a qualitative understanding of how oscillations are produced in a more complex system (as we narrated for Figure 18.1). Showing that such oscillations are *sustained*, however, requires the quantitative precision of computational models like that of Goldbeter and its successors.

8. Other visual representations relevant to mechanistic explanation

Computational modeling is not the only research approach that can contribute to mechanistic explanation. Most of the others bring their own conceptualizations and visual formats that do not dovetail so closely with mechanism diagrams, but can be interpreted in ways that point toward or elucidate an underlying mechanism. Consider Dynamical Systems Theory (DST), which offers concepts and visualization tools for better understanding patterns in the relation between variables over time. For example, the oscillations obtained by Goldbeter's (1995) differential equations can be understood as a limit cycle in state space and visualized in a phase portrait, providing a deeper understanding of the circadian mechanism's dynamics. (Chapter 20 further discusses mechanisms and DST; see especially Figure 20.1.)

Data graphics are a more general family of formats for the spatial representation of data, ranging from generic line and bar graphs to specialized formats such as spectrograms. They comprise most of the figures in published research articles, but only occasionally (as in our Figure 18.4) are integrated into a mechanism diagram. As we have discussed elsewhere (Burnston et al. 2014), although data graphics are thought of as merely describing what needs to be explained, often they serve explanatory ends as well. This is the case when the variables in a graph correspond to properties of certain parts and operations in a mechanistic explanation—measuring, for example, the changes in concentration of one type of molecule in a metabolic mechanism or the rate of one of its operations. But in his work on *explanatory relations*, Burnston (2016) emphasizes relations between variables that help explain the phenomenon of interest while not linking so simply to a mechanism diagram. Biologists might find it important, for instance, that the value of a variable peaks at a particular time, that two variables are in phase with each other, or that quantities vary proportionally to one another. Clearly, there is much more work for mechanists to do on the various roles and visualizations of data.

Finally, there are a number of research strategies that focus solely on causal, predictive, or other relations between variables. The variables might later be recognized to correspond to properties of parts and operations in a mechanistic account, but it is the variables that are named, perhaps manipulated, measured, analyzed, and reported. In structural equation modeling, for example, the strength of each pairwise relation between variables is the product of the analysis, often reported in a causal graph displaying the strongest links (see Chapter 26). As well, Coleman's boat diagram offers an interesting combination of causal and mechanistic understanding (see Chapters 30 and 32).

9. Conclusion

We have suggested that as philosophers of science turn their attention to mechanistic explanation, it is important to examine not only what scientists write but also the graphics they use as tools for their own understanding and for communication. Data graphics predominate numerically and play important descriptive and explanatory roles, but in this chapter we have focused on the special nature and advantages of mechanism diagrams. Using as exemplars several diagrams of the molecular mechanisms responsible for circadian rhythms, we have seen how different kinds of elements arranged in two-dimensional space can be used to (a) display the parts and operations of a mechanism; (b) convey their spatial, functional, and temporal organization; (c) facilitate comparison of related mechanisms; and (d) work in tandem with computational modeling. These published diagrams illustrate various uses of shapes, arrows, and lines (Tversky's glyphs) as well as color, text, iconic symbols, graphic symbols, mathematical symbols, conventional complexes, and space. These provide researchers a toolkit for representing their hypotheses about mechanisms. However, the same resources also serve crucially as visual heuristics during the research process. In section 6 we saw how researchers could exploit shape, color, and question marks to aid cross-species comparisons. It is harder to access the many versions of draft diagrams scientists produce for their own use during research or in preparing a published article, but for such case studies see Burnston et al. (2014), Sheredos (2017), and Sheredos and Bechtel (in press-a).

When researchers craft diagrams in proposing or revising a mechanistic hypothesis, innovation often takes the form of arranging glyphs and iconic symbols. The combinatorial potential of these elements can be deployed to convey novel mechanistic hypotheses and to consider components and activities in new contexts. Often the elements themselves need not change. We have not highlighted novel discoveries or disagreements about mechanisms, but they most definitely occurred over the 20 years covered here. Yet the basic types of glyphs in use have been fairly stable, as have many iconic and graphic symbols.

Despite this stability in the basic toolkit of available resources, they were deployed in a variety of ways in the diagrams we discussed. Depending on the goals of the scientist producing the diagram, different choices may be made regarding which parts and operations to show in detail, which to merge into a single element, and which to omit. Also, stylistic preferences influence whether parts are depicted by simple text labels, icons, or glyphs (labeled, color-coded, or not). Depending on appetite for innovation, a particular researcher's diagrams may tend to be conventional and consistent over time or (as in Figure 18.4) offer new, pleasing solutions satisfying ambitious goals. We would venture that just as no two scientists would write the same paragraphs, neither would they produce the same diagrams. Yet each scientist finds his or her own diagrams to be useful tools for developing, evaluating, and revising mechanistic explanations (see Chapter 19). Some diagrams—such as Figures 18.1–18.6—will also, or instead, communicate successfully with those who would not themselves produce exactly the same diagrams.

We have focused this chapter on mechanism diagrams, but conclude by noting that they are not the only tool for developing or communicating mechanistic explanations, and mechanistic explanation is not the only function that diagrams or other visualizations can serve. We have already mentioned that data graphics offer tools for displaying not only the phenomenon to be explained but also data quantifying aspects of the mechanistic explanation. We discussed how computationally annotated mechanism diagrams can serve as locality aids for computational models, which generate predicted quantifications for comparison to such data. We briefly discussed animation. Setting visual representations in motion has the potential to enhance understanding of the dynamics of a mechanism, but there is much to learn about how best to do this. For example, animation would fill in the gaps between the four snapshot diagrams in Figure 18.4, but passive viewing may be less helpful than having freeze-frame or speed-adjustment options. As a final example, visual tools providing access to augmented reality and virtual reality are moving out of engineering labs toward everywhere else, including science labs. The human drive to explore phenomena and to find, refine, and communicate explanations will continue to foster a growing visual toolbox. We invite our readers to go beyond our own work on diagrams to probe existing and future dynamic visualizations of mechanistic explanations.

References

Bechtel, W., and Abrahamsen, A. (2012) "Thinking dynamically about biological mechanisms: Networks of coupled oscillators," *Foundations of Science, 18*, 707–723.

Bechtel, W., and Abrahamsen, A. (2013) "Roles of diagrams in computational modeling of mechanisms," *Proceedings of the 35th Annual Conference of the Cognitive Science Society* (pp. 1839–1844). Austin, TX: Cognitive Science Society.

Bechtel, W., Burnston, D., Sheredos, B., and Abrahamsen, A. (2014) "Representing time in scientific diagrams," *Proceedings of the 36th Annual Conference of the Cognitive Science Society* (pp. 164–169). Austin, TX: Cognitive Science Society.

Burnston, D. C., Sheredos, B., Abrahamsen, A., and Bechtel, W. (2014) "Scientists' use of diagrams in developing mechanistic explanations: A case study from chronobiology," *Pragmatics and Cognition, 22*, 224–243.

Burnston, D. C. (2016) "Data graphs and mechanistic explanation," *Studies in History and Philosophy of Biological and Biomedical Sciences*, *57*, 1–12.

Cheng, P. C.-H. (2011) "Probably good diagrams for learning: Representational epistemic recodification of probability theory," *Topics in Cognitive Science*, *3*, 475–498.

Craver, C. F., and Darden, L. (2013) *In Search of Mechanisms: Discoveries across the Life Sciences*. Chicago, IL: University of Chicago Press.

DeWoskin, D., Myung, J., Belle, M. D., Piggins, H. D., Takumi, T., and Forger, D. B. (2015) "Distinct roles for GABA across multiple timescales in mammalian circadian timekeeping," *Proc Natl Acad Sci USA*, 112(29), E3911-9.

Glennan, S. (in press) *The New Mechanical Philosophy*. Oxford: Oxford University Press.

Goldbeter, A. (1995) "A model for circadian oscillations in the *Drosophila* period protein (PER)," *Proceedings of the Royal Society of London. B: Biological Sciences*, *261*, 319–324.

Hardin, P. E., Hall, J. C., and Rosbash, M. (1990) "Feedback of the *Drosophila period* gene product on circadian cycling of its messenger RNA levels," *Nature*, *343*, 536–540.

Harmer, S. L., Panda, S., and Kay, S. A. (2001) "Molecular bases of circadian rhythms," *Annual Review of Cell and Developmental Biology*, *17*, 215–253.

Hastings, M. H., Maywood, E. S., and O'Neill, J. S. (2008) "Cellular circadian pacemaking and the role of cytosolic rhythms," *Current Biology*, *18*, R805–R815.

Hegarty, M. (1992) "Mental animation: Inferring motion from static displays of mechanical systems," *Journal of Experimental Psychology: Learning, Memory, and Cognition*, *18*, 1084–1102.

Hegarty, M. (2011) "The cognitive science of visual-spatial displays: Implications for design," *Topics in Cognitive Science*, *3*, 446–474.

Illari, P. M., and Williamson, J. (2012) "What is a mechanism? Thinking about mechanisms across the sciences," *European Journal for Philosophy of Science*, *2*, 119–135.

Jones, N., and Wolkenhauer, O. (2012) "Diagrams as locality aids for explanation and model construction in cell biology," *Biology and Philosophy*, *27*, 705–721.

Larkin, J. H., and Simon, H. A. (1987) "Why a diagram is (sometimes) worth ten thousand words," *Cognitive Science*, *11*, 65–99.

Levy, A., and Bechtel, W. (forthcoming) "Toward mechanism 2.0. Expanding the scope of mechanistic explanation."

Nersessian, N. J. (2008) *Creating Scientific Concepts*. Cambridge, MA: MIT Press.

Perini, L. (2005) "Visual representations and confirmation," *Philosophy of Science*, *72*, 913–926.

Perini, L. (2013) "Diagrams in biology," *The Knowledge Engineering Review*, *28*, 273–286.

Reppert, S. M., and Weaver, D. R. (2001) "Molecular analyses of mammalian circadian rhythms," *Annual Review of Physiology*, *63*, 647–676.

Sheredos, B. (2017) "Communicating with scientific graphics. A descriptive inquiry into non-ideal normativity," *Studies in History and Philosophy of Science Part C: Studies in History and Philosophy of Biological and Biomedical Sciences*, *63*, 32–44.

Sheredos, B., and Bechtel, W. (in press-a) "Sketching biological phenomena and mechanisms," *Topics in Cognitive Science*.

Sheredos, B., and Bechtel, W. (in press-b) "Imagining mechanisms," in Godfrey-Smith, P. and Levy, A. (eds.) *The Scientific Imagination: Philosophical and Psychological Perspectives*. Oxford: Oxford University Press.

Tversky, B. (2005) "Prolegomenon to scientific visualizations in science education," in J. Gilbert (ed.) *Visualization in Science Education* (pp. 29–42), Dordrecht, Netherlands: Springer.

Tversky, B. (2011) "Visualizing thought," *Topics in Cognitive Science*, *3*, 499–535.

Woody, A. I. (2014) "Chemistry's periodic law: Rethinking representation and explanation after the turn to practice," in L. Soler, S. Zwart, V. Israel-Jost, and M. Lynch (eds.) *Science after the Practice Turn in Philosophy, History, and the Social Studies of Biology*. Oxford: Routledge.

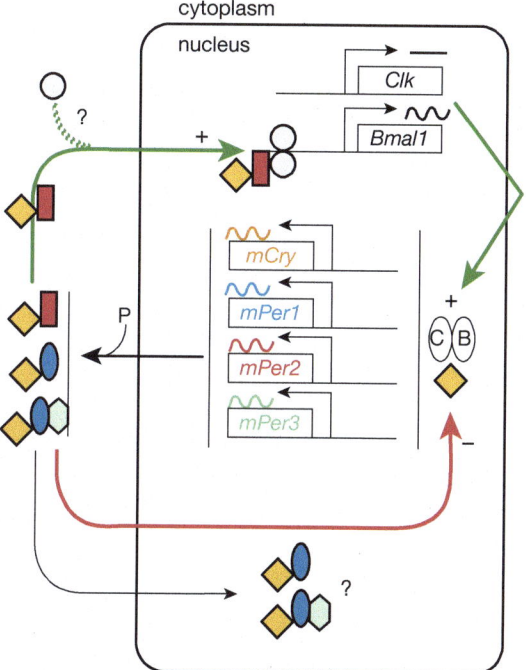

Plate 1 A diagram of the molecular clock mechanism for circadian rhythms in mammals, exemplifying several types of elements used in mechanism diagrams. From Reppert and Weaver 2001, Figure 1. Reproduced with permission of *Annual Review of Physiology*, Volume 63 © 2001 by Annual Reviews, http://www.annualreviews.org.

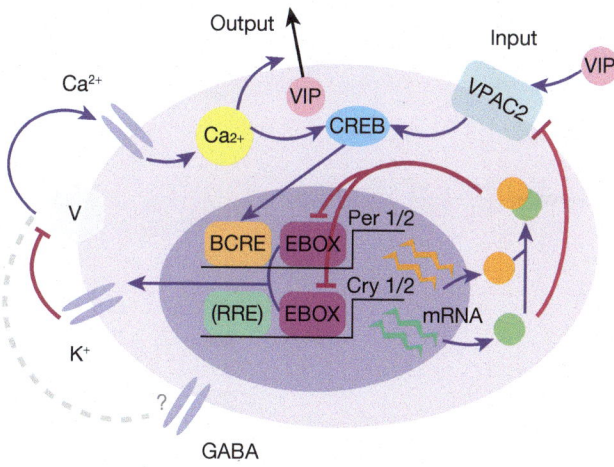

Plate 2 A mechanism diagram showing some of the key parts of the molecular clock mechanism and how, in SCN neurons, they interact with pathways extending into the extracellular environment. From DeWoskin, D., Myung, J., Belle, M. D., Piggins, H. D., Takumi, T., & Forger, D. B. (2015) "Distinct roles for GABA across multiple timescales in mammalian circadian timekeeping," *Proc Natl Acad Sci U S A.*, Figure 1.

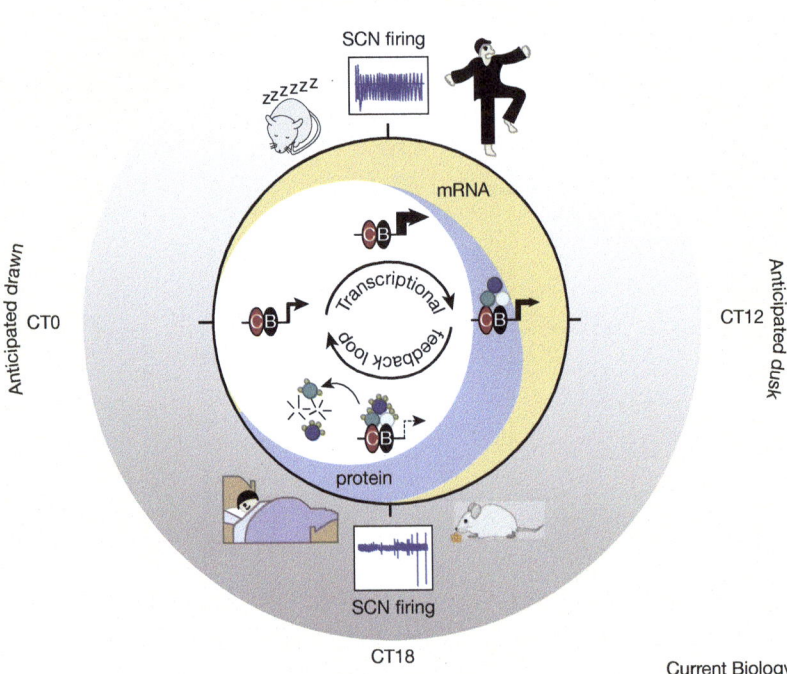

Plate 3 Using a 24-hour clock face to show the timing of behavioral and neural activity (gray region) and the underlying molecular clock (inner circle). Appropriately located iconic symbols contrast the activity of humans (diurnal) with that of mice (nocturnal). Reprinted from *Current Biology*, Vol 18, Hastings, M. H., Maywood, E. S., & O'Neill, J. S., Cellular circadian pacemaking and the role of cytosolic rhythms, Figure 2, copyright (2008), with permission from Elsevier.

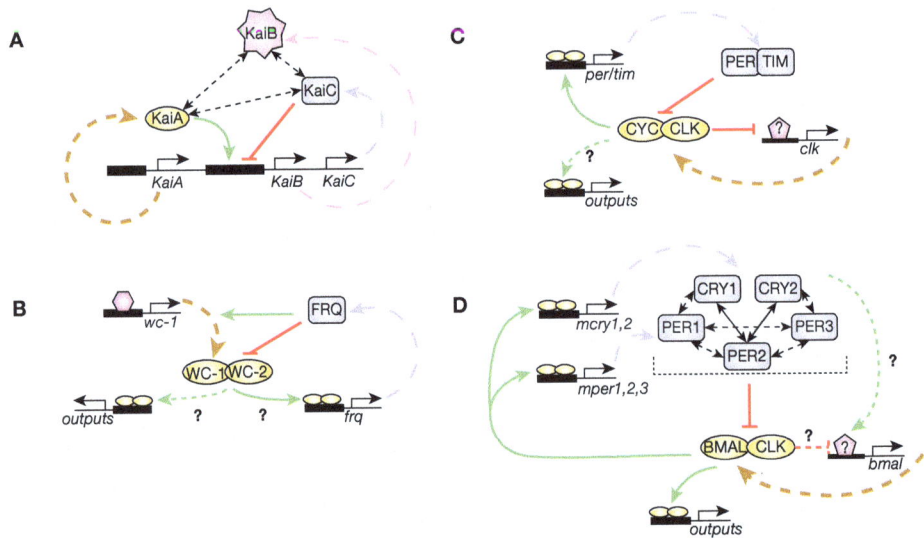

Plate 4 Comparison of the molecular clock mechanisms, as understood in 2001, in (A) cyanobacteria, (B) fungi, (C) fruit flies, and (D) mice. From Harmer, S. L., Panda, S., and Kay, S. A. 2001, Figure 2. Reproduced with permission of *Annual Review of Cell and Developmental Biology*, Volume 17 © 2001 by Annual Reviews, http://www.annualreviews.org.

19

STRATEGIES FOR DISCOVERING MECHANISMS[1]

Lindley Darden

1. Introduction

The new mechanists have applied the mechanistic perspective to numerous philosophical topics, such as explanation, reduction versus integration, function, and causation. Less discussed is the topic of reasoning in discovering mechanisms. That is the topic here.

Traditional philosophers of science distinguished the logic of discovery from the logic of justification (or falsification), arguing that there is no logic of discovery (e.g., Popper 1965). "Friends of discovery" (Gutting 1980; Nickles 1980a, 1980b; Meheus and Nickles 2009) conceded that deductive logic does not capture the nature of reasoning in discovery. Furthermore, there is no algorithm that takes data as an input and yields an explanatory theory as an output, as opposed to merely finding patterns in the data. Instead, the friends of discovery recast the task; their goal is to find heuristic problem-solving strategies for scientific discovery. Their work was influenced by Herbert Simon's heuristic problem-solving approach (Simon 1977, 1996; Nickles 1977; Wimsatt 2007). Darden (1991) argued that discovery of theories should be recast as an iterative process of cycling through stages of construction, evaluation, and revision, with heuristic reasoning strategies for each stage.

Norwood Russell Hanson was one of the few philosophers of science of the mid-twentieth century who analyzed reasoning in hypothesis construction. Following Charles Sanders Peirce, Hanson elaborated his own view of reasoning to a hypothesis. He might now be pleased that reasoning in the discovery of mechanisms follows his pattern of reasoning in discovery, a very abstract view of moving from puzzling phenomena to a hypothesis of a certain *type*:

Schematically, [retroductive reasoning] can be set out thus:

(1) Some surprising, astonishing phenomena $p_1, p_2, p_3 \ldots$ are encountered.

(2) But phenomena $p_1, p_2, p_3 \ldots$ would not be surprising were *a hypothesis of H's type* to obtain. They would follow as a matter of course from something like H and would be explained by it.

(3) Therefore there is good reason for elaborating *a hypothesis of the type of H*; for proposing it as a possible hypothesis from whose assumption phenomena $p_1, p_2, p_3 \ldots$ might be explained.

(Hanson 1961: 630; italics added)

255

In earlier versions, Hanson had simply discussed good reasons for elaborating a hypothesis H, but in this 1961 paper, he instead used "type of H." This was a step forward. Hypothesis generation and preliminary evaluation should be distinguished (Schaffner 1993: ch. 2). The ability to connect a surprising phenomenon with a type of hypothesis is a step in generation, prior to finding a specific plausible hypothesis. Using examples from physics, Hanson suggested a type might be an inverse square type, in which something varies as the distance away increases. But Hanson did not consider hypothesis types in biology and medicine. John Josephson expanded retroduction (abduction) to include systematic generation of a search space of possible types of hypotheses in medical diagnosis. He elaborated criteria for evaluating one as the most plausible and instantiating it for a specific case (Josephson and Josephson 1994).

Often in biology, the type of hypothesis to be discovered is a mechanism schema, and the search is guided and constrained by the characterization of a mechanism. The product shapes the process of discovery. One can say, in general, what counts as an adequate description of a mechanism; that will be the subject of the next section. Furthermore, heuristic reasoning strategies can guide the construction stage; those strategies are the subjects of the following sections. Because discovery consists of construction, evaluation, and revision, strategies for evaluation and revision are also part of the story, but cannot be fully discussed here (see Craver and Darden 2013: chs. 6–9). Nonetheless, final sections here discuss recent work combining construction, evaluation, and revision using computer simulations and biorobotics in mechanism discovery.

2. What is to be discovered: characterizing mechanisms and mechanism schemas

The discovery of a mechanism typically begins with a puzzling phenomenon. (For more on the nature of the phenomenon, see Garson, Chapter 8 in this volume.) Data provide evidence for the phenomenon (Bogen and Woodward 1988). When the goal is to find what produces the phenomenon, then one searches for a mechanistic type of hypothesis. That decision rules out other parts of a large search space. One is not seeking merely a set of correlated variables. One is seeking an economical equation that describes the phenomenon, although such an equation can provide a constraint in the search for a mechanism. One is not seeking a law from which a description of the phenomenon can be derived. One is not merely seeking a relation between a cause and an effect, although such a relation provides clues about mechanism components (Darden 2013; see Matthews and Tabery, Chapter 10 in this volume). Nor is one merely seeking to find a pathway, characterized by nodes and unlabeled links, although that is a very abstract way of representing some aspects of a mechanism. Rather, one is attempting to construct a mechanism schema that describes how structural and active components are spatially and temporally organized together to produce the phenomenon.

Employing a specific characterization of a mechanism provides guidance in discovery. One oft-cited mechanism characterization is this: "Mechanisms are entities and activities organized such that they are productive of regular changes from start or set up to finish or termination conditions" (Machamer, Darden, and Craver 2000: 3). (For a more minimal characterization of mechanisms, see Glennan and Illari, Chapter 1 in this volume.) The goal is to find the entities and activities, to describe how they are organized together, and to show that when they are organized together in a productively continuous way, they produce the phenomenon of interest. This characterization directs one to ask: What are the setup and finish conditions? Is there a specific, triggering start condition? What is spatially next to what? What is the temporal order of the steps? What are the entities in the mechanism? What are their structures? What are the activities that drive the mechanism? What are their range and their rate? What interacts with what? How does each step of the mechanism give rise to the next? How was each step driven by the previous one? What is the overall organization of the mechanism: does it proceed linearly or is the mechanism perhaps cyclic (with no clear start and stop), or is it organized with feedback loops, or does it have some other overall organizational motif? Where is it spatially located? In what context does the mechanism operate and how is it integrated with other mechanisms? These kinds of questions show how the product guides the process of its discovery.

Mechanism schemas are representations of mechanisms. What I call a "schema" others refer to as mechanistic "models," but "model" has many other uses, other than abstractly representing the structure of a target mechanism. (See Chapter 17.) This is an example of a very abstract schema: DNA→RNA→protein. Such schemas are often depicted in diagrams. William Bechtel and his collaborators (see Abrahamsen et al., Chapter 18 in this volume) discuss the many "visual heuristics" that diagrammatic representations of mechanism play, including a gray box with question marks to indicate gaps to be filled.

Schemas have several dimensions: sketchy to sufficiently complete, abstract to specific, small to general scope of applicability, and possible to actual (Craver and Darden 2013: ch. 3). A goal in discovering a mechanism is to convert an incomplete sketchy representation into an adequate one for the purpose at hand. Incomplete sketches indicate where black (unknown components) and gray (only functionally specified) boxes need to be filled to have a productively continuous schema. During the construction phase of discovery, moving from a sketch to a sufficiently complete schema allows one to work in a piecemeal fashion; one can work on one part of the mechanism at a time while leaving other parts as black or gray boxes. Because one is attempting to reveal the productive continuity of a mechanism from beginning to end, what one learns about one step of the mechanism places constraints on what likely has come before or what likely comes after a given step.

Abstraction comes in degrees and involves dropping details; specification involves adding details all the way to instantiation, where an instantiation is still a representation, with sufficient details to represent a productively continuous mechanism from beginning to end. A goal in discovery is to find a schema at a given degree of abstraction, from a very abstract type of schema with few specified components to a fully instantiated one for a particular case. The desired degree depends on the purpose for which the mechanism is sought. Although degree of abstraction is an independent dimension from the scope of the domain to which the schema applies, more abstract schemas (if they have any instances at all) may have a wider scope. Hence, when the goal of the discovery process is to find a very generally applicable mechanism schema, it is likely to be represented at a high degree of abstraction. The move from how possibly to how plausibly to how actually is driven by applying strategies for evaluation, such as experimental testing, and strategies for anomaly resolution, such as localizing faults and revising the schema.

Consider the example of mechanisms connecting gene to phenotype. One starts with the beginning point, e.g., a gene or gene mutation, and some characterization of the phenotype of interest. In Mendelian genetics, phenotypic characters are usually observable properties of organisms, e.g., the color of peas or eyes in fruit flies. In molecular biology the phenotype might be, e.g., the presence of a protein, the abundance of that protein, or an altered structure or abundance of a protein (see, e.g., Nachtomy et al. 2007). For medical studies, a gene mutation may be statistically associated with a disease phenotype. Between the gene or gene mutation and the phenotypic character is a black box; the goal is to fill in sufficient details so as to represent the mechanism connecting the two.

Having evidence of a beginning point (the gene or gene mutation) and the end point (the wild type or disease phenotype), the discovery task is to fill in the black box to some degree of detail. For example, if the goal is to knock out the gene or to replace the gene during gene therapy, then it may be unnecessary to find all the intervening steps in the mechanism. A highly abstract schema may be sufficient to guide the work to knock out or replace the gene. However, if the goal is to design a therapy to alter an entity or activity in a mechanism step downstream from the gene, then specific details become important: e.g., one may need to find the three-dimensional structure of a protein and identification of its active site. Rarely is one gene responsible for one phenotypic trait, so multiple beginning points may converge on the way to the trait. Or one gene mutation may play a role in multiple forking downstream steps. Or feedback loops may need to be represented. (See Abrahamsen et al., Chapter 18 in this volume.) The purpose of the study will aid in deciding how many black boxes to fill and with how much detail.

3. Stages in mechanism discovery

For convenience, we can divide the mechanism discovery process into at least four stages: characterizing the phenomenon, constructing a schema, evaluating the schema, and revising the schema (Darden 2006, ch. 12; Craver and Darden 2013, chs. 4–9). These stages are often pursued in parallel and in interaction with one another. Mechanism discovery is frequently pursued piecemeal. A scientist might work on a part of a mechanism, while leaving much else about the mechanism inside a black box. Also, the different stages of discovery frequently interact with one another. One is forced to recharacterize the phenomenon in the face of learning about the mechanism (Bechtel and Richardson 1993). Or, one is forced to re-evaluate experimental findings because one recognizes a previously unrecognized region of the space of possible mechanisms. Or, when one attempts to build a computer or biorobotic simulation, one finds previously unidentified black boxes that need to be filled.

The first stage, characterizing the phenomenon, frames the problem to be solved in the search for mechanisms. This stage involves developing a more or less precise description of the puzzling phenomenon produced by the mechanism. The nature of the phenomenon often provides clues about the kind of underlying mechanism that might possibly be responsible for it. The second stage of mechanism discovery, constructing a schema, involves generating a space of possible mechanisms for a given phenomenon. Construction strategies include employing a schema type, assembling modules, and forward and backward chaining.

The third stage of discovery, evaluation, aims at revealing the empirical and conceptual constraints on the space of possible mechanisms for a given phenomenon. The components of a mechanism—that it is composed of parts with sizes, shapes, and structures, that it has a particular temporal sequence, that it is composed of one kind of entity and not another or makes use of one kind of activity and not another—provide conceptual constraints by which scientists evaluate mechanism schemas. For example, to what extent does a proposed entity have the activity-enabling properties to carry out the following step? Empirical constraints come from observations and experiments. An experiment involves intervening to change some part of a mechanism or some background condition to learn something about how the mechanism works. Other constraints come from fitting the proposed schema into a larger context so as to cohere with other known mechanisms.

The final stage of mechanism discovery is revision. Revision is required when a favored mechanism schema is challenged by an empirical anomaly, or does not fit coherently into a larger context, or has some other kind of failure. Strategies for anomaly resolution result from the range of choices that the scientist might consider in response to an anomaly for this mechanism schema. Some of the choices concern whether the apparent challenge to the schema can be barred from being a challenge because of, for example, a failed experiment or being beyond the scope of the intended schema. When such barring fails, and the anomaly is determined to directly challenge a proposed schema, there are a number of search strategies that scientists may use to localize the site of fault within their schema and to correct it to remove the fault.

Scientists use construction strategies to populate the space of possible mechanisms. They use evaluation strategies to identify constraints that prune the space of possible mechanisms. They use revision strategies to diagnose and localize the source of error in a hypothesized mechanism and to provide clues as to how that hypothesis might fruitfully be amended.

Traditionally, "discovery" usually refers only to the construction stage, so that will be the focus of the next sections.

4. Guidance from the phenomenon

A mechanism discovery episode typically begins with at least two sources of empirical knowledge to guide the discovery process. The scientist comes to a problem with a store of knowledge about the phenomenon under investigation and about the target system in question. First, one begins knowing something, even something sketchy, about the phenomenon for which one is hoping to discover a mechanism. Second, one typically begins with prior knowledge about the kinds of entities and activities that might be involved in such phenomena in a given type of organism or system.

Because the phenomenon provides guidance in schema construction, the more that is known about it at the outset, the more clues it supplies. Work to characterize it is in order. One needs to find the setup conditions under which it occurs, any triggering conditions needed to make it come about, and inhibiting conditions that prevent its occurrence. One can investigate the phenomenon by passive observation and interventions, such as staining to find relevant structures (Kästner 2015), a technique often used to image cellular structures. Or scientists may use more deliberate experimental manipulations to activate, enhance, or ablate hypothetical components (Craver 2007), such as ablation in gene knockout experiments. If the phenomenon is the statistical association of a gene mutation with a disease phenotype, then all the standards for evaluating adequate statistical sampling apply.

If the characterization of the phenomenon shows that it is of a familiar type, the field might already have developed a library of mechanism-types through which one might search. For example, in the search for mechanisms to connect a DNA mutation to a disease phenotype, the many roles that DNA bases play provide a library of possible types of mechanisms, e.g., change in a coding region producing a change in an amino acid; change in a regulatory region leading to increased or decreased expression of the associated gene; change in a splicing region, leading to a malformed protein; or change in a region producing a microRNA, with subsequent alternation of its functional role. Standard methods exist for testing whether one of these possible types applies to a particular phenomenon.

Raoul Gervais and Erik Weber (2015) discuss a type of experiment, which they dub an "orientation experiment," that aids in finding the abstract type of mechanism schema at an early stage of the discovery process. "Orientation experiments . . . are useful in situations in which we know (next to) nothing about the mechanism, which are often (but not exclusively) the crucial early stages of an investigation into a phenomenon, namely the orientation-phase" (Gervais and Weber 2015: 49). For example, suppose the puzzling phenomenon is how an ant finds its way back to the nest. Two possible types of mechanisms are visual recognition or chemical signaling. An orientation experiment would remove light, find that the ant still finds its way back, and indicate the search should be for a chemical signal.

5. Localization

Localization is a second, often crucial, clue in the construction stage of the search for mechanisms, as William Bechtel and Robert Richardson argue (1993, 2010). It is often far from obvious at the beginning of a discovery episode where in the system under study the phenomenon takes place. In some cases it is unclear whether the phenomenon even takes place in the system under study at all, or whether, in contrast, it takes place in the interactions between the system and its environment (or perhaps entirely in the environment).

For example, the phenomenon of inheritance came to be localized to hereditary materials inside germ cells. Inheritance is the tendency of offspring to resemble their parents. In sexually breeding species, inheritance is biparental; in other words, the child has a mixture of traits from both parents. By the mid-nineteenth century (with the discovery of the mammalian egg), the locus of transmission of hereditary traits was quite plausibly the sperm and the egg, because those germ cells link the generations. However, some debate occurred as to whether influences from a previous mating could linger in the mother's womb.

Gregor Mendel's work in the 1860s on patterns of inheritance, rediscovered around 1900, provided evidence for unobservable differences among germ cells that might explain the inheritance of traits through several generations. Further evidence for localization of hypothesized hereditary particles within the germ cells occurred after the discovery in the late nineteenth century of chromosomes (thread-like bodies in the nuclei of cells) that were hypothesized to be the hereditary material. Evidence for this localization was strengthened in the early twentieth century when abnormalities in chromosomes were found to correspond to abnormalities in patterns of inheritance of traits. Biochemists noted that chromosomes are composed of two types of chemicals: proteins and deoxyribonucleic acid (DNA). Chemical analysis of DNA in the 1930s, given the sensitivity of the chemical techniques at the time, seemed to indicate that it had a simple repeating structure, whereas proteins had much more complexity. Hence in the 1930s and 1940s, the most plausible location for the hypothesized hereditary particles, the genes, was in the proteins of chromosomes. This view was dispelled with more accurate chemical analysis of the DNA molecule and the discovery of its double helical structure in 1953. The idea that the gene is a linear sequence of bases along a DNA chain localized part of the hereditary process and, in so doing, offered an entry point into the mechanism from which one could conjecture how genes could possibly produce hereditary traits (Darden 1991; Darden and Craver 2002).

Locating (and re-locating) the mechanism is often a crucial turning point in discovering a mechanism. By choosing a locus for the search for mechanisms, one places the mechanism in an appropriate context, one searches for entities and activities at that locus, one circumscribes the experimental and observational strategies that will be useful in addressing the problem, and one often has clues for choosing among an available set of mechanism schemas that might be brought to bear in the given case.

6. Employ a schema type

Characterizing the phenomenon and finding a likely location for the mechanism may provide clues for retrieving from a library one or more abstract schemas appropriate to phenomena of that type. The abstract schema provides variables, black and gray boxes, that can be filled with entities and activities to find the mechanism for the specific case (Darden 2002).

When the phenotype of interest is a disease phenotype, then the goal is to discover a disease mechanism. Thagard discusses abstract schemas for representing defective molecular mechanisms associated with diseases (e.g., Thagard 1998, 1999). He distinguishes abstract schemas for diseases caused by external agents versus internal gene defects. One searches for a different kind of mechanism if the disease is due to invading bacteria, such as pneumonia, rather than diseases due to a single gene defect, such as cystic fibrosis. Identifying the defective mechanism aids in designing therapies that target specific entities and activities to alleviate the disease (Thagard 2003; Craver and Darden 2013: ch. 11).

7. Modular subassembly

Modular subassembly is the strategy of putting together known types of parts, namely modules of entities and activities that carry out particular kinds of mechanism role function (see Chapter 8). One cobbles together different modules to construct a hypothesized how-possibly mechanism, guided by the goal of finding modules to fill all the functionally characterized gray boxes in a mechanism sketch. In doing so, scientists draw upon their knowledge of types of modules that have a particular function (Darden 2002).

Evolution itself often works by copy and edit: copies of genes can be found in mutated form, playing similar or different roles in the same or related organisms. Finding such recurrent motifs has been a powerful tool in discovery in biology. There are various types of receptors, neurotransmitters, enzymes, and gene regulatory components (e.g., inducers, repressors, and other types of transcription factors). In a given field, researchers develop a library of types of modules that perform various functions. When a functional requirement can be specified in the mechanism sketch, then appropriate types of modules can be sought to satisfy it. Experimentalists can manipulate such modules to investigate their functional roles.

For example, developmental biologists discuss modularity as a principle in evolution: components are individually modified or conserved in different lineages. One of the remarkable cases of a functional module consists of the *Pax6* gene, along with a group of genes related to it. This module plays a role in the formation of eyes in invertebrates such as fruit flies, as well as in vertebrates such as frogs and mice. This module recurs even though the eyes themselves have quite different structures (compound eyes in flies and eyes with a single lens in vertebrates). An amazing abnormality results from placing a mouse *Pax6* gene in the leg of a fruit fly, in which structures characteristic of the insect eye develop on the leg (Gilbert 2003: 413). Hence, if one is seeking the mechanism of eye development in a new species, knowledge of such types of conserved modules will aid in constructing a plausible target mechanism.

Finding a new type of module opens a new region of the space of possible mechanisms. The 1970 discovery of the enzyme reverse transcriptase was just such a new type (Temin and Mizutani 1970; Baltimore 1970; Darden 2006: ch. 10). This enzyme copies RNA back into DNA:

RNA–via reverse transcriptase–>DNA

This reverse transcriptase module plays an important role in retroviral infection. Retroviruses, such as the HIV-AIDS virus, carry this enzyme into the infected cell. Retroviruses use reverse transcriptase to copy their own viral RNA into DNA. The DNA copy of the viral genome is then integrated into the host genome and is copied as new cells are produced. Integration into the host DNA makes such retroviruses very difficult to attack with drugs. The discovery of reverse transcriptase opened up a space of possible mechanisms in which base sequences could be copied and inserted into the DNA, a hitherto unknown activity.

The discovery of this new type of module expanded the space of possible mechanism modules. Many sequences in the human genome appeared to result from reverse transcription of viral DNA in past evolutionary history. Finding recurrent motifs in biological mechanisms provides a growing store of modules of use in constructing how-possibly schemas.

8. Forward/backward chaining

Another strategy for constructing mechanism schemas involves first learning something about the mechanism or one of its components and then using that knowledge to make inferences about what came before it or what is likely to come after it (Darden 2002). In forward chaining, one uses the early steps of a mechanism to reason about the types of entities and activities that are likely to be found in later steps. In backward chaining, one reasons from the entities and activities in later steps in a mechanism to find entities and activities appearing earlier. With cyclic or feedback mechanisms, some separable step can serve as a relative starting point for reasoning about earlier or later ones. Thus, the strategy of forward/backward chaining is likely available to scientists when they know anything, or can conjecture anything, about the specific entities and activities anywhere in the hypothesized mechanism, even if no libraries of schema or module types are available.

Consider some of the ways one might use forward/backward chaining. Entities engage in activities by virtue of the fact that they have *activity-enabling properties*. One can often use general knowledge about kinds of properties and their association with particular kinds of activities in biological systems to reason forward about the activities in which the entities can or do engage. Such activity-enabling properties include three-dimensional structure, size, location, orientation, and charge. Structures can promote or inhibit the push/pull of geometrico-mechanical activities. Three-dimensional shapes can be open or closed, narrow or wide, exposing or concealing. An open entity permits movement through it more or less as its opening is narrow or wide. Entities may also have different kinds of charges, and molecules have valences, both of which affect the kinds of bonding activities in which they engage. So, finding the activity-enabling properties of entities gives clues to the kinds of activities in which they might engage in the next step of the mechanism. Melinda Fagan (2012) calls one type of such activity-enabling properties of entities "meshing" properties; they enable complexes of entities to form whenever the appropriate spatio-temporal relations between them obtain.

Conversely, in backward chaining, *activity signatures* are a source of clues as to what came before. When an activity operates, it produces an effect that changes the entities involved. Noting the properties that may have been changed, the activity signatures, allows one to conjecture what happened in a previous step. For example, if one detects a hydrogen bond between two entities in a complex in a later step of a mechanism, then one can reason that an earlier step included polar molecules, with complementary weak charges that have been neutralized in the formation of the bond. Another example is allosteric molecules. An allosteric molecule changes shape (via stresses and strains of geometrico-mechanical activities) when it bonds to an effector molecule. The new shape exposes a new active site, so that the allosteric molecule is enabled to eject or bond to a third molecule. If one detects an allosteric molecule bound to an effector (an activity signature), that indicates that effector bonding occurred in a prior step.

Forward and backward chaining both contributed to the discovery of the mechanism of protein synthesis in the 1950s and 60s. Biochemists knew that the endpoint of the protein synthesis mechanism was a string of amino acids held together by strong covalent bonds. They thus reasoned back to an antecedent step in which the amino acids were free and unbound. Because energy is required to form such strong bonds, reasoning backward suggests the existence of a high-energy intermediate in the preceding step. Biochemists isolated such a high-energy intermediate. Thus, the strong covalent bonds were activity signatures, indicating the components needed in the preceding steps to form them. Surprisingly, the activated amino acid was associated with RNA. A typical biochemical reaction schema to synthesize a molecule has the separate chemical components on one side and the newly synthesized molecule on the other. Accordingly, biochemists expected amino acids and an energy source to be required for synthesizing a protein (a chain of amino acids). However, such a schema had no role for RNA because RNA is not a component of proteins. Reasoning backward from protein to free amino acids did not suggest an RNA intermediate in the chemical reaction.

Meanwhile, molecular biologists were reasoning forward from the DNA double helix to the next step in the protein synthesis mechanism. Biochemists and cell biologists had discovered that RNA was involved in the mechanism. Molecular biologists suggested that RNA acted as a template that would guide the assembly of the protein. The order of the bases in the coding strand of DNA would be transcribed into similarly ordered bases in RNA. The RNA then serves as the template that orders the amino acids in the protein. Biologists used these tandem strategies to fill gaps in the productive continuity of the proposed mechanism. The molecular biologists reasoned forward from the DNA, while the biochemists reasoned backward from the finished protein. Their work met in the middle of the mechanism, with the discovery of the various types of RNAs and their roles. One biochemist suggested that the scientists were like workers building a tunnel: they started at opposite sides of a mountain, and eventually they met in the middle.[2]

In sum, forward and backward chaining are reciprocal strategies for reasoning about one part of a mechanism on the basis of what is known or conjectured about other parts in the mechanism. They may be used independently of or in conjunction with the other two strategies of using a schema type and assembling modules. The strategy of employing a schema is a top-down strategy that provides a how-possibly overall organizational structure for the target mechanism. Modular subassembly involves putting together functionally characterized, working subcomponents of a schema. Finally, at a finer grain, one can construct a hypothesized mechanism by reasoning forward or backward about the entities or activities themselves.

9. Computer simulations and biorobotics in mechanism discovery

William Bechtel (2016) uses case studies from circadian rhythm studies to show how computational modeling contributes to the discovery of complex mechanisms. Bacteria were found to lack what had been considered a central component of the circadian clock mechanism in other organisms and thereby indicated the nature of the crucial core of the clock mechanism in other organisms. Both computational modeling and empirical investigations played roles in the discovery of how the simplified mechanism works in bacteria. Researchers built a computational model to test whether their proposed mechanism could generate the phenomenon (a typical use of computational simulations). A state change in the model introduced a variable in need of physical interpretation. The authors proposed a hitherto unknown change in the conformation of one of the proteins, thereby leading to a newly proposed module for this mechanism requiring new experimental investigation. This case shows the interplay between constructing a mechanistic hypothesis and building a computational model to test it. If computational tractability involves adding black boxes (here, uninterpreted variables) so as to make the simulation run, then researchers are guided in adding previously undetected components to the mechanistic hypothesis. As Bechtel put it: "the modeling played a central role in the formulation of the proposed mechanistic account; it did not merely serve to demonstrate that an already advanced model could generate the phenomenon" (Bechtel 2016: 117).

Datteri and Tamburrini (2007) contrast the roles of implementing a proposed mechanism in computational simulations versus in biorobots. Many of the same issues arise in mapping a biological mechanism description into a biorobotic one as in computational modeling: what are the similarities and differences in the way the model mimics the biological activities? In biorobotics too, requirements for implementing a working model can point to previously unrecognized black boxes needed for productive continuity so that the robot works. However, a robot, placed in an appropriate environment, can mimic aspects of the biological system for which there may be no computational simulation.

Datteri and Tamburrini's example is of the investigation of the chemotaxis in lobsters, namely their capability to track turbulent chemical plumes to their sources (Grasso et al. 2000). The mechanism involves feedback analysis: lobster antennae can detect local gradients in the chemical plume, and local gradients provide lobsters with directional information to localize the plume source. Computer simulations appear to be unsuitable to test mechanism schemas that are consistent with this broad picture of lobster chemotaxis, insofar as accurate replication of plume turbulence in water is unfeasible (especially in view of the difficulty of developing accurate fluid-dynamical models at the temporal and spatial scales that are needed to simulate lobster sensorimotor loops in chemo-orientation tasks). In contrast, robots placed in real chemical plumes allow for direct comparison between robotic and biological behaviors. Two previously proposed mechanism components did not enable the robot lobster to adequately mimic the biological lobster's behavior; the experiments not only indicated which components of the mechanism needed to be changed but also provided clues as to the needed repairs.

As Datteri and Tamburrini summed up:

> Behavioral match evaluations of robotic and biological behaviors may enable one to corroborate or falsify mechanism schemata and descriptions, to identify previously overlooked interactions with the environment, to suggest the existence of entities and activities implementing some role function in the target biological system, and more generally to support the iterative specification of mechanism descriptions from mechanism sketches.
>
> *(Datteri and Tamburrini 2007: 424)*

From this computational and biorobotic work, we again see how construction, evaluation, and revision are intimately tied in mechanism discovery. A mechanistic hypothesis or hypothesis space is constructed, often in a sketchy way with numerous black boxes; evaluation of hypothesized components via experimentation or simulation provides constraints and indicates components in need of revision; and the outcome of those investigations serves to indicate the kind of fix needed during anomaly resolution. This iterative process continues to remove black boxes and find support for the hypothesized mechanism components, thereby removing incompleteness and incorrectness.

10. Conclusion

In the discovery of mechanisms, the product shapes the process of discovery. The search for mechanisms is guided by the very fact that it is a search for mechanisms rather than something else (e.g., an entity, a mathematical law). Strategies for constructing mechanism schemas include characterizing the phenomenon (to provide clues about the mechanism that produces it), localization (to find where the mechanism operates), employing an abstract schema (to provide the overall organization of the mechanism), modular subassembly (to provide functionally characterized groupings of entities and activities), and forward/backward chaining (to provide what comes next or what comes before any given step). Further evaluation and revision strategies guide the iterative process of constructing, evaluating, and revising mechanism schemas during mechanism discovery. More needs to be said about the roles played by interlevel and interfield contexts of the target mechanism. Surely other strategies will be found to add to this list as the new mechanistic philosophy of science works to understand ways of discovering mechanisms.

Notes

1 Many thanks to Carl Craver for all the work we've done together on these ideas over the years. The DC/Maryland History and Philosophy of Biology Discussion Group gave me very useful suggestions on an earlier draft. Stuart Glennan and Phyllis Illari provided very helpful guidance for rewriting.
2 For more on Mendel to Watson and Crick, see Darden (1991), Watson and Crick (1953a). On forward chaining in the discovery of the mechanism of protein synthesis, see Watson and Crick (1953a, 1953b); Crick (1988). On backward chaining by biochemists, see Hoagland (1955, 1996), Hoagland et al. (1959), and Zamecnik (1960). Zamecnik analogized their work to tunnel diggers (personal communication to author). Their work is discussed in Rheinberger (1997), who emphasized experimental systems rather than mechanisms, and in Darden and Craver (2002), who traced the discovery of the mechanism of protein synthesis.

References

Baltimore, David (1970), "Viral RNA-dependent DNA Polymerase," *Nature* 226: 1209–1211.
Bechtel, William (2016), "Using Computational Models to Discover and Understand Mechanisms," *Studies in History and Philosophy of Science Part A* 56: 113–121.
Bechtel, William and Robert C. Richardson (1993), *Discovering Complexity: Decomposition and Localization as Strategies in Scientific Research*. Princeton, NJ: Princeton University Press.
Bechtel, William and Robert C. Richardson (2010), *Discovering Complexity: Decomposition and Localization as Strategies in Scientific Research*. 2nd. ed. Cambridge, MA: MIT Press.
Bogen, James and James Woodward (1988), "Saving the Phenomena," *Philosophical Review* 97: 303–352.
Craver, Carl F. (2007), *Explaining the Brain: Mechanisms and the Mosaic Unity of Neuroscience*. New York: Oxford University Press.
Craver, Carl F. and Lindley Darden (2013), *In Search of Mechanisms: Discoveries across the Life Sciences*. Chicago, IL: University of Chicago Press.
Crick, Francis (1988), *What Mad Pursuit: A Personal View of Scientific Discovery*. New York: Basic Books.
Darden, Lindley (1991), *Theory Change in Science: Strategies from Mendelian Genetics*. New York: Oxford University Press.
Darden, Lindley (2002), "Strategies for Discovering Mechanisms: Schema Instantiation, Modular Subassembly, Forward/Backward Chaining," *Philosophy of Science* 69 (Proceedings): S354–S365.
Darden, Lindley (2006), *Reasoning in Biological Discoveries: Mechanisms, Interfield Relations, and Anomaly Resolution*. New York: Cambridge University Press.
Darden, Lindley (2013), "Mechanisms versus Causes in Biology and Medicine," in Hsiang-Ke Chao, Szu-Ting Chen, and Roberta L. Millstein (eds.), *Mechanism and Causality in Biology and Economics*. Dordrecht: Springer, pp. 19–34.
Darden, Lindley and Carl F. Craver (2002), "Strategies in the Interfield Discovery of the Mechanism of Protein Synthesis," *Studies in History and Philosophy of Biological and Biomedical Sciences* 33: 1–28. Corrected and reprinted in Darden (2006, ch. 3).
Datteri, Edoardo and Guglielmo Tamburrini (2007), "Biorobotic Experiments for the Discovery of Biological Mechanisms," *Philosophy of Science* 74: 409–430.
Fagan, Melinda Bonnie (2012) "Jointness and Mechanistic Explanation," *Philosophy of Science* 79: 448–472.
Gervais, Raoul and Erik Weber (2015), "The Role of Orientation Experiments in Discovering Mechanisms," *Studies in History and Philosophy of Science Part A* 54: 46–55.

Gilbert, Scott F. (2003), *Developmental Biology*, 7th ed. Sunderland, MA: Sinauer Associates.

Grasso, F. W., et al. (2000), "Biomimetic Robot Lobster Performs Chemo-Orientation in Turbulence Using a Pair of Spatially Separated Sensors: Progress and Challenges," *Robotics and Autonomous Systems* 30: 115–131.

Gutting, Gary (1980), "Science as Discovery," *Revue Internationale de Philosophie* 34: 26–48.

Hanson, Norwood Russell ([1961] 1970), "Is There a Logic of Scientific Discovery?" in H. Feigl and G. Maxwell (eds.), *Current Issues in the Philosophy of Science*. New York: Holt, Rinehart and Winston. Reprinted in B. Brody (ed.), *Readings in the Philosophy of Science*. Englewood Cliffs, NJ: Prentice Hall, pp. 620–633.

Hoagland, Mahlon B. (1955), "An Enzymic Mechanism for Amino Acid Activation in Animal Tissues," *Biochimica et Biophysica Acta* 16: 288–289.

Hoagland, Mahlon B. (1996), "Biochemistry or Molecular Biology? The Discovery of 'Soluble RNA'," *Trends in Biological Sciences Letters (TIBS)* 21: 77–80.

Hoagland, Mahlon B., Paul Zamecnik, and Mary L. Stephenson (1959), "A Hypothesis Concerning the Roles of Particulate and Soluble Ribonucleic Acids in Protein Synthesis," in R. E. Zirkle (ed.), *A Symposium on Molecular Biology*. Chicago, IL: Chicago University Press, pp. 105–114.

Josephson, John R. and Susan G. Josephson (eds.) (1994), *Abductive Inference: Computation, Philosophy, Technology*. New York: Cambridge University Press.

Kästner, Lena (2015), "Learning about Constitutive Relations," in U. Mäki et al. (eds.), *Recent Developments in the Philosophy of Science: EPSA13 Helsinki*, European Studies in Philosophy of Science 1, pp. 155–167. Dordrecht: Springer. DOI 10.1007/978-3-319-23015-3_12.

Machamer, Peter, Lindley Darden, and Carl F. Craver (2000), "Thinking about Mechanisms," *Philosophy of Science* 67: 1–25.

Meheus, Joke and Thomas Nickles (eds.) (2009), *Models of Discovery and Creativity*. Series: Origins: Studies in the Sources of Scientific Creativity, v. 3. Dordrecht: Springer.

Nachtomy, Ohad, A. Shavit, and A. Yakhimi (2007), "Gene Expression and the Concept of the Phenotype," *Studies in History and Philosophy of Biological and Biomedical Sciences* 38: 238–254.

Nickles, Thomas (1977), "Heuristics and Justification in Scientific Research: Comments on Shapere," in in F. Suppe (ed.), *The Structure of Scientific Theories*. Urbana: University of Illinois Press, pp. 571–589.

Nickles, Thomas (ed.) (1980a), *Scientific Discovery, Logic and Rationality*. Dordrecht: Reidel.

Nickles, Thomas (ed.) (1980b), *Scientific Discovery: Case Studies*. Dordrecht: Reidel.

Popper, Karl R. (1965), *The Logic of Scientific Discovery*. New York: Harper Torchbooks.

Rheinberger, Hans-Jörg (1997), *Towards a History of Epistemic Things: Synthesizing Proteins in the Test Tube*. Stanford: Stanford University Press.

Schaffner, Kenneth (1993), *Discovery and Explanation in Biology and Medicine*. Chicago, IL: University of Chicago Press.

Simon, Herbert A. (1977), *Models of Discovery*. Dordrecht: Reidel.

Simon, Herbert A. (1996), *The Sciences of the Artificial*. Third Edition. Cambridge, MA: MIT Press.

Thagard, Paul (1998), "Explaining Disease: Causes, Correlations, and Mechanisms," *Minds and Machines* 8: 61–78.

Thagard, Paul (1999), *How Scientists Explain Disease*. Princeton, NJ: Princeton University Press.

Thagard, Paul (2003), "Pathways to Biomedical Discovery," *Philosophy of Science* 70: 235–254.

Temin, Howard M. and Satoshi Mizutani (1970), "RNA-dependent DNA Polymerase in Virions of Rous Sarcoma Virus," *Nature* 226: 1211–1213.

Watson, James D. and F. H. C. Crick (1953a), "Molecular Structure of Nucleic Acids: A Structure for Deoxyribose Nucleic Acid," *Nature* 171: 737–738.

Watson, James D. and F. H. C. Crick (1953b), "Genetical Implications of the Structure of Deoxyribonucleic Acid," *Nature* 171: 964–967.

Wimsatt, William (2007), "Heuristics and Human Behavior," in *Re-Engineering Philosophy for Limited Beings*. Cambridge, MA: Harvard University Press, pp. 75–93.

Zamecnik, Paul C. (1960) "Historical and Current Aspects of the Problem of Protein Synthesis," *The Harvey Lectures* 1958–1959. Series 54. New York: Academic Press, pp. 256–281.

20

MECHANISMS AND DYNAMICAL SYSTEMS

David Michael Kaplan

1. Introduction

The mathematical framework of *dynamics*, or more specifically *dynamical systems theory* (hereafter DST), is having a major impact across many sectors of contemporary biology and neuroscience. The DST toolkit (described in more detail shortly), which includes differential equations, geometric state-space analyses, and other visualization techniques such as attractor landscapes and bifurcation diagrams, is playing an increasingly important role in modeling and explaining the time-varying activity of biological and neural systems ranging from single neurons and local circuits to entire brain networks, biological populations, and complex ecosystems (for book-length treatments, see Abraham and Shaw 1992; Amit 1992; Izhikevich 2005; Strogatz 1994).

The surge of scientific interest in dynamical modeling has attracted considerable philosophical attention. One recent topic of debate concerns the nature of the explanations these models provide. Some argue that the powerful mathematical framework of dynamics signals the emergence of a radically new explanatory paradigm that is importantly distinct from, and provides a competing alternative to, the dominant framework of mechanistic explanation (e.g., Chemero and Silberstein 2008; Stepp et al. 2011; Ross 2014; Silberstein and Chemero 2013). Others more friendly to the mechanistic perspective grant that DST introduces something distinctive and new—for example, a new set of mathematical techniques for describing patterns of change over time in neural systems, or even a novel theoretical framework for characterizing cognition and brain activity in non-representational or non-computational terms (Van Gelder 1995, 1998)—but reject the idea that it offers a distinct and competing explanatory framework to that of mechanism. Instead of viewing dynamical and mechanistic models as competitors, proponents of the mechanistic perspective see dynamical models as complementary to the description of mechanisms (e.g., Bechtel and Abrahamsen 2010; Kaplan and Bechtel 2011; Kaplan and Craver 2011; Kaplan 2015; Zednik 2011). In what follows, I survey the different sides of the debate over the nature of dynamical explanation with a specific focus on examples from cognitive science and neuroscience. (For related discussion of dynamical modeling in other areas of biological science including systems biology, see Chapter 27.) First, as background, I outline some basic features of the DST framework. After exploring the main positions staked out in the debate and identifying their primary strengths and limitations, I highlight the advantages of embracing a mechanistic approach to dynamical explanation. I then address several major open challenges facing the mechanistic approach and conclude by briefly discussing some important future directions for the debate.

2. Dynamics: a primer

A dynamical system is, roughly speaking, a set of variables that interact over time. Changes in these variables over time—their time series—can exhibit various patterns. Dynamics or DST comprises a highly general set of mathematical techniques for modeling, analyzing, and visualizing these patterns in time series data. At the core of every dynamical model is one or more differential equations (for the continuous time case) or difference equations (for the discrete time case), which specifies how a system changes at any given time point as a function of its state at that time. Although differential (and difference) equations are powerful modeling tools, some do not admit of exact solutions—i.e., the function or set of functions satisfying a given equation. When analytical techniques fail, numerical methods are often used to arrive at inexact or approximate solutions with arbitrarily high levels of precision (e.g., Mascagni and Sherman 1989). In other circumstances, it is more appropriate or useful to model long-term system behavior in terms of solution trajectories in a geometrically defined space.

Geometric methods and concepts are central to DST. A *state space* is defined as the set of all possible values or *states* that a given dynamical system can take or be in over time. Each system variable defines a corresponding dimension in this space and the number of dimensions of this space reflects the total number of state variables (e.g., the dynamical system in Figure 1 is defined by two variables, y_1 and y_2). Different possible trajectories of the system within that space are determined by plugging in different values to the relevant differential equation. In this manner, the differential equation serves to define a *vector field*—an assignment of an instantaneous direction and rate of change for each point in the state space (Figure 20.1a). An individual solution trajectory is a particular temporal sequence of states through this vector field, from some initial state of the system. The set of all possible solution trajectories is called a *flow* (Figure 20.1b). Many dynamical systems will eventually converge on a small subregion of the state space, called the *limit set*. If the system state passes through a limit set, the dynamics will operate to keep it there and small external perturbations will only alter the system momentarily. Afterward, it will converge back to the same state. Limit sets can either be single points (*fixed* or *equilibrium points*)

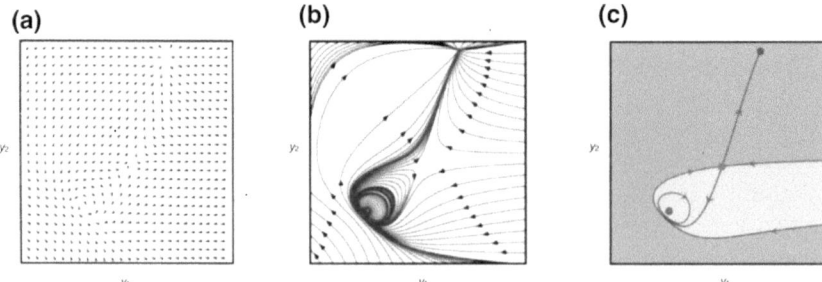

Figure 20.1 Basic constructs of DST. (a) A vector field for a two-dimensional dynamical system defined by the two differential equations, dot[y$_1$] = f$_1$(y$_1$, y$_2$) and dot[y$_2$] = f$_2$(y$_1$, y$_2$). (b) A flow diagram depicting representative solution trajectories for the same system. (c) A phase portrait depicting the system's limit sets, their stabilities, and their basins of attraction. Dots indicate equilibrium points. Other limit sets are indicated by lines. Source: Beer 2000

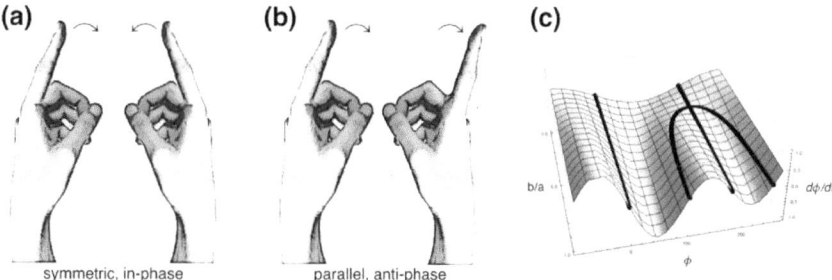

Figure 20.2   HKB experiment and model. (a) In-phase finger oscillation with relative phase (ϕ) = 0°.
(b) Anti-phase finger oscillation with relative phase (ϕ) = 180°. (c) Three-dimensional
vector field diagram. Thick lines indicate stationary or fixed points of the system (i.e.,
where $d\phi/dt = 0$). Curved surface and grayscale gradients depict the overall attractor
landscape for the system. Source: Zednik 2011; adapted from Kelso 1995

or trajectories that loop back on themselves (*limit cycles* or *oscillations*) (Figure 20.1c). For a stable
limit set, also known as an *attractor*, all nearby trajectories will converge to it. The set of points
that converge to an attractor over time is termed a *basin of attraction*. The overall characterization of all the limit sets and basins of attraction of a dynamical system is called a *phase portrait*
(Figure 20.1c). Finally, the flow and overall attractor landscape can depend on other independent parameters (called *control parameters*). Sometimes the flow field of a dynamical system changes
continuously when a control parameter is varied. Other times, a discontinuous change in the
flow pattern can result from small and continuous changes in the control parameter, called
a *bifurcation*. With this basic terminology in hand, we can move on to explore the nature of
dynamical explanation.

The well-known Haken–Kelso–Bunz (HKB) model of bimanual coordination (Haken et al.
1985) provides a useful illustration of DST in action. The HKB model was originally developed
to account for behavioral data collected in a bimanual coordination task. Subjects made repetitive
side-to-side (oscillating) movements of their index fingers in the transverse plane either in-phase
(simultaneous, mirror-symmetric movements toward the midline of the body; Figure 20.2a) or anti-
phase (simultaneous, parallel movements to the left or right of the body midline; Figure 20.2b).
Movements were completed in time with a pacing metronome, and metronome speed was an
independent variable under experimental control. Several interesting results emerged. First, subjects
could reliably perform both in-phase and anti-phase coordination patterns at low oscillation frequencies. Second, in trials where movement frequency was increased beyond a certain critical threshold,
subjects could no longer maintain the anti-phase movement pattern and switched abruptly and
involuntarily into the in-phase pattern. Third, switches never occurred on trials initiated in-phase,
even when the critical frequency was exceeded.

The core of the HKB model is the differential equation capturing the rate of change over
time in relative phase between the fingers:

$$\frac{d\phi}{dt} = -a \sin \phi - 2b \sin 2\phi \tag{1}$$

where ϕ represents relative phase (in-phase, ϕ 0°; anti-phase, $\phi \pm 180°$); a and b are empirically
fitted parameters such that the ratio b/a is inversely proportional to frequency. In the language
of DST, the ratio b/a is a control parameter. The corresponding DST analysis depicts system

behavior in terms of an attractor landscape that changes as a function of the control parameter and time (Figure 20.2c). When oscillation frequencies are relatively low ($b/a > 0.25$), two stable attractors exist corresponding to in-phase and anti-phase movement patterns. When oscillation frequencies are relatively high ($b/a < 0.25$), only in-phase coordination has a stable attractor and the anti-phase pattern becomes unstable. As the system moves in the direction of a decreasing b/a ratio (movement from top to bottom along the y-axis in Fig. 20.2c), associated with higher oscillation frequencies, the stable attractor for anti-phase coordination disappears and is subsequently transformed into an unstable equilibrium point (at $b/a = 0.25$). At this critical frequency, the system undergoes a phase transition or bifurcation. Once the system crosses this threshold, any small perturbation pushes it toward the stable fixed point attractor corresponding to the in-phase movement pattern.

The HKB model is an exemplar of the dynamical approach. It remains one of the most influential quantitative models of human motor behavior (Beek et al. 2002; Fuchs 2013; Swinnen 2002) and serves as a central case in philosophical discussions concerning dynamical explanation (Gervais and Weber 2011; Kaplan 2015; Kaplan and Craver 2011; Kelso 1995; Walmsley 2008).

3. Varieties of approaches to dynamical explanation

According to the once dominant covering law (CL) model, explanation involves (deductive or inductive) subsumption of some event or phenomenon to be explained under a law or set of laws (Hempel 1965). More precisely, a specification of the relevant laws and empirical conditions under which the phenomenon obtains (*initial* and *boundary conditions*) is supposed to provide good evidential grounds for expecting the occurrence of the explanandum-phenomenon.

According to the CL approach to dynamical explanation (hereafter, DCL), explanatory dynamical models are just special cases of CL explanations (Bechtel 1998; Bechtel and Abrahamsen 2002; Walmsley 2008). The DCL approach is relatively common in the literature, and many dynamicist researchers explicitly describe their models as involving laws (e.g., Bressler and Kelso 2001; Kelso 1995; Schöner and Kelso 1988). Zednik (2011) characterizes it as the "received view" about dynamical explanation. Despite its prominence, proponents of DCL have failed to address many of the standard challenges to the general nomological conception of explanation upon which it is based (for additional discussion, see Kaplan 2015).

The main challenge facing DCL involves showing how explanatory dynamical models either make explicit reference to a genuine law or implicitly convey information about the existence of the relevant law. If dynamical models cannot satisfy this basic requirement on CL explanations, DCL is a non-starter. Answering this challenge is difficult because it requires a sufficiently precise notion of law that can successfully distinguish laws from accidental generalizations; and consensus about general criteria for lawhood remains elusive in the philosophy of science (Hempel 1965; Salmon 2006; Woodward 2000, 2003).

Defenders of DCL can avail themselves of one of two main options for characterizing the nomological character of the generalizations appealed to in dynamical explanations. It is important to note that these are the primary options available if the defender of DCL wishes to remain within the covering law tradition. If, however, one is willing to abandon the central notion of subsumption under law, then other options (including but not limited to the mechanistic approach) come into view such as interventionist approaches to explanation (e.g., Woodward 2000, 2003). According to Woodward's account, the explanatory import of a generalization depends on the degree to which it is invariant under intervention rather than on how well it satisfies the traditional criteria for lawhood. Exploring how dynamical explanation might fit within these approaches goes beyond the scope of this chapter. Returning to the main thread, the first option is to maintain that these generalizations do in fact satisfy the standard criteria for lawhood and therefore qualify as full-blown laws, which in turn renders them suitable to feature in CL explanations. The second option is to claim that, like many explanatory generalizations found in the special sciences, dynamical generalizations are best construed as so-called *non-strict*, *qualified*, or *ceteris paribus* laws (e.g., Fodor 1991; Pietroski and Rey 1995). Both options face difficulties. (For further discussion of laws and their connection to mechanistic explanation, see Glennan (1996) and Chapters 11 and 12 of this volume.)

The main problem with the first option is that the mathematical generalizations commonly found in dynamical models do not seem to satisfy many of the standard criteria for lawhood. It is widely assumed, for example, that whatever else a law may be, it must at least be an exception-less generalization with wide scope. *Scope* describes the range of different individual systems (or types of systems) over which a given generalization holds (e.g., the gravitational inverse square law has wide scope in the sense that it applies to *all* massive bodies throughout the universe). Satisfying these criteria is important because if generalizations admit of exceptions or have scope restrictions, they will fail to play the role required of them by the CL framework. This is because deductive inference of the explanandum in a CL explanation cannot occur unless the law statement in the explanans takes the form of a universally quantified generalization (that ranges over its specified domain *without exception*). Importantly, this challenge also applies to *inductive-statistical* explanations (Hempel 1965) since exceptions in a statistical law featured in the explanans can lower the probability conferred on the occurrence of the explanandum event and thereby weaken the explanation.

Yet the mathematical generalizations typically employed in dynamical models, like most generalizations in biology, appear to be both highly restricted in scope and exception-ridden. The paradigmatic HKB model, for example, covers an impressive range of coordination patterns involving two or more oscillating components—bimanual coordination (Haken et al. 1985), coordinated movements across subjects (Schmidt et al. 1990), and even some forms of social coordination (Oullier et al. 2008). Yet its scope is restricted in important respects. For example, the model fails to apply to all rhythmic limb movements including those involved in walking or running. No discontinuous shift from walking or running (anti-phase leg movements) to hopping (in-phase leg movements) occurs when humans increase their movement speed above some critical threshold (Rosenbaum 1998).

Along similar lines, many of the mathematical regularities captured by dynamical models are exception-ridden in the sense that they only hold within a certain domain or regime of changes and break down outside of these. The HKB model holds for movement frequencies on the order of a couple of cycles per second (Haken et al. 1985), but it remains unclear whether the relationships characterized in the model remain invariant across *all* changes in this movement frequency. For example, exceptions might be expected at extreme speeds (e.g., 1000 cycles/second) approaching or exceeding the limits imposed by the musculoskeletal and nervous systems.

A number of philosophers have emphasized how the ability of dynamical generalizations to support counterfactuals might justify their status as laws (e.g., Bechtel 1998; Clark 1997; Van Gelder 1998). Clark, for instance, highlights the importance of "the way it [the system] responds to new, not-yet-encountered circumstances" (Clark 1997, 119). Bechtel claims that "DST accounts . . . are clearly designed to support counterfactuals [and] [t]his suggests that it may be appropriate to construe these DST explanations as being in the covering law tradition" (Bechtel 1998, 311).

Given the centrality of the DST notion of a state space embodying information about all possible (and actual or observed) states and trajectories that a dynamical system can take, philosophical emphasis on counterfactual support should be unsurprising. Possible state-space trajectories constitute straightforward counterfactuals about what a given dynamical system would have done if things had been different. Unfortunately, even though dynamical generalizations satisfy this criterion in spades, this criterion fails to distinguish laws from accidental generalizations. This is because many accidental generalizations support counterfactual predictions. Borrowing an example from Woodward (2003, 280), the generalization "All the coins in Clinton's pocket are dimes" is both accidental and counterfactual-supporting. Supposing as a background condition that Clinton permitted only dimes in his pocket, the above generalization supports the counterfactual: "If c were in coin in Clinton's pocket, then it would be a dime" (Woodward 2003, 280). Examples of this kind illustrate how appeals to counterfactual support appear inadequate to underwrite an account of dynamical laws.

This brings us to the second option, which involves construing dynamical generalizations as non-strict laws. This option is also fraught with difficulties. According to the general strategy, a generalization with exceptions can still play an explanatory role if it can be "completed" by specifying some further set of conditions that, together with the conditions outlined in the original generalization, are nomologically sufficient to generate the explanandum (Reutlinger and Unterhuber 2014). When supplemented with the appropriate "completer," the resulting generalization qualifies as an exceptionless law because it restricts the scope to precisely the range of circumstances where the regularity obtains without exception. The well-known problem with this proposal involves filling out the completer clause without producing generalizations that are either trivially true or false (Earman and Roberts 1999; Woodward 2002, 2003). Many philosophers remain skeptical about whether this challenge can be met (Earman et al. 2002; Reutlinger and Unterhuber 2014). Even if one assumes this problem can be handled, other difficulties arise. For example, Woodward (2002) argues that the very notion of a non-strict law incorporating qualifying clauses in this manner provides an exceedingly poor philosophical reconstruction of the kind of causal generalizations they are supposed to pick out in the special sciences.

Proponents of DCL therefore find themselves in a highly undesirable position. They must either face up to the difficult challenge of producing a satisfactory account of the nomological status of dynamical generalizations or jettison the approach.

Another approach to dynamical explanation emphasizes the tight connection between prediction and explanation (Chemero 2011; Chemero and Silberstein 2008; Stepp et al. 2011; Van Gelder 1998). This view has been termed *predictivism* (Kaplan and Craver 2011). Even though predictivism drops the problematic requirement that laws are needed for explanation, it is still closely connected to the CL approach and therefore faces many of the same difficulties.

The first problem facing predictivism as an account of the explanatory power of dynamical models is that well-known counterexamples to the CL model demonstrate how prediction is insufficient for explanation (Salmon 2006). Similar examples can be constructed to illustrate how dynamical models can be predictively adequate in the sense that they predict all the relevant aspects of the phenomenon with the desired precision and accuracy, yet fail as explanations because the model variables represent magnitudes that merely correlate with some other common cause for the phenomenon. Consider a dynamical model describing a set of three gears in an automotive transmission system. The gears are organized such that only one gear (gear C) directly connects to the motor. The other gears (A and B) rotate only as a consequence of C's rotational motion. Because the behavior of all three gears is time-locked to motor speed, their dynamics will be coupled (i.e., correlated). One could build a dynamical model that incorporates information about the speed or acceleration of gear A and thereby predicts the behavior of gear B (and vice versa). Yet it is problematic to characterize one as explaining the other because a common cause—gear C—is fundamental to explaining the correlation between them.

Similarly, one could use temporal information about A (or B) to predict the behavior of C, even though the rotational motion of C causes the motion of A (and B). This move is equally problematic from the point of view of explanation.

Explanations must respect this fundamental asymmetry between causes and effects (and mere correlates), even though effects and correlated variables can be extremely useful predictors of their causes. These and many other similar examples can be generated to illustrate how prediction is insufficient for explanation, and why the predictive force of a given model should not be confused with its explanatory force. Instead, what is needed is an account that provides an understanding of precisely *why* the regularities that constitute the phenomenon hold in the first place.

Because it assimilates explanatory and predictive power, another major problem for the predictivist view is that it is incapable of capturing the explanatory gains among predictively equivalent models where one describes the causal structure of the target system with increased accuracy. According to predictivism, the quality of an explanation can be improved primarily by increasing its predictive power. Yet there seem to be other ways—including building in more causal-mechanical details—to improve an explanation, which the predictivist view cannot accommodate. According to predictivism, unless these additional details improve the predictive power of the model, they are explanatorily inert. But this runs counter to widespread views about how scientific progress is achieved, and specifically about the kinds of refinements and model-building activities that produce better explanations.

Given the problems facing predictivism (and DCL), what is the alternative? One appealing possibility is to treat dynamical explanation as a species of mechanistic explanation. Doing so allows us to sidestep the problems outlined above. It also grounds dynamical explanations in a dominant and well-understood form of explanation that is widespread across the biological sciences.

Instead of emphasizing the gap between dynamics and mechanism, as the previous approaches do, another option is to show how the apparent gap is to be bridged. When biologists and neuroscientists put forward explanations, they frequently seek to identify the mechanism responsible for maintaining, producing, or underlying the phenomenon of interest (Bechtel and Richardson 1993/2010; Craver 2007; Machamer et al. 2000). In other words, they seek to provide *mechanistic explanations*. Mechanistic explanations specify the component parts, the operations or activities of these parts, and the (spatial and temporal) organization of the parts and their activities in the mechanism as a whole. All major accounts of mechanistic explanation identify a key role for each of these core elements. Despite the fact that lip service is frequently paid to the importance of organization, this remains the most underdeveloped aspect of many accounts of mechanistic explanation.

Brief reflection nevertheless reveals how temporal organization is critical to mechanisms and mechanistic explanations. Consider the internal combustion engine. Engines are composed of structural parts including pistons, spark plugs, and intake valves that perform activities such as sliding, sparking, and opening. Critically, these parts must interact in space and time in a highly organized manner for the engine to function properly. Engine components must bear specific spatial relationships to the other parts with which they have causal interactions. For example, pistons must be located within the cylinders, which in turn must be spatially proximate and mechanically linked to the crankshaft via connecting rods to produce torque in the axles. Precise temporal organization of the activities performed by the engine parts is no less important to engine performance. For example, spark plugs must emit their spark at the top of the compression stroke of the pistons. Intake and exhaust valves must open and close at precise times so that they are sealed shut during compression and combustion and open during the exhaust stroke. Together, these comprise the engine *dynamics*.

Although in this particular case it might not be especially illuminating, the state of each of the components and/or the global state of the engine system as a whole could be quantified (e.g., position, velocity, or acceleration for the parts; maximum power or torque for the engine as a whole) and plotted as a function of time. Each could also be subjected to a dynamical analysis according to which its evolving state is represented as a trajectory in a suitable state space. Nevertheless, for many complex natural systems as opposed to artificial or designed systems, dynamical analyses can help to reveal dynamic patterns or latent temporal structure in system activity that would otherwise be exceedingly difficult if not impossible to discern (e.g., Churchland et al. 2012).

Organization is a necessary part of most moderately complex mechanisms such that perturbing either the spatial organization or temporal dynamics of a mechanism, even while the components and their individual activities remain unchanged, can have appreciable effects on performance. Thinking about mechanistic explanation, then, it is clearly insufficient to describe only the properties and activities of the component parts in a given mechanism without giving adequate weight or attention to the spatial and/or temporal organization of those parts and activities. Often this point is underappreciated or lost when considering the nature of mechanistic explanation (for a notable exception, see Bechtel and Abrahamsen 2010).

This discussion highlights how even relatively simple mechanisms, such as internal combustion engines, can exhibit rather complex temporal organization. Consequently, understanding the dynamical "structure" of a mechanism can sometimes be just as important as understanding its physical structure. This last point rings especially true in the context of neuroscience and biology, where mechanisms often exhibit a wide range of complex dynamic patterns that are essential to their proper functioning (for additional discussion, see Kaplan 2015).

A full development and defense of the mechanistic approach to dynamical explanation sketched would ideally involve walking through some case studies in which both the framework of dynamics and that of mechanism were jointly employed in the service of building adequate explanations. Nevertheless, the view about the relationship between the frameworks of dynamics and mechanism that falls out of the above considerations is one of *subsumption*, not competition. In particular, dynamical models are usefully employed to reveal the temporal organization of activity in mechanisms. Because dynamical models are subsumed within the broader toolkit for describing mechanisms, their explanatory value can be seen as clearly depending on the presence of an associated account (however incomplete) of the parts in the mechanism (and their interactions) that support, maintain, or underlie these activity patterns. On this view, dynamical modeling approaches do not signal the emergence of a new explanatory framework that is distinct from mechanistic explanation. Instead, dynamical models with explanatory force are to be understood as playing a vital role within the mechanistic framework.

4. Open challenges for the mechanistic approach to dynamical explanation

The mechanistic approach to dynamical explanation faces challenges along several fronts. In this section, I focus on two related challenges raised by the purported non-decomposability of many dynamical systems.

It is widely acknowledged that the heuristic strategies of decomposition and localization are vitally important for discovering mechanisms and building mechanistic explanations (Bechtel and Richardson 1993/2010). Roughly, decomposition involves the identification of the component processes or activities that contribute to the overall performance of a mechanism and are recruited when the mechanism is engaged. Localization involves the identification of these different component activities with the behavior or capacities of specific physical parts of the mechanism. Silberstein and Chemero (2013) recently argue that sometimes the nature of dynamical systems prevents the application of these strategies, and that in such cases mechanistic explanations will be unavailable in principle. They identify several specific types of circumstances in which failures of decomposability or localizability will occur. I will argue that these challenges can be rebutted from a mechanistic perspective.

The first type of problematic situation Silberstein and Chemero identify is "when there are no component parts or operations that can be distinguished (such as a connectionist network), in which case one can only talk about organizational features" (Silberstein and Chemero 2013, 961–2). For ease of reference, call this the *challenge from functionally indistinguishable component parts*. The challenge seems to be that if the component parts in a system are doing functionally indistinguishable things, then the mechanistic strategies of functional decomposition and localization will fail to gain traction. Consequently, mechanistic analyses will either provide relatively thin explanations or no explanations at all. Surprisingly, arch-mechanists Bechtel and Richardson make a similar point much earlier when they claim that:

> Network models do account for the cognitive performance, but often they do so without providing an explanation of component operations that is intelligible in terms of the overall task being performed. The network is a cognitive system; the components are not. The result is that we do not explain how the overall system achieves its performance by decomposing the overall task into subtasks, or by localizing cognitive subtasks.
>
> *(Bechtel and Richardson 1993/2010, 214)*

The common idea here seems to be that when systems include parts that are functionally indistinguishable (or unintelligible)—connectionist networks being an exemplary case—mechanistic explanation cannot get off the ground. Although there are good reasons to reject the claim that mechanisms in which all the parts are essentially performing the same function are not suitable candidates for mechanistic explanation, let us grant this assumption for the sake of argument. Even under this assumption, there are problems for the view.

At a relatively coarse-grained level, it is certainly true that nodes or units in artificial neural networks do perform the same computational operations—they integrate signals and compute some function of their inputs. More specifically, the activation or transfer function (e.g., step or sigmoidal), which defines the output of a unit given some input or set of inputs, is typically implemented across all units within a layer (e.g., the hidden layer). So, there is a sense in which each network "component" is doing the same thing (i.e., has the same functional profile) and thus stands poised to block the strategies of decomposition and localization at the heart of the mechanistic approach. Nevertheless, this challenge ignores a crucial feature of such networks—their weights.

All connectionist networks contain sets of adaptive weights, i.e. numerical parameters that are precisely tuned by a learning algorithm (e.g., backpropagation) during training. It is in virtue of the systematic adjustment of weights across a network that it comes to exhibit the overall performance that it does. In the untrained state, before weight adjustment (i.e., when weight assignments are essentially random), network performance for any desired input-output function is typically extremely poor. Consequently, it might be reasonable to view untrained neural networks as lacking functionally distinguishable parts, since the weights are randomly distributed and they do not yet play any determinate functional (representational or information-processing) role. Nevertheless, once trained, this assertion is clearly false. Suitably trained networks have functionally distinguishable parts that make them susceptible to mechanistic decomposition and localization. Reinforcing this point is the fact that numerous techniques are now available to "open the hood" and quantitatively (and qualitatively) assess the diverse functional (e.g., representational or information processing) roles that individual units play in support of overall network performance. These methods include Hinton diagrams (e.g., Hinton 1986); hierarchical clustering analysis (e.g., Sejnowski and Rosenberg 1987); and even receptive-field techniques (e.g., Blohm et al. 2009; Zipser and Andersen 1988). Hence, this challenge seems to rely on an overly simplistic view of neural network function and structure.

Since artificial neural networks are simulated on digital computers, the response becomes even more forceful in the context of real biological neural networks. This is because biological circuits, in contrast to their artificial counterparts, are typically composed of a diversity of neuronal cell types (the "units") that have different functional and structural properties. Moreover, functionally distinguishable synaptic "weights" on axonal connections in real biological circuits reflect structurally distinguishable underlying parts. Differences in axonal "weights" can reflect underlying structural differences in myelination patterns that influence how well the input signal propagates through the axon, or differences in the number and/or density of presynaptic vesicles and AMPA and NMDA receptors on the cell membrane that influences the amount of neurotransmitter released into the synapse and absorbed by the postsynaptic neuron.

In contrast to the claims of Silberstein and Chemero (2013), artificial and biological neural networks possess functionally and/or structurally distinguishable parts. This opens the door to the powerful mechanistic strategies of decomposition and localization.

The second challenge amounts to the claim that the mechanistic approach and more specifically, decomposition and localization, will fail "when there are component parts and operations but their individual behaviors systematically and continuously affect one another in a nonlinear fashion" (Silberstein and Chemero 2013, 961–2). For ease of reference, call this the *challenge from nonlinearity*. Before this challenge can be addressed or rebutted, however, one must clarify what kind of nonlinearity is being invoked. It turns out that the notion of nonlinearity at the heart of Silberstein and Chemero's challenge is ambiguous and can be understood in (at least) two different ways. Consequently, there are actually two distinct challenges for the mechanist to address. In what follows, I show how both of these challenges can be answered.

One kind of nonlinearity, which has received considerably more attention in the mechanistic literature than other kinds, involves the non-sequential temporal organization of component activities or processes in a mechanism (e.g., Bechtel and Richardson 1993/2010, 202). Call this *process nonlinearity*. Bechtel and Richardson, for example, discuss "explanations in which the component tasks can be thought of as following a linear, sequential order" and go on to contrast these with explanations that invoke "cyclic" organization (1993/2010, 202). They contrast the two as follows: "[c]yclic rather than linear organization occurs when the activity of any given component is dependent on a variety of other components that, in turn, depend on it" (Bechtel and Richardson 1993/2010, 202).

Process nonlinearity poses no threat for the prospects of mechanistic explanation of dynamical systems. Many neural dynamical systems, for instance, exhibit highly non-sequential or "cyclic" temporal organization between mechanism components including parallel processing, recurrent connections, and feedback loops. Yet neural systems exhibiting these properties routinely yield to mechanistic explanation. A hallmark example is the mechanistic explanation of the action potential. The underlying neural mechanism involves a complex interaction or feedback loop between the activities of voltage-gated ion channels spanning the neuronal membrane. Briefly, depolarization of the membrane potential from its resting level initiates the rapid opening of sodium channels and a correspondingly sharp increase in sodium current into the neuron (the so-called fast positive cycle). In parallel, the initial depolarization event also induces the opening of slower potassium channels leading to a correspondingly slower increase in potassium current out of the neuron and subsequent repolarization of the membrane potential (the so-called slow negative cycle). In the case of the action potential, non-sequential temporal organization of ion channel activities (process nonlinearities) are a fundamental aspect of the underlying mechanism (for additional discussion, see Kaplan 2015). Nonlinearity in the form of non-sequential temporal organization is an unproblematic feature of (some) mechanisms and poses no problem for mechanistic explanation.

It remains possible that the anti-mechanistic challenge issued by Silberstein and Chemero (2013) involves an entirely different kind of nonlinearity and so the previous response misses its mark. In fact, there is another important kind of nonlinearity widely discussed across many sciences (especially in neuroscience) that concerns the nature of a system's input-output profile and more specifically whether a system's output is a linear function of its inputs. Call this *response nonlinearity*. Perhaps this is the kind of nonlinearity they have in mind. If dynamical neural systems are nonlinear in this sense, perhaps this is why they will evade mechanistic explanation? While it is true that many dynamical systems (including many neural systems) are nonlinear in this sense, the anti-mechanistic conclusion does not follow.

Response nonlinearities are certainly commonplace in complex dynamical systems. Computational neuroscientists Eliasmith and Anderson (2003) indicate why response nonlinearities should be expected in complex information-processing systems in the brain: "the impressively complex behavior exhibited by neural systems is unlikely to be fully explained by linear transformations alone. Rather, there is abundant evidence that nonlinear operations are central to neural transformations" (2003, 153). But this does not imply that the prospects for mechanistic explanation are dim. Understanding why requires a closer look at how this kind of nonlinearity is defined in mathematics, engineering, and signal processing theory.

Response nonlinearity is standardly defined in terms of failing to satisfy the assumptions of linearity. In mathematics, a system is said to be linear if and only if it exhibits both *homogeneity* and *additivity*. Homogeneity entails that a change in the size of the input signal to a given system produces a corresponding or proportional change in the output signal. More formally, if input signal $x[n]$ results in output signal $y[n]$, then an input of $kx[n]$ results in an output of $ky[n]$, for any input signal and constant, k. A system is said to be additive if added signals pass through it without interacting; signals added at the input produce signals that are added at the output. Again, more formally, if an input of $x_1[n]$ results in an output of $y_1[n]$, and if $x_2[n]$ results in $y_2[n]$, then $x_1[n] + x_2[n]$ results in $y_1[n] + y_2[n]$. Any system exhibiting both homogeneity and additivity is linear. Any system failing to have both of these properties is nonlinear. This is a suitably precise characterization of response nonlinearity. Now the question is whether this kind of nonlinearity raises insurmountable problems for mechanistic explanation.

Evidence from neuroscience strongly suggests a negative answer to this question. All neural dynamical systems exhibiting thresholding (e.g., spiking) behavior are nonlinear in this sense since they fail to satisfy homogeneity and additivity. Yet these systems routinely succumb to mechanistic explanation. The already discussed case of the action potential provides one clear example. Other more exotic cases involve divisive normalization in neural systems (Carandini and Heeger 2012). Normalization operations also fail to meet the strictures of homogeneity and additivity, yet neuroscientists have proceeded to develop highly refined mechanistic models for divisive normalization computation in the fly olfactory system (Olsen et al. 2010) and have produced partially worked out mechanistic explanations of comparable normalization operations in mammalian V1, among others (Carandini and Heeger 2012). Hence, response nonlinearity seems to pose no real threat to the prospects for mechanistic explanation.

5. Conclusions and future directions

I have argued that the frameworks of dynamical systems and that of mechanistic explanation are closely related and complementary. Contrary to law- and prediction-based accounts of dynamical explanation, there is no legitimate sense in which dynamical models explain phenomena independently of describing mechanisms, either by subsumption under general laws or by appealing to their predictive force alone. Although importantly distinct in many ways, these two approaches to modeling are related in terms of subsumption—dynamical models provide one important set of resources among many for discovering and describing temporal features of a mechanism. The DST framework can often reveal or predict dynamic patterns or latent temporal structure in the activity of mechanisms (and their components) that might otherwise be difficult to discern using more conventional modeling techniques. Nevertheless, the real explanatory weight of dynamical models depends on the presence of an associated description (however incomplete) of the mechanisms that support, maintain, or underlie these activity patterns. The frameworks of dynamics and mechanism are natural allies—each plays an essential part in the common enterprise of describing the parts, activities, and organization of underlying mechanisms.

In addition to the challenges for the mechanistic approach to dynamical explanation identified earlier, more hard work remains for those wishing to defend a mechanistic perspective. For instance, dynamists have also appealed to notions of emergence or downward causation in complex dynamical systems to argue for the limitations of the mechanistic perspective (e.g., Chemero and Silberstein 2008; Kelso 1995; Silberstein and Chemero 2013). To date, defenders of the mechanistic approach have not satisfactorily addressed this challenge.

Another direction for the debate over dynamical systems and mechanistic explanation involves clarifying the roles of abstraction and idealization in dynamical and mechanistic models. Although it is widely recognized that strategies of abstraction and idealization are critically important to many forms of scientific modeling (see Chapter 17), it remains unclear how these are related to mechanistic and dynamical explanation. Mathematical abstraction is clearly importantly to DST, and some see this as indicating the emergence of a distinctive non-mechanistic approach to explanation (e.g., Batterman and Rice 2014; Ross 2014). But abstraction also plays an important role in more traditional modeling of mechanisms (e.g., Levy and Bechtel 2013). This cluster of topics represents an undeniable growth area in contemporary philosophy of science that stands to shed new and valuable light on the nature of the relationship between the frameworks of dynamics and mechanism.

References

Abraham, R., Shaw, C.D. 1992. *Dynamics: The Geometry of Behavior.* Redwood City, CA: Addison-Wesley.

Ahrens, M.B., Li, J.M., Orger, M.B., Robson, D.N., Schier, A.F., Engert, F., Portugues, R. 2012. "Brain-wide neuronal dynamics during motor adaptation in zebrafish." *Nature* 485(7399):471–7.

Amit, D.J. 1992. *Modeling Brain Function: The World of Attractor Neural Networks.* Cambridge: Cambridge University Press.

Batterman, R., Rice, C. 2014. "Minimal model explanations." *Philosophy of Science* 81(3):349–76.

Bechtel, W. 1998. "Representations and Cognitive Explanations: Assessing the Dynamicist's Challenge in Cognitive Science." *Cognitive Science* 22(3):295–318.

Bechtel, W., Abrahamsen, A. 2002. *Connectionism and the Mind. Parallel Processing, Dynamics, and Evolution in Networks.* Oxford: Blackwell.

——. 2010. "Dynamic Mechanistic Explanation: Computational Modeling of Circadian Rhythms as an Exemplar for Cognitive Science." *Studies in History and Philosophy of Science Part A* 41(3):321–33.

Bechtel, W., Richardson, R.C. 1993/2010. *Discovering Complexity: Decomposition and Localization as Strategies in Scientific Research*. Cambridge, MA: MIT Press.

Beek, P.J., Peper, C.E., Daffertshofer, A. 2002. "Modeling Rhythmic Interlimb Coordination: Beyond the Haken–Kelso–Bunz Model." *Brain and Cognition* 48(1):149–65.

Beer, R.D. 2000. "Dynamical Approaches to Cognitive Science." *Trends in Cognitive Sciences* 4(3):91–9.

Blohm, G., Keith, G.P., Crawford, J.D. 2009. "Decoding the Cortical Transformations for Visually Guided Reaching in 3D Space." *Cerebral Cortex* 19(6): 1372–93.

Bressler, S.L., Kelso, J.A.S. 2001. "Cortical Coordination Dynamics and Cognition." *Trends in Cognitive Sciences* 5(1):26–36.

Carandini, M., Heeger, D.J. 2012. "Normalization as a Canonical Neural Computation." *Nature Reviews Neuroscience* 13(1):51–62.

Chemero, A. 2011. *Radical Embodied Cognitive Science*. Cambridge, MA: MIT Press.

Chemero, A., Silberstein, M. 2008. "After the Philosophy of Mind: Replacing Scholasticism with Science." *Philosophy of Science* 75(1):1–27.

Churchland, M.M., Cunningham, J.P., Kaufman, M.T., Foster, J.D., Nuyujukian, P., Ryu, S.I., Shenoy, K.V. 2012. "Neural Population Dynamics during Reaching." *Nature* 487(7405):51–6.

Clark, A. 1997. *Being There: Putting Brain, Body, and World Together Again*. Cambridge, MA: MIT Press.

Craver, C.F. 2007. *Explaining the Brain: Mechanisms and the Mosaic Unity of Neuroscience*. New York: Oxford University Press.

Earman, J., Roberts, J. 1999. "Ceteris Paribus, There is no Problem of Provisos." *Synthese* 118(3):439–78.

Earman, J., Roberts, J.T., Smith, S. 2002. "Ceteris Paribus Lost." *Erkenntnis* 57(3):281–301.

Eliasmith, C., Anderson, C.H. 2003. *Neural Engineering*. Cambridge, MA: MIT Press.

Fodor, J.A. 1991. "You Can Fool Some of the People All of the Time, Everything Else Being Equal; Hedged Laws and Psychological Explanations." *Mind* 100(397):19–34.

Fuchs, A., Jirsa, V.K. 2008. *Coordination: Neural, Behavioral and Social Dynamics*. Vol. 1. Dordrecht: Springer.

Fuchs, A. 2013. "Haken-Kelso-Bunz (HKB) Model: Nonlinear Dynamics." In *Complex Systems: Theory and Applications for the Life-, Neuro- and Natural Sciences*. Berlin: Springer-Verlag, pp. 133–45.

Gervais, R., Weber, E. 2011. "The Covering Law Model Applied to Dynamical Cognitive Science: A Comment on Joel Walmsley." *Minds and Machines* 21(1):33–9.

Glennan, S.S. 1996. "Mechanisms and the Nature of Causation." *Erkenntnis* 44(1):50–71.

Haken, H., Kelso, J.A.S., Bunz. H. 1985. "A Theoretical Model of Phase Transitions in Human Hand Movements." *Biological Cybernetics* 51(5):347–56.

Hempel, C.G. 1965. *Aspects of Scientific Explanation*. New York: The Free Press.

Hinton, G., McClelland, J.L., Rumelhart, D.E. 1986. "Distributed Representations." In *Parallel Distributed Representations: Explorations in the Microstructure of Cognition*, edited by E. Rumelhart and J.L. McClelland. Cambridge, MA: MIT Press, pp. 77–109.

Izhikevich, E.M. 2005. *Dynamical Systems in Neuroscience*. Cambridge, MA: MIT Press.

Kaplan, D.M. 2015. "Moving Parts: The Natural Alliance between Dynamical and Mechanistic Modeling Approaches." *Biology and Philosophy* 30:757–86.

Kaplan, D.M., Bechtel, W. 2011. "Dynamical Models: An Alternative or Complement to Mechanistic Explanations?" *Topics in Cognitive Science* 3(2):438–44.

Kaplan, D.M, Craver, C.F. 2011. "The Explanatory Force of Dynamical and Mathematical Models in Neuroscience: A Mechanistic Perspective." *Philosophy of Science* 78(4):601–27.

Kelso, J.A.S. 1995. *Dynamic Patterns: The Self-Organization of Brain and Behavior*. Cambridge, MA: MIT Press.

Levy, A., Bechtel, W. 2013. "Abstraction and the Organization of Mechanisms." *Philosophy of Science* 80(2):241–61.

Machamer, P., Darden, L, Craver, C.F. 2000. "Thinking about Mechanisms." *Philosophy of Science* 67(1):1–25.

Mascagni, M.V., Sherman, A.S. 1989. "Numerical Methods for Neuronal Modeling." *Methods in Neuronal Modeling*. Cambridge, MA: MIT Press.

Olsen, S.R., Bhandawat, V., Wilson, R.I. 2010. "Divisive Normalization in Olfactory Population Codes." *Neuron* 66(2):287–99.

Oullier, O., de Guzman, G.C., Jantzen, K.J., Lagarde, J., Kelso, J.A.S. 2008. "Social Coordination Dynamics: Measuring Human Bonding." *Social Neuroscience* 3(2):178–92.

Pietroski, P., Rey, G. 1995. "When Other Things Aren't Equal: Saving Ceteris Paribus Laws from Vacuity." *British Journal for the Philosophy of Science* 46(1):81–110.

Reutlinger, A., Unterhuber, M. 2014. "Thinking about Non-Universal Laws." *Erkenntnis* 79 (10): 1703–13.

Rosenbaum, D.A. 1998. "Is Dynamical Systems Modeling Just Curve Fitting?" *Motor Control* 2(2):101–4.

Ross, L.N. (2014). "Dynamical Models and Explanation in Neuroscience." *Philosophy of Science* 81(1):32–54.

Salmon, W.C. 2006. *Four Decades of Scientific Explanation*. Pittsburgh, PA: University of Pittsburgh Press.

Schmidt, R.C., Carello, C., Turvey, M.T. 1990. "Phase Transitions and Critical Fluctuations in the Visual Coordination of Rhythmic Movements between People." *Journal of Experimental Psychology: Human Perception and Performance* 16(2):227–47.

Schöner, G., Kelso, J.A.S. 1988. "Dynamic Pattern Generation in Behavioral and Neural Systems." *Science* 239(4847):1513–20.

Sejnowski, T.J., Rosenberg, C.R. 1987. "Parallel Networks that Learn to Pronounce English Text." *Complex Systems*: 145–68.

Silberstein, M., Chemero, A. 2013. "Constraints on Localization and Decomposition as Explanatory Strategies in the Biological Sciences" *Philosophy of Science* 80(5):958–70.

Stepp, N., Chemero, A., Turvey, M.T. 2011. "Philosophy for the Rest of Cognitive Science." *Topics in Cognitive Science* 3(2):425–37

Strogatz, S.H. 1994. *Nonlinear Dynamics and Chaos: With Application to Physics, Biology, Chemistry, and Engineering*. Boulder, CO: Westview Press.

Swinnen, S.P. 2002. "Intermanual Coordination: From Behavioural Principles to Neural-Network Interactions." *Nature Reviews Neuroscience* 3(5):348–59.

Van Gelder, T. 1995. "What Might Cognition Be, If Not Computation?" *The Journal of Philosophy* 92(7):345–81.

——. 1998. "The Dynamical Hypothesis in Cognitive Science." *Behavioral and Brain Sciences* 21(5):615–28.

Walmsley, J. 2008. "Explanation in Dynamical Cognitive Science." *Minds and Machines* 18(3):331–48.

Woodward, J. 2000. "Explanation and Invariance in the Special Sciences." *British Journal for the Philosophy of Science* 51(2):197–254.

——. 2002. "There Is No Such Thing as a Ceteris Paribus Law." *Erkenntnis* 57(3):303–28.

——. 2003. *Making Things Happen: A Theory of Causal Explanation*. Oxford: Oxford University Press.

Zednik, C. 2011. "The Nature of Dynamical Explanation." *Philosophy of Science* 78(2):238–63.

Zipser, D., Andersen, R.A. 1988. "A Back-Propagation Programmed Network that Simulates Response Properties of a Subset of Posterior Parietal Neurons." *Nature* 331(6158): 679–84.

PART IV

Disciplinary perspectives on mechanisms

21

MECHANISMS IN PHYSICS[1]

Meinard Kuhlmann

1. Introduction

Since mechanisms became a central topic in contemporary philosophy of science some 15 years ago, physics has never been in the focus of larger debates. Prima facie, this is understandable in the face of one major motivation for today's mechanistic program: The traditional twentieth-century philosophy of science with its roots in logical empiricism was dominated by physics as the paradigmatic science. In physics, theories with universal laws seem to be center stage. They are the basis for predictions, explanations, and our general understanding of nature. However, this law-focused philosophy of science seems much less suited when it comes to special sciences such as biology, neuro-science, and psychology—leaving aside the question of whether it is appropriate for physics itself. In the special sciences, mechanisms play a far greater role than universal laws, if indeed there are any such laws at all. With the growing attention paid to special sciences in the philosophy of science, mechanisms have accordingly become a central issue. Today, with the ascent of the new mechanistic program in philosophy of science, the question has almost turned: How does physics fit into the picture?

When thinking about mechanisms in physics, the most interesting question is different than for other fields. The primary concern is not how the notion of mechanisms can be captured and how exactly mechanistic explanations work. Rather, the main question is to what extent physics, in particular fundamental physics, deals with mechanisms in the first place. A second question concerns whether the character of the physical processes that underlie all natural and social phenomena may even endanger the tenability of mechanistic reasoning in the special sciences.

Physics can be important for mechanisms on higher levels, like those in chemistry or biology. However, is physics itself concerned with mechanisms? Here are some reasons why physics or parts of physics may not be dealing with mechanisms:

a) One may argue that fundamental physics rests on the brute fact that one observes certain regularities, which physical theories organize in a systematic way. Thus no explanation of the observed regularities is given, and thus *a fortiori* no mechanistic explanation. If this should be true then it seems that there is no room for mechanistic explanations in fundamental physics.

b) At least some explanations in physics are non-causal and thereby arguably not mechanistic. Examples are explanations based on energy conservation considerations or dimensional analysis. These cases may not be representative, however. Moreover, arguably these often very elegant explanations are only convincing provided that one also has (more complicated) micro explanations of the same type of phenomena.

c) Only very recently, a number of philosophers of science have argued that the range of non-causal explanations may be greater than has been thought previously. In particular, it may also include important classes of micro explanations in physics, such as explanations in statistical physics and specifically so-called renormalization group explanations. In our context this debate is of particular importance because it concerns explanations which seem mechanistic. If true, this line of thought could mean that mechanistic explanations are not only absent in fundamental physics but also in higher levels.

d) To make things worse for the new mechanists, one could even argue that the specific nature of quantum physics undermines the mechanical philosophy concerning all higher levels, i.e. even regarding phenomena that are not studied in physics but in the special sciences.

These potential limitations to mechanistic reasoning in physics bring us to three interesting questions:

1) Do mechanisms play any role in fundamental physics?
2) To what extent are explanations in the non-fundamental part of physics mechanistic?
3) Does quantum physics undermine mechanistic reasoning on higher levels?

As we will see, none of these three questions has an easy, straightforward answer. Since the answers depend crucially on which parts of physics one is looking at, the structure of this chapter largely follows the structure of physics as a discipline.

2. Fundamental physics: general reflections

Presupposing for now that there is a fundamental physics,[2] it is natural to assume that it comprises those theories that state what the fundamental physical entities and the fundamental laws are. The laws reflect how the entities behave, in isolation as well as when they interact with each other. Advanced accounts usually describe the fundamental entities in terms of permanent properties, which are given by the invariants under certain symmetry transformations, i.e. changes that only affect how we describe the world but not what is described. Which symmetry transformations are actually supposed to be the ones that leave the physical reality unchanged is itself a vital part of our theories: The symmetries that characterize Classical Mechanics differ from the ones for Special Relativity Theory.

We do not know the true fundamental physics yet, at least not in full, and we may never do so. The considerations discussed in this section are, however, general enough that they should not depend much on the state of the art in current fundamental physics. This changes when we consider specific theories, as we will do in the subsequent sections. There, one must keep in mind that we are just dealing with previous or current candidates for fundamental physics. Although these theories are provisional or even proven wrong, it is still worth exploring how they fit into the mechanistic program. First, these theories have shaped our worldview, and unveiling the role of mechanisms in them helps us to clarify this picture. Second, exploring the occurrence or non-occurrence of mechanisms in today's physics may contribute to finding or evaluating candidates for future fundamental physics.

Within the modern debate on mechanisms, the special status of fundamental physics for the mechanistic program was put on record early on. Among the advocates of this program, Glennan is arguably the one who has most persistently pointed to this issue. Glennan (1996: 50) notes:

> [That] two bodies . . . gravitationally attract each other . . . is just a "brute fact" about the world in which we live. . . . [W]e cannot explain the interaction by reference to any mechanism. . . . I suggest that there should be a dichotomy in our understanding of causation between the case of fundamental physics and that of other sciences (including much of physics itself).

Glennan (2002: S348) adds:

> While most laws are mechanically explicable, inevitably there must be some laws that are not. For instance . . . Maxwell's equations are *not mechanically explicable*. There is not, for instance, a mechanical ether consisting of particles whose interactions could explain the propagation of electromagnetic waves. Laws such as these, which I call *fundamental laws*, represent *brute nomological facts of our universe*.
>
> *(My emphases, MK)*

This view is close to the one I listed as argument a) above.[3] The lack of mechanistic explanations for fundamental interactions has nothing to do with the nature of the specific theories in place. If there were a mechanistic explanation for a supposedly fundamental law, then the laws governing the interacting parts of this mechanism would be the truly fundamental laws and not the one we started with. These laws in turn have no lower mechanistic basis, for otherwise one could invoke the same argument again. The upshot is that there are no mechanisms in fundamental physics, but only in higher-level sciences, including the non-fundamental parts of physics.[4]

Recently, Glennan has adopted a somewhat different take on this issue:

> Our discussion of this problem will be facilitated by some new terminology. Let us call a mechanism (that is, a set of entities engaging in activities and interactions) that does not depend upon other mechanisms, a fundamental (or basic) mechanism. All other mechanisms are mechanism-dependent or compound mechanisms.
>
> *(Glennan forthcoming, sec. 7.5)*

While in the original view, mechanisms played no role in fundamental physics, now it abounds with mechanisms. Glennan comments on this change as follows:

> In past work (Glennan, 1996, 2002, 2011) I have simply spoken of fundamental interactions rather than saying they are interactions produced by fundamental mechanisms, so this language represents a terminological shift. This shift does not affect the main substantive claim, which is that if there are such fundamental mechanisms, there is a base case of mechanisms whose productive capacities are not themselves mechanism-dependent. . . . The advantage of saying these interactions are produced by fundamental mechanisms is that then every causally related pair of events will be connected by a mechanism.
>
> *(Glennan forthcoming, chap. 7, footnote 8)*

One may question whether this is really just a "terminological shift"; it seems at least a shift in emphasis from epistemology toward ontology. But, in any case, it gives us an attractive coherent picture: The interactions described by the laws of fundamental physics cannot be *explained* mechanistically. However, this does not mean that there is no room for mechanisms in fundamental physics, because the fundamental interactions *are* themselves mechanisms. The only aspect that distinguishes them from mechanisms in higher-level sciences is that they are basic mechanisms, since they do not rest on further mechanisms on a lower level.

If we accept this line of reasoning, one crucial question then is which basic mechanisms there are in fundamental physics. Let us start with a straightforward, somewhat naïve list of—old or current—candidates of basic mechanisms in fundamental physics:

1) Newtonian physics: gravitational attraction as action-at-a-distance
2) Special Relativity Theory (SRT): length contraction and time dilation
3) General Relativity Theory (GRT): gravitational attraction in terms of the curvature of space-time
4) Classical Electrodynamics: electromagnetic interaction
5) Fermi theory: weak interaction
6) Quantum Electrodynamics (QED): quantum version of electromagnetic interaction
7) "Standard Model" of elementary particle physics:[5]

   (7.1) Salam-Glashow-Weinberg theory: electroweak interaction (unified weak and electromagnetic interaction)
   (7.2) Quantum Chromodynamics (QCD): strong interaction

8) Quantum Gravity (String theory or Loop Quantum Gravity): all four forces in a unified way.

That some interactions appear repeatedly in the above list is mainly because they are dealt with in different ways in subsequent theories, such as Newtonian physics and Relativity Theory, or classical and quantum theories. Also in the background is the concept of "effective field theories," which I will discuss in more detail below.

We have seen above that there can be no mechanistic explanations for fundamental interactions. However, for a number of independent reasons, mechanistic reasoning may nevertheless play a role in specific parts of fundamental physics. To see why this possibility is not in contradiction with the above general line of argument, I propose to distinguish four sorts of entities that physical theory seeks to characterize:

i) *Fundamental laws* (at their time): Newton's law of gravitation, Maxwell's equations, Einstein's field equations in General Relativity Theory, the Schrödinger equation, the field equation in Quantum Electrodynamics.
ii) *Phenomena that are immediate consequences of fundamental laws*: Electromagnetic waves (e.g. visible light) in Maxwell's Theory, Heisenberg's uncertainty relations in Quantum Mechanics, black holes in General Relativity Theory, pair creation in Quantum Field Theory.
iii) *Basic composite systems*: Composite systems whose immediate constituents are fundamental.
iv) *Higher-order composite systems*: Composite systems whose constituents are themselves composite systems.

In actual cases, the distinction between (i) and (ii) as well as that between (iii) and (iv) may be controversial. Nevertheless, I think in principle it is comprehensible and justified and it has an important bearing on our present question, since immediate consequences of fundamental laws, i.e. (ii), make up large parts of fundamental physics, and the treatment of higher-order composite systems (iv) is also partly a topic in fundamental physics. In the following I will consider to what extent mechanistic reasoning is in fact relevant in the different parts of fundamental physics.

3. Fundamental physics: relativity theory and quantum physics

In Classical Mechanics, forces remain largely unspecified. Gravitation is the only force for which we have a specific fundamental law. We argued above that gravitational attraction *is*—ontologically—a basic mechanism, but it cannot be *explained* mechanistically. The same applies to electromagnetic forces. In Special and General Relativity the situation is less straightforward. Length contraction and time dilation of speeding objects are two closely connected phenomena, which are crucial for the development of twentieth-century physics. What is the nature of the explanation of these phenomena? Is there a mechanism that leads to the decreased length of a speeding spaceship? The received view today is that the correct explanation in terms of Einstein's Special Theory of Relativity from 1905 is not mechanistic. However, historically this was not the only answer. The Dutch physicist and 1902 Nobel Prize winner Hendrik Lorentz famously proposed a very different theory. Moreover, reconsidering this alternative proposal, one may question the orthodox view of Special Relativity, too (see Brown 2005). This turns the question of whether the correct explanation of length contraction and time dilation is mechanistic into a controversial one. Even if there should not be any doubt about the correct answer in the end, the discussion is still worth pursuing because it sheds an interesting light on the general question concerning the scope of mechanistic explanations in physics.

The crucial aspect about the Lorentz theory in our context is that it supplies a *dynamical* account of length contraction. The length of a speeding spaceship has really decreased. So why does it not get tight in the spaceship?—Because everything in the spaceship got smaller, too. In particular, all measuring devices like rods will also be smaller. It is like a big conspiracy: Everything has changed but we do not notice this change because everything we could use to detect this change has itself changed exactly such that measurements will not tell us of any change. The conspiracy flavor is one reason why most physicists have rejected the Lorentz theory. Nevertheless, it has its attraction and that is probably because it tells a dynamical and thereby potentially a *mechanistic* story of why relativistic effects occur. Spaceships and rods shrink because of how their micro constituents behave and of course interact.

In Einstein's Special Theory of Relativity the explanation of length contraction and time dilation is of a completely different nature. It is *kinematical*, i.e. it has to do with the very *geometry* of space-time and not with anything specific about the behavior of objects due to their properties and their interaction with physical fields. As Janssen (2009: 28) puts it:

> Special relativity is completely agnostic about what inhabits . . . Minkowski space-time. All the theory has to say about systems inhabiting/carrying Minkowski space-time is that their spatio-temporal behavior must be in accordance with the rules it encodes. Special relativity thus imposes the kinematical constraint that all dynamical laws must be Lorentz invariant. This property transcends the individual laws.

Janssen nicely compares the geometry of space-time with the rectangularity of paintings, which puts certain restrictions on paintings and which is realized in paintings, but which has nothing to do with the nature of the individual paintings. The upshot is that there are certain kinds of explanations in fundamental physics that genuinely show why certain phenomena occur (as opposed to just describing what occurs), but which are not in any way mechanistic. Special Relativity really *explains* length contraction and time dilation but the explanation does not rest on any physical processes in space and time but instead on the geometrical properties of space-time itself, regardless of any specific objects in space-time. Felline (2015) proposes to rate this as a "structural explanation" because its rests on the geometrical *structure* of space-time rather than on any mechanisms in space-time. Moreover, Felline argues, this shows that a lack of mechanistic explanations in fundamental physics does not entail a lack of explanations *tout court*.

Examples such as the one we just discussed, and many others, are of paramount importance in a recent debate on non-causal explanations because it is surprisingly hard to draw a clear line between causal and non-causal explanations, and the way they work, in a general way (see Baker 2005; Batterman 2010; Lange 2013; Skow 2014; Pincock 2015). Counting only explanations as causal that explicitly cite causes would arguably be too restrictive. Therefore, Lange (2013: 493) proposes the broader notion that an explanation is causal if "it explains by virtue of describing contextually relevant features of the result's causal history or, more broadly, of the world's network of causal relations." In the light of this characterization, the way in which SRT explains why the Lorentz transformations hold, or more specifically, the occurrence of length contraction and time dilation is clearly non-causal. The explanation rests on spatiotemporal symmetries and the principle of relativity, and these do not describe "the world's network of causal relations," but rather impose constraints on the (causal) laws of nature (Lange 2013: 494). This is exactly what Janssen metaphorically compared with the rectangularity of paintings.

The situation seems different when we consider gravitation in Einstein's General Theory of Relativity (GRT): Masses *affect* the curvature of space-time. This curvature in turn *causes* bodies to move in a certain way. This behavior appears to be due to gravitational forces between bodies but in reality they just follow their geodesics, i.e. the shortest way from A to B, given the curved space-time in which they live. Thus it seems reasonable to say that GRT explains the apparently gravitational behavior of massive bodies in a mechanistic fashion, namely by the interaction between large masses, the bendable space-time and massive bodies in question.

Now let us turn our attention to quantum physics. The most serious obstacle for the interpretation of quantum mechanics is the fact that the standard formalism is completely silent about the dynamical process that leads to an observable measurement outcome. Apparently, such a process would have to be one that leads from an initial state of the composite system of measurement apparatus and object to be measured via their interaction to a final determinate state of (at least) the pointer of the measurement apparatus. Therefore, one could rephrase our first statement by saying it is the lack of a *mechanism* leading to determinate measurement outcomes that blocks the interpretation of standard quantum mechanics. Note that physically there is nothing specific about measurements in a laboratory. The problem is generic. In general the dynamical law of QM does not lead to determinate values, in contrast to the apparent occurrence of classical properties everywhere in nature, be it in the laboratory or in the wild. However, it is customary to phrase the problem in terms of measurements. Alternatively, one could say that quantum measurements happen everywhere and all the time, i.e. not only in the laboratory. Using the famous Schrödinger cat example: The cat itself is the measurement apparatus already. There is no need for an experimenter who observes the result. In particular, it has nothing to do with a *conscious* observer.[6]

Unfortunately, whatever the sought-after mechanism that leads to determinate "measurement" outcomes would look like, it seems inevitable that it is in conflict with the standard assumption that the Schrödinger equation specifies the universally valid dynamical law for the time evolution of quantum systems. The reason is that there is usually a whole spectrum of different possible measurement outcomes and the *linear* Schrödinger time evolution cannot accommodate the fact that in a measurement process the spectrum of possibilities seems to collapse into one actual outcome.

Thus in the context of the standard Hilbert space formalism of quantum mechanics, the prediction of certain measurement outcomes does not have any mechanistic explanation. However, this is one of the issues in modern physics that count as not completely understood. And every approach that is a candidate for solving the measurement problem either supplies a mechanistic explanation of definite measurement outcomes—where mechanism is understood in the sense of, e.g., Machamer, Darden, and Craver (2000) and Glennan (1996, 2002, forthcoming)—or shows that there is some physical process that leads to it—which fits with Dowe's (2000) notion of mechanisms.

The oldest proposal for solving this conflict is to simply postulate that there is a second kind of dynamical law that is only valid for the apparent collapse in a measurement process. The disadvantage of this move is that it remains unclear why measurement processes should differ so fundamentally from other physical processes. Other proposals are more creative, but they all agree in that it is inevitable to give up some dear belief (see Maudlin 1995). Ghirardi, Rimini, and Weber give up the standard *linear* Schrödinger equation and replace it with a stochastic version—paying the price of introducing *ad hoc* two new natural constants, which cannot be measured independently. Bohm gives up the assumption that the quantum state is *complete* and uses the always determinate particle positions as the additional hidden parameters—but in doing so uses a preferred frame, which causes a conflict with Special Relativity. The many-worlds interpretation takes the plurality of possible measurement outcomes as reality, but only in different causally separated worlds. Thus it gives up the assumption that a measurement process actually leads to one determinate measurement outcome. The many-worlds interpretation, too, has big downsides, however: the way it tries to account for the probabilities for different measurement outcomes in terms of decision theory is highly controversial. Moreover, it severely violates the principle of ontological parsimony.

What all these proposals have in common is that they try to give—in very different ways—a mechanistic explanation for the occurrence of determinate measurement outcomes. Since each of these proposals has its merits and advocates but also serious drawbacks, we currently have a certain stalemate. However, there is at least something that most people agree upon as being at least a part of the solution for the conflict between quantum mechanics and the classical appearance of our world: namely, decoherence. In Kuhlmann and Glennan (2014) we argue that decoherence is of crucial importance for showing that quantum mechanics does not generally undermine the "new mechanical philosophy."

One of the outstanding features of decoherence and arguably one main reason why it attracts attention from so many very different camps is that it is a real physical process. Schlosshauer (2007) offers a comprehensive exposition of the theoretical and experimental aspects of decoherence. It is not a postulate, a story, or an *ad hoc* addition but an observable and analyzable mechanism. Moreover, one can fully describe this mechanism within the limits of QM. To be sure, decoherence does not solve the quantum measurement problem, but it is discernible how it may contribute at least something to its solution—in fact, whereas it is essential for the many-worlds interpretation, the role of decoherence is not so straightforward for other interpretations. The main idea of decoherence is that, because of a suitable interaction of a quantum system with a macroscopic environment, the troublesome superpositions (e.g. of a dead and a live cat) are effectively no longer observable even though they do not in fact disappear. This non-observability is not just an illusion; it is grounded in solid physical circumstances. In principle, the superpositions are not only still there but they are even enhanced by decoherence. However, they are spread outside of the system in question such that it is virtually impossible to recover them in a detectable manner.

4. Fundamental physics of today

The list of candidates for basic mechanisms in fundamental physics (in section 1) contained numerous duplications. As we saw, many of these duplications are due to the transitions either from Newtonian physics to Relativity Theory or from classical to quantum theories. However, there is also another explanation of these duplications, which is less known to non-physicists but which is crucial for contemporary physics. Most of the above theories are regarded as so-called "effective field theories" today. Hartmann (2001) and Castellani (2002) discuss effective field theories from a philosophical point of view, in particular in the context of reduction. Very generally, physicists call some quantity X "effective" if it is what we actually observe, while there is some underlying quantity, let's call it Y, that can have completely different characteristics. For instance, the underlying contributions to a potential can be of a very different kind than the overall potential we effectively measure.

One particularly important "mechanism"[7] that brings us from the underlying hidden Y to the observable X is so-called "spontaneous symmetry breaking"—a notion that was explicitly introduced only in the 1960s and turned out to be crucial in a huge spectrum of otherwise largely unrelated contexts. The key idea of spontaneous symmetry breaking is that we can have underlying laws (our "Y") that are symmetric, whereas the ground state, which we observe (i.e. the "X"), is not symmetric.

In effective field theories the X we "observe"[8] is one particular field theory, e.g. Quantum Electrodynamics (QED) of the Standard Model, and the underlying Y can be something that is not even a field theory at all, like string theory. So why do we "see" the electromagnetic quantum field and not the supposedly underlying theory? The reason is the energy regime that is relevant in the world we are dealing with today. The conditions in our world are quite different from the ones after the big bang because the universe has cooled down considerably. In this cooling spontaneous symmetry breaking occurred repeatedly, so that we do not directly see the "true physics" any more. So the idea is that the laws as such have not changed, but we effectively see different laws for different temperatures.

Obviously these considerations have an immediate bearing for our search of basic mechanisms. The effective field theory lesson would be that there are no basic mechanisms simpliciter but only basic mechanisms for a given energy regime. For anyone but physicists working on fundamental theories, this is good enough, of course, because we have no handle on the energy regime we live in today anyway.

Another important—although yet speculative—case of an effective field is General Relativity Theory (GRT). Famously GRT is a field theory, i.e. space and time are at the root of the theory. However, space and time may prove to be only "effective" or emergent entities: in a future theory of Quantum Gravity, space and time may no longer be basic notions but rather macroscopic notions that emerge in an approximate way from the underlying microscopic theory. If this speculation should turn out to be true and if we think that mechanisms are spatiotemporal entities, then it seems possible that there is no room for basic mechanisms in fundamental physics.

Thus, effective field theories may pose two threats to the mechanistic program. The first threat is that there is no most fundamental theory and therefore no basic mechanisms. In reaction to this threat one may simply welcome the infinite tower of effective field theories and say that *every* theory has a mechanistic grounding, namely in terms of the respective underlying theory on another energy regime.

The second threat is that space and time no longer appear in fundamental physics but emerge as macroscopic notions. Therefore, fundamental physics would not be dealing with any mechanisms—provided one understands mechanisms as spatiotemporal entities. Two possible reactions by mechanists come to mind. One possibility would be to reconceptualize mechanisms in a way that they do not need to be spatiotemporal. An alternative reaction would be that emergent space-time is not really a threat for the very notion of basic mechanisms because basic *mechanisms* simply appear only on one level higher where space-time is already manifest.

5. Physics of composite systems

Two things may block mechanistic explanations for composite systems. First, the composition laws of quantum systems could preclude mechanistic explanations for the behavior of composite systems. Second, complex systems dynamics may be incompatible with the micro-reductive nature of mechanistic explanations, roughly to the same extent that complex systems exhibit emergent behavior. I will argue that both concerns should disappear on closer scrutiny.

Let us turn to the first concern. Although classical and not quantum physics is used for building cars and refrigerators, the universal validity of quantum physics casts doubts on the theoretical legitimacy of mechanistic explanations even on macroscopic scales. Of course, any consideration on this issue ultimately hinges on the pending solution of the quantum measurement problem. However, even before this is accomplished, one may partially alleviate the concern that QM undermines mechanistic reasoning. In Kuhlmann and Glennan (2014) we identify three features of quantum mechanics that could conflict with the mechanistic approach, namely (A) indeterminacy of properties, (B) non-localizability of quantum objects, and (C) quantum holism. Although (B) is just a special case of (A), it deserves separate attention, first, because well-localized parts in particular may appear prerequisite for mechanisms and, second, because one can dissolve worry (B) even without a general solution to (A). The main point is that causal and not spatiotemporal organization is crucial for mechanisms.[9]

Quantum holism (C) is another issue that may endanger a mechanistic approach to composite systems. Since the mechanistic conception of explanation rests on the reductionist idea that the behavior of a composite system is reducible to the activities and interactions of its parts, it may seem that the failure of reductionism due to quantum holism may infect the mechanistic program too. However, this is not the case because mechanistic explanations are concerned with the *dynamics* of compound systems and not with the question whether the states of the subsystems determine the state of the compound system at a given time (cf. Hüttemann 2005). To determine how a compound quantum system evolves in time, all we need to know are the relevant terms for the subsystems and their respective interactions, and these terms are simply added up.

We have argued in section 3 that as we approach macroscopic scales, mechanistic explanations usually become less problematic. However, here too, things are not as straightforward as one might expect. Arguably the most interesting issue is dynamical complexity, which occurs in two flavors (see Strevens 2016). First, simple macro behavior can result from *convoluted* micro dynamics. Second, *simple* micro dynamics can produce surprisingly intricate macro behavior.[10] Both cases do not appear to be amenable to mechanistic reasoning because micro dynamics and macro behavior do not fit with each other: In the first case, the interactions between the system's parts seem to be too chaotic to explain the simple behavior of the macro variables. In the second case, the reverse obtains, since the interactions between the system's parts seem to be too simple to explain the non-trivial macro behavior.

Let us take a closer look at both cases. The first case characterizes quite generally the way in which statistical mechanics underlies thermodynamics: thermodynamics deals with the relation of various macroscopic quantities such as pressure, density, and temperature. One crucial difference between (classical) thermodynamics and statistical mechanics is that only the latter deals with probabilities. This fact not only represents a qualitative difference between micro dynamics (statistical mechanics) and macro behavior (thermodynamics), but also opens a venue for bridging the gap: convoluted micro dynamics can lead to simple macro behavior because only certain probabilistic features of the micro dynamics matter.[11] Whether statistical mechanics (micro dynamics) actually does the trick of explaining thermodynamical macro behavior depends on whether thermodynamics can be reduced to statistical mechanics.[12] Strevens (2016, section 2.3) proposes a so-called *enion probability analysis* which shows how "probability distributions over microlevel properties . . . can be used to explain stabilities or simplicities in a gas's macrolevel properties."

But is this a *mechanistic* explanation? While Strevens is silent on this matter, Weber and Lefevere (2014) propose a negative answer. They rate the above explanation (not specifically Strevens' account of it) as an "aggregation explanation," which they see as a different type of micro-explanation than mechanistic explanation:

> We use the term *aggregation explanation* to refer to explanations which explain the behaviour of the macro-systems . . . by means of the component parts, the activities of these parts, organisation and *random distribution of behaviour of components*. This means that, compared to mechanistic explanations, aggregation explanations have an extra ingredient [so that they should be rated as] non-mechanistic micro-explanations.
>
> *(2014: 43f)*

However, if randomness is merely an *extra* ingredient, then why should aggregation explanations no longer be mechanistic? It seems more natural to say that they are a sub-class of mechanistic explanations (see Chapters 7 and 13).

Finally, let us consider complex systems that have non-trivial macro behavior despite their simple micro dynamics (see Chapter 20 for biological examples). One example is the formation of macroscopic convection cells in a container with a viscous fluid that is uniformly heated from below. Not a few rate such self-organized coordination as *emergent* behavior, and then argue that this makes mechanistic reasoning inapplicable.[13] For complex systems, the occurrence of emergent behavior has been claimed in various overlapping contexts. The most important ones involve the non-linear connection between micro causes and macro behavior (resulting in chaotic behavior), feedback with top-down causation, "enslavement" in *Synergetics*, the dynamical role of the so-called order parameter (e.g. in superconductivity), and universal behavior in phase transitions across diverse materials.[14] In claims that these phenomena in complex systems are emergent and therefore defy mechanistic explanation, the general idea is roughly this: It is not the arrangement of the micro details that accounts for the behavior of the whole system but rather something that only emerges when the composite system evolves, and which acquires a certain life of its own. To show what this can mean in more concrete terms, I want to pick out two issues, namely universal behavior (or short "universality") and enslavement.

Universal behavior arises in phase transitions, for instance from a liquid to a vaporous state (or phase) or from an unmagnetized to a ferromagnetic state of a piece of metal. Phase transitions in radically diverse materials can exhibit remarkably similar or even identical properties. Numerically, universality shows up in (almost) identical "critical exponents," which characterize a system's behavior in phase transitions. Whereas similar behavior of materially diverse tables is easy to explain by reference to their geometry, universal behavior in phase transitions is more intricate. A satisfactory explanation was only given in the 1970s by Kenneth Wilson's "renormalization group method." Batterman (2002, 2010) argues that renormalization group explanations are a genuine new type of explanation (involving so-called "infinite idealizations"), which are not mechanistic (at least not in the traditional sense) because their essence consists in "systematic methods for throwing details away" (Batterman 2002: 18), i.e. exactly those things that seem essential for a mechanistic approach. This claim prompted a number of critical responses (Butterfield 2011; Norton 2012) as well as largely sympathetic ones (Morrison 2012). In any case, it initiated a fruitful debate that is by no means finished. My own view is that renormalization group explanations should be rated as causal and, more specifically, as mechanistic because showing that *most* micro details are irrelevant does not mean that *none* are relevant—in accordance with Lange's (2013: 493) above-mentioned broader notion of causal explanation. The procedure shows us exactly which (structural) aspects of the micro details do matter. In Kuhlmann (2014) I then argue—in a related context—that one needs a notion of *structural mechanisms* to account for this fact. In this account, the explanatory weight rests on certain structural features of the dynamical interactive organization and not on the details of the initial set-up.

6. Summing up

One naturally assumes that physics gives us the ultimate basis for mechanistic reasoning in the special sciences. However, it is not evident to what extent physics itself deals with mechanisms. One reason for the potential lack of mechanisms is that on the fundamental level mechanistic explanations come to a natural end. Arguably the most coherent way to deal with this fact is to say that fundamental physics gives us the basic mechanisms, which have no further mechanistic grounding. One recent development in physics that could endanger this view is the notion of effective field theories, according to which there may be no most fundamental level. Moreover, space and time, which seem required for mechanisms to operate in, may be emergent rather than basic features of the universe. Apart from these more recent developments, conventional quantum physics already may undermine mechanistic reasoning for a number of reasons. Finally, the mechanistic approach seems to be at odds with the behavior of complex systems, because either the convolution of micro details prevents the explanation of simple macro behavior, or the simple organization on the micro level leaves the intricate macro behavior inexplicable. I have shown that one can dispel or at least alleviate each of these worries.

Notes

1 I wish to thank Elena Castellani, Roman Frigg, and Carl Hoefer, as well as the editors Stuart Glennan and Phyllis Illari for many valuable comments on an earlier draft.
2 As we will see below, this is controversial. The threat are so-called "effective field theories" which could result in an infinite cascade of ever more fundamental theories.
3 The difference is that Glennan explicitly rejects the regularity account of laws and causation. In our context, the inexplicability of fundamental laws, this difference does not matter.
4 Felline (forthcoming) argues that the lack of a mechanistic underpinning for fundamental phenomena does not necessitate an ancillary account of causation for the fundamental level because according to her—contrary to widespread opinion—the causal anti-fundamentalism in Glennan's approach is in fact compatible with the idea that there is no causality at the fundamental level.

5 The so-called standard model of particle physics from the 1980s is still the valid theory of elementary particles and interactions. In this model the "Higgs mechanism" plays a crucial parts because—according to common gloss—it explains why particles have mass. However, on closer scrutiny the "Higgs mechanism" does not fulfill the requirements for being a mechanism in several respects. Instead, the "Higgs mechanism" rather seems to be the core ingredient of a *non-causal explanation* (see Kuhlmann forthcoming).

6 In the early period of thinking about the quantum measurement problem it was quite common to see the act of a *conscious* observer as an indispensable ingredient. Partly because of the formulation of various potential solutions that do "Quantum Theory Without Observers" (BI), there is hardly anyone today in the relevant scientific community who assigns any physical significance to a conscious observer.

7 Below I will address the question of whether—or under which conditions—this is really a mechanism.

8 One cannot *observe* a theory, of course. What I mean by this way of talking is that the world appears to us as if this was the correct underlying theory.

9 In Kuhlmann (2015) I show in detail what this means for the quantum-mechanical explanation of laser light.

10 Note that often only this second case goes under the title "complex systems."

11 Note that the role of probabilities depends crucially on the framework. While they are essential in the Gibbsian approach, they are something like epiphenomena in the Boltzmannian approach. See Frigg (2008) for details.

12 See Sklar (2015, section 6) for a non-technical introduction, and Dizadji-Bahmani, Frigg, and Hartmann (2010) for a defense of the reductive account.

13 To my knowledge Broad (1925, chapter 2) introduced the distinction of mechanistic versus emergent explanations. In any case, his discussion was arguably the most influential one.

14 Batterman (2002, 2010) and Morrison (2012, 2014) are among the most active contributors to this debate.

References

Baker, A. (2005) "Are There Genuine Mathematical Explanations of Physical Phenomena?," *Mind* 114: 223–38.

Batterman, R. W. (2002) *The Devil in the Details*, Oxford: Oxford University Press.

Batterman, R. W. (2010) "On the Explanatory Role of Mathematics in Empirical Science," *British Journal for the Philosophy of Science* 61: 1–225.

Broad, C. D. (1925) *Mind and Its Place in Nature*, London: Kegan Paul.

Brown, H. (2005) *Physical Relativity: Space-Time Structure from a Dynamical Perspective*, Oxford: Oxford University Press.

Butterfield, J. (2011) "Less Is Different: Emergence and Reduction Reconciled," *Foundations of Physics* 41: 1065–135.

Castellani, E. (2002) "Reductionism, Emergence, and Effective Field Theories," *Studies in History and Philosophy of Modern Physics* 33: 251–67.

Dizadji-Bahmani, F., R. Frigg and S. Hartmann (2010) "Who's Afraid of Nagelian Reduction?," *Erkenntnis* 73: 393–412.

Dorato, M., and L. Felline (2011) "Scientific Explanation and Scientific Structuralism," in A. Bokulich and P. Bokulich (eds): *Scientific Structuralism*, Dordrecht: Springer, pp. 161–76.

Dowe, P. (2000) *Physical Causation*, Cambridge: Cambridge University Press.

Falkenburg, B., and M. Morrison (eds) (2015) *Why More Is Different. Philosophical Issues in Condensed Matter Physics and Complex Systems*, Berlin: Springer.

Felline, L. (2015) "Mechanisms Meet Structural Explanation," *Synthese*, First online: 30 May 2015: 1–16.

Felline, L. (forthcoming) "Mechanistic Causality and the Bottoming-out Problem," in Felline et al. (eds.) *New Developments in Logic and Philosophy of Science*, London: College Publications.

Frigg, R. (2008) "A Field Guide to Recent Work on the Foundations of Statistical Mechanics," in D. Rickles (ed.): *The Ashgate Companion to Contemporary Philosophy of Physics*, London: Ashgate, pp. 99–196.

Glennan, S. S. (1996) "Mechanisms and the Nature of Causation," *Erkenntnis* 44: 49–71.

Glennan, S. S. (2002) "Rethinking Mechanistic Explanation," *Philosophy of Science* 69: S342–S353.

Glennan, S. S. (forthcoming) *The New Mechanical Philosophy*, Oxford: Oxford University Press.

Hartmann, S. (2001) "Effective Field Theories, Reductionism, and Explanation," *Studies in History and Philosophy of Modern Physics* 32: 267–304.

Hüttemann, A. (2005) "Explanation, Emergence and Quantum-Entanglement," *Philosophy of Science* 72: 114–27.

Janssen, M. (2009) "Drawing the Line Between Kinematics and Dynamics in Special Relativity," *Studies In History and Philosophy of Modern Physics* 40: 26–52.

Kuhlmann, M. (2014) "Mechanisms in Dynamically Complex Systems," in P. McKay Illari, F. Russo, and J. Williamson (eds.): *Causality in the Sciences*, Oxford: Oxford University Press, 2011, pp. 880–906.

Kuhlmann, M. (2015) "A Mechanistic Reading of Quantum Laser Theory," in B. Falkenburg and M. Morrison (2015), pp. 251–71.

Kuhlmann (forthcoming) "Is the Higgs Mechanism a Mechanism?"

Kuhlmann, M. and S. Glennan (2014) "On the Relation Between Quantum Mechanical and Neo-Mechanistic Ontologies and Explanatory Strategies," *European Journal for Philosophy of Science* 4: 337–59.

Lange, M. (2013) "What Makes a Scientific Explanation Distinctively Mathematical?," *British Journal for the Philosophy of Science* 64: 485–511.

Machamer, P., L. Darden, and C. F. Craver (2000) "Thinking about Mechanisms," *Philosophy of Science* 67: 1–25.

Maudlin, T. (1995) "Three Measurement Problems," *Topoi* 14: 7–15.

Morrison, M. (2012) "Emergent Physics and Micro-Ontology," *Philosophy of Science* 79: 141–66.

Morrison, M. (2014) "Complex Systems and Renormalization Group Explanations," *Philosophy of Science* 81: 1144–156.

Norton, J. (2012) "Approximation and Idealization: Why the Difference Matters," *Philosophy of Science* 79: 207–32.

Pincock, C. (2015) "Abstract Explanations in Science," *The British Journal for the Philosophy of Science* 66: 857–82.

Schlosshauer, M. (2007) *Decoherence and the Quantum-to-Classical Transition*, Heidelberg: Springer.

Sklar, L. (2015) "Philosophy of Statistical Mechanics," in E. N. Zalta (ed.): *Stanford Encyclopedia of Philosophy* (Fall 2015 Edition), available at https://plato.stanford.edu/entries/statphys-statmech.

Skow, B. (2014) "Are There Non-Causal Explanations (of Particular Events)?," *British Journal for the Philosophy of Science* 65: 445–67.

Strevens, M. (2016) "Complexity Theory", in P. Humphreys (ed.): *Oxford Handbook of the Philosophy of Science*, doi: 10.1093/oxfordhb/9780199368815.013.35.

Weber, E., and M. Lefevere (2014) "The Role of Unification in Micro-Explanations of Physical Laws," *Theoria* 79: 41–56.

22

MECHANISMS IN EVOLUTIONARY BIOLOGY

Lane DesAutels

1. Introduction

In 1961, renowned Harvard biologist Ernst Mayr asked himself why the warbler he had been observing on the grounds of his New Hampshire summer home began its southward migration on the night of the 25th of August. The answer, he came to realize, was not a simple one. There were ecological causes: being an insect-eater, the warbler would starve in the New Hampshire winter. There were physiological causes: the warbler had an intrinsic capacity to sense the dwindling number of daylight hours. And there were genetic causes: the warbler's special sensitivity to environmental stimuli indicating the approach of colder climate was programmed into its very DNA. But on Mayr's view, these myriad reasons for the warbler's migration were really just of two kinds: the *proximate causes* dealing with the physiology of the warbler as it related to the photoperiodicity and air conditions in its environment, and the *ultimate causes* dealing with the bird's evolutionary history, the way its genetic constitution had been molded by natural selection over many thousands of generations (Mayr 1961).

Well beyond understanding the migration habits of the common warbler, Mayr's distinction between proximate and ultimate causes was to serve as grounds for demarcating the distinct explanatory magisteria of two different kinds of biology. On the one hand, *functional biology* seeks to understand proximate causes: "the functional biologist is vitally concerned with the operation and interaction of structural elements, from molecules up to organs and whole individuals. His ever-repeated question is 'How?'" (Mayr 1961, 1502). *Evolutionary biology*, on the other hand, lives squarely in the domain of ultimate causes:

> The animal or plant or micro-organism . . . [the evolutionary biologist] is working with is but a link in an evolutionary chain of changing forms, none of which has any permanent validity. There is hardly any structure or function in an organism that can be fully understood unless it is studied against this historical background.
>
> *(Mayr 1961, 1502)*

Rather than asking "How?", the evolutionary biologist's perpetual question is "Why?"

Mayr's ideas regarding functional vs. evolutionary biology have been the subject of renewed interest in the philosophy of biology literature. Some have argued that they have significant flaws (Laland et al. 2011, 2013; Calcott 2013); while others have attempted to support the spirit, if not the letter, of Mayr's conclusions (Ariew 2003; Scholl and Pigliucci 2014). I will not weigh in on the specifics of this ongoing debate, but I bring up Mayr's framework because it represents an influential and pervasive way of thinking about the explanatory scope of evolutionary biology—one that, if sound, has important implications for the role of mechanistic philosophy of science in evolutionary biology. Consider, for example, the following recent characterization of evolutionary biology from Futuyma's foundational text, *Evolution*:

> Evolutionary biology is a more historical science than most other biological disciplines, for it seeks to determine what the history of life has been and what has caused those historical events. It complements studies of the PROXIMATE CAUSES (immediate, mechanical causes) of biological phenomena—the subject of cell biology, neurobiology, and many other biological disciplines—with an analysis of the ULTIMATE CAUSES (the historical causes, especially the action of natural selection) of those phenomena.
>
> *(Futuyma 2013, 13)*

As seen here, evolutionary biology is still today conceived of as an essentially historical science devoted to understanding ultimate causes. And it is still conceived of as distinct from fields like cell biology, molecular biology, developmental biology, or genetics whose job is to delineate immediate, mechanical, proximate causes.

What does any of this have to do with the philosophy of mechanisms and its role in evolutionary biology? Here is the rub. Implicit in both Mayr's and Futuyma's ideas are the following two claims:

i) Evolutionary biology takes ultimate rather than proximate causes as its subject matter.

and

ii) Mechanistic causes are proximate rather than ultimate.

But if we accept (i) and (ii), it would seem to follow that

iii) Mechanistic causes are not the subject matter of evolutionary biology.

Ergo, we might extrapolate,

iv) *The mechanistic approach to philosophy of science has nothing* (at least directly) *to offer the study of evolution.*

The purpose of this chapter is to explore whether these conclusions are warranted—that is, to explore whether, and to what extent, the mechanistic philosophical framework has any purchase in evolutionary biology.

There are a number of ways we might go about this. One way would be to question whether there *are* any immediate mechanistic causes driving evolution. Is natural selection aptly understood as a mechanism? Is drift? Mutation? Let us refer to this as the *metaphysical question* (MQ):

(MQ) Are any of the primary causal drivers of evolution aptly understood as mechanisms?

If any of the primary causal drivers of evolution are aptly understood as mechanisms, and the study of the primary causes of evolution is within the subject matter of evolutionary biology, then it would seem that—contra (iii)—there *is* a straightforward sense in which evolutionary biology should take mechanistic causes as (at least partly constitutive of) its subject matter.

Beyond the metaphysical question, however, there is also an important *epistemological question* (EQ):

(EQ) When applied to evolutionary biology, does the mechanistic philosophical framework help add to our understanding of evolution?

The answer to this question, I suggest, may be quite independent from the first.[1] Regardless of whether there are any mechanisms of evolution in the robust metaphysical sense, it might still be the case that the mechanistic philosophical framework can offer some important explanatory, pragmatic, or otherwise strategic resources that, when applied to the field of evolutionary biology, help *illuminate* the phenomenon of evolution. If so, then—contra (iv)—there may indeed be a valuable role for the philosophy of mechanisms in the study of evolution.

In what follows, I will consider responses to both the metaphysical and epistemological questions. In section 2, I briefly lay out some of the central metaphysical features of mechanisms as they have been characterized in recent literature. In section 3, I examine whether these features of mechanisms are aptly understood as being instantiated by three of the primary causal drivers of evolution: natural selection, drift, and mutation. In section 4, I expound a few of the strategic roles played by mechanistic thinking and hint at the ways they might apply to the field of evolutionary biology regardless of whether there turn out to be any mechanisms driving evolution. I conclude, in section 5, that—on *at least some* philosophical characterizations of mechanisms, and with regard *to at least some* of the central processes of evolutionary biology—there is room for affirmative answers to both the metaphysical and epistemological questions. And thus, there may indeed be an important place for mechanisms in evolutionary biology.

2. Central metaphysical features of mechanisms

In this section, I highlight five central metaphysical features of mechanisms: components, operations, organization, function, and regularity.[2] Many of these aspects of mechanisms have been discussed by other authors in this volume (see especially Chapters 7, 8, 9, 10, 12), so I will not spend much time here. The point is to say just enough about what makes mechanisms the kinds of things that they are to be able to arbitrate the question of whether natural selection, drift, or mutation might qualify.

F1: components

Mechanisms have *components* or *constitutive parts*. Machamer, Darden, and Craver (MDC) call these "entities" and claim that, along with their "activities," they form the two aspects of the "dualist ontology" of mechanisms (MDC 2000, 3). Stuart Glennan refers to these as "parts" (Glennan 1996, 52, 2002, S344) and Bechtel and Abrahamsen as "component parts" (Bechtel and Abrahamsen 2005, 47) (see Chapter 9 for further discussion about mechanism components).

F2: operations

Mechanisms' component parts perform *operations*; they *do* things. In MDC's terms, these are the "activities" performed by the entities—the second aspect of mechanisms' dualist ontology (MDC 2000, 3). Glennan characterizes this aspect of mechanisms as the "interactions" between parts (Glennan 1996, 52, 2002, S344) and Bechtel and Abrahamsen as "component operations" (Bechtel and Abrahamsen 2005, 47).

F3: organization

Mechanisms must be—in some sense—*organized*. For MDC, this means that mechanisms have "start-up" and "termination" conditions, and that mechanisms' entities must be (1) located, (2) structured, (3) oriented; and a mechanism's activities must have (4) temporal order, (5) rate, (6) duration (MDC 2000, 3). Bechtel and Abrahamsen speak of a mechanism as a "structure" performing a task by virtue of its component parts, operations, and their "organization" (Bechtel and Abrahamsen 2005, 47). And Glennan speaks of mechanisms as "complex systems" underlying a given phenomenon (Glennan 1996, 52, 2002, S344).

Unlike the merely nominal disagreements mentioned in (F1) and (F2), however, there appear to be substantive differences in the organizational requirements placed on mechanisms between contemporary mechanists. Bechtel and Abrahamsen (2009, 2010, 2013) argue that the MDC organizational requirements are untenable if we wish to characterize feedback mechanisms (e.g., the mechanisms responsible for circadian rhythms) or for mechanisms for the maintenance of equilibrium (e.g., regulatory mechanisms)—as these phenomena have no clear start or termination conditions. To accommodate single causal chains as mechanisms (e.g., the 1980 death of French literary critic Roland Barthes when he was struck by a laundry truck while crossing a Paris street), Glennan goes even further toward loosening the organizational requirement by arguing for a notion of "ephemeral mechanism" according to which the configuration of a mechanism's parts may be "short-lived" and "non-stable" (Glennan 2010, 260).

F4: function

Mechanisms have to be *set up to do something*; they must carry out a *function*. Glennan (1996) points out that "one cannot even identify a mechanism without saying what it is that the mechanism does" (Glennan 1996, 52). For MDC, mechanisms have a function "to the extent that the activity of a mechanism as a whole contributes to something in the context that is taken to be antecedently important" (MDC 2000, 6). And Bechtel and Abrahamsen state explicitly that "a mechanism is a structure performing a function" (Bechtel and Abrahamsen 2005, 47).

Once again, however, there appear to be substantive disagreements about how to understand mechanistic function. As in the more general debate about functions in biology, there seem to be roughly two camps: proponents of *normative* notions of function (e.g., Millikan's (1989) and Neander's (1991) notion of "proper function"—notably defended in the context of mechanisms by Garson (2011)) vs. *non-normative* causal notions of function (e.g., Cummins' (1975) idea of "causal-role function"—notably applied to mechanisms by Craver (2001)). (See Chapter 8 for a detailed discussion of mechanistic function.)

F5: regularity

Mechanisms carry out their function in a *regular* fashion. For MDC, the entities and activities constitutive of a given mechanism "work always or for the most part in the same way under the same conditions" (MDC 2000, 3). Bechtel and Abrahamsen (2005) say that mechanisms are structures responsible for a given "phenomenon"—where "phenomenon" is here understood in the regularist sense advocated by Bogen and Woodward (1988).

As with (F3) and (F4), there is again substantive disagreement between mechanists about the degree to which regularity should be conceived of as a metaphysical prerequisite for mechanisms. Machamer (2004) as well as Bogen (2005) have argued that regularity should be struck from the MDC characterization of mechanism on the grounds that many of the phenomena targeted for mechanistic explanation (e.g., especially synaptic transmission) fail to achieve termination conditions much more often than they succeed. DesAutels (2011, 2015) argues that neither a fully regularist nor a fully irregularist characterization of mechanism is tenable, and a notion of stochastic mechanism must be developed. Holly Andersen (2012) defends a version of the MDC regularity requirement but develops a helpful taxonomy according to which mechanistic regularity comes in a variety of strengths and may be located in a variety of loci within a given mechanism. (See Chapters 11, 12, and 13 for further exposition of some of the issues surrounding mechanistic regularity.)

3. Do the primary causal drivers of evolution instantiate (F1)–(F5)?

Having now laid out the above central metaphysical features of mechanisms (F1)–(F5) as well as having drawn attention to a few places where contemporary mechanists disagree on the nature and strength of these various requirements, we can get on with exploring whether they are instantiated by some of the primary causal drivers of evolution: natural selection, drift, and mutation.

Natural selection

Provided there is heritable variation of fitness-relative traits among a population as well as competition for limited resources, nature tends to preserve those characteristics that afford their possessors the greatest chance to survive and reproduce, and it tends to reject those that do not. The result is that, over time, species become increasingly well matched to their respective environments—or else they go extinct. In its basic form, this is *natural selection*. Understood in this way, does natural selection instantiate (F1)–(F5)?

Regarding (F1) and (F2), I suggest that natural selection fares pretty well. Consider Darwin's finches as a paradigmatic case. In this example, natural selection seems clearly made up of *components*: finches with differing beak-lengths and seeds that are harder or easier to forage depending on beak-shape. And these componential entities undertake *operations*: foraging for seeds, differential reproduction, and predator avoidance among fitter finches. Barring general skepticism about the nature of part–whole relations—regarding which mechanistic mereology would hardly seem to be a special case—it seems entirely reasonable to conceive of natural selection as being composed of active entities.

Regarding (F3)–(F5), the story is more complicated. Whether natural selection is taken to meet the organization criterion set forth in (F3) depends on how strict we take the requirement to be. On the original MDC characterization of mechanism according to which mechanisms are to have definitive set-up/start and termination conditions, natural selection seems to fall short.[3] When would we say of Darwin's finches that the mechanism of natural selection begins or ends? It seems that any attempt to delineate such temporal boundaries would be arbitrary or ad hoc. Furthermore, unlike human-made mechanisms like clocks or toasters, there is not an obvious physical boundary around the mechanism of natural selection—which makes it hard to determine what its internal structure might be. That said, on more lenient versions of the organizational requirement, there do not seem to be any obvious organizational impediments for natural selection. If, like Bechtel and Abrahamson (2009, 2010, 2013), we allow for feedback mechanisms or mechanisms for the maintenance of equilibrium, neither of which have obvious start or termination conditions, then it seems we should not disallow natural selection from being a mechanism on the grounds that it lacks such conditions. Or if, like Illari and Williamson (2010), we cash out mechanistic organization in terms of functional individuation, then natural selection meets the organizational requirement just as well as paradigmatic mechanisms from molecular biology (e.g., protein synthesis). Or if, like Glennan (2010), we allow for any single causal chain to be a short-lived, non-stable, ephemeral mechanism, then it is even easier to see that natural selection passes muster.

What about (F4)? Does natural selection have a function? Is it set up to do something? Once again, these questions do not have straightforward answers. Whether there is sense to be made of teleology in evolution is an issue with a long and complicated history.[4] That said, even if we agree (as most do) that there is not teleological directionality to evolution—no end-goal it sets out to achieve—there may still be a sense in which natural selection has a function. Namely, *it is that which brings about adaptation*. And if we take on something like the causal-role understanding of biological function (à la Cummins (1975) and Craver (2001)), and grant that natural selection is that which plays the causal role of bringing about adaptation, then natural selection might well satisfy (F4).[5]

So how about (F5)? Does natural selection evince the regularity that is required for being a mechanism? Once again, whether we answer this in the affirmative depends on how one understands mechanistic regularity. If one requires fully deterministic output conditions from a process for it to count as a mechanism, then natural selection surely does not pass. For as Gould (1990) famously argued, if the tape of evolution were played back again and again, it would never turn out the same way twice. In this way, natural selection should be understood as probabilistic rather than deterministic. And Skipper and Millstein (2005) have argued that this probabilistic character of natural selection precludes it from meeting the regularity requirements set forth by MDC (2000). On the other hand, failure to behave fully deterministically does not preclude all notions of mechanistic regularity from being met. Barros (2008) argues that natural selection is a biased stochastic mechanism: one that is regular in that it succeeds in producing predicted outcomes over 50 percent of the time. DesAutels (2016) argues that, when we distinguish between process vs. product, internal vs. external, and abstract vs. concrete regularity, then natural selection escapes the Skipper and Millstein regularity critique just fine.

Drift

Drift happens when a population changes over time—where these changes are *not* the result of natural selection.[6] Consider, for example, a population of snails, half of which are pink and half are which are yellow.[7] Imagine further that the yellow snails are twice as fit as the pink ones because of their greater resistance to the sun. Scientists observing these snails expect, therefore, that the population of yellow to pink snails should increase from one half of the population to two thirds in the subsequent generation. However, suppose that after observing these snails for one generation, they find that the population of yellow snails actually decreases from one half to two fifths of the snail population. Because this change in the population is not due to natural selection (selection would have increased the relative proportion of yellow snails), they attribute this unexpected result to drift. Understood in this way, how does drift fare with regard to (F1)–(F5)?

Regarding (F1) and (F2), I suggest it remains reasonable to consider drift as comprising both components and operations. In our above example, there are pink and yellow snails engaged in usual survival and reproductive activities. There are also whatever entities and activities were responsible for the unexpected decrease in the population of yellow snails—perhaps a dangerous fungus to which only the yellow snails were vulnerable. So once again, modulo general mereological skepticism, there seems to be little problem with conceiving of this instance of drift as being composed of active entities.

Unfortunately, the story regarding (F3)–(F5) is even more difficult than it was with natural selection. For instance, it seems particularly difficult to conceive of a sense in which drift is *organized*—or that it has any particular *structure*. At least with natural selection, there are some commonalities between the sorts of relationships that occur among its participants (e.g., competition for limited resources or differential reproduction favoring fitter members of the population), and the participants must meet certain preconditions for natural selection to occur (e.g., variation in the population and heritability of these variations). On the other hand, because drift is an umbrella term catching all non-selective instances of population change over time, it seems much more difficult to delineate any common characteristics shared among all of its instances. And this, I suggest, makes it exceedingly difficult to understand drift as meeting any kind of organizational constraint. That said, there may still be some sense to be made of drift as a *causal* process.[8] (The decrease in population size of the yellow snails, though not caused by natural selection, was surely caused.) And if we follow Glennan (2010) in allowing for a conception of ephemeral mechanism that covers all instances of single causal chains, then it may well be that we could conceive of instances of drift as ephemeral mechanisms: i.e., mechanisms that are "short-lived" and "non-stable" like the one responsible for the death of Roland Barthes.

But does drift have a *function*? Is there something that drift is *set up to do*? Once again, this is tough to answer affirmatively. Unlike with natural selection which (at least on a causal-role notion of function) can be understood as that which brings about adaptation, drift does not seem to do any specific thing; it does not serve any particular unified causal role. Unless we allow for the function of drift to be the fixation of a given trait absent selection (a function that risks vacuity given the very definition of drift), it seems drift cannot meet (F4).

Which leaves (F5): *regularity*. Is there any sense to be made of drift operating regularly? Here, I suggest that the prospects are once again somewhat dim. Because of the aforementioned nature of drift as a catch-all for every instance of non-selective population change, it is near impossible to conceive of it as operating in a regular fashion. A population might drift because of disease, human encroachment, natural disasters, or just plain bad luck. And if mechanistic regularity requires, as MDC suggest, that mechanisms operate "always or for the most part in the same way under the same conditions," then we might think that drift appears to fall woefully short. That said, there may be more space for understanding drift as operating regularly than all of this might portend. For one thing, drift (at least according to many biologists) *tends to occur more often among smaller populations*. So in this sense, there are at least some conditions we can specify under which drift is more likely to occur. And if such conditions can be specified, then there may at least be a sense in which drift operates more regularly in some circumstances than others.

Mutation

Deoxyribonucleic acid (DNA) is composed of two polymers made up of nucleotides and a backbone of sugars and phosphate groups—all organized into the shape of a double helix. DNA replication begins when the parent molecule gets unzipped as the hydrogen bonds between the base pairs are broken. Once separated, the sequence of bases on each of the unzipped strands becomes a template for the insertion of a complementary set of bases. Deoxynucleoside triphosphates assemble the new strands in the order that complements the order of bases on the strand serving as the template. When the process is complete, two DNA molecules have been formed identical to each other and to the parent molecule. However, there are several ways that the process of DNA replication can (and does) go wrong—resulting in *mutation*. One such way is when a base is changed by the repositioning of a hydrogen atom, altering the hydrogen bonding pattern of that base, and resulting in incorrect base pairing during replication. Another way is when there is a loss of a purine base (A or G) to form an apurinic site (AP site). There can also be denaturation of the new strand from the template during replication, followed by renaturation in a different spot. This can lead to insertions or deletions.

Despite the fact that mutations are essentially *mistakes* in DNA replication, it is a good thing they occur. Without them, there would be no evolution. Mutations are the source of the variation on which natural selection acts. As such, mutation is an indisputable causal driver of evolution. So how does mutation fare with regard to (F1)–(F5)?

Here, again, there are no serious obstacles to meeting (F1) or (F2): we have component entities: DNA, a purine base, individual molecules, etc., and we have operations: unzipping, separating, inserting, etc.

Regarding (F3)–(F5), the situation with mutation seems a good measure improved from that of both natural selection and drift. Mutation appears to have plausible start and set-up conditions: the presence of DNA in a living organism and the initiation of hydrogen bond separation. And it has reasonable termination conditions: the existence of a mutated strand of DNA. Additionally, unlike either natural selection or drift, the process responsible for mutation has a readily discernible molecular structure common among all of its instances. Furthermore, much like natural selection, mutation can easily be seen to carry out a unified causal-role function: it is that which brings about genetic variation at the population level as a result of copying errors at the individual level. Finally, more so than either selection or drift, mutation seems to occur in a regular fashion. It is well known, for example, that errors during nucleotide substitution are biased and regularly occur more often at certain locations on the chromosome.

So how should we sum up the metaphysical question? To what extent is it the case that these primary causal drivers of evolution (natural selection, drift, and mutation) are aptly understood as mechanisms? What we have just seen is that the answer to the metaphysical question is complicated. Among the primary causal drivers of evolution, whether they are aptly understood as meeting (F1)–(F5) depends crucially on how liberal we understand these requirements to be. With strict enough constraints on organization, function, and regularity, it may well be that none of these processes end up counting as mechanisms. On highly liberal versions of these constraints, they all may well qualify. So the painfully residual question, then, is: *how liberal should we understand these metaphysical requirements to be*? The answer to *this* question, I am afraid I cannot offer in this short chapter. But I will say this. Liberalized metaphysical requirements on mechanisms have the advantage of allowing more physical processes to be understood as mechanisms, and this may have several pragmatic benefits (to be briefly expounded in the following section). However, highly liberalized metaphysical requirements have drawbacks as well. Namely, the more physical processes we allow to count as mechanisms, the less interesting and distinctive mechanistic explanations will end up being. As with any tradeoff between theoretic virtues, the right balance can only be arbitrated by extra-theoretic values.

4. Does the mechanistic framework add to our understanding of evolution?

Let us turn our attention to the epistemological question. That is, let us consider whether, *regardless* of whether any of the primary causal drivers of evolution are aptly understood as mechanisms, the mechanistic philosophical framework might help add to our understanding of evolution.

But how could this possibly be? How could mechanistic philosophy of science help illuminate the study of evolution even if there are no mechanisms driving it? In pondering this question, it will be helpful to consider the notion of "strategic mechanism" recently developed by Levy (2013). According to Levy, one way of approaching the philosophy of mechanisms is to conceive of it as constituting a *strategy*: "a way of doing science, a framework for representing and reasoning about complex systems" (Levy 2013, 105). Regardless of the metaphysical status of mechanisms out there in the world, the idea is that the mechanist "sees mechanistic methods as having particular cognitive and epistemic features" (ibid., 105) and maintains that "certain phenomena are best handled mechanistically" (ibid., 100).

In keeping with Levy, I suggest that there are at least three significant strategic roles that mechanisms might play in our understanding of science regardless of their metaphysical status: *reductionist explanation, manipulation,* and *discovery.* And when applied to the field of evolutionary biology, it seems to me that each of these strategic roles has the potential for adding to our understanding of evolution.

The strategic role of reductionist explanation is emphasized most convincingly by Bechtel and Richardson (1993) in their discussion of "complex localization." They write,

> Complex localization requires a decomposition of systemic tasks into subtasks, localizing each of these in a distinct component. Showing how systemic functions are, or at least could be, a consequence of these subtasks is an important element in a fully mechanistic explanation.
>
> *(Bechtel and Richardson 1993, 125)*

Thinking about nature mechanistically leads naturally to explanation by reduction and decomposition. When thinking mechanistically, we can more readily come to an understanding of *why* some phenomenon occurs by reducing it to its component parts and operations and by showing *how* the phenomenon in question is brought about by these parts and operations. Evolution is no exception. To understand why some trait (or individual with a given trait) was selected for, it seems a perfectly good strategy to decompose the larger ecological system in which the individual is a member into its component parts and operations—and in doing so, gain insight as to the advantages the trait confers. A similar story can be told of the strategic role of mechanisms for manipulation. When we model an instance of selection, drift, or mutation as a mechanism (often this is done using mechanism schemata made up of boxes and arrows representing causal relationships), it may become more salient where to intervene on the system in question to test various hypotheses about the nature of these causal interactions.[9] Finally, as emphasized in their recent book, Craver and Darden (2013) convincingly show that mechanistic thinking has tremendous benefits viz. scientific discovery: by progressively filling in the black boxes in our incomplete mechanism schemata, scientists are supplied with an invaluable heuristic for moving from *how-possibly* explanations to *how-actually* explanations of the natural world. This benefit seems especially apropos in the study of evolution where so many of our discoveries (e.g., fossils of intermediate forms like *Archaeopteryx* or *Tiktaalik*) originated as the product of speculative hypotheses.

Of course, it would require much more argument to establish that these strategic advantages of mechanistic thinking *actually do* pay off in the field of evolutionary biology. However, since I cannot undertake this work here in any detail, I am content for the time being to establish that an affirmative answer to the epistemological question has a reasonably high degree of plausibility. And this plausibility remains even if no affirmative answer can be given to the metaphysical question. Levy's notion of strategic mechanism helps to do just that.

5. Conclusion

In this chapter, I have explored answers to both the metaphysical and epistemological questions regarding the status of mechanisms in evolutionary biology. Regarding the metaphysical question, I briefly laid out some of the central metaphysical features of mechanisms and examined whether they are aptly understood as being instantiated by three of the primary causal drivers of evolution: natural selection, drift, and mutation. Provided we adopt relatively lenient versions of these constraints, I suggest the answer is yes. Regarding the epistemological question, I briefly expounded Lévy's notion of strategic mechanism and hinted at ways the strategic benefits of mechanistic thinking might apply to the field of evolutionary biology regardless of whether there turn out to be any mechanisms driving evolution. I conclude, therefore, that Mayr and Futuyma may have been a bit hasty in their implicit claim that evolutionary biology be confined to the realm of ultimate causes at the exclusion of mechanistic ones. Contrary to this view, I have shown that—on *at least some* philosophical characterizations of mechanisms, and with regard *to at least some* of the central processes of evolutionary biology—there is room for affirmative answers to both the metaphysical and epistemological questions. And thus, there may indeed be an important place for mechanisms in evolutionary biology.

Notes

1 The distinction between metaphysical and epistemological questions regarding the appropriateness of mechanistic thinking as applied to a given domain echoes one made by Glennan and Illari in Chapter 7 of this volume.
2 See Illari and Williamson (2012) and Chapter 1 for arguments that these requirements can be expressed by a common conception of mechanism that applies widely across the sciences.
3 See Skipper and Millstein (2005) for an argument to this effect.
4 See Ariew (2007) for a nice survey of some of these issues.
5 Not everyone agrees that natural selection is a causal process (cf. especially Matthen and Ariew 2002). Following Millstein (2006) and more recently Ramsey (forthcoming), I presume that natural selection can be given a causalist interpretation.
6 More detailed philosophical analyses of drift vary greatly. Some argue that drift cannot be distinguished from selection and so cannot be conceived of as a distinct evolutionary process (for instance, Matthen and Ariew 2002; Matthen 2009). And there are some (for example, Millstein 2002, 2008) who argue that drift is distinct from selection by virtue of the kind of process it is (indiscriminate as opposed to discriminate sampling of a population).
7 Example comes from Roberta Millstein (1996, S15).
8 As with natural selection, there are some who deny that drift is causal. See especially Lange (2013) as an example.
9 See especially Craver (2007) for a detailed exposition of this advantage.

References

Andersen, H. (2012) "The Case for Regularity in Mechanistic Causal Explanation," *Synthese* 189 (3): 415–432.
Ariew, André (2003) "Ernst Mayr's 'Ultimate/Proximate' Distinction Reconsidered and Reconstructed," *Biology and Philosophy* 18 (4): 553–565.
Ariew, André (2007) "Teleology," in D. L. Hull and M. Ruse (eds.), *The Cambridge Companion to the Philosophy of Biology*, Cambridge: Cambridge University Press, 160–181.
Barros, Benjamin (2008) "Natural Selection as a Mechanism," *Philosophy of Science* 74: 306–322.
Bechtel, W. and Abrahamsen, A. (2005) "Explanation: A Mechanist Alternative," in C. F. Craver and L. Darden (eds.), Special Issue: "Mechanisms in Biology," *Studies in History and Philosophy of Biological and Biomedical Sciences* 36: 421–441.

Bechtel, W. and Abrahamsen, A. (2009) "Decomposing, Recomposing, and Situating Circadian Mechanisms: Three Tasks in Developing Mechanistic Explanations," in H. Leitgeb and A. Hieke (eds.), "Reduction: Between the Mind and the Brain," *Ontos* 12–177.

Bechtel, W. and Abrahamsen, A. (2010) "Dynamic Mechanistic Explanation: Computational Modeling of Circadian Rhythms as an Exemplar for Cognitive Science," *Studies in History and Philosophy of Science Part A* 41 (3): 321–333.

Bechtel, W. and Abrahamsen, A. (2013) "Thinking Dynamically about Biological Mechanisms: Networks of Coupled Oscillators," *Foundations of Science* 18 (4): 707–723.

Bechtel, W. and Richardson, R. C. (1993) *Discovering Complexity: Decomposition and Localization as Strategies in Scientific Research*, Princeton, NJ: Princeton University Press.

Bogen, J. (2005) "Regularities and Causality; Generalizations and Causal Explanations," in Special Issue: "Mechanisms in Biology," in C. Craver and L. Darden (eds.), *Studies in History and Philosophy of Biological and Biomedical Sciences* 36 (2): 397–420.

Bogen, J. and Woodward, J. (1988) "Saving the Phenomena," *The Philosophical Review* 97 (3): 305–352.

Calcott, B. (2013) "Why How and Why Aren't Enough: More Problems with Mayr's Proximate-Ultimate Distinction," *Biology and Philosophy* 28 (5): 767–780.

Craver, C. F. (2001) "Role Functions, Mechanisms, and Hierarchy," *Philosophy of Science* 68: 53–74.

Craver, C. F. (2007) *Explaining the Brain: Mechanisms and the Mosaic Unity of Neuroscience*, New York: Oxford University Press.

Craver, C. F. and Darden, L. (2013) *In Search of Mechanisms: Discoveries across the Life Sciences*, Chicago, IL: University of Chicago Press.

Cummins, R. (1975) "Functional Analysis," *Journal of Philosophy* 72: 741–764.

DesAutels, L. (2011) "Against Regular and Irregular Characterizations of Mechanisms," *Philosophy of Science* 78 (5): 914–925.

DesAutels, L. (2015) "Toward a Propensity Interpretation of Stochastic Mechanism for the Life Sciences" *Synthese* 192: 2921–2953.

DesAutels, L. (2016) "Natural Selection and Mechanistic Regularity," *Studies in History and Philosophy of Biological and Biomedical Sciences* 57: 13–23.

Futuyma, D. J. (2013) *Evolution* (3rd Edition). Sunderland, MA: Sinauer Associates.

Garson, J. (2011) "Selected Effects and Causal Role Functions in the Brain: The Case for an Etiological Approach to Neuroscience," *Biology and Philosophy* 26 (4): 547–565.

Glennan, S. (1996) "Mechanisms and the Nature of Causation," *Erkenntnis* 44 (1): 49–71.

Glennan, S. (2002) "Rethinking Mechanistic Explanation," *Proceedings of the Philosophy of Science Association* 3: S342–S353.

Glennan, S. (2010) "Ephemeral Mechanisms and Historical Explanation," *Erkenntnis* 72 (2): 251–266.

Gould, S. (1990) *Wonderful Life: The Burgess Shale and the Nature of History*, New York: W. W. Norton and Company.

Illari, P. and Williamson, J. (2010) "Functional Organization: Comparing the Mechanisms of Protein Synthesis and Natural Selection," *Studies in History and Philosophy of Biological and Biomedical Sciences* 41: 279–291.

Illari, P. and Williamson, J. (2012) "What Is a Mechanism: Thinking about Mechanisms across the Sciences," *European Journal for Philosophy of Science* 2: 119–135.

Laland, K. N., Sterelny, K., Odling-Smee, J., Hoppitt, W., and Uller T. (2011) "Cause and Effect in Biology Revisited: Is Mayr's Proximate–Ultimate Dichotomy Still Useful?" *Science* 334: 1512–1516.

Laland, K. N., Odling-Smee, J., Hoppitt, W., and Uller, T. (2013) "More on How and Why: Cause and Effect in Biology Revisited," *Biology and Philosophy* 28 (5): 719–745.

Lange, M. (2013) "Really Statistical Explanations and Genetic Drift," *Philosophy of Science* 80 (2): 169–188.

Levy, A. (2013) "Three Kinds of New Mechanism," *Biology and Philosophy* 28: 99–114.

Machamer, P., Darden, L., and Craver, C. (2000) "Thinking about Mechanisms," *Philosophy of Science* 67 (1): 1–25.

Machamer, P. (2004) "Activities and Causation: The Metaphysics and Epistemology of Mechanisms," *International Studies in the Philosophy of Science* 18 (1): 27–39.

Matthen, M. (2009) "Drift and Statistically Abstractive Explanation," *Philosophy of Science* 76: 464–487.

Matthen, M. and Ariew, A. (2002) "Two Ways of Thinking about Fitness and Natural Selection," *Journal of Philosophy* 49: 55–83.

Mayr, E. (1961) "Cause and Effect in Biology," *Science* 134: 1501–1506.

Millikan, R. G. (1989) "In Defense of Proper Functions," *Philosophy of Science* 56: 288–302.

Millstein, R. L. (1996) "Random Drift and the Omniscient Viewpoint," *Philosophy of Science* 63 (3): S10–S18.

Millstein, R. L. (2002) "Are Random Drift and Natural Selection Conceptually Distinct?" *Biology and Philosophy* 17: 33–53.

Millstein, R. L. (2006) "Natural Selection as a Population-Level Causal Process," *British Journal for the Philosophy of Science* 57: 627–653.

Millstein, R. L. (2008) "Distinguishing Drift and Selection Empirically: 'The Great Snail Debate' of the 1950s," *Journal of the History of Biology* 41 (2): 339–367.

Neander, K. (1991) "Functions as Selected Effects: The Conceptual Analyst's Defense," *Philosophy of Science* 58: 168–184.

Ramsey, G. (forthcoming) "The Causal Structure of Evolutionary Theory," *Australasian Journal of Philosophy*: DOI: 10.1080/00048402.2015.1111398.

Scholl, R. and Pigliucci, M. (2014) "The Proximate–Ultimate Distinction and Evolutionary Developmental Biology: Causal Irrelevance Versus Explanatory Abstraction," *Biology and Philosophy* 30: 653–670.

Skipper, R. A., Jr. and Millstein, R. A. (2005) "Thinking about Evolutionary Mechanisms: Natural Selection," in C. F. Craver and L. Darden (eds.), Special Issue: "Mechanisms in Biology," *Studies in History and Philosophy of Biological and Biomedical Sciences* 36: 327–347.

23

MECHANISMS IN MOLECULAR BIOLOGY

Tudor M. Baetu

1. Introduction

While mechanistic explanations are by no means a novelty in biology (see Part I on historical perspectives on mechanisms), their appearance dating back to Harvey's discovery of the blood's circulation and Descartes' mechanistic manifesto touted in *The Treatise of Man*, it is only in the second half of the twentieth century, with the rise of molecular biology, that mechanistic thinking overtakes the whole of biology, becoming the predominant type of explanation. Explaining biological phenomena as the effects of molecular mechanisms turned out to have a marked influence on philosophy of science on two accounts. First, it motivated a renewed interest in mechanistic explanation, providing philosophers with a wealth of examples that didn't quite fit the deductive-nomological approach promoted by the logical positivists (Wimsatt 1976). Second, molecular biology motivated a shift in the way we think about mechanisms, fostering a "new" mechanistic philosophy. Unlike the "old" mechanistic philosophy, which was closely linked to the theory of classical mechanics and the clockwork view of the world, the "new" mechanistic philosophy had to come to terms with the notion that most biological mechanisms do not look and behave like eighteenth-century automata, but are much more complex and "noisier" systems composed of hundreds, thousands, and even millions of non-fixed, non-rigid parts whose behavior is nevertheless sufficiently constrained both by the properties of the parts and by the spatio-temporal organizational features of the system as to reliably produce and sustain biological phenomena and, ultimately, life itself.

2. What is molecular biology?

Molecular biology is the field of scientific investigation concerned with the molecular basis of biological activity. Its most significant achievements are the elucidation of the mechanisms of replication, transcription, and translation in the 1950s and 60s, providing an explanation of how cells replicate their genetic material and how genes are expressed as proteins that contribute to the phenotype of the organism. Molecular biologists quickly extended their inquiries to other biological phenomena, most notably gene expression regulation, the cell cycle, and cellular signaling. By the 1970s, molecular biology gained a firm footing in many other fields of biology, with developmental biology, immunology, neurology, and microbiology "going molecular." In this respect, molecular biology can be viewed as providing the guidelines of a general explanatory approach which I shall explore in more detail in the subsequent sections of this chapter.

Historically, molecular biology was born from the convergence of work in genetics, biochemistry, and physical chemistry in an attempt to figure out how genes determine phenotypes (Carlson 1967; Darden 1991, 2006a; Fox Keller 2000; Kay 1993; Morange 1998; Schaffner 1993).[1] Such questions fall outside the immediate explanatory scope of classical genetics, which is mainly concerned with the transmission of inherited traits (Morgan 1935; Moss 2003; Waters 1994). They also fall outside the immediate scope of biochemistry, which focuses predominantly on metabolic activities (e.g., the chemical reactions taking place during glycolysis or protein synthesis), as well as that of physical chemistry, which is concerned with the inter- and intramolecular forces shaping the tridimensional structure and the physicochemical properties of macromolecules (Baetu 2012a; Darden 2006b; Morange 2002; Olby 1994).

It is customary to distinguish molecular biology from related fields on the basis of its specific techniques of investigation revolving around the "cloning" of genetic material (creating copies of DNA fragments), along with techniques required to detect (e.g., electrophoresis), sequence (e.g., chain-termination sequencing), amplify (e.g., polymerase chain reaction, or PCR), and manipulate (e.g., site-directed mutagenesis) genetic material, many of which rely on the understanding of the mechanisms of replication, transcription, and translation which are at the core of molecular biology (Astbury 1961; Waters 2008). These techniques involve experimental interventions at the resolution of individual nucleotides—and, via the alteration of codon sequences, of individual amino-acids in a peptide sequence—thus tracking the flow of information within the cell, revealing the functional role of sequence motifs such as promoters, codons, and zinc-fingers relative to the operation of molecular mechanisms, and providing an understanding of how changes in the sequence of various molecular components result in changes in the operation of mechanisms and their ability to produce or sustain biological phenomena.

It should be noted, however, that molecular biology was and continues to be part of a highly integrated cluster of fields. Molecular biologists routinely rely on data, theoretical assumptions, and formal and experimental techniques from genetics, biochemistry, physical chemistry, cell biology, microbiology, statistics, epidemiology, systems biology, systematics, and many others. At the same time, scientists in other fields, from evolutionary biology to psychiatry, rely on explanatory strategies and experimental techniques inspired from molecular biology. As a result, our current understanding of the molecular basis of biological activity integrates findings from a wide variety of fields within and outside biology.

3. The nature of molecular mechanisms

In his celebrated essay *What Is Life?*, Erwin Schrödinger tackles the difficult question of the origin of order in biological systems. He argues that biological systems escape entropy because there is information in the system telling it how to assemble itself in an organized fashion. On this account, information is a source of order in what would have otherwise been a thermodynamically disordered system. Genes would be the repositories of information, which is propagated throughout the cell by a series of deterministic mechanisms that preserve order. This view, predating the major discoveries of molecular biology, endures until today, to the point that some philosophers define molecular biology as the study of the mechanisms of information propagation within cells (Darden 2006a, 2006b; Morange 1998).

Schrödinger's order-generating information was eventually identified with genetic information, while its propagation throughout the cell by means of deterministic mechanisms came to be known as genetic determinism. The best illustration is the "genetic program" view popularized by François Jacob (1976) and Jacques Monod (1972); for a philosophical discussion, see Baetu (2012b). According to this view, information is unidirectionally propagated via the mechanisms of transcription and translation to peptide sequences, which in turn determine the tridimensional shape of proteins via the folding of α-helices, ß-sheets, and other secondary structures, itself determining the specificity of binding to other molecules according to a lock-and-key or induced fit model, thus directing the flow of chemical reactions underpinning metabolic and signaling pathways as well as the self-assembly of the various supramolecular structures that constitute the living cell.

While immensely popular in the 1970s and still echoing today, this view is in fact a special case of a more general concept in molecular biology, namely that of specific binding (Kupiec 2009; Morange 1998). Specificity of binding is assumed to determine not only the preferential pairing of nucleotides in complementary strands of nucleic acids during replication and transcription or between codons and aminoacyl-tRNAs during translation, but also virtually any single activity related to the operation of molecular mechanisms, including enzyme-substrate interactions, the recognition of extracellular ligands by cell-surface receptors, the binding of transcriptional factors to particular DNA sequences, the self-assembly of microtubules and ribosomes, etc. Specificity is determined by chemical affinity, which is typically measured as the average life span of a supramolecular complex in a given chemical environment. When molecules form stable, long-lasting complexes likely to have a marked impact on biological activity, their binding is said to be specific; when the complexes are short-lived, their binding is said to be non-specific. Thus, order-generating information is not restricted to the genome, but is in fact manifest in every single specific molecular interaction taking place in the cell.

Less appreciated by the general public is the fact that binding specificity is an "analog," or stochastic concept (Rao et al. 2002). Any given molecule always interacts with many other molecules, for some with stronger and for others with weaker affinity, such that specific binding invariably occurs against the "noisy" background of myriad non-specific interactions. Furthermore, it is not uncommon that the same macromolecule can bind with relatively high specificity not only one, but many other molecules (e.g., most transcriptional factors can bind with variable, but relatively high specificity several related DNA sequences).[2] By contrast, genetic determinism assumes an idealized limit case, a "noise-free" or "digital" specificity propagating itself throughout the cell without significant distortion as the inevitable result of the expression of genomic sequences.

"Digital" specificity idealizations are still ubiquitous in the qualitative descriptions of molecular biology. While there are epistemic benefits associated with such idealizations, most notably increased intelligibility, the truth is that, until very recently, molecular biologists had no other choice but to idealize. Traditionally, molecular techniques rely on the amplification of the properties, states, and activities of molecular components up to an unequivocal threshold of detection. This is achieved by studying millions of cells in bulk over (chemically speaking) long periods of time, which means ignoring both time fluctuations, as well as differences from one cell to another. Yet the underlying biochemistry dictates that most molecular activities are chemical reactions controlled by the concentrations, states, and locations of the various molecules involved. If these parameters are not identical in all cells, there will be fluctuations from one cell to another with respect to quantitative and dynamic aspects of the operation and output of mechanisms. Furthermore, chemical reactions are intrinsically stochastic since they rely on random microscopic events that govern how fast reactions occur and in what order. For years, it was assumed that although stochastic fluctuations are bound to occur, they are biologically irrelevant "background noise" cells keep to a minimum. Thus, it was and often still is customary to assume that what is true of a large population of cells can be safely extrapolated to individual cells within that population. In turn, this justifies the belief that, on average, the same mechanism, following the sequence of chemical events, is synchronously operating in all cells of the same type subjected to the same conditions.

In contrast with this view, recent single-cell experiments revealed that stochastic fluctuations are non-negligible, meaning that cells must rely on noise-suppressing mechanisms, such as negative feedback and DNA "proof-reading" (Elowitz et al. 2002). It also turned out that stochasticity itself can be biologically relevant and is in fact used as a means for generating diversity and histological-level patterns. For example, probabilistic biases in the distribution of adhesion proteins suffice to generate the right amount of twisting in the developing gut, while stochastic gene expression is responsible for generating blue- to ultraviolet-sensitive cells ratios in the *Drosophila* eye; for discussion, see Baetu (2015b), Heams (2011), and Merlin (2011). The growing realization that many mechanisms in biology are stochastic motivated a more careful investigation of the extent to which regularity is a distinctive characteristic of mechanisms (Andersen 2012; Darden 2008; DesAutels 2011), with some authors emphasizing the highly irregular (Bogen 2005) and even the singular nature of some mechanisms (Glennan 2010, 2011). The issues of stochasticity, regularity, and singular causation are discussed in more detail in Chapters 10–13. More general implications for theories of causation can be found in Hall (2004) and Psillos (2004).

Specific binding is clearly responsible for generating some supramolecular structures, from the recruitment of polymerase complexes to the self-assembly of proteasomes and microtubules; furthermore, techniques like in vitro translation or PCR would be impossible without specific binding. Nevertheless, specificity is sensitive to the effective concentrations of molecular components, with the "background noise" of non-specific interactions increasing for low copy numbers of molecules. This observation led to a questioning about whether specific binding is the exclusive origin of order in biological systems.

The traditional model, inherited from early twentieth-century biochemistry, is the cell as a "bag of enzymes." According to this admittedly idealized model, the spatio-temporal organization of molecular mechanisms, from the dynamics of activities to the assembly of supramolecular structures, is driven by specificity of binding in the context of a free diffusion solution chemistry. In short, were it not for differential binding specificities, the molecules inside a cell would display the same amount of order as sugar dissolved in a glass of water. This model turned out to be unsatisfactory. Molecular mechanisms operate within an intracellular environment filled with other molecules. Macromolecular crowding functions as an excluded volume effect, favoring the aggregation of macromolecules while drastically decreasing the rate of any diffusion-dependent molecular activity (Ellis 2001). This strongly suggests that the intracellular environment cannot consist of a disordered collection of molecules, as this would result in the proliferation of noise-generating non-specific interactions at the expense of order-generating specific ones. Instead, it must be structured in such a way as to bring in close proximity proteins and their ligands, thus favoring the specific chemical interactions required for the operation of molecular mechanisms (Hochachka 1999; Mathews 1993). However, if this is the case—and recent evidence supports this conclusion, e.g. the effects of nuclear architecture on transcription (Cremer and Cremer 2001)—then the tridimensional structure of the cell and tissue organization are also a source of order, functioning as a scaffold constraining the behavior of molecular mechanisms by favoring some activities while suppressing others. This "return to holism" in molecular biology did not go unnoticed in philosophy of biology, motivating a renewed attack on genetic determinism on the grounds that gene expression is a stochastic process (Kupiec 2009), that information is distributed throughout the cell (Jablonka 2001), and that genes alone—or for that matter the properties of individual molecules, such as their binding affinities—cannot fully account for phenotype and development (Dupré 2010; Griffiths and Stotz 2006, 2013; Oyama et al. 2001).

One final twist in the story of the molecular basis of biological activity is the "systemic turn" in biology. Its impact on the way we think about mechanisms is discussed in Chapter 27. The present discussion will focus on one of the many findings that prompted this turn, namely the realization that molecular mechanisms operate in a less modular fashion than initially thought. Molecular biologists often work under the assumption that mechanisms amount to discrete functional modules organized hierarchically, serially or in parallel. While useful as a heuristic of discovery, the modularity assumption came to be questioned by a growing body of evidence revealing that distinct mechanisms responsible for distinct phenomena share mechanistic components. Notoriously, intracellular signaling pathways and gene regulatory mechanisms invariably intersect at some point, forming widespread molecular networks (Davidson and Levine 2005). It has therefore been proposed that many molecular mechanisms are in fact inextricably interlocked into vast networks of partially overlapping mechanisms, where the sharing of mechanistic components is thought to play a role in the fine-tuning of quantitative-dynamic aspects of the phenomena produced by these mechanisms (Barabási and Oltvai 2004). In some cases, mathematical models were able to account for minute discrepancies between the predicted and observed outcomes by taking into consideration interference from overlapping mechanisms, thus providing some evidence for the biological significance of non-modular modes of organization; for philosophical discussion and references to the original scientific literature, see Baetu (2015b); Bechtel and Abrahamsen (2010, 2011).

In recent philosophical debates, modularity is often understood along the lines of "independent disruptability" stating that the overall effect of a causal system can be decomposed into a set of independent causal contributions attributable to each of the constituents of the system (Steel 2007; Woodward 2002). The systemic turn in biology prompted some authors to reject causal modularity (Bogen 2004; Mitchell 2009), and even interpret it as a vindication of a more holistic approach defended by the old organicist school of thought (Nicholson 2013); for a historical discussion of mechanism vs. organicism, see Chapter 5. It is interesting, however, to note that if molecular mechanisms are constrained by higher-level cellular and histological structures, then significant forms of modularity must prevail in virtue of the physical partitioning of biochemical reactions (Callebaut and Rasskin-Gutman 2005; Hartwell et al. 1999). If so, the relevant issue is not whether cells and organisms are organized in a modular fashion, but the extent to which molecular mechanisms, as described in biology textbooks, amount to functional or causal modules (Baetu 2016; Woodward 2010).

4. Mechanistic explanations in molecular biology

One of the explicit aims of molecular biology is to provide a reductive explanation of biological phenomena in terms of molecular mechanisms. It is therefore crucial to investigate more closely the notion of mechanistic explanation: how do mechanisms explain?

According to the ontic view, mechanistic explanations are objective features of the world. To explain a phenomenon is to fit it into the causal structure of the world (Craver 2007). The advantages of this view is that it eliminates the subjective notion of understanding while emphasizing the tight link between explanation and the ability to control effects by intervening on their causes. Notwithstanding, most philosophers prefer an epistemic view, according to which mechanistic explanations are step-by-step descriptions of how mechanisms produce the phenomena for which they are responsible, although some authors point out that both ontic and epistemic considerations contribute to a successful mechanistic explanation (Illari 2013). (For treatment of ontic vs. epistemic conceptions, see Chapter 16.)

In one version of the epistemic view, descriptions of mechanisms, in words or by means of diagrams, convey an intuitive understanding consisting in simulating the working of mechanisms in our imagination (Bechtel and Abrahamsen 2005) or by analogy with more common types of activities (Machamer 2004). While imagination is indispensable for our ability to learn about and understand molecular mechanisms, equating mechanistic explanation with intuitive understanding can be problematic. For one thing, the simulations we perform in our minds are heavily idealized. Detailed quantitative aspects are absent, while known facts are distorted to outline a deterministic sequence of events tracking the fate of single molecules (a more accurate biochemical description would be that of a series of back-and-forth equilibria involving populations of molecules bumping into each other), each assumed to be rigidly structured (ignoring the fact that molecules "breathe," vibrating and cycling through various configurations), as they modify one another to bring about a change from an initial to a final state of affairs (thus masking an underlying variety of chemical pathways that contribute to the same final state, as well as ignoring alternate pathways where mechanisms fail to contribute to the output).

Such idealizations are due to the fact that intuitive understanding relies on analogies with macroscopic mechanisms more familiar to us. In contrast, examples of mechanistic explanations from the sciences show that the working entities of a mechanism—that is, the parts of the mechanism engaged in the operation of the mechanism (Machamer et al. 2000)—span multiple levels of composition, thus supporting the notion that biological mechanisms combine a wide variety of activities associated with different levels of composition, from push-pull mechanical interactions to chemical reactions and thermodynamic processes (Craver 2007; Craver and Darden 2013; Darden 2006b; also Chapter 14 in the present volume). There is even evidence that some molecular mechanisms harness the quirks of quantum mechanics (Ball 2011); for a philosophical discussion, see Barwich (2015). Arguably, an intuitive understanding of chemical equilibria, thermodynamic processes, and quantum mechanics by means of analogies with macroscopic mechanisms is problematic, as many of these concepts defy our imagination and are best captured by a rather complex mathematical formalism.

This brings us to a third view meant to reflect a relatively recent quantitative turn in molecular biology, namely an understanding mediated by evidence from the testing of mathematical models aimed at demonstrating that mechanisms can in principle produce the phenomena for which they are responsible in close approximation of detailed quantitative measurements (Baetu 2015b; Bechtel 2012; Bechtel and Abrahamsen 2011; Braillard 2010; Brigandt 2013; Gross 2015; also Chapters 20 and 27). This view also qualifies as epistemic, albeit the understanding is more of a theoretical than an intuitive-imaginative nature, as it involves showing how phenomena are consequences of explicit rules and assumptions about the operation of mechanisms. (A more comprehensive treatment of representations of mechanisms in general can be found in Chapter 18. Models of mechanisms are discussed in more detail in Chapter 17.)

It is interesting to note that mechanistic explanations do not automatically preclude a role for generalizations and regularities in biology. The ability of some mechanisms to regularly produce a phenomenon accounts for some of the generalizations observable in the biological world (Glennan 2002; Illari and Williamson 2012; Machamer et al. 2000). For example, mechanistic constraints can explain why some evolutionary outcomes are more probable than others, and allow for predictions in specific lineages (Baetu 2012c), while the reliance on exemplar organism models supports the notion that general patterns shared by a large number of species exist and play an important role in guiding research (Ankeny 2001; Bolker 1995; Schaffner 2001; Weber 2005).

Conversely, the behavior of mechanistic parts is often characterized by invariant change-relating generalizations (Glennan 1996, 2002; Woodward 2002, 2010). Likewise, mathematical modeling presupposes a set of rules specifying how mechanisms change from one state to another, where some of these rules are laws borrowed from chemistry and physics (Schaffner 1993; Weber 2005). The current tendency to complement experimental research with mathematical modeling is also responsible for reviving the notion that mathematical or, more often, computational derivation contributes to the mechanistic explanation by demonstrating that certain aspects of a phenomenon are the consequences of the rules governing the operation of the mechanism. These considerations suggest a rather complicated relationship between mechanisms and regularities in molecular biology, whereby mechanisms both rely on regularities governing the behavior of their components and are themselves responsible for generating novel regularities, typically under the form of recurrent or reproducible phenomena.

Another issue that requires clarification concerns the completeness of mechanistic explanations. This issue is closely linked to the more general problem of how mechanisms are discovered (Chapter 19). In this chapter, I will focus on questions about the levels of composition at which mechanistic explanations bottom-out and top-off—that is, the optimal resolution of detail at which mechanisms should be described and the extent to which mechanisms act as independent modules that can produce (and therefore explain) the phenomena for which they are responsible when separated from the systems in which they are embedded.[3] The decomposition of biological systems reveals a hierarchical organization, with lower-level components organized as mechanisms underlying higher-level components (Bechtel 2006; Bechtel and Richardson 2010; Craver 2007). It might therefore be tempting to conclude that a more complete explanation can be achieved by investigating the components of a mechanism to reveal the finer-grained sub-mechanisms responsible for their properties and activities, as well as understanding how the mechanism fits in the context of progressively more comprehensive systems of mechanisms. Nonetheless, even though such investigations are bound to generate new knowledge—for instance, by explaining why mechanistic components have the properties they have or by providing the structural details necessary to intervene on these components and their properties—they cannot guarantee explanatory completeness. Assuming that the primary objective of a mechanistic explanation is to demonstrate causation—that is, show how an organized system of parts produces a specific phenomenon—then whether a mechanistic explanation is complete is a matter of providing evidence that parts having the properties and organization specified in the explanation can and do produce the phenomenon of interest.[4] This kind of evidence is generated by attempts to build the mechanism in vitro using reconstitution experiments, in vivo using the techniques of genetic engineering and synthetic biology, and in silico by testing mathematical models, as well as more indirectly, by assessing the ability to correctly foresee and correct side-effects of treatments and other technological applications based on mechanistic explanations (Baetu 2015a; Craver and Darden 2013; Morange 2009; Weber 2005).

Another way of looking at the issue of explanatory completeness is from a pragmatic standpoint. There are many ways in which a mechanistic explanation is useful to the scientist, and the purpose will determine when the explanation is deemed to be complete (Craver and Darden 2013). The main advantage of a pragmatic stance is that it can accommodate seemingly conflicting evaluations. For instance, if one seeks to gain control over the phenomenon of interest, then the explanation should focus on the manipulable components of the actual mechanism responsible for the phenomenon, such as gene and protein sequences (Craver 2006; Waters 2007); if prediction is the main goal, then the focus will fall on showing how changes in the components of a mechanism result in changes in the phenomenon produced by the mechanism (Cartwright 2002; Woodward 2010); if one seeks intelligibility, be it for didactic purposes or to reveal general patterns, then abstracting or even idealizing may be necessary (Levy 2014).

5. Conclusion

Molecular biology was undoubtedly the most influential field of biological research in the second half of the twentieth century, with hardly any branch of biology left untouched by the long-reaching arm of the molecular revolution. At the same time, it is equally important to realize that molecular biology itself changed over time as a result of interactions with other fields, from its origins as interdisciplinary research in genetics, biochemistry, and physical chemistry, to its subsequent integration of ideas from cell and developmental biology, to its current interactions with synthetic, systems biology, and nanoscience. From a philosophical point of view, advances in molecular biology prompted an inquiry about scientific explanation and the nature of biological mechanisms. Under the impetus of the elucidation of the mechanisms of genome replication and expression, molecular mechanisms were initially idealized as thoroughly deterministic devices propagating genetic information. This deterministic conception with a strong reductionistic flavor gave way to a more realistic interpretation endorsed by most molecular biologists today, namely that of molecular mechanisms as deterministic systems with noise, where the spatio-temporal organization of mechanisms is generated not only by specificity of binding (an important part of which is genetic information), but also by cellular and other supramolecular organization constraints.

Notes

1 Among these authors, some take the official date of birth of molecular biology to coincide with the coining of the term by Warren Weaver in 1938 (Kay 1993; Olby 1994), while others postpone it until the elucidation of the structure of DNA in 1953 and the emergence of the modern concept of genetic information (Darden 2006a, 2006b).

2 There are many reasons why molecular interactions are "noisy." Organic molecules can assume multiple stable configurations, each characterized by its distinct affinities; for instance, spontaneous mutations can occur via the tautomerization of nucleotides. Such variability is expected to be even more pronounced in macromolecules such as proteins, which can fold in a variety of configurations. Binding itself deforms molecules, thus altering their affinities and generating rather hard-to-predict phenomena such as allosterism and cooperative binding. Finally, new discoveries suggest that some proteins are unstructured and assume different folding configurations depending on the partners with which they interact. For references and philosophical discussion, see Kupiec (2009).

3 In many respects, these issues are a continuation of an earlier debate about reductionism in biology. While some philosophers have argued that the best explanations in biology are those bottoming out at a molecular level (Rosenberg 2006; Waters 1994), others replied that descriptions at higher levels of composition are needed to account for organizational features of biological systems (Laubichler and Wagner 2001) or to enhance intelligibility (Kitcher 1984).

4 Some physicists proposed that biological explanations are likely to bottom-out at a molecular level because at this size scale the mechanical, electrostatic, chemical, and thermal energies of objects have similar magnitudes. The convergence of energy values means that mechanical, electrical, and chemical energy can be converted into one another, which can explain some of the most fundamental properties of living things, such as their ability to convert food (chemical energy) into motion (mechanical energy). Furthermore, thermal energy (random molecular collisions) may quite literally "kick start" molecular mechanisms, thus explaining their ability to work autonomously and spontaneously. In contrast, mechanical forces at macroscopic scales or binding forces stabilizing sub-molecular structures such as atoms are largely insensitive to thermal fluctuations at room/body temperature, and therefore phenomena such as spontaneous activation, self-assembly, or change of shape are impossible (Hoffmann 2012; Philips and Quake 2006). This suggests that there is something objectively special about the molecular level in the sense that only molecules, as opposed to atoms or macroscopic objects, have the kind of properties required by mechanistic explanations in biology.

References

Andersen, H. 2012. "The Case for Regularity in Mechanistic Causal Explanation." *Synthese* 189 (3):415–32.

Ankeny, R. 2001. "Model Organisms as Models: Understanding the 'Lingua Franca' of the Human Genome Project." *Philosophy of Science* 68:S251–S61.

Astbury, W. 1961. "Molecular Biology or Ultrastructural Biology?" *Nature* 190 (4781):1124.

Baetu, T. M. 2012a. "Emergence, Therefore Antireductionism? A Critique of Emergent Antireductionism." *Biology and Philosophy* 27 (3):433–48.

——. 2012b. "Genomic Programs as Mechanism Schemas: A Non-Reductionist Interpretation." *British Journal for the Philosophy of Science* 63 (3):649–71.

——. 2012c. "Mechanistic Constraints on Evolutionary Outcomes." *Philosophy of Science* 79 (2):276–94.

——. 2015a. "The Completeness of Mechanistic Explanations." *Philosophy of Science* 82 (5):775–86.

——. 2015b. "From Mechanisms to Mathematical Models and Back to Mechanisms: Quantitative Mechanistic Explanations." In *Explanation in Biology: An Enquiry into the Diversity of Explanatory Patterns in the Life Sciences*, ed. P.-A. Braillard and C. Malaterre, 345–63. Dordrecht: Springer.

——. 2016. "From Interventions to Mechanistic Explanations." *Synthese* 193: 3311–27

Ball, P. 2011. "Physics of Life: The Dawn of Quantum Biology." *Nature* 474 (7351):272–74.

Barabási, A.-L., and Z. N. Oltvai. 2004. "Network Biology: Understanding the Cell's Functional Organization." *Nature Reviews Genetics* 5 (2):101–13.

Barwich, A.-S. 2015. "Bending Molecules or Bending the Rules: The Application of Theoretical Models in Fragrance Chemistry." *Perspectives on Science* 23 (4):1–23.

Bechtel, W. 2006. *Discovering Cell Mechanisms: The Creation of Modern Cell Biology*. Cambridge: Cambridge University Press.

Bechtel, W. 2012. "Understanding Endogenously Active Mechanisms: A Scientific and Philosophical Challenge." *European Journal for Philosophy of Science* 2:233–48.

Bechtel, W., and A. Abrahamsen. 2005. "Explanation: A Mechanist Alternative." *Studies in History and Philosophy of Biological and Biomedical Sciences* 36:421–41.

——. 2010. "Dynamic Mechanistic Explanation: Computational Modeling of Circadian Rhythms as an Exemplar for Cognitive Science." *Studies in History and Philosophy of Science Part A* 41:321–33.

——. 2011. "Complex Biological Mechanisms: Cyclic, Oscillatory, and Autonomous." In *Philosophy of Complex Systems*, ed. C. A. Hooker, 257–85. New York: Elsevier.

Bechtel, W., and R. Richardson. 2010. *Discovering Complexity: Decomposition and Localization as Strategies in Scientific Research*. Cambridge, MA: MIT Press.

Bogen, J. 2004. "Analyzing Causality: The Opposite of Counterfactual Is Factual." *International Studies in the Philosophy of Science* 18:3–26.

——. 2005. "Regularities and Causality: Generalizations and Causal Explanations." *Studies in History and Philosophy of Biological and Biomedical Sciences* 36:397–420.

Bolker, J. 1995. "Model Systems in Developmental Biology." *BioEssays* 17 (5):451–5.

Braillard, P.-A. 2010. "Systems Biology and the Mechanistic Framework." *History and Philosophy of Life Sciences* 32:43–62.

Brigandt, I. 2013. "Systems Biology and the Integration of Mechanistic Explanation and Mathematical Explanation." *Studies in History and Philosophy of Biological and Biomedical Sciences* 44 (4):477–92.

Callebaut, W., and D. Rasskin-Gutman. 2005. *Modularity: Understanding the Development and Evolution of Natural Complex Systems*. Cambridge, MA: MIT Press.

Carlson, E. O. 1967. *The Gene: A Critical History*. Philadelphia: W.B. Saunders.

Cartwright, N. 2002. "Against Modularity, the Causal Markov Condition and any Link between the Two: Comments on Hausman and Woodward." *British Journal for the Philosophy of Science* 53 (3):411–53.

Craver, C. 2006. "When Mechanistic Models Explain." *Synthese* 153:355–76.

——. 2007. *Explaining the Brain: Mechanisms and the Mosaic Unity of Neuroscience*. Oxford: Clarendon Press.

Craver, C., and L. Darden. 2013. *In Search of Biological Mechanisms: Discoveries across the Life Sciences*. Chicago, IL: University of Chicago Press.

Cremer, T., and C. Cremer. 2001. "Chromosome Territories, Nuclear Architecture and Gene Regulation in Mammalian Cells." *Nature Reviews Genetics* 2:292–301.

Darden, L. 1991. *Theory Change in Science: Strategies from Mendelian Genetics*. New York: Oxford University Press.

——. 2006a. "Flow of Information in Molecular Biological Mechanisms." *Biological Theory* 1 (3):280–7.

———. 2006b. *Reasoning in Biological Discoveries: Essays on Mechanisms, Interfield Relations, and Anomaly Resolution*. Cambridge: Cambridge University Press.

———. 2008. "Thinking Again about Biological Mechanisms." *Philosophy of Science* 75:958–69.

Davidson, E., and M. Levine. 2005. "Gene Regulatory Networks." *Proceedings of the National Academy of Science* 102 (14):4935.

DesAutels, L. 2011. "Against Regular and Irregular Characterizations of Mechanisms." *Philosophy of Science* 78 (5):914–25.

Dupré, J. 2010. "It Is Not Possible to Reduce Explanations in Biology to Explanations in Chemistry and/ or Physics." In *Contemporary Debates in Philosophy of Biology*, ed. F. Ayala and R. Arp, ch. 2. Chichester, UK: Wiley-Blackwell.

Ellis, J. 2001. "Macromolecular Crowding: Obvious but Underappreciated." *Trends in Biochemical Sciences* 26 (10):597–604.

Elowitz, M., A. J. Levine, E. D. Siggia, and P. S. Swain. 2002. "Stochastic Gene Expression in a Single Cell." *Science* 297 (5584):1183–6.

Fox Keller, E. 2000. *The Century of the Gene*. Cambridge, MA: Harvard University Press.

Glennan, S. 1996. "Mechanisms and the Nature of Causation." *Erkenntnis* 44:49–71.

———. 2002. "Rethinking Mechanistic Explanation." *Philosophy of Science* 69:S342–S53.

———. 2010. "Ephemeral Mechanisms and Historical Explanation." *Erkenntnis* 72:251–66.

———. 2011. "Singular and General Causal Relations: A Mechanist Perspective." In *Causality in the Sciences*, ed. P. McKay, J. Williamson and F. Russo, 789–817. Oxford: Oxford University Press.

Griffiths, P., and K. Stotz. 2006. "Genes in the Postgenomic Era." *Theoretical Medicine and Bioethics* 27 (6):499–521.

———. 2013. *Genetics and Philosophy: An Introduction*. Cambridge: Cambridge University Press.

Gross, F. 2015. "The Relevance of Irrelevance: Explanation in Systems Biology." In *Explanation in Biology: An Enquiry into the Diversity of Explanatory Patterns in the Life Sciences*, ed. P.-A. Braillard and C. Malaterre. Dordrecht: Springer.

Hall, N. 2004. "Two Concepts of Causation." In *Causation and Counterfactuals*, ed. J. Collins, N. Hall and L. Paul, 225–76. Cambridge, MA: MIT Press.

Hartwell, L. H., J. J. Hopfield, S. Leibler, and A. W. Murray. 1999. "From Molecular to Modular Cell Biology." *Nature* 6761supp (C47–52).

Heams, T. 2011. "Expression stochastique des gènes et différenciation cellulaire." In *Le hasard au coeur de la cellule: Probabilités, déterminisme, génétique*, ed. J.-J. Kupiec, O. Gandrillon, M. Morange and M. Silberstein, 31–59. Paris: Éditions Matériologiques.

Hochachka, P. W. 1999. "The Metabolic Implications of Intracellular Circulation." *Proceedings of the National Academy of Sciences* 96 (22):12233–9.

Hoffmann, P. M. 2012. *Life's Ratchet: How Molecular Machines Extract Order from Chaos*. New York: Basic Books.

Illari, P. 2013. "Mechanistic Explanation: Integrating the Ontic and Epistemic." *Erkenntnis* 78 (2):237–55.

Illari, P., and J. Williamson. 2012. "What is a Mechanism? Thinking about Mechanisms *across* the Sciences." *European Journal for Philosophy of Science* 2 (1):119–35.

Jablonka, E. 2001. "The Systems of Inheritance." In *Cycles of Contingency: Developmental Systems and Evolution*, ed. S. Oyama, P. Griffiths and R. Gray. Cambridge, MA: MIT Press.

Jacob, F. 1976. *The Logic of Life*. New York: Pantheon.

Kay, L. E. 1993. *The Molecular Vision of Life: Caltech, the Rockefeller Foundation, and the Rise of the New Biology*. New York: Oxford University Press.

Kitcher, P. 1984. "1953 and All That: A Tale of Two Sciences." *Philosophical Review* 93:335–73.

Kupiec, J.-J. 2009. *The Origin of Individuals*. Singapore: World Scientific Publishing.

Laubichler, M., and G. Wagner. 2001. "How Molecular is Molecular Developmental Biology? A Reply to Alex Rosenberg's Reductionism Redux: Computing the Embryo." *Biology and Philosophy* 16:53–68.

Levy, A. 2014. "What was Hodgkin and Huxley's Achievement?" *British Journal for the Philosophy of Science* 65 (3):469–92.

Machamer, P. 2004. "Activities and Causation: The Metaphysics and Epistemology of Mechanisms." *International Studies in the Philosophy of Science* 18 (1):27–39.

Machamer, P., L. Darden, and C. Craver. 2000. "Thinking about Mechanisms." *Philosophy of Science* 67:1–25.

Mathews, C. K. 1993. "The Cell-Bag of Enzymes or Network of Channels?" *Journal of Bacteriology* 175 (20):6377–81.

Merlin, F. 2011. "Pour une interprétation objective des probabilités dans les modèles stochastiques de l'expression génétique." In *Le hasard au coeur de la cellule: Probabilités, déterminisme, génétique*, ed. J.-J. Kupiec, O. Gandrillon, M. Morange and M. Silberstein, 31–59. Paris: Éditions Matériologiques.

Mitchell, S. 2009. *Unsimple Truths: Science, Complexity, and Policy*. Chicago: University of Chicago Press.

Monod, J. 1972. *Chance and Necessity: An Essay on the Natural Philosophy of Modern Biology*. London: Vintage Books.

Morange, M. 1998. *A History of Molecular Biology*. Cambridge, MA: Harvard University Press.

——. 2002. "The Gene: Between Holism and Generalism." In *Promises and Limits of Reductionism in the Biomedical Sciences*, ed. M. Van Regenmortel and D. Hull. Chichester: John Wiley & Sons. DOI: 10.1002/0470854189.ch9.

——. 2009. "Synthetic Biology: A Bridge Between Functional and Evolutionary Biology." *Biological Theory* 4 (4):368–77.

Morgan, T. H. 1935. "The Relation of Genetics to Physiology and Medicine." *Les prix Nobel en 1933*. *Imprimerie Royale*: 1–16.

Moss, L. 2003. *What Genes Can't Do*. Cambridge, MA: MIT Press.

Nicholson, D. 2013. "Organisms ≠ Machines." *Studies in History and Philosophy of Biological and Biomedical Sciences* 44 (4):669–78.

Olby, R. 1994. *The Path to the Double Helix: The Discovery of DNA*. Mineola, NY: Dover.

Oyama, S., P. Griffiths, and R. Gray. 2001. "Introduction: What Is Developmental Systems Theory?", In *Cycles of Contingency: Developmental Systems and Evolution*, ed. S. Oyama, P. Griffiths and R. Gray, 1–11. Cambridge, MA: MIT Press.

Philips, R., and S. R. Quake. 2006. "The Biological Frontier of Physics." *Physics Today* 59:38–43.

Psillos, S. 2004. "A Glimpse of the Secret Connexion: Harmonizing Mechanisms with Counterfactuals." *Perspectives on Science* 12:288–391.

Rao, C. V., D. M. Wolf, and A. P. Arkin. 2002. "Control, Exploitation and Tolerance of Intracellular Noise." *Nature* 420 (6912):231–7.

Rosenberg, A. 2006. *Darwinian Reductionism, or, How to Stop Worrying and Love Molecular Biology*. Chicago: University of Chicago Press.

Schaffner, K. F. 1993. *Discovery and Explanation in Biology and Medicine*. Chicago: University of Chicago Press.

——. 2001. "Extrapolation from Animal Models: Social Life, Sex, and Super Models." In *Theory and Method in the Neurosciences*, ed. P. Machamer, R. Grush and P. McLaughlin, 231–49. Pittsburgh: University of Pittsburgh Press.

Steel, D. 2007. *Across the Boundaries: Extrapolation in Biology and Social Science*. Oxford: Oxford University Press.

Waters, C. K. 1994. "Genes Made Molecular." *Philosophy of Science* 61:163–85.

——. 2007. "The Nature and Context of Exploratory Experimentation: An Introduction to Three Case Studies of Exploratory Research." *History and Philosophy of Life Sciences* 29 (3):275–84.

——. 2008. "Beyond Theoretical Reduction and Layer-Cake Antireduction: How DNA Retooled Genetics and Transformed Biological Practice." In *The Oxford Handbook of Philosophy of Biology*, ed. M. Ruse, 238–62. Oxford: Oxford University Press.

Weber, M. 2005. *Philosophy of Experimental Biology*. Cambridge: Cambridge University Press.

Wimsatt, W. C. 1976. "Reductive Explanation: A Functional Account." In *Boston Studies in the Philosophy of Science*, ed. A. C. Michalos, 671–710. Dordrecht: Reidel.

Woodward, J. 2002. "What Is a Mechanism? A Counterfactual Account." *Philosophy of Science* 69:S366–S77.

——. 2010. "Causation in Biology: Stability, Specificity, and the Choice of Levels of Explanation." *Biology and Philosophy* 25:287–318.

24

MECHANISMS AND BIOMEDICINE[1]

Brendan Clarke and Federica Russo

1. Introduction

Our aim in this chapter is to give a "functionalist account" of the ways that mechanisms are sought, formulated, and used in medicine. Rather than giving a single analytic account of mechanism, or a review of ways that existing accounts of mechanism fail to describe one or other aspects of medical practice, we instead work from a starting position that one of us has previously called the "mosaic view" of causality (Illari and Russo 2014). According to this mosaic view, the objective of research by practically and historically engaged philosophers of science is not to find *The-One* definition of causality, but instead to understand what role related notions play in our epistemologies and methodologies. In a similar vein, we here focus on exploring the use of mechanisms in the methodologies and epistemologies of medicine. In this we are motivated by (amongst others) Andrea Woody's functionalist account of explanation. Her approach is to "think about where and when explanations are sought and formulated, and subsequently to consider what role(s) they might play in practice" (Woody 2015: 81).

Our aim is to provide a related inquiry, concentrating on the ways that mechanisms are sought, formulated, and used in medicine. We take the main aims of medicine to be to understand and intervene on the health of individuals and of populations, where those interventions seek to cure, mitigate, or prevent disruptions to human health.

Mechanisms, we submit, are found to the point of ubiquity in medicine. Because of this, we will use six "episodes" to draw out some of the role(s) that mechanisms play in making sense of medicine. In section 2, we introduce these episodes in a fairly descriptive way. We then, in section 3, analyze these episodes to draw lessons about mechanisms in medicine. Finally, in sections 4 and 5, we reach some tentative conclusions about mechanisms in medicine. Here, a major job is to answer the following question: if the mechanisms project often or usually looks to the sciences for inspiration, why is there such a mis-match between the high prominence of mechanisms in medical practice, and the much lower level of attention that medical mechanisms have received from philosophers?

2. Examples of mechanisms in medicine

We begin with what seems a simple fact of the matter: talk of mechanisms is nearly ubiquitous in medical practice. This is not to claim that mechanisms are either necessary or sufficient to establish the causes and effects of health and disease, but just to notice that mechanisms enter, very, very frequently, into *several* inferential practices in the medical sciences. The interesting question then becomes what these different uses of mechanisms have in common or in which respects they differ. This is in line with Woody's functionalist approach.

To build our argument for this diversity, we present some "episodes" of medical mechanisms at work. We borrow from Chang (2011: 110ff) the term "episode," rather than "case study," to emphasize these are selected as exemplary cases of the numerous uses of mechanisms in medicine, rather than unique instances chosen *ad hoc* for the purpose of the present discussion. To develop this point slightly, we intend to tell a diverse group of stories about the practices of the medical sciences, present and historical. Yet we hope that the term "episode" will suggest that, while each tells a different story about medicine, these stories have something in common from the perspective of researchers interested in mechanisms.

Recall that we take the main aims of medicine to concern understanding and intervening on the health of individuals and of populations. This means we use the term "medicine" to include all clinical, scientific, and even political forms of engagement with health and disease (for a detailed discussion, see Clarke and Russo 2016). We therefore use "medicine" as an umbrella term with the aim of moving toward a broad and inclusive understanding of medicine which—we hope—will foster a thorough discussion of the role of mechanisms in this broad-church medicine. It is in this sense that, in this chapter, we explore the *applied* epistemology of mechanisms in medicine. We submit that mechanisms are powerful tools for understanding, establishing, and intervening on causal relations in medicine. However, we worry that insufficient attention has been paid to the *different* ways that an understanding of mechanisms can contribute to this applied epistemology. Hence our six episodes.

From the discussion of the episodes it will become clear (i) that mechanisms *contribute to*, rather than constitute, several inferential practices in medicine and (ii) that the effects of these contributions are extremely heterogeneous. We will go on to think more analytically about these various contributions in section 3.

As a first case, consider aspirin, which has been widely used as analgesic, antipyretic, and anti-inflammatory since its synthesis by Hoffmann in 1897 (Schrör 1997: 349). Its effects are so well known that the example of aspirin is often used by philosophers of medicine to argue that mechanisms are not needed to establish causal relations (e.g. Howick 2011: 930). At first sight, the argument seems compelling: the efficacy of aspirin as a painkiller was known for many decades before its mechanism of action was understood. Many of us, too, will have taken and trusted aspirin to (say) relieve our headache despite (we venture) few of us possessing the knowledge of the relevant mechanisms that explained why it was effective. Yet the analgesic, antipyretic, and anti-inflammatory effects are not the only effects of aspirin of interest to medical practice. Aspirin is widely used because of its effect on platelet function. Taking regular low-dose aspirin improves cardiovascular outcomes (Antithrombotic Trialists' Collaboration 2002). Yet, we argue, this role of aspirin is not as perspicuous as its painkilling one. Instead, knowledge of this effect was intimately linked to understanding the mechanism by which the drug worked.

We summarize here the main steps of how the effect of aspirin on platelet function was established, based on Schrör's presentation (1997: 349–50). Quick reported that aspirin increased bleeding time (which is a measure of the overall rate at which blood clots) in 1967. This was shortly followed by several reports that low-dose aspirin appeared to inhibit platelet aggregation—in itself, an important part of blood clotting (Weiss, Aledort, and Kochawa 1968; O'Brien 1968). This inhibition began within two hours of taking the aspirin, and lasted for several days. The investigation of this impressive degree of platelet aggregation in turn led to the finding that aspirin inhibited prostaglandin synthesis (Vane 1971). Incidentally, this inhibition of prostaglandin biosynthesis was later to provide the roots of the mechanistic explanations of the other actions of aspirin. Smith and Willis (1971) then identified that the inhibition of prostaglandin biosynthesis was "the mechanism of the antiplatelet action of aspirin" (Schrör 1997: 350), which was traced specifically to inhibition of the cyclooxygenase (or COX) enzyme in 1975, and to a specific amino acid residue in the COX-1 enzyme in 1991. Further mechanistic research on COX would reveal routes for future interventions that could capitalize on aspirin's beneficial effects, while hopefully avoiding its adverse effects.

Our second example is high-density lipoprotein (HDL)-raising drugs. Observational evidence suggests high levels of HDL in the blood are inversely correlated with heart disease. But is this correlation due to an underlying causal (preventive) relationship between heart disease and HDL?

How about investigating this question by using HDL-raising as an intervention designed to prevent heart disease? A recently developed class of drugs called cholesteryl ester transfer protein (CETP) inhibitors seemed to promise just such an investigation. Clinical trials of the drug torcetrapib seemed effective at raising HDL levels (Brousseau et al. 2004). However, the drug did not improve clinical outcomes. In fact, it severely worsened them, leading to the abandoning of a large phase III clinical trial (known as ILLUMINATE) in 2006, owing to excess deaths in patients receiving torcetrapib. This result was unexpected—so much so that one commentator suggested that this result should lead us to conclude that "We know so much about the cholesterol pathway, but we never seem to know what matters" (Lehrer 2011). The trial authors (Barter et al.) were unsurprisingly more measured in their analysis of the trial result, and responded with some educated speculation as to its cause(s):

> Clinical trials such as ours are not designed to elucidate mechanisms of either benefit or harm associated with the use of a drug. However, they may provide clues that have the potential to inform future research. . . . There are at least two possible explanations for the observation of increased mortality and morbidity associated with the use of torcetrapib in our study: an off-target effect of torcetrapib, unrelated to CETP inhibition, and an adverse effect of CETP inhibition per se, with the possible generation of dysfunctional or even proatherogenic HDL cholesterol.
>
> *(Barter et al. 2007)*

Further research has suggested that the "off-target" mechanism suggested by Barter et al. may be correct. It appears that torcetrapib increases blood pressure (Tall, Yvan-Charvet, and Wang 2007)—itself a well-known cause of cardiac death. If that is the case, then the increased mortality and morbidity found in the ILLUMINATE trial appear to be adverse consequences of torcetrapib specifically, and therefore unlikely to be replicated in clinical research on alternative CETP inhibitors (such as evacetrapib and anacetrapib).

The third example is that asbestos is known to be responsible for fatal diseases such as asbestosis and mesothelioma. The biochemical mechanisms that connect exposure to asbestos fibers with the development of cancer have been studied and examined carefully (see e.g. IARC Working Group 2012). However, work remains to be done studying the (largely social) mechanisms by which exposure to asbestos occurs. For example, occupational medicine researchers have been studying the disease from the perspective of the workplace. This led to the study of populations living close to asbestos factories. Examples abound across different geographical locations. We might mention, for instance, Barking in the United Kingdom (Greenberg 2003) and Eternit in Italy, for which a memorable sentence was issued in 2009 after a long and difficult trial (Mossano 2011; Allen and Kazan-Allen 2012): the owners of the asbestos multinational were deemed guilty of fraudulent environmental disaster and omission. In this context, and because of their bearing on legal and policy questions, some aspects of the mechanisms underlying asbestos exposure remain disputed, for instance latency (see, e.g., Terracini et al. 2014; La Vecchia and Boffetta 2014). The point here is that mechanisms investigated via occupational and environmental epidemiology largely do not concern the way that asbestos causes mesothelioma. This, as we shall further discuss later, poses a question for the "overbiologization" of diseases, namely the reduction of disease causation to biochemical reactions in the body.

In the philosophy of medicine our fourth example, the discovery of *Helicobacter pylori* as a cause of gastric ulcer, has been used to illustrate how hypotheses are generated in medicine (Gillies 2005; Hutton 2012; Thagard 1998a, 1998b). The same episode has also been used to illustrate the mutual need for evidence of difference-making *and* of mechanisms in establishing causal claims in medicine (Russo and Williamson 2007). This body of literature showcased how the use of mechanisms in given inferential practices *also* depends on the available theoretical framework. In this case, available background knowledge had it that bacteria could not live in acid environments such as the stomach. This tended to preclude the hypothesis that *H. pylori* could be a cause of gastric ulcer, and until the mechanism by which this bacteria could survive in low-pH environments had been investigated, the causal claim could not be properly assessed.

Another stock example in the philosophy of medicine, and our fifth, is the story of Ignaz Semmelweis, a doctor active in nineteenth-century Vienna. His notoriety is due to his hypothesis that puerperal fever was due to some infection. The mechanisms had not been clarified at that time, but his suggested preventive intervention was remarkably simple: doctors should wash their hands after performing autopsies and before assisting women in labor. Part of the debate in philosophy of medicine concerns the question of whether the scientific community was right or wrong in rejecting Semmelweis' precautionary measure on the basis of available evidence and of the theoretical framework supporting the intervention.

The sixth and final episode we want to present is the *comparison* between approaches to health and disease. This becomes highly relevant in times where the public is showing skepticism and mistrust for so-called "Western medicine" and increasingly seeking advice from "alternative" approaches. "Alternative" or "complementary" medicine is, however, a basket where too often anything that is "non-Western" is placed. Instead, care is needed in making comparisons, and non-Western medical traditions ought not to be conflated.

To illustrate the relevance of our point, we rely on the contribution of Hugh Shapiro in the volume *Medicine across Cultures* (Shapiro 2003). Shapiro provides an invaluable contextualization of what is usually called "Chinese medicine," including a brief history of its relation to Western approaches. Shapiro then explains that from a practice of forensic medicine and of dissection, which was common to both Chinese and Western approaches, very different conceptualizations of the body (anatomy and physiology) and of pathology derived. Moreover, this led to very different developments: the study of the living body in Chinese medicine and the study of the dead body in Western medicine. Shapiro illustrates these claims with the case of "nerves."

At the beginning of the last century the medical community gathered to translate and standardize terms coming from the West. It became clear that there wasn't a one-to-one correspondence. For instance, Chinese lacked both the concept and the word for "nerve." Since the Renaissance, attempts to translate Western medical concepts into Chinese ones had to make recourse to periphrases and to neologisms that would make sense in their conceptual framework. Thus, for instance, nerve was used often as synonymous with sinew, and its function was to transmit the "vital power." Chinese doctors have known for a long time how to intervene on this vital power, using the technique of acupuncture. The lack of a corresponding term for "nerve" in Chinese testifies, argues Shapiro, to the different conceptions and understanding of the body and of the phenomena of health and disease. This, it should be noted, holds for Western culture too. In fact, Shapiro continues, in the West the understanding of what nerves are and of how they function has been related to the character trait of "volition," an idea that traces back to Greek medicine. Moreover, this is intimately connected with action and especially volitional actions, which are defining features of identity.

3. Mechanisms contribute to inferences in medicine

After this overview of episodes of mechanisms in medicine, it would be tempting to try and pin down The-One definition that fits them all. In accordance with the "mosaic view" and the "functionalist approach" we espoused (see section 1), we will not pursue this objective here. For our purposes, we can safely rely on the working characterization by Jon Williamson and Phyllis Illari: "A mechanism for a phenomenon consists of entities and activities organized in such a way that they are responsible for the phenomenon" (Illari and Williamson 2012: 120).

This kind of view is now routinely referred to as "minimal mechanism" (see also Chapter 1). This characterization avoids equating mechanisms with Cartesian deterministic machines (see Chapter 3). Also, this characterization is flexible enough to accommodate mechanisms that travel across a generic or population-level case and a single, individual case. It allows us to deal with biochemical, pharmacological, or social mechanisms of health and disease. These and other aspects of mechanisms are discussed next.

Returning to our first episode, aspirin, examples like this are instructive in that they illustrate, in a concrete way, the difficulties that can be experienced in finding effects in medicine. This is hard—effects are complex, and we think that researchers need to know about mechanisms to reliably find and measure many effects. While there are exceptions—you need no detailed knowledge to know that an anesthetic is effective in sending someone to sleep—it can be extremely hard to discover and understand all the clinically relevant effects of an anesthetic (or, in the aspirin case, analgesic) drug. This is also the case when it comes to finding side effects (such as gastric and duodenal ulcers, Reye's syndrome, and exacerbations of asthma in the case of aspirin).

The second example, HDL-raising drugs, shows that medical effects of interest may result from complex systems of mechanisms that are difficult to separate (like the hypertensive effect of torcetrapid). The complexity comes from the interaction of drug(s) with a pathology, or rather with *several* pathologies. In fact, cases of co-morbidity (i.e., a patient that has more than one pathological condition at one time) are the norm rather than the exception. Knowledge of mechanisms is necessary to explain, predict, and treat, especially in these cases. Differently put, the situation of one-drug-one-pathology rarely occurs. Knowledge of mechanisms doesn't come from large clinical trials *alone* (as Barter et al. noted in their quote given above), but is instead to be complemented with other types of studies, for instance lab experiments.

Cases like the third, asbestos, point to the question of whether, in the light of the stunning advancement of biomedicine, disease ought to be reduced to biochemical reactions in the body. On the one hand, in the medical sciences, attention *is* given to social mechanisms of health and disease, but these are often considered "distant," as "adjunct" descriptions of how biological mechanisms of health and disease correlate with socio-economic differences or inequalities across individuals. On the other hand, the sociology of health and the behavioral sciences describe the role that social, psychological, economic, or behavioral factors play in the etiology of disease. To be sure, this is an area where evidence of difference-making (*that* social factors make a difference to the occurrence of disease) exceeds evidence of mechanisms (*how* social factors intervene in the mechanisms of disease causation). The philosophy of mechanisms has a chance there in developing an account of *mixed mechanisms*, where both biological and social factors play an active role (Kelly et al. 2014). This is not just to elucidate the etiology of disease, but also to better plan public health interventions. In fact, while most (non-communicable) diseases have a proper explanation in terms of biochemical mechanisms, successful public health interventions in the past targeted social determinates, lifestyles, and basic hygiene or sanitary measures; to be sure, this is still the case nowadays, as evidenced by major public health interventions in Brazil (see e.g. Barreto et al. 2010; Barreto and Aquino 2009). In a nutshell, bio-social mechanisms of health and disease are needed to plan, research, and implement interventions. This is because successful interventions need a complex view of what counts as relevantly beneficial or harmful.

The fourth case, *H. pylori* and gastric ulcer, is interesting in that it shows that mechanisms do not function as an "experimentum crucis" for a theory, nor are they "deductively" inferred from available theories. The search for explanatory and therapeutic mechanisms requires constant interplay with available background knowledge. Hunting for specific mechanisms, like that linking *H. pylori* infection with stomach ulcers, might confirm or disconfirm existing background knowledge. It is undeniable that experiments depend on mechanisms; more precisely, on *knowledge* of mechanisms. However, without standards to assess the quality of knowledge of mechanisms, poor-quality assertions and speculation ("the stomach is sterile") appear to fill the gaps that should instead be populated with empirically based, carefully researched knowledge about mechanisms. This allows us to introduce a *normative* dimension of the mechanism project in medicine, next to the descriptive one carried out in the previous section. Mechanisms support the infrastructure of medical inference. To be even more precise, they support a *variety* of medical inferences at different stages of the scientific process. They aren't just a way of expressing existing knowledge, but instead actively participate in producing new knowledge. Mechanisms have a vital role in the setup of trials or of lab experiments. And the results of these scientific practices may lend further support to these mechanisms, or they may lead us to significantly revise our knowledge base (as Craver and Darden 2013 suggest).

The fifth case, puerperal fever, is an instructive case for several reasons. One is that we should avoid "presentism": it is wrong to assess the past with the theories available today. The recent debate on evidential pluralism (see Broadbent 2011) focuses on whether the scientific community, at the time of Semmelweis, was right or wrong in rejecting his preventive measure. The point here is that this measure was not supported by a solid theoretical background, as this was happening well before the germ theory of disease had been developed. Given what we know *today* about bacteria and so on, it is too easy a judgment to say that the scientific community at that time was wrong in rejecting Semmelweis. The point here is of course broader than just *this* specific episode. It is a point of philosophical methodology and of how to do philosophy of science, properly informed by the history of science. Yet much is historically disputed—indeed, "misleading" (Tulodziecki 2013: 1074) in the Semmelweis case. Another reason why this episode is relevant to us is that, even if the historical quarrel cannot be settled, this can provide important lessons about the present. In particular, examples like these should encourage reflection on what to do when knowledge of mechanisms is incomplete or highly uncertain. What actions are nonetheless justified? This is where proper attention to mechanisms in philosophy of medicine pushes back the frontier of epistemology and of methodology. In fact, exquisitely epistemological questions about the (un)certainty of current medical knowledge quickly turn into ethico-political questions about what justifies the implementation (or refusal) of interventions or preventive measures. A case in point here is the current controversy over vaccination programs. Apart from the disputed question of the correlation of vaccines and autism, another question concerns the tension between the individual freedom not to vaccinate versus the public health duty to protect the population at large by promoting vaccination to all individuals.

Health, disease, and worldviews, the sixth episode, is an area where *much* work is needed, from a philosophical, historical, and also socio-anthropological perspective. These sorts of comparison are instructive in that they remind us that mechanisms, their entities and activities, are not *given*. They are instead products of our investigations into the phenomena of health and disease. From an epistemological point of view, this means that mechanisms are part of our inferential systems—a point that already arose earlier in the reflection on the case of *Helicobacter pylori*. But mechanisms are also a part of our inferential system in another sense: different understandings of the phenomena of health and disease may lead to *incommensurable* mechanisms, or to syndromes that are culture-bound (Guarnaccia and Rogler 1999). Thus all these phenomena, as well as the investigations and explanations thereof, are deeply embedded in cultural factors, in the East as well as in the West. In our sixth episode, this was shown in the way the nervous system was conceptualized differently in Western and in Chinese medicine, and in the different terms used to refer to parts of the body and to its functioning. This should warn us about the perils of adopting a naïvely scientist attitude and imperialistic imposition of Western standards in an uncritical way.

4. Evidence and contrastive focus

The sextet presented in section 2, and then analyzed in section 3, reveals many issues of interest to those thinking about mechanisms. While we would like to follow the episodes with a detailed and exhaustive theoretical discussions of the issues, we will restrict ourselves to a discussion of just two of them. The first—evidence of mechanism—is a hotly disputed issue in the current debate. We will develop our account of evidence of mechanism in medicine from the six episodes, as a useful contribution. The second—contrastive focus—has not (we think) received sufficient attention from philosophers interested in mechanisms. We therefore highlight it here by way of raising interest in it as a research problem more widely.

We begin by considering evidence, reasoning, and narratives in medicine. As the aspirin example suggests, mechanisms do not typically appear as ready-made pieces of knowledge. A great deal of work is needed to acquire and establish knowledge of mechanism. Here, the notion of *evidence* plays a crucial role. We borrow from Illari the definition of *evidence of mechanism*. She takes evidence of mechanisms as being "evidence of the actual existence of the postulated mechanism linking cause and effect" (Illari 2011: 120), and we share her position.

While schematic, this definition is sufficiently clear to warrant some initial clarification regarding our thoughts about the use of mechanisms in medicine as evidence. What is at stake, in fact, is a distinction between *evidence* of mechanism and mechanistic *reasoning*, around which substantial disagreement exists. For instance, Howick et al. define mechanistic reasoning as follows: "the inference from mechanisms to claims that an intervention produced a patient-relevant outcome. Such reasoning will involve an inferential chain linking the intervention (such as antiarrhythmic drugs) with the outcome (such as mortality)" (Howick et al. 2010: 434).[2]

We take this to mean that mechanistic reasoning would involve making a clinical decision about the likely efficacy or harms of a medical intervention *just* by thinking about mechanisms. We do not endorse this way of using mechanisms to make medical decisions as a normative aspiration. Nor do we think it is a good description of the generality of medical practice (it seems hard to connect to our episodes). We instead think that reasoning about mechanisms *alone* is highly unlikely to make for reliable clinical inferences. As we have argued in other places (Clarke et al. 2013, 2014), a kind of *pragmatic* evidential pluralism featuring some contribution from evidence of relevant mechanisms may instead help make better medical inferences. According to this view, evidence of mechanism participates, along with *other* kinds of relevant evidence such as that produced by randomized clinical trials (RCTs), in various inferential practices. For the purposes of this chapter, though, it suffices to note that our *evidence of mechanism* is just one kind of evidence (amongst many), rather than the ambitious and free-standing *mechanistic reasoning* discussed by Howick et al. Along with Solomon (2015: 123–4), we therefore note that *evidence of mechanism* and *mechanistic reasoning* cannot be used interchangeably.[3]

A further distinction that we wish to draw is between evidence of mechanism and *narratives*. While the role of narratives in medicine (see Kleinman 1989; Greenhalgh 1999; Montgomery 2006) and in science more generally (Roth 1988, 1989) is a topic that deserves more philosophical attention than it gets, it is important to note that we think there is a difference between evidence of mechanism and of the *narrative(s) that may be used to* describe, present, discuss, or criticize said evidence of mechanism.

It's possible to use (empirically grounded) narratives to describe mechanisms, but that is not the same thing as saying that mechanisms just *are* narratives (see also Chapter 31). We think that any scientific result that is expressed in ordinary language can be understood as a narrative, without the converse assertion that all scientific results are just narratives. More positively, we think that there are standardized rules that govern the relation between a result and its narratives, which supply rigid structure on the presentation of those narratives. Consider how a scientific article must present a research question, the data, the analyses, and the results.

We turn now to our second main point. One aspect that many mechanisms in the episodes above share is *contrastive focus*. Mechanisms, as also discussed elsewhere in this volume (see Chapter 16), help answer the question "*How* does C cause E?" Medicine is no exception to this use of mechanisms. Typically, emphasis is given to those characteristics of mechanisms (most notably, *organization*) that participate in explaining *how* C causes (in the sense of *producing*) E.

In this section, though, we want instead to draw attention to another way that we might interpret the question "How does C cause E?" Rather than give an answer that highlights the productive continuity that obtains between C and E, we might instead give an answer that explains why our C causes E rather than E_1, or that C, rather than C_1, causes E. This is the question of *contrastive focus*, which has been fairly extensively discussed in relation to causality (Dretske 1972; Schaffer 2005; Northcott 2008). Here, we aim to show that attention to contrastive focus is key to several inferential practices in medicine, and that dealing with questions of contrastive focus often involves thinking about mechanisms, in different ways, also illustrated by the sextet of episodes. We highlight here two specific aspects of contrastive focus as applied to medicine: (i) normal vs pathological and (ii) biological vs social. Simply put, the idea is that the aspects of contrastive focus we discuss here concern (i) different mechanisms describing normal or pathological behavior and (ii) the factors—biological or social—that are at work in such mechanisms.

To begin with, mechanisms that deal with the etiology of pathology typically do so in a contrastive way with normal physiology (see also Chapter 8). For example, the mechanism that explains how infection with *H. pylori* leads to ulcers does so in a contrastive way with normal physiology (see the extremely thorough review by Kusters, van Vliet, and Kuipers 2006 for further details). This means that we need a *generic* mechanism about normal physiology *and* a suitable contrastive model of a mechanism that explains pathogenic behavior. Many experimental studies in biomedicine are based on this idea—they effectively compare subjects that do and do not receive some intervention. The very same idea is at the basis of extrapolation from animal studies to humans (see e.g. Steel 2007).

This means that contrastive focus is at the basis of RCTs too. In fact, one way to use RCTs is precisely to differentiate between the effect E_1 rather than E_2, or between the role of C_1 rather than C_2. This adds to the recent debates on the role of mechanisms in RCTs. Specifically, according to evidence-based medicine (EBM)-theorizers, mechanisms play little or no role at all in establishing causal relations—a view challenged by supporters of "evidential pluralism," who hold the view that to establish causal relations one typically needs multifarious evidence (for a discussion, see Clarke et al. 2014).

The contrastive focus can also participate in establishing causal relations at the individual, single-case level, precisely using considerations about what is (thought to be) normal and what is not. This, however, is far from being an easy task, as testified by the several challenges of diagnosis. We shall focus here on just one challenge. To make a diagnosis in the single case, we need to contrast it with a generic mechanism. But how *generic* should a generic mechanism be?

There is no principled answer to this question. One problem has to do with finding mechanisms that are stable in sufficiently homogeneous reference classes. A reference class is homogenous when all its members have the same characteristics. This is usually not the case in medicine—we expect those in the reference class of "patients with heart disease" to be suffering from a range of different conditions. Homogeneous reference classes—where all individuals have exactly the same conditions—are either very scarce or very hard to find in medicine. A notable example in this respect is the study of pathologies that frequently co-occur with other pathologies. As the HDL-raising drug example above shows, it can be extremely hard to unpick the contributions of different conditions with similar effects. In the HDL-raising case, we can presumably assume that this isn't just one mechanism at work, but several nested and interconnected mechanisms. This is because people using HDL-raising drugs have very different conditions, and most of the time *multiple* pathologies, not just heart disease. Consequently, the contrastive focus of mechanisms should not simply be in terms of differentiating the effect of a drug against a placebo. Instead, drugs should be tested taking these other mechanisms into account, including interactions with other drugs.

The second aspect of the issue of contrastive focus is whether only biological factors should contribute to answering the question "How does C cause E?" We encountered this issue in the asbestos example, where a relevant question is whether workplace or other social factors are appropriate to be included in the etiology of the disease. So, when we check for appropriate contrasts between reference classes, we shouldn't just consider biochemical factors (i.e. those biochemical factors marking normal vs pathological behavior), but also psycho-demo-socio-economic ones. Asbestos is of course not the only case. Another example is epigenetics, which attempts to reconstruct the effect of the environment, including *in utero* events, at the genetic level even one generation later. The contrastive focus here might be with events that happened much earlier on in our lives, if not generations before. So the mechanisms of health and disease may be acting over long time spans, and the causes may reside in events that cannot be assessed at the time of diagnoses. Studies on the "Dutch famine" of 1944–5 are a good case in point. According to these studies, the children of women that were pregnant during the famine are more susceptible to develop diseases such as obesity, diabetes, cardiovascular conditions, etc. Epigenetics attempts to shed light on the long-term mechanisms that, by affecting women during pregnancy, also affected their offspring even decades after birth (see Roseboom, de Rooij, and Painter 2006 for an introduction).

The health sciences (including sociology of health) have recognized for a long time that health and disease are not independent of socio-economic factors (for an overview, see Kelly et al. 2014). But this recognition is typically explicated in two ways. On the one hand, socio-economic factors are merely correlated with health and disease, and they are used to map which parts of the population are healthier and which are more ill. But here socio-economic factors do not actively participate in the production of health and disease. On the other hand, the action of psycho-socio-demo-economic factors is totally explained away by reducing them to the action of biochemical factors. Here, a proper understanding of how these factors *actively contribute* to health and disease is missing. Differently put, the mechanisms of health and disease are not just biochemical or just socio-eco-psychological. They are most often *mixed*, a blend of those categories, and in need of further investigation within medicine and public health, and philosophy too. Most non-communicable diseases, from obesity to type-2 diabetes, and alcohol-related diseases arguably require mixed mechanisms both for understanding and for intervening on their causes and effect (see also Kelly et al. 2014). Admittedly, these (and many others) are cases where what we *don't* know exceeds by far what we know. And, precisely for this reason, thinking in terms of how mechanisms contribute to several inferential practices (rather than just working out The-One definition of mechanism) may be of great help.

5. Conclusions

As shown by the above sextet of examples, we have a very broad and inclusive understanding of what medicine is and does. We hope that this will help redress part of the existing debates in philosophy of medicine, particularly for what we call the "narrow view of medicine," held by a number of authors, both from medicine and from philosophy.

For many philosophers in recent years, medicine has been largely synonymous with EBM. This is perhaps an artifact, due to how philosophers of science came to pay attention to medicine. The establishment of EBM as a dominant paradigm prompted a peculiar reaction in the philosophical community, which was captured by the iconic paper by John Worrall (2002). He asked: "What on earth was medicine based on before?" This might explain why so much attention was, since then, devoted to this *sub*-field of medicine. There is much more to medicine than evaluating the efficacy and harms of drugs. At the same time, particularly in the way that philosophical work on EBM intersects with other interests of analytic philosophers of science (particularly the philosophy of statistics), we see just why this narrowing might be both useful and comforting. However, when it comes to thinking about mechanisms, we worry about this narrowness, largely because (we think) it has led to the pre-eminence of a narrow set of related questions that are ostensibly about medicine, but are really about statistical inference in the context of the clinical trial: evidence, bias, and so on. In turn, this has led to a caricature of medical practice such that mechanisms are either (a) described as playing a negligibly minor evidential role or (b) treated as a simple (but faulty!) alternative method of inference to the clinical trial. Both of these caricatures, we submit, are inaccurate. It is high time to broaden the scope of the debate, by looking at the very many practices in medicine, including, but not restricted to, EBM.

The approach taken in this chapter points to a *pluralistic* account of medicine, and of mechanisms in medicine. This stems from the different ways in which mechanisms contribute to medical practices. It is in this sense that we have explored the *applied* epistemology of mechanisms in medicine. We submit that mechanisms can be used as powerful conceptual tools for establishing and intervening on causal relations in medicine. However, we worry that insufficient attention has been paid to the *different* ways that an understanding of mechanisms can contribute to this applied epistemology. This is all the more important, because the philosophy of mechanisms itself became somewhat specialized, discussing mechanisms in specific, isolated domains: in biology, in psychology, or in sociology. But mechanisms, in medicine or elsewhere, are seldom intrinsically biological, psychological, or sociological. We hope that our approach to mechanisms in medicine will be an encouragement to start crossing the borders of these sub-fields in the philosophy of mechanisms.

Notes

1 The authors gratefully acknowledge financial support from the Arts and Humanities Research Council (AHRC) (UK) for the research done under grant number AH/M005917/1 ("Evaluating Evidence in Medicine").

2 Howick also gave a similar formulation in a later solo article: "Mechanistic reasoning is an inferential chain (or web) linking the intervention (such as HRT) with a patient-relevant outcome, via relevant mechanisms" (Howick 2011: 929).

3 A brief clarification is in order here. We do not agree that our previous work (Clarke et al. 2013, 2014) endorsed the equivalence of mechanistic reasoning and mechanistic evidence, as Miriam Solomon seems to suggest: "Although there are differences between Howick (2011b), Andersen (2012), and Clarke et al. (2013), they all share the assumption that mechanistic reasoning should be regarded as mechanistic evidence, moreover, evidence that has a place in the hierarchy. This is an assumption that I challenge" (Solomon 2015: 120).

References

Allen, D., and L. Kazan-Allen. 2012. *Eternit and the Great Asbestos Trial*. London: The International Ban Asbestos Secretariat. http://www.ibasecretariat.org/eternit-great-asbestos-trial-toc.htm.

Andersen, H. 2012. Mechanisms: What are they evidence for in evidence-based medicine? *Journal of Evaluation in Clinical Practice*. 18(5): 992–9.

Antithrombotic Trialists' Collaboration. 2002. Collaborative meta-analysis of randomised trials of antiplatelet therapy for prevention of death, myocardial infarction, and stroke in high risk patients. *British Medical Journal*. 324(7329): 71–86.

Barreto, M. L., and Aquino, R. (2009). Recent positive developments in the Brazilian health system. *American Journal of Public Health*, 99(1): 8. doi:10.2105/AJPH.2008.153791.

Barreto, M. L., Genser, B., Strina, A., Teixeira, M. G., Assis, A. M. O., Rego, R. F., . . . Cairncross, S. 2010. Impact of a citywide sanitation program in northeast Brazil on intestinal parasites infection in young children. *Environmental Health Perspectives*. 118(11): 1637–42. doi:10.1289/ehp.1002058.

Barter, Philip J., Caulfield, M., Erikson, M. . . . Brewer, B. 2007. Effects of torcetrapib in patients at high risk for coronary events. *New England Journal of Medicine*. 357:2109–22. doi:10.1056/NEJMoa0706628.

Broadbent, A. 2011. Inferring causation in epidemiology: Mechanisms, black boxes, and contrasts. In Illari, P. M., Russo, F. and Williamson, J. (eds). *Causality in the Sciences*. Oxford University Press. 45–69.

Brousseau, M. E., Schaefer, E. J., Wolfe, M. L., Bloedon, L. T., Digenio, A. G., Clark, R. W., Mancuso, J. P., and Rader, D. J. 2004. Effects of an inhibitor of cholesteryl ester transfer protein on HDL cholesterol. *New England Journal of Medicine*. 350(15): 1505–15. doi:10.1056/NEJMoa031766.

Chang, H. 2011. Beyond case-studies: History as philosophy. In Mauskopf, Seymour and Schmaltz, Tad (eds). *Integrating History and Philosophy of Science*. Springer. 109–24.

Clarke, B., Gillies, D., Illari, P., Russo, F., and Williamson, J., 2013. The evidence that evidence-based medicine omits. *Preventive Medicine*. 57(6): 745–7.

Clarke, B., Gillies, D., Illari, P., Russo, F., and Williamson, J. 2014. Mechanisms and the evidence hierarchy. *Topoi*. 33(2): 339–60. doi:10.1007/s11245-013-9220-9.

Clarke, B., and Russo, F., 2016. Causation in medicine. In J. Marcum (ed.). *Companion to Contemporary Philosophy of Medicine*. Bloomsbury. 297–322.

Craver, C. F., and Darden, L. 2013. *In Search of Mechanisms: Discoveries across the Life Sciences*. University of Chicago Press.

Dretske, F. I. 1972. Contrastive statements. *The Philosophical Review*. 81(4): 411–37.

Gillies, D. 2005. Hempelian and Kuhnian approaches in the philosophy of medicine: The Semmelweis case. *Studies in History and Philosophy of Science Part C*. 36(1): 159–81. doi:10.1016/j.shpsc.2004.12.003.

Greenberg, M. 2003. Cape asbestos, Barking, health and environment; 1928–1946. *American Journal of Industrial Medicine*. 43(2): 109–19.

Greenhalgh, T. 1999. Narrative-based medicine in an evidence-based world. *BMJ*. 318(7179): 323–5.

Guarnaccia, P. J., and Rogler, L. H. 1999. Research on culture-bound syndromes: New directions. *American Journal of Psychiatry*. 156(9): 1322–7. doi:10.1176/ajp.156.9.1322.

Howick, J. 2011. Exposing the vanities – and a qualified defence – of mechanistic evidence in clinical decision-making. *Philosophy of Science*. 78(5): 926–40. Proceedings of the Biennial PSA 2010.

Howick, J., Glasziou, P., and Aronson, J. K. 2010. Evidence-based mechanistic reasoning. *Journal of the Royal Society of Medicine*. 103(11): 433–41. doi:10.1258/jrsm.2010.100146.

Hutton, J. 2012. Composite paradigms in medicine: Analysing Gillies' claim of reclassification of disease without paradigm shift in the case of Helicobacter pylori. *Studies in History and Philosophy of Science Part C*. 43(3): 643–54. doi:10.1016/j.shpsc.2012.05.003.

IARC Working Group. 2012. *Arsenic, metals, fibres and dusts*. International Agency for Research on Cancer. http://monographs.iarc.fr/ENG/Monographs/vol100C/index.php.

Illari, P. M. 2011. Mechanistic evidence: Disambiguating the Russo-Williamson Thesis. *International Studies in the Philosophy of Science*. 25(2): 139–57.

Illari, P. and Russo, F., 2014. *Causality: Philosophical Theory Meets Scientific Practice*. Oxford: Oxford University Press.

Illari, P. M., and Williamson, J. 2012. What is a mechanism?: Thinking about mechanisms across the sciences. *European Journal of the Philosophy of Science*. 2: 119–35.

Kelly, M. P., Kelly, R. S., and Russo, F. 2014. The integration of social, behavioural, and biological mechanisms in models of pathogenesis. *Perspectives in Biology and Medicine*. 57(3): 308–28.

Kleinman, Arthur. 1989. *Illness Narratives: Suffering, Healing and the Human Condition*. Basic Books.

Kusters, J. G., van Vliet, A. H. M., and Kuipers, E. J. 2006. Pathogenesis of *Helicobacter pylori* infection. *Clinical Microbiology Reviews*. 19(3): 449–90 doi:10.1128/CMR.00054-05.

La Vecchia, C., and Boffetta, P. 2014. A critique of a review on the relationship between asbestos exposure and the risk of mesothelioma: Reply." *European Journal of Cancer Prevention*. 23(5): 494–6. doi:10.1097/CEJ.0000000000000051.

Lehrer, Jonah. 2011. Trials and errors: Why science is failing us. *Wired*. https://www.wired.com/2011/12/ff_causation/all/1.

Montgomery, K. 2006. *How Doctors Think: Clinical Judgment and the Practice of Medicine*. Oxford University Press.

Mossano, Silvana. 2011. The Eternit trial: The verdict is close. *Epidemiologia E Prevensione*. 35(3–4): 175–7.

Northcott, R. 2008. Causation and contrast classes. *Philosophical Studies*. 139(1): 111–23.

O'Brien, J. R. 1968. Effects of salicylates on human platelets. *Lancet* 1: 779–83.

Quick, A. J. 1967. Salicylates and bleeding: The aspirin tolerance test. *The American Journal of the Medical Sciences*. 252(3): 265–9.

Roseboom, T., de Rooij, S., and Painter, R. 2006. The Dutch famine and its long-term consequences for adult health. *Early Human Development*. 82(8): 485–91.

Roth, P. A. 1988. Narrative explanations: The case of history. *History and Theory*. 27(1): 1–13.

Roth, P. A. 1989. How narratives explain. *Social Research*. 51(2): 449–78.

Russo, F., and Williamson, J. 2007. Interpreting causality in the health sciences. *International Studies in the Philosophy of Science*. 21(2): 157–70.

Schaffer, J. 2005. Contrastive causation. *The Philosophical Review*. 114(3): 297–328.

Schrör, Karsten. 1997. Aspirin and platelets: The antiplatelet action of aspirin and its role in thrombosis treatment and prophylaxis. *Seminars in Thrombosis and Hemostasis*. 23(4): 349–56. doi:10.1055/s-2007-996108.

Shapiro, H. 2003. How different are Western and Chinese medicine? The case of nerves. In Selin, Helaine (ed.). *Medicine across Cultures: History and Practice of Medicine in Non-Western Cultures*. Springer. 351–72.

Smith, J. B., and Willis, A. L. 1971. Aspirin selectively inhibits prostaglandin production in human platelets. *Nature New Biol* 231: 235–7.

Solomon, M. 2015. *Making Medical Knowledge*. Oxford University Press.

Steel, D. 2007. *Across the Boundaries: Extrapolation in Biology and Social Science*. Oxford University Press.

Tall, Alan R., Yvan-Charvet, Laurent, and Wang, Nan. 2007. The failure of torcetrapib. *Arteriosclerosis, Thrombosis, and Vascular Biology*. 27(2): 257–60. doi:10.1161/01.ATV.0000256728.60226.77.

Terracini, B., Mirabelli, D., Magnani, C., Ferrante, D., Barone-Adesi, F., and Bertolotti, M. 2014. A critique to a review on the relationship between asbestos exposure and the risk of mesothelioma. *European Journal of Cancer Prevention*. 23(5): 492–4. doi:10.1097/CEJ.0000000000000057.

Thagard, P. 1998a. Ulcers and bacteria I: Discovery and acceptance. *Studies in History and Philosophy of Science Part C*. 29(1): 107–36. doi:10.1016/S1369-8486(98)00006-5.

Thagard, P. 1998b. Ulcers and bacteria II: Instruments, experiments, and social interactions. *Studies in History and Philosophy of Science Part C*. 29(2): 317–42. doi:10.1016/S1369-8486(98)00024-7.

Tulodziecki, D. 2013. Shattering the myth of Semmelweis. *Philosophy of Science*. 80(5): 1065–75. doi:10.1086/673935.

Vane, J. R. 1971. Inhibition of prostaglandin biosynthesis as a mechanism of action of aspirin-like drugs. *Nature New Biology*. 231: 232–5.

Weiss, H. J., Aledort, L. M., and Kochawa, S. 1968. The effect of salicylates on the hemostatic properties of platelets in man. *Journal of Clinical Investigation*. 47: 2169–80.

Woody, Andrea I. 2015. Re-orienting discussions of scientific explanation: A functional perspective. *Studies in History and Philosophy of Science Part A*. 52: 79–87. doi:10.1016/j.shpsa.2015.03.005.

Worrall, J. 2002. What evidence in evidence-based medicine? *Philosophy of Science*. 69(S3): S316–S330.

25

DEVELOPMENTAL MECHANISMS[1]

Alan C. Love

1. Mechanisms in developmental biology

A developmental mechanism is a mechanism that operates during development or ontogeny. Typically, the scope of development is narrowed to one of four broad problem domains that constitute the majority of contemporary developmental biology: pattern formation, growth, morphogenesis, and cellular differentiation (Love 2014, 2015). We can observe this in the following introductory discussion from *Mechanisms of Morphogenesis*:

> The purpose of this book is to bring together in one place some of the most significant advances that have been made in identifying mechanisms of morphogenesis from a variety of species, systems and scales. . . . The morphogenetic mechanisms described in later chapters tend to invoke, and 'take for granted' the fine-scale mechanisms described in earlier chapters.
>
> *(Davies 2013, 5)*

This passage assumes that there are different *types* of mechanisms operating in development: morphogenetic mechanisms and fine-scale mechanisms. The latter is shorthand for *molecular genetic* mechanisms and Davies thinks developmental biologists have fixated on them.

> [Older descriptive embryology] books, which provided the foundations on which programs of molecular biology could stand, have been succeeded in recent years by large and solid works full of well-established mechanisms of signalling, pattern formation and gene control. . . . The spotlight of developmental books and review volumes is usually focused on the molecular biology of gene control and pattern formation . . . current texts [focus on] molecular mechanisms of morphogenesis, taking for granted . . . morphogenetic mechanisms.
>
> *(Davies 2013, 4)*

Given that molecular mechanisms of morphogenesis (signaling or gene regulatory networks) are distinguished from morphogenetic mechanisms (cell migration or epithelial invagination), there is value in marking the distinction explicitly: molecular genetic mechanisms (MGMs) and cellular-physical mechanisms (CPMs). This distinction plays a significant role in reasoning about mechanisms that operate during ontogeny and explaining how animals and plants take shape. It also serves exegetically to accent two conceptual issues: (1) the conservation or stable identity of MGMs across phylogenetically disparate taxa compared with the non-conservation of CPMs, and (2) the explanatory generality of MGMs compared with the explanatory completeness of integrating these with CPMs to comprehend more fully how complex interactions yield morphological outcomes in ontogeny.

The conservation of MGMs is central to the reasoning practices of contemporary developmental biology and a major source of its recent success.

> The past two decades have brought major breakthroughs in our understanding of the molecular and genetic circuits that control a myriad of developmental events in vertebrates and invertebrates. These detailed studies have revealed surprisingly deep similarities in the mechanisms underlying developmental processes across a wide range of bilaterally symmetric metazoans . . . [these] comparisons have defined a common core of genetic pathways guiding development.
>
> *(Bier and McGinnis 2003, 25)*

Despite dramatic differences in the *phenomena* of development ("myriad of developmental events"), the "deep similarities" of genetic *mechanisms* underlying these phenomena are striking. These "conserved mechanisms" serve as a foundation for reasoning from a small set of model organisms to explanatory generalizations about how all or most animals develop (Love and Travisano 2013; Ankeny and Leonelli 2011; Halina and Bechtel 2013). However, conserved MGMs are not identical and therefore a question arises about how deep the similarities must be to license these inferences. Additionally, mechanisms are individuated by the outcomes they produce. Since the claim of conservation is a judgment of homology, which is based typically on structure rather than function or behavior (Love 2007), what constitutes the individuation conditions for a conserved MGM? And why is it the case that many CPMs, such as those associated with morphogenesis, appear not to be conserved?

The explanatory generality of MGMs is constitutive of much contemporary developmental biology. Many take this generality as a fundamental point of departure: "Developmental complexity is the direct output of the spatially specific expression of particular gene sets and it is at this level that we can address causality in development" (Davidson and Peter 2015, 2). However, others disagree: "One might be tempted to think of development simply in terms of mechanisms for controlling gene expression. But this would be highly misleading. . . . To think only in terms of genes is to ignore crucial aspects of cell biology" (Wolpert et al. 1998, 15). The concern expressed pertains to explanatory completeness; something is missing if we only focus on MGMs. CPMs add crucial detail to mechanistic explanations. What are the prospects for generating integrated explanations of development using both types of mechanisms? Could generality and completeness exhibit a tradeoff as explanatory values in the context of developmental biology (Weisberg 2013; see also Chapter 17, this volume)?

To approach these two conceptual issues, I first provide a framework of interpretive categories and characterize examples of both types of mechanisms. Then I probe what is involved in claims about the phylogenetic conservation of MGMs, as well as the difference between the explanatory generality of MGMs compared with the potential explanatory completeness derived from integrating these with CPMs. In closing, I identify one further conceptual issue—the dynamic constitution and organization of MGMs—and provisionally conclude with two broad lessons derived from my analysis of developmental mechanisms.

2. Mechanisms: philosophical background and developmental examples

Philosophical explorations of mechanisms and mechanistic explanation have grown dramatically over the past two decades and encompass a wide variety of sciences (Craver and Darden 2013; Craver and Tabery 2016; Illari and Williamson 2012). Although a number of different accounts of mechanisms have been offered, four shared elements are discernable (Craver and Tabery 2016; Illari and Russo 2014, ch. 12; see Chapter 1, this volume): (1) what a mechanism is for, (2) its constituents, (3) its organization, and (4) the spatiotemporal context of its operation. From these we can generate an ecumenical characterization to serve as a template for guidance in our analysis.

> A mechanism is constituted by a number of parts and activities or component operations that are organized into patterns of interacting relationships within a particular spatiotemporal context so as to produce a specific behavior or phenomenon (or set thereof).[2]

Mechanistic explanation works by decomposing systems into their constituent parts, localizing their characteristic activities, and articulating how they are organized to produce a particular effect at or within a specific time or place. These explanations illustrate and display the generation of specific phenomena at particular times and places by describing the organization of a system's constituent components and activities.

Consider an established MGM: the initial formation of segments in *Drosophila* due to the segment polarity network of gene expression (Wolpert et al. 2010, 70–81; Damen 2007). By Stage 8 of development (~3 hours post-fertilization), *Drosophila* embryos have 14 parasegment units that were defined by pair-rule gene expression in earlier stages. The transcription factor Engrailed accumulates in the anterior portion of each parasegment. This initiates a cascade of gene activity that defines the boundaries of each compartment of cells that will eventually become a segment. One element of this activity is the expression of *hedgehog*, a secreted signaling protein, in cells anterior to the band of cells expressing *engrailed*, which marks the posterior boundary of each nascent segment. This, in turn, activates the expression of *wingless*, another secreted signaling protein, which maintains the expression of both *engrailed* and *hedgehog* in a feedback loop so that segment boundaries persist.

This mechanism description can be expanded to illustrate the productive continuity of interactions involved in the feedback loop. On one side, the Hedgehog signaling pathway is activated when Hedgehog binds to the membrane protein Patched (Lum and Beachy 2004). In the absence of Hedgehog, Patched inhibits another membrane protein (Smoothened). Once Hedgehog binds to Patched, Smoothened is able to block the production and operation of repressors of the transcription factor Cubitus interruptus, which then turns on the expression of *wingless*. On the other side, Wingless (a member of the Wnt protein family) jointly binds to a member of the Frizzled family and lipoprotein receptor-related membrane protein, which disrupts a complex of proteins that continually degrade β-catenin in the cytoplasm. The phosphoprotein Dishevelled is also activated, further blocking this complex from operating by anchoring it to the plasma membrane. β-catenin then accumulates and is able to reach the nucleus in sufficient concentrations to initiate transcription and expression of *engrailed* and *hedgehog*.

The segment polarity network is a MGM that exhibits the features of our ecumenical characterization. It is constituted by a number of parts (e.g., Engrailed, Wingless, Hedgehog) and activities or component operations (e.g., signaling proteins bind receptors, transcription factors bind to DNA and initiate gene expression), which are organized into patterns of interacting relationships (the positive feedback loop, the Wnt and Hedgehog signaling pathways) within a spatiotemporal context (in parasegments of the *Drosophila* embryo, ~3 hours post-fertilization) so as to produce a specific behavior or phenomenon (a set of distinct segments with well-defined boundaries). The Wnt and Hedgehog signaling pathways are also distinct mechanisms (Craver and Tabery 2016; Lum and Beachy 2004; van Amerongen and Nusse 2009). The positive feedback loop of the segment polarity network of gene expression has these as component mechanisms. However, whereas these component mechanisms are stable and present throughout the organism's life history, the segment polarity network is transient and operative for only a specific period of ontogeny. The expression of *hedgehog* and *wingless* becomes decoupled later in embryogenesis so that the feedback loop is no longer present. There is a change in the organization of the mechanism as the spatiotemporal context changes subsequent to the production of a particular outcome.

Having observed a MGM, consider the CPM of branching morphogenesis, which refers to combinations of cellular proliferation and movement that yield branch-like structures in kidneys, lungs, glands, or blood vessels. There are many types of branching morphogenesis, but one primary mechanism is epithelial folding, which involves cells invaginating at different locations on a structure to yield branches (Davies 2013, ch. 20). Different CPMs can produce invaginations that lead to branching structures (Varner and Nelson 2014): the constriction of one end of a subset of columnar cells in an epithelium ("apical constriction"); increased cell proliferation of one epithelial sheet in relation to another ("differential growth"); and compression of an epithelium leading to periodic invaginations ("mechanical buckling"). That different mechanisms can lead to the same morphological outcome means it can be difficult to discern which mechanism is operating in an embryonic context. Sometimes divergent requirements of different CPMs can be isolated, such as the stiffness of the extracellular matrix, and divergent predictions can be identified for cellular properties, such as the dynamics of cell shape changes, which permit different CPMs to be teased apart and confirmed to apply to a particular system (Davidson et al. 1995).

CPMs like apical constriction bear complex relations to MGMs. The transcription factor Trachealess initiates a signaling cascade during *Drosophila* embryogenesis that leads to apical constriction and thereby epithelial invagination to form the reticulate system of branching morphology that is characteristic of the trachea (Affolter and Caussinus 2008). However, CPMs are a distinct type of mechanism and exhibit the features of our ecumenical characterization differently. The parts constituting the CPM are no longer transcription factors or signaling molecules but cells and tissues. The activities or component operations (e.g., apical constriction, differential growth, mechanical buckling) are organized into patterns of interacting relationships (apical constriction leading to epithelial invagination) within a spatiotemporal context (in tracheal precursors within the *Drosophila* embryo around Stage 7 and 8). This organization produces a specific behavior or phenomenon (a set of branching structures—the trachea).

It is tempting to interpret the complex relations between MGMs and CPMs in terms of nested mechanisms. The MGMs generating apical constriction could be considered component mechanisms of the CPM of branching morphogenesis. However, caution is required because MGMs are stable over evolutionary time whereas associated CPMs vary significantly.

> The molecular signaling pathways [MGMs] that control branching morphogenesis appear to be conserved across organs and species. However, despite this molecular homology, recent advances in cell lineage analysis and real-time imaging have uncovered surprising differences in the [cellular-physical] mechanisms that build these diverse tissues.
>
> *(Varner and Nelson 2014, 2750)*

This evolutionary decoupling of the different types of mechanisms brings us to our first issue: what does it mean to label MGMs as "highly conserved"?

3. Conserved genetic mechanisms of development

The Wnt signaling pathway is found across animal phyla (van Amerongen and Nusse 2009; Komiya and Habas 2008). Whether biologists look in fruit flies, soil worms, or humans, this pathway is present in some form; it is a *conserved* MGM. The claim of conservation is a judgment of homology. The core idea behind homology is identifying the same trait in different taxa under every variety of form and function, where the sameness derives from a common evolutionary heritage due to descent with modification. Similarity is neither necessary nor sufficient for determining whether two traits are homologous (Ghiselin 2005). Many morphological features are similar because of natural selection (i.e., analogous), but not as a result of common descent (i.e., homologous).

The application of homology reasoning in the context of conserved mechanisms has not received sustained attention (but see Halina and Bechtel 2013; Love 2007). Halina and Bechtel (2013) offer the following definition: "A biological mechanism is conserved when it can be identified as the product of evolutionary descent" (1201). To some degree, this is too liberal. All biological mechanisms are the product of evolutionary descent. The suppressed premise concerns the comparative nature of the judgment. A biological mechanism cannot be "conserved" except with respect to another mechanism in another taxon. Thus, the Wnt signaling pathway is designated as conserved because it is found in fruit flies *and* soil worms *and* humans. Yet why do we think the "same" Wnt signaling pathway is present in these taxa? Is it because of similarities in the pathways? These could be due to natural selection operating at the molecular level in these taxa, leading to similar (but not conserved) mechanisms (Halina and Bechtel 2013). This highlights a critical link between our understanding of what a conserved mechanism is and the criteria for determining whether a mechanism is genuinely conserved. Standard criteria for identifying homologs include relative position in the body with respect to other traits, similarity of structural detail, special quality, and embryological origin. Each of these is defeasible since they can manifest because of other factors (e.g., similarity due to natural selection) or fail despite homology obtaining (e.g., many homologous traits do not share the same embryological origin because of developmental system drift; Haag 2014).

What might count as criteria for a conserved mechanism? One thing to note is that phylogenetic reconstruction is necessary but not sufficient. Thus, the claim that "conservation is ultimately established through phylogenetic analyses" (Halina and Bechtel 2013, 1201) is not strictly true because reconstructing a phylogeny does not necessarily involve applications of particular criteria of homology for the trait (or mechanism) in question. (Methodologically, one should explicitly avoid including the trait in question when constructing a phylogeny to test for conservation.) More constructively, we can use our ecumenical characterization and explicitly map its four features onto the standard criteria for homology.

The first mapping is between the constituents of a mechanism and the criterion of special quality. If the same constituents are identified as standard components (e.g., Wnt proteins), then this is a particular feature (e.g., an initiating signaling molecule) that is pertinent to determining whether something is conserved. This accounts for why many conserved mechanisms are named in terms of constituent molecules that have these special qualities (e.g., Wnt, Hedgehog, Hox). The second mapping is between the organization of a mechanism and similarity of structural detail. If a mechanism is organized in similar ways in terms of which families of molecules interact with one another (e.g., in the segment polarity network), then this is another potential indicator of conservation. This supplements the first criterion because many of the molecules whose name labels a conserved mechanism are involved in other molecular processes that do not exhibit the same organization among participating constituents. The third mapping obtains between the criterion of relative position and the spatiotemporal context of a mechanism. The conserved mechanism of establishing anterior-posterior axes in the embryo using combinations of *Hox* gene expression occurs at distinct temporal junctures and spatial regions during development (e.g., early on in the entire embryo, but later only in appendages).

The fourth mapping introduces a twist into our analysis. A stereotypical behavior or phenomenon produced by a mechanism does not correspond to any of the criteria, and for a good reason. If a homolog is the same under every variety of form and function, then a stereotypical behavior or function that does not vary clashes with typical homology judgments based on structure. Unlike most traits evaluated for homology, the individuation of mechanisms is functional, not structural:

> The boundaries of a mechanism—what is in the mechanism and what is not—are fixed by reference to the phenomenon that the mechanism explains. The components in a mechanism are components in virtue of being relevant to the phenomenon.
> *(Craver and Tabery 2016)*

In essence, change the phenomenon or behavior and you change what counts as the mechanism. Yet similarity of function is a problematic criterion of homology (Abouheif et al. 1997); what a trait does typically should not enter into an evaluation of homolog correspondence because similarity of function is often the result of adaptation via natural selection to common environmental demands, not common ancestry. As a consequence, the idea of "functional homology" has long been thought suspect, though there are ways to recover a conceptually coherent notion of homology of function (Love 2007). However, this strategy assumes that a functional trait can be defined in terms of its activity (what it is) rather than use (what it is for). Is the notion of a "conserved genetic mechanism" inherently problematic because of its reliance on functional individuation?

Given that conserved MGMs play important roles in the reasoning of developmental biologists (Love and Travisano 2013; Ankeny and Leonelli 2011; Halina and Bechtel 2013), we should be hesitant to jettison it outright and instead attempt to account for these roles. One plausible strategy for understanding the situation is to recognize that claims about conserved mechanisms are not simply claims about homology. A judgment of correspondence for mechanisms across diverse taxa does not hold under every variety of form and function. Instead, we can use our analogies between the traditional criteria for homology and those for conserved genetic mechanisms to recognize a richer characterization:

> Conserved genetic mechanisms are shared, derived traits composed of particular constituents, organized in a specific way, and found in delimitable spatiotemporal contexts where they manifest a stereotypical behavior or phenomenon.

This is more circumscribed than the typical understanding of homology because sameness is not maintained under every variety of form and function; substantive claims about the sameness of constituents, organization, context, and function are involved. Importantly, this characterization underwrites the role of conserved MGMs in securing explanatory generality from investigating model organisms: "researchers assume deep commonalities between the mechanisms such that when a part or operation is found within a model mechanism, they expect to find it and search for it within the target mechanism" (Halina and Bechtel 2013, 1202). A typical judgment of homology would not necessarily license this kind of inference.[3]

This characterization retains several advantages of homology judgments, such as the ability to talk about sameness in the midst of evolutionary modifications across taxa. One can still accommodate variation in a mechanism's composition (form) or what it is for (function). First, although particular types of constituents are necessary for a conserved MGM, there can be variation in the number of components and intensity of activities for each type (Halina and Bechtel 2013). For example, the number of Wnt proteins involved in segment formation is variable across protostomes even though the interactions between Wnt and Hedgehog pathways are maintained (Janssen et al. 2010). Second, the requirement of specific organization need not imply identical organization across taxa. One well-known example is the diverse rearrangement of gene interactions in conserved mechanisms related to the initial patterning of insect embryos (Chipman 2015). Third, spatiotemporal contexts and functional individuation can be treated at different levels of abstraction to establish correspondence. One reproductive signaling pathway is considered conserved even though it produces outcomes of larval formation in soil worms, metamorphosis in insects, and puberty in mammals (Antebi 2013). At a higher level of abstraction, larval formation, metamorphosis, and puberty are all outcomes related to reproduction in a life cycle. Similar reasoning permits identifying conserved mechanisms across spatiotemporal contexts, such as the coordinated expression of Hox genes in establishing major body axes early in development and individual appendage axes later in development.

With this picture of MGM conservation in view, we need to ask why many CPMs (e.g., branching morphogenesis) appear *not* to be conserved. The short answer is that they exhibit a form of multiple realization (Bickle 2016). Constraints on what counts as a conserved MGM prevent it from being defined solely in terms of an outcome. However, the same is not true for many CPMs. There are numerous different ways to achieve epithelial invagination to generate branching morphologies. Another illustration comes from the developmental stage labeled "gastrulation," which refers to when a single-layered ball of animal cells is transformed into a three-layered embryo of endoderm, mesoderm, and ectoderm. The MGMs involved are highly conserved (Tam and Loebel 2007), but a plethora of CPMs manifest during gastrulation in different taxa, including invagination, ingression, involution, delamination, epiboly, intercalation, and convergent extension (Stern 2004). Similar studies on diverse anatomical features in ontogeny corroborate this theme: "a conserved molecular mechanism can control convergent morphogenesis through different cell behaviours" (Steventon et al. 2016, 1732).

This evolutionary decoupling of MGMs from CPMs has challenged the thinking of developmental biologists.

> Despite the fact that many of the same signaling molecules are used during branching morphogenesis of the *Drosophila* trachea and vertebrate lung, the physical mechanisms that generate branches are distinct . . . how do homologous signaling pathways yield such a diversity of physical mechanisms? . . . Is the molecular homology simply misleading our search for blueprints that govern the development of branching epithelia? These questions are not limited to branching systems. . . . Many of the transcription factors involved in early cardiogenesis . . . are conserved across species, but the tissue deformations that drive heart development can be very different.
>
> *(Varner and Nelson 2014, 2756)*

In some ways this recapitulates what happened two decades earlier when developmental genetic expression was overextended to make problematic claims of homology for many embryological and anatomical features across metazoans (Abouheif et al. 1997). In that situation, the molecular homologies were misleading. Another question that naturally emerges about the relation between MGMs and CPMs pertains to the explanatory expectations of developmental biologists. Do diverse CPMs bear commensurate explanatory weight with conserved MGMs? Do they individually or in combination provide answers in the "search for blueprints that govern development"?

4. Integrating genetic and physical developmental mechanisms?

Our discussion of how MGMs are evolutionarily decoupled from CPMs vindicates the caution in interpreting the former simply as component mechanisms of the latter. However, the significance of this decoupling can be probed further. Building explanatory models of complex phenomena (like developing embryos) is a central challenge in diverse sciences. The problem intensifies when the aim is to offer an *integrated* account of how different types of causes or mechanisms make contributions to complex biological phenomena (Mitchell 2003). How are different types of causes "combined" to explain how an effect results from their joint operation? Is the integration a relationship of realization, where some mechanisms are parts of other mechanisms, or a relationship of causation, where productive continuity is established between diverse causal factors in a single mechanistic description?

This framing arises directly out of scientific practice. It is manifested in questions about how to combine causes associated with CPMs and causes associated with MGMs that operate during embryogenesis (Miller and Davidson 2013). Genetic mechanism explanations appeal to changes in the expression of genes and interactions among their RNA and protein products to causally explain how processes of differentiation, pattern formation, growth, and morphogenesis produce anatomical structures. Physical mechanism explanations appeal to mechanical forces (e.g., mechanical buckling) resulting from the geometrical arrangements of soft condensed materials within the embryo to causally explain the same effects. The issue is not *whether* they both operate—"both the physics and biochemical signaling pathways of the embryo contribute to the form of the organism" (Von Dassow et al. 2010, 1)—but *how* to combine different types of causes to understand their joint contribution to the effect of organismal form.

A degree of pessimism about achieving this type of integration among causal contributions can be found in prior philosophical discussion. One analysis claims that apportioning causal responsibility (i.e., how much a cause contributes to an effect) is impossible for developmental phenomena because there is no common currency for measuring the contributions (Sober 1988). Sober distinguishes questions of relative contribution from questions of whether a cause makes a difference. Although there are sciences where questions of relative contribution can be addressed (e.g., Newtonian mechanics), only difference-making questions can be asked in developmental biology. Careful experimental methodology can answer whether one type of difference-maker accounts for more of the variation in an effect variable for a particular population (e.g., vary the genetic constitution while growing a plant in identical environments or *vice versa*), but the measure of interaction between variables says only *that* there is a joint effect, not *how* they jointly bring about their effect. Since the latter is a primary aim of mechanistic explanation, this paints a bleak picture for building integrated explanatory models of the morphological outcomes of embryogenesis that include both genetics (the presence, absence, or change in frequency of RNA molecules or proteins) and physics (stretching, compression, fluid shear stress).

Representations of time are an underappreciated dimension of explanatory models. Difference-making accounts of causal reasoning typically have no representation of time apart from an ordinal relation of causes preceding their effects: "dependence concerns a relation between cause and effect, without concern about what happens in between cause and effect" (Illari and Russo 2014, 252). This is a common idealization in these accounts and facilitates dissecting causal relationships within complex phenomena. (Idealizations are reasoning strategies that purposefully depart from features known to be present in nature to achieve prediction, explanation, or control.) Production accounts of causal reasoning, including those associated with mechanisms, do not share this idealization and explicitly attend to the representation of time, especially to illustrate the productive continuity of causal factors in generating an outcome: "production concerns the linking between cause and effect" (Illari and Russo 2014, 252; compare Craver and Darden 2013). One can distinguish the contributions of some causes from others because they do not operate instantaneously. Combinations of causes can be conceptualized not only as "this *and* that," but also as "this *then* that." However, even though descriptions of mechanisms can incorporate temporal duration between difference-makers or stages of a mechanism, there is a question about whether the forms of integration common to mechanistic explanation are suited to combining genetic and physical causes (see Chapters 10 and 11, this volume).

Consider an account of the origin of aortic arch asymmetry (Yashiro et al. 2007). Although the morphological outcome results from both gene expression and physical dynamics, the organization of these causal factors within a representation of developmental time permits a dissection of how contributions are made. Gene expression at an earlier time makes a difference in the structure of the outflow tract, which leads to a differential distribution of blood flow on the left and right. The decreased blood supply on the right down-regulates genes normally expressed on the left, which causes the branchial arch artery to disappear from the right and leave only a left-sided aortic arch. While this type of explanatory model does not permit a quantitative evaluation of the relative contribution of each type of cause, it does illuminate how types of causes are compounded to bring about an effect. It also aligns with the reasoning of developmental biologists: "an increasing number of examples point to the existence of a reciprocal interplay between expression of some developmental genes and the mechanical forces that are associated with morphogenetic movements or with hydrodynamic flows during development" (Brouzés and Farge 2004, 372). This reciprocal interplay reinforces the need to conceptualize MGMs and CPMs as causally interacting, possibly within a single mechanism.

Mechanistic explanations involve decomposing systems into their constituent parts, localizing their characteristic activities, and articulating how they are organized to produce a particular effect at a specific place and time (Darden 2006; see section 2 of this chapter). The element of organization appears crucial to building integrated explanatory models of how an effect is produced by genetic and physical mechanisms during development. A natural strategy is to integrate these two types of mechanism according to tactics that have applied in other contexts. Craver and Darden (2013, ch. 10) describe three ways that integration can occur in mechanistic explanation. The first is "simple mechanistic integration" where different fields are working at the same level or spatial scale of a mechanism, such as the role of RNAs in protein synthesis. Molecular biologists, focusing on an earlier time in protein synthesis, elucidated how messenger RNAs act as a template to guide the incorporation of amino acids via transfer RNAs into a polypeptide that corresponds with the triplet codon derived from a DNA sequence. Biochemists, focusing on a later time, elucidated how individual amino acids attached to transfer RNAs bonded to one another to form a stable polypeptide chain. Although different methods and experimental systems were required to ascertain how these aspects of the mechanism operate, they eventually were integrated into a single mechanistic description of continuous organization.

The second form of mechanistic integration is "interlevel integration"—exploring mechanisms at different spatial scales or hierarchical levels understood as part–whole relations. For example, to comprehend mechanisms of learning and memory, component mechanisms at the level of the hippocampus need to be integrated with component mechanisms for how long-term potentiation occurs in individual neurons. These, in turn, need to be integrated with component mechanisms for how specific protein receptors are activated by signaling molecules. Integrating these nested, component mechanisms yields a more robust explanatory model of the mechanisms of learning and memory.

A third form of mechanistic integration is "sequential intertemporal integration," such as different research communities investigating distinct steps of a complex mechanism where the steps are temporally disjointed, sometimes across generations. The mechanism of heredity involves a complex series of events (e.g., gene replication and expression), each of which contain different relevant component mechanisms (e.g., meiosis or transcription factor binding). Although interlevel integration helps to dissect various steps in this series, one also needs intralevel integration to bridge temporal gaps and purchase productive continuity of organization for the mechanistic description. Sequential intertemporal integration requires combining the other two modes of integration across spans of time; otherwise, there are black boxes between various steps in the series.

Although integrated explanatory models exhibiting these three types can be found in biological practice, it is unclear whether they are suitable for building integrated explanatory models of genetic and physical causes in development. Simple mechanistic integration is inappropriate because the causal dynamics occurring through the temporal duration are *inter*level (e.g., gene expression and fluid flow). Interlevel integration is inadequate because the mode of explanatory reasoning does not involve conceptualizing MGMs as hierarchically organized parts of CPMs. This is the justified caution identified earlier of not simply interpreting MGMs as nested components of CPMs. By implication, sequential intertemporal integration, which combines features of the other two types of integration across substantial temporal gaps, is not applicable either. None of these appear suitable for combining genetics and physics in explanations of development, even though the causal reasoning associated with mechanisms foregrounds the temporal dimension that is crucial in overcoming the idealization about time inherent to difference-making approaches.

Why are these approaches inadequate to the task of building integrated explanatory models of genetic and physical factors? First, the reciprocal interaction between genetic and physical causes does not conform to the expectation that mechanism descriptions "bottom-out" in lowest-level activities of molecular entities (Darden 2006). The interlevel nature of the causal dynamics between genetic and physical factors runs counter to this expectation and is not amenable to an interpretation in terms of nested mechanisms realizing another mechanism. Second, the reciprocal interaction between genetic and physical causes does not require stable, compositional organization, which is a key criterion for mechanisms (Craver and Darden 2013). The productive continuity of a sequence of genetic and physical difference-makers can be maintained despite changes in the number and types of elements in a mechanism. Although compositional differences can alter relationships of physical causation (fluid flow or tension), these relationships do not require the specificity of genetic interaction predominant in most mechanistic explanations from molecular biology. (The multiple realizability of CPM outcomes is central to this conclusion.) Standard mechanistic strategies of representation and explanation appear inadequate to capture these mechanisms.

An alternative form of integration that could illuminate how different genetic and physical causes jointly operate involves departing from a standard assumption about time and mechanisms.

Temporal duration for a mechanisms approach is typically time "in" the mechanism. In contrast, most developmental explanations use standardized representations of times—periodizations— that are measured and calibrated apart from specific mechanisms. The most ubiquitous of these in developmental biology are normal stages (Hopwood 2005; DiTeresi 2010). Stages facilitate the study of diverse mechanisms, with different characteristic rates and durations, within a common framework for a model organism or its parts. They also permit the study of conserved MGMs in different species because the mechanism description is not anchored to a particular periodization. An example of explanatory modeling with genetics and physics using a standardized periodization illustrates this point.

Experiments have shown that gene expression is initiated by mechanical deformations of tissue structure in *Drosophila* embryos (Farge 2003). As germ-band extension occurs dur- ing early ontogeny,[4] the associated physical motions of morphogenesis induce expression of the gene *twist* in the anterior foregut and stomodeal primordium (precursor cells of the mouth). Although the protein Dorsal regulates *twist*, three experiments showed that *twist* is also controlled by mechanical movements: (a) a transient lateral deformation induces the ectopic expression of *twist* in the dorsal region of the embryo independent of *dorsal* expression; (b) in mutant *Drosophila* where mechanical movements are blocked, normal expression of *twist* in foregut and stomodeal primordium fails to occur, and artificial compression can rescue the mutant phenotype; and (c) the laser ablation of cells in the dorsal region, which reduces mechanical compression, leads to decreased expression of *twist* in the stomodeal primordium. In the absence of the protein Dorsal, these mechanical movements permit an accumulation of the regulatory protein Armadillo, which subsequently translocates into the nucleus and—in associa- tion with other proteins—activates *twist*.

In this account of how aspects of the gut develops, a standard periodization for *Drosophila* frames the description of causal interactions between physical and genetic factors (see also Desprat et al. 2008). The pertinent events (e.g., mesoderm invagination or germ-band exten- sion) correspond to the formal structure of normal stages: Stage 6 (mesoderm invagination), Stage 7 (early germ-band extension/endodermal anterior midgut invagination), and Stage 10 (stomodeal foregut invagination). These stages, which are calibrated independently of the mechanisms of gut formation, permit the linkage of difference-makers into chains of produc- tive continuity. The difference-makers are not ranked according to how much variation in the effect variable they account for, but are combined in a periodization that orders different kinds of causal factors that contribute to an effect. Researchers capture the dependency rela- tions between genetic and physical factors by mapping them onto a temporal sequence during which the mediation of causes and effect occurs. The same set of stages can accomplish a similar integration with respect to different sets of causes pertinent to other developmental processes. Although no standard composition of causes emerges from this strategy (i.e., quantified relative contributions), we do achieve an understanding of *how* multiple causes combine mechanistically to yield a morphological outcome.

Is this integrated explanatory model best understood as a single mechanism that incorporates both physical and biochemical interactions? We have the linkage of difference-makers into chains of productive continuity—a standard criterion for mechanistic descriptions (Craver and Darden 2013). This interpretation is possible, though biologists don't adopt it and continue to distinguish molecular and physical mechanisms. The reason why they might not adopt the perspective of a single mechanism returns us to a question of explanatory values: should "inte- grated" mechanisms be valued over separate mechanisms that are more widely generalizable, especially conserved MGMs? A consequence of using standardized periodizations to combine genetic and physical causation in mechanistic explanations of embryogenesis is a reduction in the scope of resulting generalizations. To achieve an integrated or more complete explanation for two modes of causation, we must give up the explanatory generality secured for one mode of causation treated separately. The periodizations used to assign responsibility to genetic and physical difference-makers in chains of productive continuity are relatively specific to a model organism. Additionally, "there are no one-to-one correspondence principles between gene functions and the mechanical events that they affect" (Miller and Davidson 2013, 741); this is the diversity of CPMs that can yield the same outcome in different contexts.

There is a tradeoff between the specificity of diverse causal interactions in an integrated explanatory model and the generality found in non-integrated explanatory models, especially those constructed from conserved MGMs. It is not a small thing to give up on the explanatory generality of the latter since it is a large part of the rationale for the use of model organisms in developmental biology. Since integrated explanatory models of genetic and physical mechanisms remain rare, explanatory generalizations of wide scope based on conserved MGMs appear more valued by developmental biologists than combined models of genetic and physical causes. A preference for generality fits with the standard genetic orientation of developmental biology as a discipline, but also has a wider application. Building integrated mechanistic models is frequently the result of interdisciplinary efforts, where differences in explanatory standards are exposed; generalization within a discipline is likely to take priority over integration across disciplines. This implies that the tradeoff between explanatory generalization and explanatory integration (or completeness) identified for genetics and physics in developmental biology is broadly applicable across the sciences.

5. Further issues and provisional conclusions

The two issues analyzed in this chapter do not exhaust the space of conceptual questions related to developmental mechanisms. To see this, let us briefly describe one more issue: the dynamic constitution and organization of MGMs. This issue is observable in an overview of the Developmental Mechanisms Section of the National Institute of Dental and Craniofacial Research.[5] The institute's research focuses on "how signaling interactions control regulatory state transitions during embryogenesis" and it "emphasizes that signals not only activate new gene regulatory networks, but also eliminate or alter pre-existing ones." MGMs, such as gene regulatory networks, come into existence, transform, and go out of existence during the processes of embryogenesis.

The contrast between the stability of the Hedgehog and Wnt signaling mechanisms and the transient nature of the segment polarity network of gene expression illustrates this issue concretely. Various components and activities are still present but their patterns of interaction within a particular spatiotemporal context have changed. Additionally, MGMs can exhibit changes in constitution. The Wnt signaling pathway has a "noncanonical" version that is involved in planar cell polarity during ontogeny, the establishment of directionality in epithelial tissue (e.g., hairs pointing in the same direction). Again, a Wnt protein jointly binds to a Frizzled transmembrane protein and another receptor, which activates Dishevelled. However, β-catenin and its degradation complex are not involved. Instead, Dishevelled forms a complex with different elements to activate proteins that regulate actin polymerization in the cytoskeleton to facilitate the establishment of polarity (Komiya and Habas 2008).

A natural response to changes in constitution or organization is that these are no longer the same mechanism. And, since the outcome of noncanonical Wnt signaling is different, this change in the phenomenon being produced could indicate that these are different mechanisms. Whether this is true remains an open question among scientific practitioners; some researchers argue for a unified view of Wnt signaling as a mechanism that includes both canonical and non-canonical dimensions (van Amerongen and Nusse 2009). However, the point is that changes of constituents and organization for these mechanisms, and hence their causal capacities, differ from what philosophers have analyzed in molecular biological mechanisms (but see McManus 2012 and Parkkinen 2014). Most of the biological mechanisms analyzed to date (e.g., DNA replication) are stable and recurring throughout an organism's life cycle. Questions about the individuation of mechanisms that change through developmental time require further analysis.

We began our discussion with the distinction between MGMs and CPMs. Then our ecumenical characterization of mechanisms illuminated concrete examples of MGMs involved in insect segment formation and CPMs of branching morphogenesis. This set the stage for exploring two central issues about developmental mechanisms: the conservation of MGMs and the explanatory integration of genetic and physical mechanisms to understand complex interactions and morphological outcomes in contrast to the explanatory generality of conserved MGMs alone. For the former, it was necessary to incorporate comparative reasoning from evolutionary biology because "conservation" is a form of homology for mechanisms. However, it is a special kind of homology and its characteristics help account for the role that model organisms play in developmental biological research. At the same time, delineating this circumscribed homology judgment for conserved MGMs facilitated a better understanding of why CPMs are not conserved.

For questions surrounding integration and generality, we isolated the significant role that representations of time play in providing a template for combining different kinds of causes into chains of productive continuity. Although strategies for integrating mechanisms exist, none were adequate for combining genetic and physical causes in accord with reasoning patterns in developmental biology. Instead, standardized periodizations offered one possibility that could serve this integrative role, but at a cost of generality. The local nature of these periodizations in conjunction with the diversity of CPMs that are decoupled from conserved MGMs meant that an integrated explanatory model sacrificed the generality secured by reasoning with conserved MGMs alone. There is a tradeoff between the localized explanatory completeness achieved in an integrated mechanistic model and the explanatory generality achieved in a mechanistic model based on molecular genetics alone. The ubiquity of the latter and rarity of the former suggest that generality is preferred over integration by most developmental biologists.

Two provisional conclusions can be drawn from the preceding analysis. First, scrutinizing conserved MGMs showed the importance of incorporating reasoning strategies from evolutionary biology with those commonly used in molecular and developmental biology. To date, most philosophers interested in mechanistic explanation have neglected these dimensions of reasoning and therefore questions surrounding the meaning of "conservation" critical for licensing inferences from model organisms were insufficiently analyzed. Second, the tradeoff between integration and generality, displayed when comparing mechanistic models that combine genetic and physical causes with those that rely on MGMs alone, is a fault line in how biologists conceptualize adequate explanations. The preference for explanatory generality over integration, with an attendant loss of completeness in our understanding of how an outcome is actually produced mechanistically, reveals new terrain for philosophical analysis.

A better understanding of the relative significance of explanatory values such as generality, specificity, and completeness is necessary. Although philosophers have highlighted the importance of abstraction and the role of mathematical modeling (Levy and Bechtel 2013; Brigandt 2013), as well as idealization (Love and Nathan 2015), less work has been done on how different explanatory values obtain across diverse biological disciplines. Although our analysis of developmental biology brings some of these issues into view, more work is needed that explores mechanisms and mechanistic explanation in particular areas of biological (and other scientific) research with special attention to these differences in explanatory expectations. Just as the distinction between MGMs and CPMs was crucial to identifying issues in developmental biology, so also the differences in reasoning about mechanisms in other areas of science, rather than commonalities, hold promise for a deeper comprehension of how mechanistic explanation operates across the sciences, especially under conditions of interdisciplinarity.

Notes

1 I received helpful feedback on an earlier version of this chapter from Stuart Glennan, Phyllis Illari, Tom Stewart, and Yoshinari Yoshida. The research and writing of this article were supported in part by a grant from the John Templeton Foundation (Integrating Generic and Genetic Explanations of Biological Phenomena; ID 46919).
2 This is similar to other ecumenical versions: "A mechanism for a phenomenon consists of entities and activities organized in such a way that they are responsible for the phenomenon" (Illari and Williamson 2012, 120).
3 A justification for the deep commonalities in constituents and organization of conserved mechanisms derives from the concept of generative entrenchment, which refers to how much a complex system or behavior depends on these features and thereby maintains them (see Wimsatt 2015).
4 The germ-band is a coordinated group of cells that develops into the segmented trunk of the embryo. During embryogenesis it extends along the anterior-posterior axis and narrows along the dorsal-ventral axis.
5 http://www.nidcr.nih.gov/Research/NIDCRLaboratories/DevelopmentalMechanisms.

References

Abouheif, E., M. Akam, W.J. Dickinson, P.W.H. Holland, A. Meyer, N.H. Patel, R.A. Raff, V.L. Roth, and G.A. Wray. 1997. Homology and developmental genes. *Trends in Genetics* 13:432–433.

Affolter, M., and E. Caussinus. 2008. Tracheal branching morphogenesis in *Drosophila*: new insights into cell behaviour and organ architecture. *Development* 135:2055–2064.

Ankeny, R., and S. Leonelli. 2011. What's so special about model organisms? *Studies in History and Philosophy of Science* 42:313–323.

Antebi, A. 2013. Steroid regulation of *C. elegans* diapause, developmental timing, and longevity. In *Current Topics in Developmental Biology*, eds. A.E. Rougvie and M.B. O'Connor, 181–212. San Diego: Academic Press.

Bickle, J. 2016. Multiple realizability. In *The Stanford Encyclopedia of Philosophy*, ed. E.N. Zalta, http://plato.stanford.edu/archives/spr2016/entries/multiple-realizability/.

Bier, E., and W. McGinnis. 2003. Model organisms in the study of development and disease. In *Molecular Basis of Inborn Errors of Development*, eds. C.J. Epstein, R.P. Erickson, and A. Wynshaw-Boris, 25–45. New York: Oxford University Press.

Brigandt, I. 2013. Systems biology and the integration of mechanistic explanation and mathematical explanation. *Studies in the History and Philosophy of Biological and Biomedical Sciences* 44:477–492.

Brouzés, E., and E. Farge. 2004. Interplay of mechanical deformation and patterned gene expression in developing embryos. *Current Opinion in Genetics & Development* 14:367–374.

Chipman, A.D. 2015. Hexapoda: comparative aspects of early development. In *Evolutionary Developmental Biology of Invertebrates 5: Ecdysozoa III: Hexapoda*, ed. A. Wanninger, 93–110. Vienna: Springer.

Craver, C.F., and L. Darden. 2013. *In Search of Mechanisms: Discoveries across the Life Sciences*. Chicago and London: University of Chicago Press.

Craver, C.F., and J. Tabery. 2016. Mechanisms in science. In *The Stanford Encyclopedia of Philosophy*, ed. E.N. Zalta, http://plato.stanford.edu/archives/spr2016/entries/science-mechanisms/.

Damen, W.G.M. 2007. Evolutionary conservation and divergence of the segmentation process in arthropods. *Developmental Dynamics* 236:1379–1391.

Darden, L. 2006. *Reasoning in Biological Discoveries: Essays on Mechanisms, Interfield Relations, and Anomaly Resolution*. New York: Cambridge University Press.

Davidson, E.H., and I.S. Peter. 2015. *Genomic Control Process: Development and Evolution.* San Diego: Academic Press.

Davidson, L.A., M.A.R. Koehl, R. Keller, and G.F. Oster. 1995. How do sea urchins invaginate? Using biomechanics to distinguish between mechanisms of primary invagination. *Development* 121:2005–2018.

Davies, J.A. 2013. *Mechanisms of Morphogenesis: The Creation of Biological Form.* 2nd ed. Amsterdam: Academic Press, Elsevier.

Desprat, N., W. Supatto, P.A. Pouille, E. Beaurepaire, and E. Farge. 2008. Tissue deformation modulates Twist expression to determine anterior midgut differentiation in *Drosophila* embryos. *Developmental Cell* 15:470–477.

DiTeresi, C.A. 2010. Taming Variation: Typological Thinking and Scientific Practice in Developmental Biology. PhD dissertation. University of Chicago, Chicago.

Farge, E. 2003. Mechanical induction of Twist in the *Drosophila* foregut/stomodeal primordium. *Current Biology* 13:1365–1377.

Ghiselin, M.T. 2005. Homology as a relation of correspondence between parts of individuals. *Theory in Biosciences* 124:91–103.

Haag, E.S. 2014. The same but different: worms reveal the pervasiveness of developmental system drift. *PLoS Genetics* 10:e1004150.

Halina, M., and W. Bechtel. 2013. Mechanism, conserved. In *Encyclopedia of Systems Biology*, eds. W. Dubitzky, O. Wolkenhauer, K.-H. Cho, and H. Yokota, 1201–1204. New York: Springer.

Hopwood, N. 2005. Visual standards and disciplinary change: normal plates, tables and stages in embryology. *History of Science* 43:239–303.

Illari, P., and F. Russo. 2014. *Causality: Philosophical Theory Meets Scientific Practice.* New York: Oxford University Press.

Illari, P., and J. Williamson. 2012. What is a mechanism? Thinking about mechanisms across the sciences. *European Journal of the Philosophy of Science* 2:119–135.

Janssen, R., M. Le Gouar, M. Pechmann, F. Poulin, R. Bolognesi, E.E. Schwager, C. Hopfen, J.K. Colbourne, G.E. Budd, S.J. Brown, N.-M. Prpic, C. Kosiol, M. Vervoort, W.G.M. Damen, G. Balavoine, and A.P. McGregor. 2010. Conservation, loss, and redeployment of Wnt ligands in protostomes: implications for understanding the evolution of segment formation. *BMC Evolutionary Biology* 10:1–21.

Komiya, Y., and R. Habas. 2008. Wnt signal transduction pathways. *Organogenesis* 4:68–75.

Levy, A., and W. Bechtel. 2013. Abstraction and the organization of mechanisms. *Philosophy of Science* 80:241–261.

Love, A.C. 2007. Functional homology and homology of function: biological concepts and philosophical consequences. *Biology & Philosophy* 22:691–708.

Love, A.C. 2014. The erotetic organization of developmental biology. In *Towards a Theory of Development*, eds. A. Minelli, and T. Pradeu, 33–55. Oxford: Oxford University Press.

Love, A.C. 2015. Developmental biology. In *The Stanford Encyclopedia of Philosophy*, ed. E.N. Zalta, http://plato.stanford.edu/archives/fall2015/entries/biology-developmental/.

Love, A.C., and M. Nathan. 2015. The idealization of causation in mechanistic explanation. *Philosophy of Science* 82:761–774.

Love, A.C., and M. Travisano. 2013. Microbes modeling ontogeny. *Biology & Philosophy* 28:161–188.

Lum, L., and P.A. Beachy. 2004. The Hedgehog response network: sensors, switches, and routers. *Science* 304: 1755–1759.

McManus, F. 2012. Development and mechanistic explanation. *Studies in the History and Philosophy of Biological and Biomedical Sciences* 43:532–541.

Miller, C.J., and L.A. Davidson. 2013. The interplay between cell signalling and mechanics in developmental processes. *Nature Reviews Genetics* 14:733–744.

Mitchell, S.D. 2003. *Biological Complexity and Integrative Pluralism.* New York: Cambridge University Press.

Parkkinen, V-P. 2014. Developmental explanations. In *New Directions in the Philosophy of Science: The Philosophy of Science in a European Perspective, Vol. 5*, eds. M.C. Galavotti, D. Dieks, W.J. Gonzalez, S. Hartmann, T. Uebel, and M. Weber, 157–172. Berlin: Springer.

Sober, E. 1988. Apportioning causal responsibility. *Journal of Philosophy* 85:303–318.

Stern, C. 2004. *Gastrulation.* Cold Spring Harbor, NY: Cold Spring Harbor Press.

Steventon, B., F. Duarte, R. Lagadec, S. Mazan, J.-F. Nicolas, and E. Hirsinger. 2016. Species-specific contribution of volumetric growth and tissue convergence to posterior body elongation in vertebrates. *Development* 143:1732–1741.

Tam, P.P.L., and D.A.F. Loebel. 2007. Gene function in mouse embryogenesis: get set for gastrulation. *Nature Reviews Genetics* 8:368–381.

van Amerongen, R., and R. Nusse. 2009. Towards an integrated view of Wnt signaling in development. *Development* 136:3205–3214.

Varner, V.D., and C.M. Nelson. 2014. Cellular and physical mechanisms of branching morphogenesis. *Development* 141:2750–2759.

Von Dassow, M., J. Strother, and L.A. Davidson. 2010. Surprisingly simple mechanical behavior of a complex embryonic tissue. *PLoS ONE* 5:e15359.

Weisberg, M. 2013. *Simulation and Similarity: Using Models to Understand the World*. New York: Oxford University Press.

Wimsatt, W.C. 2015. Entrenchment as a theoretical tool in evolutionary developmental biology. In *Conceptual Change in Biology: Scientific and Philosophical Perspectives on Evolution and Development*, ed. A.C. Love, 365–402. Dordrecht: Springer.

Wolpert, L., R. Beddington, J. Brockes, T. Jessell, P.A. Lawrence, and E.M. Meyerowitz. 1998. *Principles of Development*. New York: Oxford University Press.

Wolpert, L., C. Tickle, T. Jessell, P.A. Lawrence, E.M. Meyerowitz, E. Robertson, and J. Smith. 2010. *Principles of Development*. 4th ed. New York: Oxford University Press.

Yashiro, K., H. Shiratori, and H. Hamada. 2007. Haemodynamics determined by a genetic programme govern asymmetric development of the aortic arch. *Nature* 450:285–288.

26

MECHANISMS IN ECOLOGY[1]

Viorel Pâslaru

1. Introduction

Genetics, cell biology, and molecular biology offered to philosophers of science sufficient examples to challenge the logical empiricist view on scientific inquiry as a search for laws of nature and scientific explanation as subsumption of phenomena under laws. Those areas of biology indicated instead that scientific inquiry is a search for mechanisms and explanation is a matter of describing them (see Chapters 1 and 23). The resulting new mechanistic philosophy of science offered various accounts of mechanisms whose common assumptions are articulated by this formulation of a minimal mechanism: "A mechanism for a phenomenon consists of entities (or parts) whose activities and interactions are organized so as to be responsible for the phenomenon" (see Chapter 1).

Mechanisms are real and local complex systems or processes that are responsible for the phenomena scientists study. Mechanisms produce phenomena in virtue of their parts that are organized and engage in activities or interact. For example, a heart is a mechanism for pumping blood. It produces this phenomenon because of its parts: right and left ventricles, right and left aortas, and the valves. They perform activities, such as the right ventricle contracting and pushing the blood into the pulmonary trunk, while the valves open and close. Parts and activities are organized. Blood is collected in the right atrium first, but not in the left atrium, and the tricuspid valve opens toward the apex of the heart, but not toward its top. The mechanistic approach is reductionist in that it explains properties and activities of the whole in terms of properties and activities of its lower-level parts. That walls of the heart contract is explained by referring to three layers of cardiac muscles and to the intercalated disks that enable direct transmission of electrical impulses between cells, which also shows that mechanisms are hierarchical. The mechanistic approach acknowledges the complexity of real systems and examines mechanisms in their contexts. Accordingly, operation of the heart mechanism is considered in the context of the thorax, its interaction with the diaphragm, the lungs, and the system of blood vessels.

Ecologists frequently describe mechanisms for purposes of explanation and prediction. However, philosophers of science have not examined ecological mechanisms, even though they might be different from the biological mechanisms that have so far received philosophical scrutiny. I use the case of the invasive shrub *Lonicera maackii* (Amur honeysuckle) to examine ecological mechanisms at the level of individual organisms. I then explore the case of mechanisms involving groups of organisms. But first, I offer a brief overview of the field of ecology.

2. What is ecology?

Ecology is not environmentalism and its focus is not environmental problems and the effects of human impact on the environment. Ecology is a discipline of biology that studies the world of plants and animals, including humans, and is widely understood as "the scientific study of the interactions that define the distribution and abundance of organisms" (Krebs 2008, 5). Distribution is the place in nature where organisms are found, while abundance is their number in an area. Distribution and abundance are two facets of the same phenomenon and the factors that affect the former could also influence the latter. The factors that determine the distribution and abundance of organisms fall into two categories: biotic and abiotic. Biotic factors comprise interactions between organisms, such as predation, or competition. Abiotic factors comprise influences of the environment on organisms, such as temperature, availability of light or water, its pH, or the nature of the ground.

Distribution and abundance are examined at various levels of increasing complexity and integration, but of decreasing understanding: populations, communities, ecosystems, landscapes, and biosphere. To explain phenomena of distribution and abundance at one level, ecologists typically seek explanatory mechanisms at lower levels. For example, behavioral and physiological mechanisms operating at the level of individual organisms are sought to explain changes in a population. Because its area of investigation spans so many levels, ecology neighbors with multiple sciences. It draws on physiological and behavioral studies of individual organisms, as well as on meteorology, geology, and geochemistry for its explanations (Krebs 2008). Ernest Haeckel introduced the term *ecology* in 1869, but ecological research and interest predate this date, making it one of the oldest disciplines of biology.

3. An example of individual-level ecological mechanisms

Research on the competitive success of the invasive shrub *Lonicera maackii* (Amur honeysuckle)[2] against native plants illustrates how ecologists examine and conceive of mechanisms at the individual level and their effects at the macro scale.

Specimens of the shrub were introduced to the US in 1898 to the New York Botanical Garden. From there, humans further dispersed shrub specimens to serve as ornaments or to stabilize soil. Furthermore, birds, deer, and small mammals consume the seeds of the shrub and drop them with feces at various locations, which helped the invasive plant establish throughout vast areas of the Midwestern US.

L. maackii dominates native plants in competition for light and suppresses their growth because its canopy (its uppermost leaves and branches) is dense, and because it is able to produce numerous stem shoots and grow rapidly in habitats with different light regimens. When grown in understory, i.e., beneath the canopy of other plants, *L. maackii* maximizes height gain and allocates energy to continuous production of long shoots growing from the base; it consequently has a lower canopy width, fewer shoot ends, and a smaller diameter of basal stems. Should more light become available, which happens when a patch of forest is cleared, say, by a fallen tree, the shrub stops producing basal long shoots and starts producing long shoots at higher levels in the canopy, develops wider canopies, and produces more leaves. Compared with indigenous shrub species, *L. maackii* is able to equal or exceed the maximum stem growth, higher stomatal density, and thickness of leaves, higher leaf number, area, and mass in low-light conditions, but exceeds such a performance in high-light conditions that occur following disturbances in forests. Under

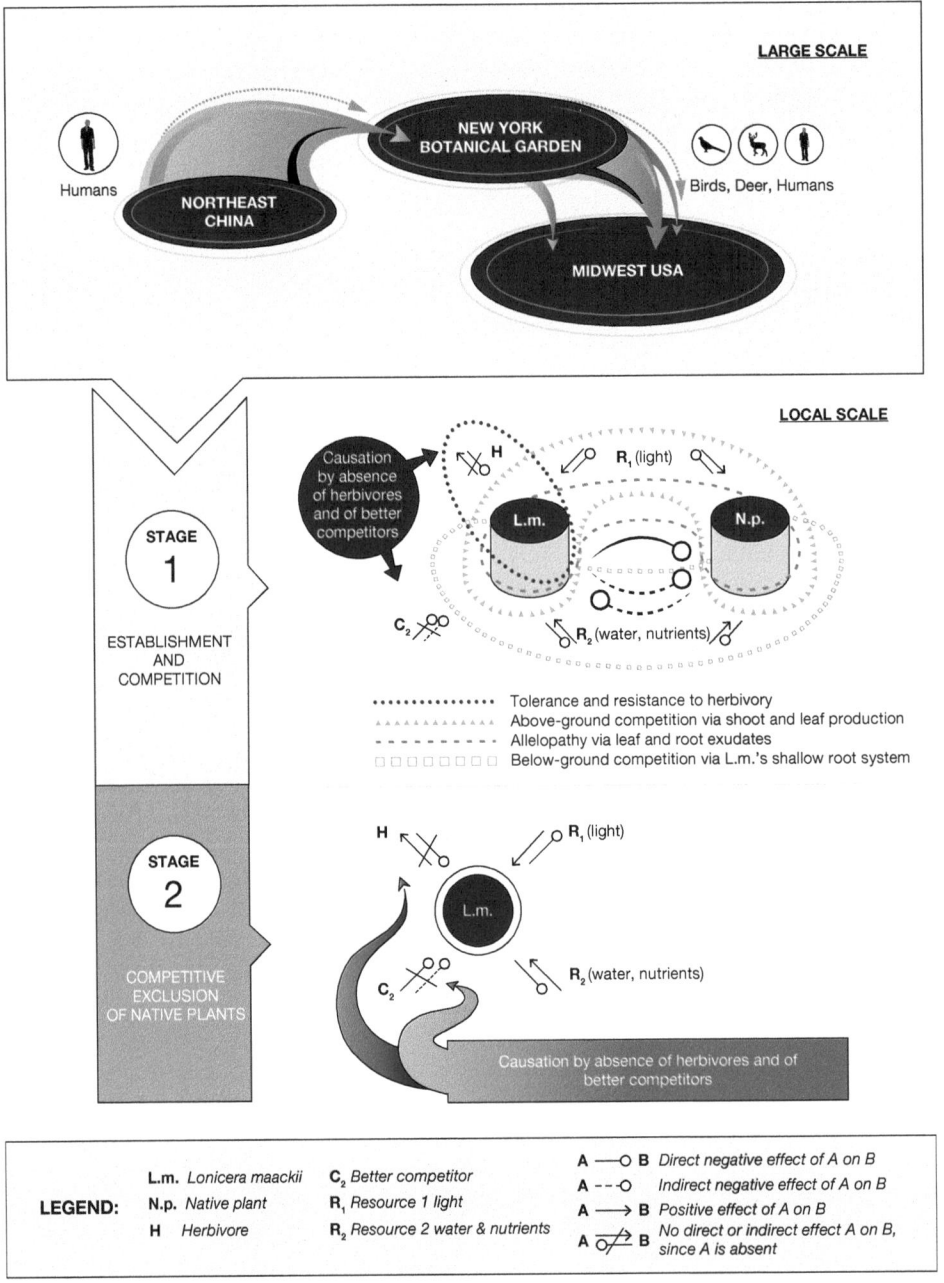

Figure 26.1 Integrated representation of the mechanisms of competitive success of *L. maackii*. Drawing by the author

both light conditions, *L. maackii* allocates more energy to branch and leaf mass growth and this allows it to overgrow the slow-growing native shrubs. The shrub is fully able to regenerate after the first clipping both in open-growth and in forested areas. It leafs two to three weeks before the native plants and drops the leaves four to six weeks later than the native plants, and can be observed as late as December. Its leaves are also freeze-resistant and are not damaged by spring bouts of cold, in contrast to the native species. Because of all these features, *L. maackii* is able, under various conditions and despite various environmental stresses, to get more light than the native species and deprive them of this precious resource, suppressing their growth. Seedlings establish throughout a variety of light conditions, and even those that were suppressed because of clumped seed input can survive and gradually replace adult *L. maackii* and native shrubs.

In addition to outcompeting native plants for light, *L. maackii* dominates below-ground competition for nutrients and water. Its roots cover a wider area close to the surface, which allows it to deprive neighboring plants of water and nutrients, suppressing their growth. Additionally, it suppresses the growth of native plants by means of chemicals in a process called allelopathy. Leaves produce thirteen mildly acidic toxic compounds that inhibit or delay germination of seeds, or alter size, survival, and architecture (number of branches and bolts) of native plants and deter generalist herbivores from consuming its leaves. Roots produce similar chemicals, but they have milder inhibiting effects on neighboring plants. Chemicals released into the ground from decomposing leaves and by roots modify soil and leaf microbial communities, altering ecosystem function and processes.

Ecologists treat above-ground competition, below-ground competition, and allelopathy as separate mechanisms, yet none of them is solely responsible for the competitive success of *L. maackii*. Since they act jointly, I represent the separate descriptions in an integrated account (Figure 26.1), which will also facilitate the forthcoming philosophical examination of these mechanisms.

4. Individual-level mechanisms and the minimal account

Individual-level ecological mechanisms satisfy most, but not all features of the minimal account of mechanisms and their particularities contribute toward the general examination of mechanisms.

Several decades after the introduction of *L. maackii* to the US, it became clear that it was not an innocent shrub that beautifies landscapes, but a shrub that efficiently suppresses and eliminates native plants. This phenomenon of suppression and elimination of native flora required explanation that ecologists formulated from the beginning, not in terms of laws, ecological or otherwise, but in terms of mechanisms. And so they articulated the mechanism of competition for light, of below-ground competition for water and nutrients, and the mechanism of allelopathy, which were meant to account entirely for the elimination of native flora or offer partial explanations of that phenomenon. Accordingly, ecological mechanisms, just like in other areas of biology, are mechanisms for a phenomenon and are formulated to explain it.

Ecologists often characterize individual-level mechanisms as consisting of individual organisms with certain properties doing certain activities. In Gause's (1934) account of the mechanism of competition, the components are the individual cells of yeast differentiated by the property tolerance to alcohol. Alcohol produced by *S. kephir* is more toxic to *S. cerevisiae* than vice versa, the latter being eliminated. *L. maackii* is representative of cases in which the phenomenon cannot be explained by reference to a property of the individual organism and its activity, but to several properties and activities that engage different parts of the same organism. Accordingly, it is parts of organisms that are entities in various mechanisms. The roots of *L. maackii* are components of the mechanism for below-ground competition, while its shoots and leaves are parts of the mechanism for above-ground competition, not the individual plant, although it is necessary for the roots, shoots, and leaves to function. The case of *L. maackii* helps stress that entities that appear simple are in fact composites of many parts that function in different mechanisms.

The composite nature of the organism *L. maackii* makes the mechanism of competition between this invasive plant and native species hierarchical. Its lower-level components are the mechanism of competition for light, the below-ground mechanism of competition for nutrients, the mechanism of allelopathy, and the mechanism of tolerance and resistance to herbivores. Each of these mechanisms is made up of parts at lower levels that are also investigated. The mechanism of competition for light consists of leaves and branches, with specific properties and performing activities. Shrubs that grow in understory have long shoots of a small diameter, fewer shoot ends, and a low canopy width. Since leaves are a key component of this mechanism, their properties are closely studied. Ecologists examine their higher stomatal density and thickness, their number per stem, area, mass, and their photosynthetic performance in low-light conditions compared with native species. The mechanism of photosynthesis that allows *L. maackii* to thrive in low-light conditions and the mechanism for longer leaf duration in *L. maackii* are lower-level components of the mechanism of competition for light.

Hierarchical organization of ecological mechanisms shows some entities can be components in several mechanisms. Leaves and roots, in addition to being parts of the mechanisms of above-ground and respectively below-ground competition, are also components in the mechanism of allelopathy. Leaves produce toxic compounds that, once they reach the ground with falling leaves, inhibit the germination of seeds of native species, while roots produce metabolites that inhibit the growth of roots of native species. The outcome is the same in both cases: native species are prevented from growing. Descriptions of the mechanism of allelopathy include chemical analyses of the allelochemicals that leaves and roots produce.

Furthermore, ecological mechanisms do not have clear boundaries, and what identifies them is the operational unification of components of different sizes. For example, the boundaries of the mechanism of competition for light are set by the operational unification of the parts of different sizes: shoots and leaves of individuals of species involved and light.

The explanatory role played by absent entities is another feature that, while not unique to ecological mechanisms, is very important. According to the Enemy Release Hypothesis, one of the hypotheses formulated to explain the success of *L. maackii*, the shrub spreads well in new environments *because* its traditional predators are absent. Similarly, it is not subject to intense competition, for there are no competitors that could challenge it. Ecologists consider the absent predators and competitors in light of the assumption that all species are parts of food webs and have predators and competitors that control their growth where the absence of either of them is a cause of a species' growth. Had either a predator or competitor been present, the dynamics of the species would have been different. Ecologists treat absent predators and competitors as possible causes that could influence the phenomenon under scrutiny, and as such they are part of the mechanistic account of the phenomenon along with the detailed descriptions of actual entities. This is not a departure from the realism of mechanistic philosophy, since the absent predators and competitors are not abstractions or fictions, but real organisms that happen not to be present in the habitat under scrutiny. The list of absent causes could also include other instances of causation by absence involving abiotic or biotic factors. Absence of sunlight explains the absence of shade-intolerant plants, and absence of water explains the presence of drought-resistant plants and the absence of drought-intolerant ones. The absence of an entity is not sufficient by itself to account for the phenomenon. The other components of the mechanism are required. Mechanisms in ecology involve absences and preventions, and a mechanistic perspective must explain their causal relevance. (See also Chapter 11 and Glennan (forthcoming).)

While the mechanisms described above seem generally to fit the characterization of minimal mechanisms, there are two features of ecological mechanisms that seem to distinguish them from many of the mechanisms discussed by New Mechanists. The first is that ecological mechanisms are often described in terms of properties rather than activities or interactions, and the second is that ecological mechanisms are often less dependent on organization than many exemplars of biological mechanisms.

In the characterization of mechanisms by Machamer, Darden, and Craver (2000), entities and activities take center stage, while properties seem subordinated and only briefly acknowledged as bases of activities. In Glennan's latest work causal powers or capacities perform the metaphysical role that properties have relative to activities in the characterization of Machamer, Darden, and Craver (Glennan forthcoming). The minimal account of mechanisms does not either thoroughly examine the role of properties in mechanisms. However, if the philosophy of mechanisms is to do justice to ecology, one has to emphasize the properties of entities. Ecologists are explicit about the importance of properties in characterizing mechanisms:

> Perhaps I should clarify what I mean by mechanism. The mechanistic basis for population ecology is provided by the properties of entities one hierarchical level lower than populations, that is, by the behavior and physiology of individual organisms.
>
> *(Turchin 1999, 156)*

And it is this focus on traits that defines ecology's subdisciplines of functional ecology and physiological ecology (Shipley 2010). In ecological context, properties of entities are the actual traits of organs, individuals, populations, and communities that ecologists measure, rather than the not-yet-manifested capacities. A focus on properties is necessary to better understand the activities of organisms. *L. maackii* chemically suppresses the growth of native plants. To understand this activity, ecologists performed an analysis at the molecular level of the chemicals and confirmed experimentally their inhibitory effect (Cipollini et al. 2008). In other instances, description of properties amounts to a description of causes of an organism's impact, or of distribution and abundance. Variety of canopy shapes, i.e., leaf-forms, growth forms, and heights of plants, accounts for better use of the three-dimensional space and, consequently, better light interception (Tremmel and Bazzaz 1993). Tolerance to high soil salinity and to deficiency of oxygen in water-logged soils is a cause of the presence of the grass *Juncus gerardi* in those areas of marshes that are flooded regularly, while intolerance of the shrub *Iva frutescens* to those conditions explains its presence on the terrestrial borders of marshes that are not flooded (Bertness and Hacker 1994).

"Organization is the least controversial element in any characterization of mechanisms," yet how to understand organization is not a trivial matter and it has not been discussed extensively (Illari and Williamson 2012, 127). The consensus in the literature is that along with entities and activities, organization is a necessary condition for a mechanism to be able to produce the phenomenon for which it is responsible. As a condition different from entities and activities, organization is generally treated as something that can be changed actually or in principle by a human or natural agent: should the same entities and activities be organized differently, they would produce a different phenomenon (Illari and Williamson 2012, 127). This view on organization is a version of what Glennan (forthcoming; see also Chapter 7) calls induced organization, which he distinguishes from affinitive organization. This distinction between induced and affinitive organization helps address the nature of organization in ecological mechanisms.

Induced organization is imposed upon entities and/or activities by an agent, human or otherwise, determining the organization of the mechanism. Affinitive organization depends upon affinities, or properties of entities and/or activities. The latter are organized as they are because of their properties. Two ways to organize a dinner party illustrate the two types of organization. The host could induce the organization of the party and assign the seats of guests at the tables, determining thereby who talks to whom. The second way is to let the organization of the party to be determined by accident (where open seats are available) and by the affinities of people attending the party—foodies, football fans, children, etc. (Glennan forthcoming).

Induced organization of mechanisms can be found in biogeography. A human or natural agent can change the distance between an island and its source region or change the arrangement of islands in an archipelago. Nature accomplishes this by means of tectonic changes, while humans do so by modifying the landscape, as in the case of land-based islands. Changing the distance between islands and their source region modifies the number of species present on an island (Pâslaru 2014). One could further analyze induced organization and add that it can be changed by either adding or removing components. That is, the host could invite additional guests, or disinvite some if they fail to behave properly. Humans induced the organization of entities—plant species—at the New York Botanical Garden when they introduced a new component: *L. maackii*. Likewise, humans induce the organization of ecological communities when they decide to remove *L. maackii* because it behaves as an invasive species and suppresses native plants.

An illustration of affinitive organization is the mechanism of dispersal of seeds of *L. maackii*. It involves the interaction between birds, deer, and small mammals with the shrub's seeds and it was not induced, but resulted from affinities of the entities and their activities. Birds consume seeds because of their nutritious value and taste, and spread them to other places, where they can germinate even after they have passed through the digestive tract of birds. No agent induced the interaction between birds, seeds, and germination of the latter. Their properties determined them to interact and organization emerged as a result. The mechanism of competition for light appears organized affinitively: *L. maackii* leafs out before native plants and suppresses their growth by depriving them of light. Leafing at a certain time in the spring is a property of *L. maackii* that it will show regardless of where and in what plant community it is planted. Similarly, succession, another central ecological process, shows affinitive organization. An area is colonized in a certain order by species belonging to certain functional types. Properties of species determine the order of colonization. Soil-fixing plants establish first, and then trees, herbivores, and carnivores. Even if an herbivore were introduced first, it could not get established in the absence of plants suitable for consumption. These examples indicate that affinitive organization is restrictive. It restricts an entity's interactions. They also indicate that properties of components limit induced organization and, more generally, "[a]ll mechanistic organization depends to some degree upon the existence of affinities—as parts must have the capacities to interact with other parts" (Glennan forthcoming). One cannot just organize differently the same entities and activities to get a different phenomenon. Trading the location of *L. maackii*'s canopy with that of its roots will produce no meaningful ecological phenomenon. The shrub will simply die. Whether induced organizations persist also depends on the properties of entities. *L. maackii* was able to establish in New York, as well as throughout the Midwestern US, because of its properties and activities. Had it not been a good competitor for light, or a producer of allelochemicals, it would have been limited to the grounds of botanical gardens.

As in the case of non-hierarchical organization, properties determine hierarchical organization. Toxic compounds produced by leaves and roots of *L. maackii* cannot be a component of the mechanism of competition for light and occupy the same level with leaves of native species, because the compounds do not have the geometrical and material properties necessary to diminish the amount of light reaching native plants. And roots cannot be part of this mechanism, since their properties are such that they can only subsist in the ground. Therefore, the hierarchical organization is a product of the properties of entities and activities and one cannot induce a different hierarchical organization of a mechanism by reshuffling its components. Instead, one has to introduce different entities which, however, could destroy the mechanism.

The stronger metaphysical finding of the foregoing examination is that induced organization is on a par with properties in determining the functioning of mechanisms and it requires an inducer, human or natural. Affinitive organization does not require an inducer and is a product of entities interacting in a certain manner in virtue of their properties, and so it appears to be an epiphenomenon. In mechanisms embodying affinitive organization, it is not the central element that determines how entities and activities produce phenomena. Instead, it is the properties of entities and activities that do this work. While affinitive organization is a form of organizing ecological mechanisms, it does not have the causal power that properties of entities and activities have. This implies that to change the organization of a mechanism, one has to change the entities and their activities. Only changing the spatial or temporal organization, while preserving the components unaltered, is rarely possible.

5. Experimental methods for mechanism discovery

Experimental techniques in ecology exemplify a variety of experimental designs that have been identified by New Mechanists as strategies of mechanism discovery. Some experiments test the causal relevance of an entity, property, activity, or organizational characteristic. Others examine the activity or the entity that mediates between a cause and its effects, and are called by-what-activity and by-what-entity experiments, respectively (Craver and Darden 2013; see also Chapter 19). Experiments examining multilevel mechanisms are bottom-up or top-down experiments, using the analytic or the synthetic strategies, respectively (Bechtel and Richardson 1993, 20). Alternatively, one intervenes on the start conditions to produce the explanandum phenomenon and examines the behavior of the putative components (Craver and Darden 2013, 128). Ecologists' quest to identify mechanisms illustrates all of the foregoing experiments.

Cipollini et al. (2008) did by-what-entity experiments when they identified the thirteen phenolic compounds contained in the leaves of *L. maackii*, and then singled out the four chemicals (including their formulas) that have an allelopathic effect. They did by-what-activity experiments when they determined how the identified chemicals impact the growth of other plants by inhibiting seed germination, and deter the foraging behavior of herbivorous insects. Additionally, one could identify in ecology by-what-property experiments that detect the property of an entity or activity that is solely responsible for or is necessary for the production of a phenomenon. Such is the experiment of Luken and Mattimiro (1991) of clipping off repeatedly for four years all stems at the top of the base of forest- and open-grown *L. maackii*. The experiment determined the shrub has the property of being resilient under stress, i.e. it is able to resprout in different habitats.

Ecologists also perform experiments to test for causal relevance. Luken et al. (1997) created large gaps in the shrub thicket to determine whether increasing light availability by removing the shrub enhances establishment of understory plants. This is an experiment that tests the causal relevance of shrub removal to enhancing abundance of understory plants. Similarly, Cipollini et al. (2008) tested the causal relevance of leaf metabolites in suppressing the growth of other plants. The experiment of Luken et al. (1997) illustrates bottom-up stimulation experiments. They exposed *L. maackii* and *L. benzoin* to increasing levels of light (0 percent, 25 percent, 100 percent) to simulate disturbance regimes that increase light availability. Typically, stimulation experiments examine putative components. However, light was not a putative component, but its level of causal efficacy, that Luken et al. examined, was. Bottom-up interference experiments in ecology decrease to various levels an abiotic or biotic component. An entire population of organisms could be removed, or parts of the organism, or the amount of an abiotic component could be decreased. For example, *L. maackii* was removed in some experiments to examine how its absence increases species richness (Gould and Gorchov 2000); in others only half of a plants' leaves were removed, and the strength of nitrogen fertilization was halved (Lieurance and Cipollini 2013).

Top-down experiments are also common. Based on the hypothesis that leachate of *L. maackii* affects other species, Watling et al. (2011) created artificial pools in which tadpoles were exposed to water containing leachate and to non-contaminated water. This experimental strategy was also used to investigate the mechanistic relationship between various levels of species diversity and ecosystem processes. In light of hypotheses about the role of species diversity in ecosystem processes, Naeem et al. (1994) created various communities in laboratory conditions, while Tilman et al. (2006) assembled them in field experiments. Instead of examining the behavior of specific components as the system is activated, ecologists focused on how species diversity produces ecosystem processes and compared them with ecosystem processes that are produced by an actual system of similarly varying species diversity.

It is not always possible to manipulate components of a mechanism, and manipulation might not change the component in the same way as natural factors do. Additionally, manipulation might not be sufficient to assess the magnitude of the indirect effects correlated with

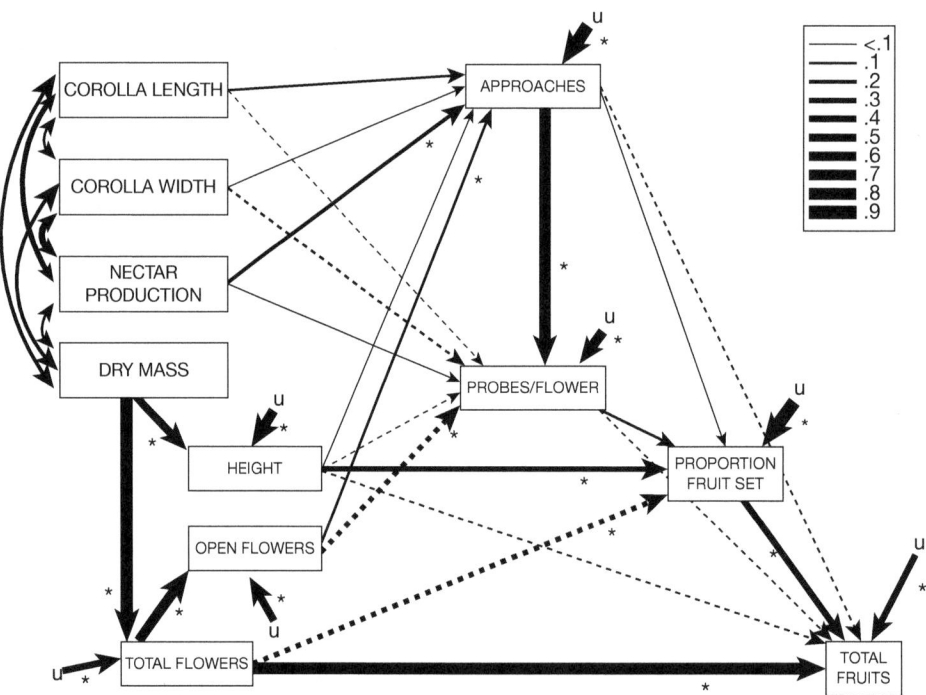

Figure 26.2   Final causal model of the causal relationships between reproductive success of flowering plants given plant traits and pollinator behavior. From Mitchell (1994, 879). Reprinted by permission of the University of Chicago Press

the component from its direct ones. In such cases, some ecologists use path analysis and structural equation modeling (SEM) to identify causal relationships making up mechanisms (Pâslaru 2015). Mitchell (1992, 1994) used these methods to identify the "underlying causal mechanisms" responsible for the reproductive success of flowering plants given plant traits and pollinator behavior. The first step in the use of SEM is to conjecture causal relationships among variables based on available knowledge about the phenomenon in question and to formulate a path model incorporating hypothesized causal relationships. The proposed model is then tested for fit with available data. Should the model not fit the data, one can change some paths of the model in light of a new hypothesis about the underlying causal mechanism and re-test the model (Figure 26.2). If deemed necessary, one could study the lower-level mechanisms that produce the causal relations between two variables.

6. Mechanisms above the individual level

Ecologists generally accept individual-level mechanisms, but some question the existence of mechanisms involving populations. A frequent objection states that explaining population dynamics in terms of abundances or density-dependence is a phenomenological approach, and, therefore, unsatisfactory (Tilman 1987; Turchin 1999). According to this view, mechanisms are found exclusively at the individual level; at the population level one finds only summaries of interactions among individuals, which are necessary, since it is not possible to follow the interactions among individual organisms (Turchin 1999). However, other ecologists affirm the existence of mechanisms at various levels beyond individuals, as I show below. Given ecologists' disagreement on this issue, further research is required. The following examination serves this goal.

Ecologists describe mechanisms that involve groups of organisms of different species, or of the same species, which are populations. Because of this, I prefer the term group-level mechanisms, instead of population-level, to cover both cases. Groups have properties that are different from those of individuals. The latter have properties that do not apply to groups: they have skeletons, physiological systems, organs, and engage in specific behaviors, e.g., feeding or mating. Groups are characterized by frequency, density, growth rate, generation, age structure, and diversity, properties that arise only when there is a collection of individuals. Group-level mechanisms contain causal relations that are best understood by analogy to natural selection being a population-level causal process (Millstein 2006) and according to the difference-making account of causation. For example, "If one systematically manipulates the density of a population to change the annual fecundity, then variation in population density is the cause of variation in annual fecundity." Groups can be viewed as individuals, based on an argument by Millstein (2009), and so they are the components of such mechanisms. They are organized spatially and temporally and changes in the properties of one group bring about changes in the properties of another group.

An example of group-level mechanisms is what Palmer et al. (2000) call the hypothetical mechanism by which diverse terrestrial and aquatic plant communities may increase biodiversity in aquatic sediments (see Figure 26.3). Its components are not individual organisms, but groups of organisms of various species. For example, component microbial diversity groups together populations of microbes of different species. Diversity means here the number of species to which microbes belong and the functional identity of those species. In addition to groups of organisms as components, this mechanism features a property as a component—temporal stability of detrital pool. Components of the mechanism are linked causally in the sense of the manipulability perspective on causation. Changing the number of

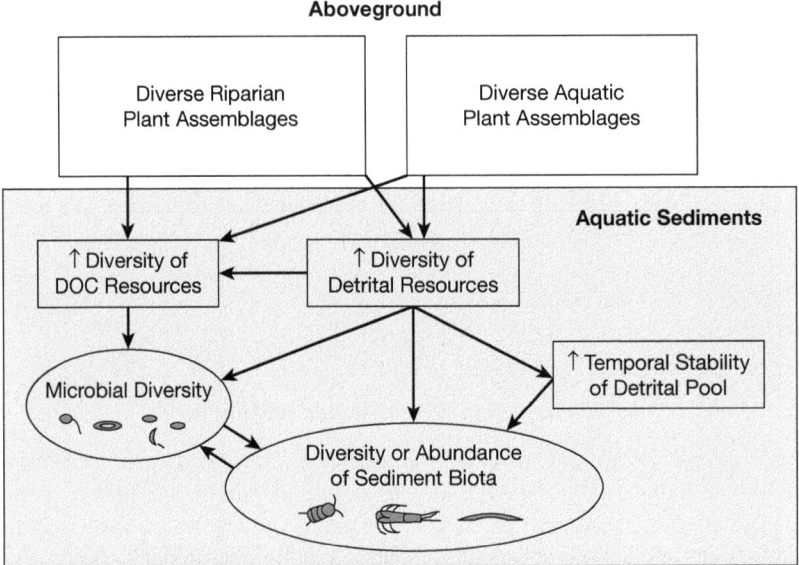

Figure 26.3 Graph of a group-level mechanism. From Palmer et al. (2000, 1069). Reprinted by permission of Oxford University Press

species, say, by decreasing it, or eliminating certain species of functional importance, changes the number and functional identity of species present in the sediment. And the presence of a component is responsible for the presence of another one, for, e.g., diverse riparian assemblages are partly responsible for the presence of dissolved organic carbon resources. While components appear organized in how they are causally interrelated, they do not perform any activity. This example and ecologists' use of the term *mechanism* suggest viewing group-level mechanisms as causal networks.

Description of this group-level mechanism is not sufficient to provide a satisfactory explanation of the phenomenon biodiversity in aquatic sediments, because Palmer et al. require a description of the lower-level mechanisms underlying the group-level mechanism. Even if the group-level mechanism is explanatorily insufficient, it appears to be necessary to provide the context for individual-level mechanisms producing the upper-level causal relations.

Another example of purported group-level mechanism is the increased population size mechanism that was proposed to explain the macro-scale relationship between the amount of energy that an assemblage of species receives and the number of species it contains. Adding energy increases population size that in turn reduces its extinction risk, and since populations are of different species, this increases the number of species of the assemblage (Evans, Warren, and Gaston 2005). This seems to be a group-level mechanism in which components, energy, and populations are linked causally such that they produce the phenomenon—the number of species in an assemblage. However, I propose viewing this as a phenomenological description of changes at the group level. The mechanism is at the individual level, as follows. Increased amounts of energy delivered in the form of nutrients allow individual organisms to grow and reproduce, which amounts to increases in population size. Because organisms are of different species, increasing the amount of energy increases the population size of all species, which results in the community containing a greater number of species compared with the situation when less energy is available. This interpretation of the increased population size mechanism indicates that at least some purported group-level mechanisms are reducible to individual-level mechanisms. Ecologists' reason for speaking about mechanisms is methodological: "[i]t is neither profitable nor necessary . . . to try to explain all macroscopic phenomena in terms of the mechanisms at lower levels" (Brown 1995, 155). (This case is analogous to the statistical description of evolutionary theory that sees genuine forces of evolution as taking place at the level of individuals. See Chapter 22.)

Elements of group-level mechanisms irreducible to individual-level mechanism appear in the mechanisms responsible for population regulation. This phenomenon occurs in populations fluctuating around a mean. Regulation happens when the per capita growth rate of the population is influenced by density-dependence. If the density of a population in an area is high, population growth is restricted, but it is amplified if the density is low (Rockwood 2015, 70–1). One of the proposed mechanisms to account for population regulation is crowding. In it, the frequency or intensity of interactions among individuals or with predators or parasites increases as population density increases, resulting in a lower fecundity for the population (Rodenhouse et al. 2003). As population density decreases, the frequency of interactions decreases, resulting in higher fecundity followed by population increase. And so the population fluctuates around a mean. However, interactions among individuals, which constitute the individual-level mechanism of crowding, are insufficient to explain population regulation, because they occur at various densities. To account for population regulation, they have to be related to changes in population density, yet it is a property of a group-level component. This indicates that contrary to what ecologists claim, crowding is not a strictly individual-level mechanism but one that involves both individual- and group-level components.

The foregoing examination points toward further work needed to clarify the nature of mechanisms above the individual level. Tentatively, there are at least three types of such mechanisms: (i) group-level mechanisms as causal networks that require individual-level mechanisms for a satisfactory explanation; (ii) group-level mechanistic descriptions of phenomena produced by individual-level mechanisms; and (iii) group-level mechanisms containing individual-level components.

7. Conclusion

Describing mechanisms is a central way of delivering explanations in ecology, especially at the individual level, where they could be characterized as follows:

> An individual-level mechanism for a phenomenon in ecology consists of entities, biotic or abiotic, that perform specific activities and are organized in certain ways by virtue of their properties such that they are responsible for the phenomenon.

These mechanisms show the importance of properties in determining activities and organization. The category of absent entities could be extended to comprise not just absent components, but also absent properties, activities, and organization. Such an extended category of absent entities allows for a more general account of background conditions necessary for the operation of certain mechanisms.

While the nature of group-level mechanisms is less clear, and additional research is needed, it can be concluded based on ecologists' use of the term mechanism that some of them are causal networks of group-level components, properties, and activities. Others offer useful methodological approaches despite being reducible to individual-level mechanisms, and yet others are group-level mechanisms.

Since ecology studies phenomena on various levels of integration above individuals and groups, further research should address the nature of mechanisms ecologists describe at those levels and their relationship with mechanisms at other levels, and answer this question: are there mechanisms specific for each level, or are they just variations of the individual- and group-level mechanisms examined above? I conjecture the latter to be the case.

Notes

1 I am grateful to Stuart Glennan and Phyllis Illari for helpful suggestions and comments on earlier drafts of the chapter. I am also thankful to Dragoş Popa-Miu for helping me improve Figure 26.1 and to Marilyn Marx for refining the English expression of this text. Hanley Sustainability Institute at the University of Dayton financially supported part of the research for this project.
2 Summary of the research is based on the review by McNeish and McEwan (2016).

References

Bechtel, William, and Robert C. Richardson. 1993. *Discovering Complexity: Decomposition and Localization as Strategies in Scientific Research*. Princeton, NJ: Princeton University Press.

Bertness, Mark D., and Sally D. Hacker. 1994. "Physical stress and positive associations among marsh plants." *The American Naturalist* 144 (3):363–372.

Brown, James H. 1995. *Macroecology*. Chicago: University of Chicago Press.

Cipollini, Don, Randall Stevenson, Stephanie Enright, Alieta Eyles, and Pierluigi Bonello. 2008. "Phenolic metabolites in leaves of the invasive shrub, Lonicera maackii, and their potential phytotoxic and anti-herbivore effects." *Journal of Chemical Ecology* 34(2):144–152.

Craver, Carl F., and Lindley Darden. 2013. *In Search of Mechanisms: Discoveries across the Life Sciences*. Chicago: University of Chicago Press.

Evans, Karl L., Philip H. Warren, and Kevin J. Gaston. 2005. "Species–energy relationships at the macroecological scale: a review of the mechanisms." *Biological Reviews of the Cambridge Philosophical Society* 80:1–25.

Gause, G. F. 1934. *The Struggle for Existence*. Baltimore: The Williams & Wilkins Company.

Glennan, Stuart. forthcoming. *The New Mechanical Philosophy*: Oxford: Oxford University Press.

Gould, Andrew M. A., and David L. Gorchov. 2000. "Effects of the exotic invasive shrub Lonicera maackii on the survival and fecundity of three species of native annuals." *The American Midland Naturalist* 144(1):36–50.

Illari, Phyllis McKay, and Jon Williamson. 2012. "What is a mechanism? Thinking about mechanisms across the sciences." *European Journal for Philosophy of Science* 2(1):119–135.

Krebs, Charles J. 2008. *Ecology: The Experimental Analysis of Distribution and Abundance*. 6 ed. San Francisco: Benjamin Cummings.

Lieurance, Deah, and Don Cipollini. 2013. "Environmental influences on growth and defence responses of the invasive shrub, Lonicera maackii, to simulated and real herbivory in the juvenile stage." *Annals of Botany* 112(4):741–749.

Luken, J. O., L. M. Kuddes, T. C. Tholemeier, and D. M. Haller. 1997. "Comparative responses of Lonicera maackii (Amur Honeysuckle) and Lindera benzoin (Spicebush) to increased light." *American Midland Naturalist* 138(2):331–343.

Luken, James O., and Daniel T. Mattimiro. 1991. "Habitat-specific resilience of the invasive shrub Amur honeysuckle (Lonicera maackii) during repeated clipping." *Ecological Applications* 1(1):104–109.

Machamer, Peter, Lindley Darden, and Carl F. Craver. 2000. "Thinking about mechanisms." *Philosophy of Science* 67(1):1–25.

McNeish, Rachael E., and Ryan W. McEwan. 2016. "Invasion ecology of Amur honeysuckle (*Lonicera maackii*), a case study of impacts at multiple ecological scales." *The Journal of the Torrey Botanical Society* 143(4):367–385.

Millstein, Roberta L. 2006. "Natural selection as a population-level causal process." *The British Journal for the Philosophy of Science* 57(4):627–653.

Millstein, Roberta L. 2009. "Populations as individuals." *Biological Theory* 4(3):267–273.

Mitchell, Randall J. 1992. "Testing evolutionary and ecological hypotheses using path analysis and structural equation modelling." *Functional Ecology* 6(2):123–129.

Mitchell, Randall J. 1994. "Effects of floral traits, pollinator visitation, and plant size on *Ipomopsis aggregata* fruit production." *The American Naturalist* 143(5):870–889.

Naeem, Shahid, Lindsey J. Thompson, Sharon P. Lawler, John H. Lawton, and Richard M. Woodfin. 1994. "Declining biodiversity can alter the performance of ecosystems." *Nature* 368:734–737.

Palmer, Margaret A., Alan P. Covich, Sam Lake, Peter Biro, Jacqui J. Brooks, Jonathan Cole, Cliff Dahm, Janine Gibert, Willem Goedkoop, and Koen Martens. 2000. "Linkages between aquatic sediment biota and life above sediments as potential drivers of biodiversity and ecological processes." *BioScience* 50(12):1062–1075.

Pâslaru, Viorel. 2014. "The mechanistic approach of *The Theory of Island Biogeography* and its current relevance." *Studies in History and Philosophy of Science Part C: Studies in History and Philosophy of Biological and Biomedical Sciences* 45:22–33.

Pâslaru, Viorel. 2015. "Causal and mechanistic explanations, and a lesson from ecology." In *Romanian Studies in Philosophy of Science*, edited by Ilie Pârvu, Gabriel Sandu, and Iulian D. Toader, 269–289. Cham, Switzerland: Springer.

Rockwood, Larry L. 2015. *Introduction to Population Ecology*. 2nd ed. Chichester, West Sussex: John Wiley & Sons.

Rodenhouse, Nicholas L., T. Scott Sillett, Patrick J. Doran, and Richard T. Holmes. 2003. "Multiple density–dependence mechanisms regulate a migratory bird population during the breeding season." *Proceedings of the Royal Society of London B: Biological Sciences* 270(1529):2105–2110.

Shipley, Bill. 2010. *From Plant Traits to Vegetation Structure: Chance and Selection in the Assembly of Ecological Communities*. Cambridge: Cambridge University Press.

Tilman, David. 1987. "The importance of the mechanisms of interspecific competition." *The American Naturalist* 129(5):769–774.

Tilman, David, Peter B. Reich, and Johannes M. H. Knops. 2006. "Biodiversity and ecosystem stability in a decade-long grassland experiment." *Nature* 441(7093):629–632.

Tremmel, D. C., and F. A. Bazzaz. 1993. "How neighbor canopy architecture affects target plant performance." *Ecology* 74(7):2114–2124.

Turchin, Peter. 1999. "Population regulation: a synthetic view." *Oikos* 84(1):153–159.

Watling, James I., Caleb R. Hickman, and John L. Orrock. 2011. "Predators and invasive plants affect performance of amphibian larvae." *Oikos* 120(5):735–739.

27

SYSTEMS BIOLOGY AND MECHANISTIC EXPLANATION

Ingo Brigandt, Sara Green, and Maureen A. O'Malley

Systems biology is a new and highly interdisciplinary field that combines elements from molecular biology and physiology with quantitative modeling approaches from disciplines such as engineering, physics, computer science, and mathematics. The term "systems biology" was used originally in 1968 by Mesarović to urge the use of systems theory to understand biological systems (Mesarović 1968); some commentators would trace the historical roots even further back (Green and Wolkenhauer 2013). But when the term is used in the context of contemporary bioscience it typically refers to a much more recent approach, initiated in the late 1990s as a response to the new experimental techniques and fast computers that allowed the rapid sequencing of DNA and automated measurements of molecular interactions (Ideker et al. 2001; Kitano 2001). These innovations afforded major new initiatives in the life sciences but also produced unforeseen challenges. Systems biology addresses one of these, namely the interpretation of extensive quantitative data via mathematical and computational modeling (Alberghina and Westerhoff 2005; Boogerd et al. 2007).

Research in systems biology is driven by complex problems that require multidisciplinary integration (Carusi 2014; MacLeod and Nersessian 2014; O'Malley and Soyer 2012). Consequently, it is a diverse field. Some proponents pursue strategies that extend molecular biology with sequence-based tools (see Chapter 23), while others explore the relevance of abstract mathematical systems theory to molecular interactions (O'Malley and Dupré 2005). Common to all branches of systems biology is the willingness to borrow reasoning and representation tools from engineering and the physical sciences, including network diagrams and graph-theory, other types of mathematical modeling (primarily ordinary differential equations), and computational simulations. We focus on just one of the many possible questions about systems biology: To what extent can the *modeling strategies and explanations* in systems biology be characterized as mechanistic?

1. Dynamic mechanistic explanation and other modeling aims

A hallmark of mechanistic research is to understand a complex whole by decomposing it into component parts, and by localizing phenomena of interest to certain parts of the system (Bechtel and Richardson 1993; Craver 2007; see Chapters 9 and 19). Models in systems biology are similarly based on empirically measured molecular entities and interactions. Given the abundance of different molecules and pathways in every cell, modeling involves the selection of components relevant to the system being investigated (Donaghy 2014). But the role of mathematical models and computational tools—as distinctive aspects of systems biology—was not addressed in original philosophical accounts of mechanistic explanations (see Chapter 16), primarily because molecular systems biology is so new. Lately, the relationship between models in systems biology and mechanistic accounts has become an important philosophical topic of debate, with some commentators arguing that a traditional mechanistic account is sufficient to describe research in systems biology (e.g., Richardson and Stephan 2007), and others instead stressing the need for a more pluralistic perspective of explanatory integration (e.g., Braillard 2010; Fagan 2016; Mekios 2015).

Although it is possible to focus on differences between dynamic models in systems biology and mechanistic explanations in general (Issad and Malaterre 2015; Théry 2015), Bechtel and Abrahamsen (2010) instead highlight the continuity between the two by introducing the notion of *dynamic mechanistic explanation*. Dynamic mechanistic explanations are also based on concrete entities and interactions, but they extend mechanistic explanation by mathematically or computationally capturing the dynamical operation of the system and its parts across time. In fact, in the case of complex systems, mathematical models are *required* for the purpose of mechanistic explanation (Baetu 2015; Bechtel 2012; Brigandt 2013; see Chapter 20). How is this the case?

In addition to decomposition and localization as strategies for discovering mechanism components, mechanistic explanations must reassemble those components and specify epistemically how their organization and operation result in the overall features of the mechanism to be explained (Bechtel and Abrahamsen 2005). Bechtel illustrates the importance of mathematical models with circadian rhythms, which are endogenous oscillations of about 24 hours present in most organisms. A mechanism diagram can depict various components of the underlying mechanism, including specific genes and proteins, some of which have oscillating expression levels. The diagram can also represent the activation or inhibition of interactions among proteins and other entities, thereby depicting positive as well as negative feedback loops (see Chapter 18). For some simple mechanisms, mental simulation (using a mechanism diagram) suffices to show how the phenomenon to be explained is generated (Bechtel and Abrahamsen 2005). However, in the case of circadian rhythms, only mathematical models are able to reveal that over time the component interactions—which involve changing protein concentrations, several feedback loops, and time-delaying gene expression pathways—actually produce sustained periodic oscillations (see also Brigandt 2015).

Computational modeling strategies in systems biology can thus extend mechanistic accounts in various ways. By providing mathematical tools for modeling the dynamics of large systems of nonlinear organization, computational models can help researchers *recompose* knowledge about subsystems that have been taken apart for functional analysis. It is well known that there is cross-talk between different mechanisms, and computational tools can afford a better understanding of how mechanisms relate to one another dynamically (Bechtel 2016; Fagan 2012). Among the many examples are computer simulations of whole cells (and even organs) that explore dynamic interactions between different functional subsystems (Bassingthwaighte et al. 2009; Karr et al. 2012). If a mechanistic explanation is characterized as accounting for how a system behavior is causally generated by the organized interaction of its particular parts, computational models of circadian rhythms or large-scale simulations can indeed be interpreted as vehicles for mechanistic explanations.

However, we should be careful about assuming that modeling in systems biology is *always* geared toward mechanistic explanations. First, ethnographic studies of research practices in systems biology show that scientists may not even frame their modeling aims in terms of explanation, but instead in terms of control or prediction (MacLeod and Nersessian 2015). Practical concerns (application) and pragmatic constraints (not all parameters can be measured and modeled mathematically) direct particular modeling aims. MacLeod and Nersessian make this point with the example of how reducing the amount of cell-wall hardening lignin in plants is highly desirable for biofuel production. But because the relevant model of the lignin synthesis pathway depends on estimated parameters, it indicates primarily a robust relation between particular system components and the lignin output. The model's main achievement, therefore, is to reveal an angle of technological control. Although it does give partial insight into how the system works, this model might not yield a mechanistic explanation in the sense of accounting for the behavior of the whole in terms of its parts and their properties (MacLeod and Nersessian 2015). Second, even when systems biologists have explanation as an explicit target, they may not be offering causal explanations of *specific* systems. Instead, they may intend to provide more abstract functional classifications of all the variants of system organization that could possibly realize a particular function (see section 3).

The lesson we draw from these observations is that while systems biology can fruitfully extend philosophical accounts of mechanistic explanation to include dynamical and quantitative aspects by means of mathematical models and simulations, philosophers investigating systems biology also need to pay attention to the diverse practices across this field, and to the actual, context-dependent aims of the modelers and experimental researchers. We will take this finding into account as we explore mechanisms in systems biology further via the use of network analysis.

2. Network models: from motifs to global topologies

Because cellular systems are highly complex webs of molecular interactions, one approach in systems biology involves the investigation of *networks*. A network can be represented and studied computationally as a graph, in which the nodes correspond to molecular entities, while an edge between two nodes represents an interaction between them. A graph can be undirected, such as a protein-interaction network that depicts all the interactions in which the protein types inside a cell engage, or it can be directed. The latter category includes metabolic reaction networks, signal transduction networks, and gene regulatory networks that depict genes regulating other genes.

While decomposition and localization have proven to be useful strategies of mechanistic research, additional methods (e.g., graph-theoretic and computational) are needed in systems biology to process large data sets and analyze highly integrated systems. One particular approach is to screen larger networks for the repeated occurrence of the same type of small connectivity patterns called *network motifs* (Figure 27.1; Alon 2007). The functionality of any network motif can then be investigated computationally. Consider, for example, a feedforward loop, in which X has a direct as well as a mediated input on Z (Figure 27.1A). Using engineering language, systems biologists might say that Z processes its two potential inputs as an AND-gate, which is when both inputs are needed for activation. In this case, the motif will function as a persistence detector. In other words, output Z will be activated only upon sustained activation of X, which can be turned on by some external signal. The reason is that when receiving an input by means of the time-delayed pathway via Y (involving two activation steps), Z would not receive a second input (directly from X) unless X has already been active for some time. Such a persistence-detector design makes biological sense when it is energetically costly for an organism to synthesize an enzyme that processes a particular substrate. In that case, synthesis is best

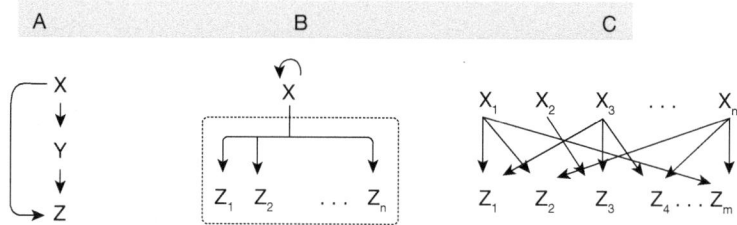

Figure 27.1   Examples of network motifs. A: Feedforward loop; B: Single-input motif; C: Dense overlapping regulons. Adapted with permission from Alon (2007), copyright Chapman & Hall/CRC Press

initiated only if the substrate is reliably present. Particular design motifs are expected to function in the same general way, whatever the particular biological and environmental contexts of their implementation (but see section 4).

Network motifs abstract away from a good deal of molecular detail. They neither specify what kinds of entities the nodes are, nor do they indicate the actual means by which one entity would activate another (e.g., that a eukaryotic gene is transcribed to RNA, which when transported outside the nucleus is translated to a protein, which later binds to a different gene so as to activate it). Despite this loss of mechanistic detail, Levy and Bechtel (2013) argue that the analysis of a network motif is still a dynamic mechanistic explanation. This is because once abstracted, the network account points directly to the *organization* of the mechanism that is responsible for the phenomenon to be explained. Generally, this sort of abstraction occurs widely in systems-biological modeling, including the examples already mentioned in section 1 (see also Chapters 17 and 34).

Even though the analysis of an individual network motif's functionality might be largely mechanistic, what makes research on network motifs distinctively systems-biological stems from the fact that large networks are screened to determine the frequency with which motifs occur (e.g., the feedforward loop is known to be highly abundant). Doing so reveals both common and uncommon elements of biological design, and draws attention to the former, which are likely to be more biologically important. Moreover, different kinds of large networks, from gene regulatory to neural networks, can be screened for the same design element. This generalizability also applies to networks from different taxa, whether prokaryotes or eukaryotes. These strategies indicate that abstract organizational schemes, which systems biologists call *design principles*, transcend the organization of a single mechanism, and even a single species (Green 2015b). We will elaborate on design principles in section 3.

While the scrutiny of an individual motif pertains to a very small network, research at the other end of the spectrum investigates large networks for their *global properties*, also via graph-theoretical means. Earlier work initially addressed regular networks (where each node has the same number of edges) and random networks (where a certain proportion of nodes is randomly connected by edges; Figure 27.2A). In the last fifteen years, however, small-world and scale-free networks have gained prominence because of their interesting properties and widespread occurrence among real biological systems. A *small-world network* is defined in terms of the global property of the average path length between two nodes—averaged across all pairs of the network's nodes—which grows logarithmically as the number of nodes increases. This means that for two randomly chosen nodes, the shortest distance between them (in terms of a path of intermediate nodes connected by edges) will be small relative to the size of the network. This global property usually entails that a signal propagates quickly from one part of the network to another. For a biological system, this can have the advantage of enabling rapid reaction times. Many protein-interaction networks are small-world for this reason (Albert 2005).

A network's degree distribution *P(k)* is the network-wide proportion of edges that is connected to *k* other nodes (i.e., the network's proportion of nodes connected to only one other node, the proportion of nodes connected to two other nodes, and so on). The degree distribution is thus a global characteristic. A *scale-free network* is defined as a network that has a degree distribution that follows a power law of the form $P(k) = c \cdot k^{-\gamma}$. This exponentially declining function means that across any scale-free network there are many nodes that are connected to only one or a few other nodes, while only few nodes are so-called *hubs*, which are connected to many other nodes (Figure 27.2B). From this global property, predictions can be made about the network's functionality. One is that it will exhibit *robustness*, which is the biologically important feature that a system will maintain its functionality despite perturbations. While the elimination of a node that is connected to one or only a few other nodes is unlikely to affect a network's functionality, eliminating a node that is a hub may seriously impact how the network functions. But in a scale-free network there are comparatively few hubs, meaning that such a network is generally robust. A variety of actual biological networks of interest to systems biology are approximately scale-free, including metabolic networks and the gene regulatory networks of prokaryotes and eukaryotes (Albert 2005).

Huneman (2010) coined the term *topological explanation* for explanations of phenomena in terms of topological properties (including the structural properties of a graph). Although his focus is on ecological systems and evolutionary contexts, one important explanandum he addresses is equally relevant in systems biology: namely, robustness. Huneman argues that an explanation of a system's general robustness to random node elimination in terms of its scale-free network structure is a topological explanation. A topological explanation appeals to a system's basic organization, in the same way the network motif explanations mentioned above do (Levy and Bechtel 2013), but it abstracts away from even the generic interactions or temporal features

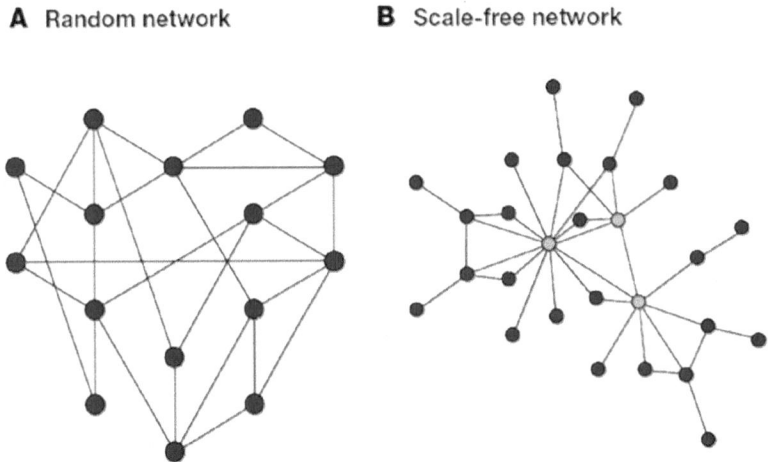

A Random network **B** Scale-free network

Figure 27.2 An illustration of two kinds of large-scale networks. In the scale-free network, highly connected hub nodes are visualized in lighter gray. Reprinted from Barabási and Oltvai (2004) with permission from *Nature Reviews Genetics*, Macmillan Publishers Ltd, copyright 2004

seen in motifs. This is at odds with mechanistic explanation, if the latter is to include significant physical detail (Craver 2007; Kaplan and Craver 2011), or if a mechanistic account's explanatory status is taken to increase when more detail is added (Kaplan 2011; see Chapter 20). In any case, the fact that topological explanations neither list specific activities nor trace their operation from set-up to termination conditions is Huneman's primary reason for contrasting this type of explanation against mechanistic explanation.

Another case of topological explanation in the context of systems biology is the explanation of vulnerability (the opposite of robustness) in terms of bowtie structures (Jones 2014). A bowtie structure is a molecular network with the shape of a bowtie (Figure 27.3), in which it is obvious that the bowtie's core is the weakest link because its deactivation (compared to any other node) will probably damage the whole network's functionality. A concrete example is the explanation of why the human immune system is vulnerable to attacks on CD4+ T-cells (by HIV). The reason is that the molecular network of intercellular interactions and signaling pathways forms a bowtie that has the CD4+ T-cell type as its core (Figure 27.3; Kitano and Oda 2006). Generally, then, it holds for scale-free networks that they are robust to random perturbations but vulnerable to attacks on the highly connected nodes (hubs or bowtie cores) that participate in a large number of interactions. Topological explanations such as these may well be instances of what some philosophers have discussed as distinctively mathematical explanations in natural science, which have even been claimed to offer non-causal explanations of physical phenomena (Lange 2013). Our reason for invoking topological explanation, however, is simply to show how it contrasts with classic mechanistic accounts (see Bechtel 2015; Woodward 2013).

Despite these insights, the value of graph-theoretical analysis for biological research is a contested issue, and critics have pointed to problems with the generalizations made about such networks (Arita 2004). For instance, whether gene regulatory networks are scale-free has been

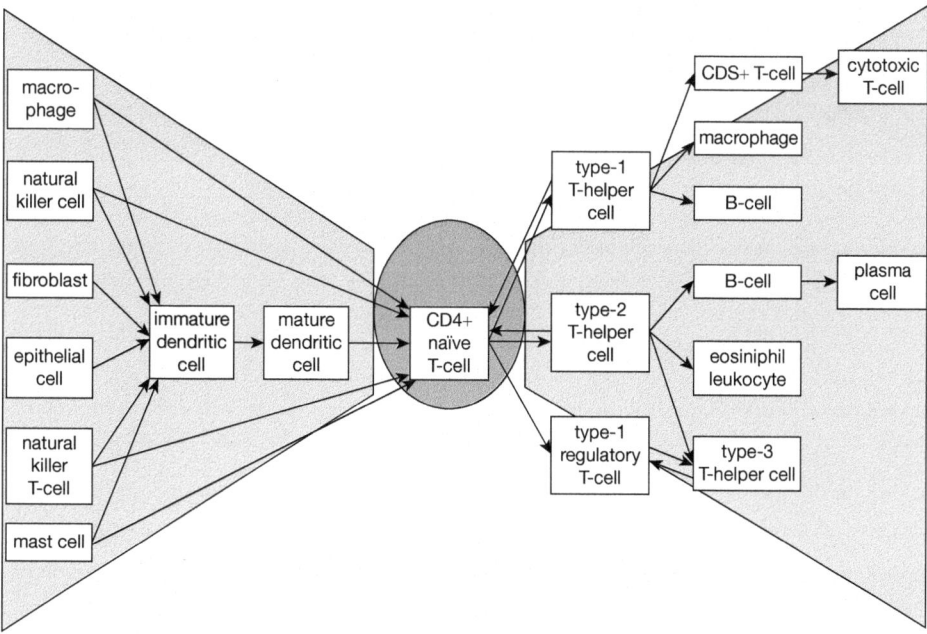

Figure 27.3   Bowtie network of CD4+ T-cells. Reproduced from Jones (2014) with permission from *Erkenntnis*, Springer, copyright 2014

disputed, because some biological networks also exhibit properties similar to random networks (Barabási and Oltvai 2004; Keller 2005). Another common challenge is that networks usually provide a static picture of cellular systems, because the data that network edges are based on combine the totality of interactions that have been measured. However, all the edges represented need not be active at the same time or in the same location *in vivo*. A recent development, therefore, is to include temporal change when constructing topological mappings. When time-course data are used, distinctions can be made between a "party hub," which interacts with many entities at the same time, and a "date hub," which interacts with only a few other entities despite having many overall connections and interaction partners (Han et al. 2004). In yeast metabolism research, protein-interaction and gene expression data have been used to develop a time-dependent network that is sensitive to which proteins interact at a particular phase of the cell cycle (de Lichtenberg et al. 2005). This has provided new insight into the processes underlying the periodization of protein synthesis.

As section 1 discussed, research methods in systems biology have huge potential not only for extending mechanistic research but also for providing novel insights into how and why biological systems are organized into generalizable schemes with broad applications. As we have demonstrated, graph-theoretical analysis affords a quantitative understanding of biological function and makes possible a comparison of organizational schemes in different functional systems. As well as cellular systems, neuronal, ecological, and even non-biological communication and transport networks are often scale-free or small-world networks, and can be analyzed accordingly. Now we will show how network and systems analysis can be taken even further epistemically, in a way that provides additional philosophical insight into the relationship between systems biology and mechanistic research.

3. Searching for and using design principles

An important research question in systems biology is the extent to which biological functions rely on general *design principles* that are largely independent of specific causal details and particular contexts of implementation (Poyatos 2012). Design principles are abstractions that describe characteristic organizational features of importance for a system's functionality, such as negative feedback control, network motif configurations, or common architectures of biological and engineered networks. Aside from understanding how these design principles are causally instantiated in specific biological systems, an important explanatory question is why the same basic principles can describe the functioning of so many different systems. Some philosophers have recently argued that certain abstract models in systems biology, when answering that question, provide non-mechanistic *design explanations* that focus on generalizable constraints for biological functions (Braillard 2010; Green 2015b). In contrast, discussions of mechanisms have typically interpreted abstract models solely as heuristic tools or as mechanism schemas that guide the formulation of more realistic models (Matthiessen 2017; see Chapter 19). This resonates with the perspective of many experimental biologists, but has long been opposed by proponents of systems theory (Green and Wolkenhauer 2013). Rather than assuming that a model is useful only insofar as it explains a biological system in concrete detail, current philosophical investigations of design explanations (and of topological explanations) are motivated by the goal of making sense of why some scientists rely on abstract models even in situations where more detailed models exist.

One illuminating example is how biologists investigate systems exhibiting *robust perfect adaptation* (RPA) from an engineering perspective. RPA is the capacity of a system with sensors to return to the exact pre-stimulus activity after a stimulus-response reaction. This is important because it maintains the responsiveness of sensors. Creating designs with RPA is a goal in engineering human-made systems. Biological systems also exhibit RPA. Examples include the regulation of calcium homeostasis in mammals and membrane turgor pressure in yeast (Briat et al. 2017). In bacterial chemotaxis (movement in response to external chemical stimuli), RPA pertains to the regained responsiveness of transmembrane receptors (i.e., sensors) that detect changes in the concentration of chemicals in the environment. Adding a repellent to the bacterial environment leads to changes in the bacterial tumbling frequency (and thereby to random reorientations in space), but the receptor system returns very quickly to its equilibrium value. This enables the receptors to become sensitive to new changes, even if the repellent concentration continues increasing, which occurs when the bacterium swims along a chemical gradient. In the case of the bacterium *E. coli*, the mechanistic basis of its chemotactic RPA is known. It consists of transmembrane receptors, a signal transduction pathway inside the bacterium, and its connection to the flagellar motor. A feedback loop from the intracellular process back to the transmembrane receptor is important for achieving RPA (Barkai and Leibler 1997).

The explanatory issue we are highlighting with bacterial chemotaxis is the question of what *generic properties* (abstract organizational features) make it possible for any system—not just *E. coli*—to exhibit RPA. The answer is integral feedback control (Yi et al. 2000). Used in engineering, integral feedback control is known in mathematical control theory as a special case of the internal model principle. When the environmental input u changes (see Figure 27.4), the difference between the actual output y_1 and the desired output y_0—the equilibrium value of the receptor—is fed back into the system as the integral of the system error. This feedback functions as a signal for the renormalization of the receptor, so that integral feedback control is sufficient for RPA. A crucial insight is provided by a theorem of Yi et al. (2000), which shows that (at least in linear systems) integral feedback control is *necessary* for achieving RPA. This explains why any system that exhibits RPA has to have an organization that instantiates integral feedback control (Iglesias 2013). *E. coli* should be no exception in this regard, and Yi et al. (2000) point out that an influential dynamic mechanistic model of RPA in bacteria (Barkai and Leibler 1997) does indeed embody the basic principle of integral feedback control.

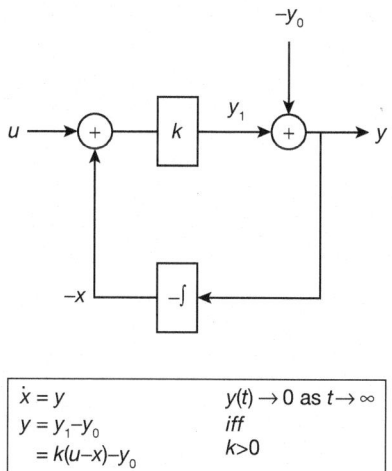

Figure 27.4 Diagram showing the abstract principle of integral feedback control. Reproduced from Yi et al. (2000), with permission from *Proceedings of the National Academy of Sciences*, copyright (2000) National Academy of Sciences, USA

We can compare this account of chemotaxis with a standard mechanistic explanation. The latter would show how a particular structural organization causally generates and thus explains some function (e.g., RPA). In contrast, what Wouters (2007) calls a design explanation proceeds in the *opposite direction*, as the function to be performed (RPA in our case) explains the presence of some structural organization (integral feedback control). Using examples from physiology and functional anatomy, Wouters argues that such an explanation is non-causal, because it is based on law-like dependency relations between structures and functions. It maps out the possible structural realizers of a certain function, without going into a diachronic account of how the realizer or the need for the function came about causally. But regardless of where one stands on the status of non-causal explanations, in the case of bacterial chemotaxis, the design explanation does not just offer a list of the various concrete mechanisms that perform RPA (e.g., a transmembrane receptor, six Che proteins, and other details in *E. coli*). This is a non-mechanistic explanation in that it points to the abstract organizational feature of integral feedback control as a generic property that *any* system exhibiting RPA must instantiate (Braillard 2010). This explanatory aim addresses a *why*-question that is distinct from the aim of explaining *how* a behavior is mechanistically produced in some specific system.

The example of design principles underpinning RPA in engineering and biology also provides more general philosophical lessons about the theoretical relevance of delineating the space of biological possibility. Mechanistic accounts have typically taken how-possibly models to have less explanatory power than how-actually models (Craver 2007; Kaplan 2011; Kaplan and Craver 2011; see Chapter 19). Yet understanding the wider constraints on biological variation can in some contexts be of higher importance than describing how a specific function is causally produced in any specific system. Design principles can help researchers understand *why* the same structural patterns are found across different contexts: as a result of the constraints on possible architectures that can realize a given function. Importantly, this is not to be understood as a question that presupposes natural selection as the answer. Rather, the why-question addressed here is about the physically determined boundaries of the design space for a given function.

Design principles do, however, have significance for evolutionary research as well as functional biology. Investigations of the constraints on evolutionary and developmental trajectories have often been associated with rather speculative "structuralist" accounts, but some of these ideas have gained new relevance in the context of *evolutionary systems biology* (Green et al. 2015; Jaeger and Crombach 2012). Evolutionary systems biology is an umbrella term for very diverse approaches (O'Malley 2012), but one important aim is to investigate why certain general patterns arise in evolution. This is often done via models that represent the *in silico* evolution of gene regulatory networks. Structures like the network motifs discussed in section 2 are often assumed to be common because of regulatory functions favored by natural selection (Alon 2007). Yet evolutionary simulation studies suggest that common structural patterns of networks, such as feedforward loops, may also result from constraints on genome evolution. These constraints are inherent in the mutational dynamics of gene duplication, deletion, and recombination (Cordero and Hogeweg 2006). Research on evolutionary design principles, when understood as general patterns occurring from evolutionary trajectories, can thus generate insight into the potential and limits of biological variation.

Design principles also identify the generic features that unite diverse systems exhibiting similar functional patterns (Green 2015b). By relating specific systems to general functional types, such as signal amplifiers, filters, or homeostatic regulators, these abstract principles facilitate the transfer of theoretical frameworks across disciplinary borders. Aside from this epistemic role, such structure-function mappings can serve practical aims. Similarly to MacLeod and Nersessian's (2015) emphasis on practical purposes such as control and prediction of modeling in systems biology, research on possibility spaces for biological structures can have practical goals such as templates for synthetic biology designs. Synthetic biology is the biological construction of material models, usually guided by mathematical modeling. "How-possibly models" can in this context be more important than "how-actually models" because they elucidate the necessary structures for a certain function, like RPA, or reveal whether there are simpler possible designs than the ones found in nature (Briat et al. 2017; Ma et al. 2009).

The upshot of this discussion is that abstract models are not always stepping stones toward more detailed mechanistic models. Aside from the practical purposes of control and technological implementation, abstract design principles afford an understanding of why causally different systems in biology and engineering share certain organizational features, and how they are situated within larger spaces of physically possible designs. Consequently, an exclusive philosophical focus on mechanistic explanation (and even on dynamic mechanistic explanation) risks missing out on these diverse epistemic activities in systems biology.

4. Discussion and outlook

Research in systems biology shows how strategies of abstraction are used in biology not only to simplify the task of identifying mechanisms but also to elucidate system-level patterns of organization that may not be visible at the level of the molecular details. Mechanistic accounts have usually been framed in opposition to explanatory unification, understood as the subsumption of the particular to general laws or explanatory schemas. But network modeling and the quest for design principles suggest an alternative way of thinking about the role of *unification* in biology: not via reduction of the particular to the general, but through abstraction from causal details for the purpose of identifying generic organizational patterns.

Mathematical models (including network models and design principles) serve various roles in systems biology. Generally, mathematical frameworks provide a more rigorous way of exploring the extent to which biological functions are underpinned by characteristic organizational structures. Mathematical frameworks can also make engineering analogies more precise. Section 2 mentioned the identification of functional network motifs based on mathematically guided screening for overabundant circuit types. This search is inspired by an analogy to design principles in electronic networks, and the structural decomposition of the network preceded the functional analysis of the modules of the network (see Chapter 34). In other cases, the biological function is known and systems biologists set out to explore the extent to which the function is similarly realized in engineered systems (e.g., robustness). Mathematical abstractions and design principles can articulate constraints that delimit the search space for an analysis. Delineating search space may serve the development of mechanistic explanations. At the same time, network models and design principles provide an understanding complementary to mechanistic explanation. Although an important virtue of mechanistic explanations is to make sense of biological diversity through attention to specific causal difference-makers and material composition, abstraction strategies can help scientists see similarities in the way functional systems—from airplanes to organisms—are organized.

Generally, a focus by philosophers on the issue of mechanistic explanation has left many aspects of systems biology unexplored. We have pointed to the use of models for the purpose of prediction, control, or the creation of simple and efficient designs that can be implemented in synthetic organisms. Another topic of major interest to systems biologists that is philosophically rich is robustness. In many cases when a system maintains its functioning despite noise and even major perturbations, this is due to *dynamic reorganization*, where the organismal system responds flexibly by changing interaction patterns and levels, including establishing new interactions (Wagner 2005). This puts pressure on the assumption that systems biology always investigates mechanisms (on a machine-like conception), or that all explanations about systems exhibiting dynamic reorganization or robustness are mechanistic (in the sense of referencing the mechanism's specific organization; Brigandt 2015; Gross 2015; Woodward 2013).

Some of the questions that deserve more attention by philosophers pertain to issues that are currently controversial within the systems biology community itself. One is the question of whether complex living systems can be understood in terms of engineering notions (Braillard 2015; Green 2015a), and particularly whether the heuristic assumption of *modularity* is warranted. Research on network motifs is often predicated on the idea that an individual motif is modular, meaning that its functionality is unaffected by the system context in which it occurs (section 2). The traditional mechanistic strategies of decomposition into distinct components and the localization of some function to a certain component also resonate with the assumption that biological systems are modular. However, many systems biologists observe highly integrated functionality across large-scale networks, which suggests that systems need to be investigated not in terms of modularity but via more connectivist perspectives that can capture features emerging from system-wide dynamics (Huang 2004; see also Bassingthwaighte et al. 2009; Bechtel 2015).

Our discussion draws attention to a wide range of explanatory and modeling strategies in systems biology. We have shown how some explanatory aims and outputs are not mechanistic according to standard philosophical interpretations of mechanistic explanations, and indeed, that some of the practices in systems biology lie outside existing philosophical frameworks. But well beyond these negative insights, we have depicted the wealth of modeling approaches at work in systems biology, and how further philosophical scrutiny of them will enhance the investigation of biological systems and philosophical accounts of mechanistic explanation and explanation in general.

References

Alberghina, L. and H. V. Westerhoff (eds) (2005) *Systems Biology: Definitions and Perspectives*, Berlin: Springer.

Albert, R. (2005) "Scale-Free Networks in Cell Biology," *Journal of Cell Science* 118: 4947–57.

Alon, U. (2007) *An Introduction to Systems Biology: Design Principles of Biological Circuits*, Boca Raton, FL: Chapman & Hall / CRC Press.

Arita, M. (2004) "The Metabolic World of *Escherichia coli* Is Not Small," *Proceedings of the National Academy of Sciences USA* 101: 1543–7.

Baetu, T. (2015) "From Mechanisms to Mathematical Models and Back to Mechanisms: Quantitative Mechanistic Explanations," in P.-A. Braillard and C. Malaterre (eds.), *Explanation in Biology: An Enquiry into the Diversity of Explanatory Patterns in the Life Sciences*, Dordrecht: Springer, pp. 345–63.

Barabási, A.-L. and Z. N. Oltvai (2004) "Network Biology: Understanding the Cell's Functional Organization," *Nature Reviews Genetics* 5: 101–13.

Barkai, N. and S. Leibler (1997) "Robustness in Simple Biochemical Networks," *Nature* 387: 913–17.

Bassingthwaighte, J., P. Hunter and D. Noble (2009) "The Cardiac Physiome: Perspectives for the Future," *Experimental Physiology* 94: 597–605.

Bechtel, W. (2012) "Understanding Endogenously Active Mechanisms: A Scientific and Philosophical Challenge," *European Journal for Philosophy of Science* 2: 233–48.

—— (2015) "Can Mechanistic Explanation Be Reconciled with Scale-Free Constitution and Dynamics?" *Studies in History and Philosophy of Biological and Biomedical Sciences* 53: 84–93.

—— (2016) "Using Computational Models to Discover and Understand Mechanisms," *Studies in History and Philosophy of Science* 56: 113–121.

Bechtel, W. and A. Abrahamsen (2005) "Explanation: A Mechanist Alternative," *Studies in History and Philosophy of Biological and Biomedical Sciences* 36: 421–41.

—— (2010) "Dynamic Mechanistic Explanation: Computational Modeling of Circadian Rhythms as an Exemplar for Cognitive Science," *Studies in History and Philosophy of Science* 41: 321–33.

Bechtel, W. and R. C. Richardson (1993) *Discovering Complexity: Decomposition and Localization as Strategies in Scientific Research*, Princeton, NJ: Princeton University Press.

Boogerd, F. C., F. J. Bruggeman, J.-H. S. Hofmeyr and H. V. Westerhoff (eds) (2007) *Systems Biology: Philosophical Foundations*, Amsterdam: Elsevier.

Braillard, P.-A. (2010) "Systems Biology and the Mechanistic Framework," *History and Philosophy of the Life Sciences* 32: 43–62.

—— (2015) "Prospect and Limits of Explaining Biological Systems in Engineering Terms," in P.-A. Braillard and C. Malaterre (eds.), *Explanation in Biology: An Enquiry into the Diversity of Explanatory Patterns in the Life Sciences*, Dordrecht: Springer, pp. 319–44.

Briat, C., A. Gupta and M. Khammash (2017) "Antithetic Integral Feedback Ensures Robust Perfect Adaptation in Noisy Biomolecular Networks," *arXiv.org e-Print Archive* arXiv:1410.6064v7 [math.OC].

Brigandt, I. (2013) "Systems Biology and the Integration of Mechanistic Explanation and Mathematical Explanation," *Studies in History and Philosophy of Biological and Biomedical Sciences* 44: 477–92.

——(2015) "Evolutionary Developmental Biology and the Limits of Philosophical Accounts of Mechanistic Explanation," in P.-A. Braillard and C. Malaterre (eds.), *Explanation in Biology: An Enquiry into the Diversity of Explanatory Patterns in the Life Sciences*, Dordrecht: Springer, pp. 135–73.

Carusi, A. (2014) "Validation and Variability: Dual Challenges on the Path from Systems Biology to Systems Medicine," *Studies in History and Philosophy of Biological and Biomedical Sciences* 48: 28–37.

Cordero, O. X. and P. Hogeweg (2006) "Feed-Forward Loop Circuits as a Side Effect of Genome Evolution," *Molecular Biology and Evolution* 23: 1931–6.

Craver, C. F. (2007) *Explaining the Brain: Mechanisms and the Mosaic Unity of Neuroscience*, Oxford: Oxford University Press.

Donaghy, J. (2014) "Temporal Decomposition: A Strategy for Building Mathematical Models of Complex Metabolic Systems," *Studies in History and Philosophy of Biological and Biomedical Sciences* 48: 1–11.

Fagan, M. B. (2012) "Waddington Redux: Models and Explanation in Stem Cell and Systems Biology," *Biology & Philosophy* 27: 179–213.

—— (2016) "Stem Cells and Systems Models: Clashing Views of Explanation," *Synthese* 193: 873–907.

Green, S. (2015a) "Can Biological Complexity Be Reverse Engineered?" *Studies in History and Philosophy of Biological and Biomedical Sciences* 53: 73–83.

—— (2015b) "Revisiting Generality in Biology: Systems Biology and the Quest for Design Principles," *Biology & Philosophy* 30: 629–52.

Green, S., M. Fagan and J. Jaeger (2015) "Explanatory Integration Challenges in Evolutionary Systems Biology," *Biological Theory* 10: 18–35.

Green, S. and O. Wolkenhauer (2013) "Tracing Organizing Principles: Learning from the History of Systems Biology," *History and Philosophy of the Life Sciences* 35: 553–76.

Gross, F. (2015) "The Relevance of Irrelevance: Explanation in Systems Biology," in P.-A. Braillard and C. Malaterre (eds.), *Explanation in Biology: An Enquiry into the Diversity of Explanatory Patterns in the Life Sciences*, Dordrecht: Springer, pp. 175–98.

Han, J.-D. J., N. Bertin, T. Hao, D. S. Goldberg, G. F. Berriz, L. V. Zhang, D. Dupuy, et al. (2004) "Evidence for Dynamically Organized Modularity in the Yeast Protein-Protein Interaction Network," *Nature* 430: 88–93.

Huang, S. (2004) "Back to the Biology in Systems Biology: What Can We Learn from Biomolecular Networks?" *Briefings in Functional Genomics and Proteomics* 2: 279–97.

Huneman, P. (2010) "Topological Explanations and Robustness in Biological Sciences," *Synthese* 177: 213–45.

Ideker, T., T. Galitski and L. Hood (2001) "A New Approach to Decoding Life: Systems Biology," *Annual Review of Genomics and Human Genetics* 2: 343–72.

Iglesias, P. (2013) "Systems Biology: The Role of Engineering in the Reverse Engineering of Biological Signaling," *Cells* 2: 393–413.

Issad, T. and C. Malaterre (2015) "Are Dynamic Mechanistic Explanations Still Mechanistic?" in P.-A. Braillard and C. Malaterre (eds.), *Explanation in Biology: An Enquiry into the Diversity of Explanatory Patterns in the Life Sciences*, Dordrecht: Springer, pp. 265–92.

Jaeger, J. and A. Crombach (2012) "Life's Attractors: Understanding Developmental Systems through Reverse Engineering and in Silico Evolution," in O. S. Soyer (ed.), *Evolutionary Systems Biology*, New York: Springer, pp. 93–119.

Jones, N. (2014) "Bowtie Structures, Pathway Diagrams, and Topological Explanation," *Erkenntnis* 79: 1135–55.

Kaplan, D. M. (2011) "Explanation and Description in Computational Neuroscience," *Synthese* 183: 339–73.

Kaplan, D. M. and C. F. Craver (2011) "The Explanatory Force of Dynamical and Mathematical Models in Neuroscience: A Mechanistic Perspective," *Philosophy of Science* 78: 601–27.

Karr, J. R., J. C. Sanghvi, D. N. Macklin, M. V. Gutschow, J. M. Jacobs, B. Bolival, N. Assad-Garcia, J. I. Glass and M. W. Covert (2012) "A Whole-Cell Computational Model Predicts Phenotype from Genotype," *Cell* 150: 389–401.

Keller, E. F. (2005) "Revisiting 'Scale-Free' Networks," *BioEssays* 27: 1060–8.

Kitano, H. (ed.) (2001) *Foundations of Systems Biology*, Cambridge, MA: MIT Press.

Kitano, H. and K. Oda (2006) "Robustness Trade-Offs and Host–Microbial Symbiosis in the Immune System," *Molecular Systems Biology* 2: 22.

Lange, M. (2013) "What Makes a Scientific Explanation Distinctively Mathematical?" *British Journal for the Philosophy of Science* 64: 485–511.

Levy, A. and W. Bechtel (2013) "Abstraction and the Organization of Mechanisms," *Philosophy of Science* 80: 241–61.

de Lichtenberg, U., L. J. Jensen, S. Brunak and P. Bork (2005) "Dynamic Complex Formation During the Yeast Cell Cycle," *Science* 307: 724–7.

Ma, W., A. Trusina, H. El-Samad, W. A. Lim and C. Tang (2009) "Defining Network Topologies That Can Achieve Biochemical Adaptation," *Cell* 138: 760–73.

MacLeod, M. and N. J. Nersessian (2014) "Strategies for Coordinating Experimentation and Modeling in Integrative Systems Biology," *Journal of Experimental Zoology Part B: Molecular and Developmental Evolution* 322: 230–9.

—— (2015) "Modeling Systems-Level Dynamics: Understanding without Mechanistic Explanation in Integrative Systems Biology," *Studies in History and Philosophy of Biological and Biomedical Sciences* 49: 1–11.

Matthiessen, D. (2017) "Mechanistic Explanation in Systems Biology: Cellular Networks," *British Journal for the Philosophy of Science* 68: 1–25.

Mekios, C. (2015) "Explanation in Systems Biology: Is It All about Mechanisms?" in P.-A. Braillard and C. Malaterre (eds.), *Explanation in Biology: An Enquiry into the Diversity of Explanatory Patterns in the Life Sciences*, Dordrecht: Springer, pp. 41–72.

Mesarović, M. D. (1968) "Systems Theory and Biology—View of a Theoretician," in M. D. Mesarović (ed.), *Systems Theory and Biology: Proceedings of the III Systems Symposium at Case Institute of Technology*, Berlin: Springer, pp. 59–87.

O'Malley, M. A. (2012) "Evolutionary Systems Biology: Historical and Philosophical Perspectives on an Emerging Synthesis," in O. S. Soyer (ed.), *Evolutionary Systems Biology*, New York: Springer, pp. 1–28.

O'Malley, M. A. and J. Dupré (2005) "Fundamental Issues in Systems Biology," *BioEssays* 27: 1270–6.

O'Malley, M. A. and O. S. Soyer (2012) "The Roles of Integration in Molecular Systems Biology," *Studies in History and Philosophy of Biological and Biomedical Sciences* 43: 58–68.

Poyatos, J. (2012) "On the Search for Design Principles in Biological Systems," in O. S. Soyer (ed.), *Evolutionary Systems Biology*, New York: Springer, pp. 183–93.

Richardson, R. C. and A. Stephan (2007) "Mechanism and Mechanical Explanation in Systems Biology," in F. C. Boogerd, F. J. Bruggeman, J.-H. S. Hofmeyr and H. V. Westerhoff (eds.), *Systems Biology: Philosophical Foundations*, Amsterdam: Elsevier, pp. 123–44.

Théry, F. (2015) "Explaining in Contemporary Molecular Biology: Beyond Mechanisms," in P.-A. Braillard and C. Malaterre (eds.), *Explanation in Biology: An Enquiry into the Diversity of Explanatory Patterns in the Life Sciences*, Dordrecht: Springer, pp. 113–33.

Wagner, A. (2005) *Robustness and Evolvability in Living Systems*, Princeton, NJ: Princeton University Press.

Woodward, J. (2013) "Mechanistic Explanation: Its Scope and Limits," *Aristotelian Society Supplementary Volume* 87: 39–65.

Wouters, A. (2007) "Design Explanation: Determining the Constraints on What Can Be Alive," *Erkenntnis* 67: 65–80.

Yi, T.-M., Y. Huang, M. I. Simon and J. Doyle (2000) "Robust Perfect Adaptation in Bacterial Chemotaxis through Integral Feedback Control," *Proceedings of the National Academy of Sciences USA* 97: 4649–53.

28

MECHANISTIC EXPLANATION IN NEUROSCIENCE[1]

Catherine Stinson and Jacqueline Sullivan

1. Introduction

Perhaps the most striking thing you notice when thumbing through the pages of a neuroscience textbook like *Principles of Neural Science* (Kandel et al. 2012) are the elaborate diagrams of the central nervous system, brain, spinal cord, synapses, neurons, and molecules. It is equally striking that whatever topic you look at, whether the action potential, synaptic transmission, cognition, or perception, it is inevitably described and explained using the word "mechanism." You can become so accustomed to seeing and hearing about mechanisms in neuroscience that it never occurs to you to question what mechanisms in fact are, or what that choice of terminology implies. As it turns out, there are tricky philosophical problems lurking beneath the surface of mechanism talk in neuroscience.

In this chapter we explore some of the ways that mechanisms are invoked in neuroscience, and look at a selection of the philosophical problems that arise when trying to understand mechanistic explanations (several chapters in this volume go into more detail about particular philosophical problems encountered in mechanistic explanation in neuroscience, and Chapter 6 also describes some of the history we discuss below). We begin in section 2 by introducing a series of historical case studies that illustrate how neuroscientists have depended on mechanistic metaphors in their efforts to understand the mind and brain, and how their mechanistic explanations have developed over time. We revisit these examples throughout the remainder of the chapter. In section 3, we use these case studies to highlight what contemporary philosophers have identified as the fundamental features of mechanisms and mechanistic explanation. In section 4, we consider some of the methodological issues that arise in neuroscience including (1) how to integrate psychological with neural models, (2) how to generalize findings in model organisms like the sea slug *Aplysia* to human learning and memory, and (3) whether to favor top-down or bottom-up methods.

2. A short history of neural mechanisms of learning and memory

The historical examples we focus on are episodes in the search for the neural mechanisms of learning and memory, which are among the most important cognitive phenomena studied in neuroscience. These examples illustrate how mechanisms are discovered, reasoned about, represented, and how they figure in explanations.

In the seventeenth century, in *Treatise on Man*, the French philosopher René Descartes likened human beings to machines. Descartes drew an analogy between the movements of human beings and the movements of the automated figures in the fountains of the Royal Gardens at St. Germain. Descartes described how when visitors to the gardens step on certain tiles, statues of Roman Gods, Goddesses, and other mythical creatures move, gesture, play music, spray water, and speak. Pressing on the tiles triggers a flow of water from storage tanks beneath the fountains through a network of hidden pipes. The flow of water then causes the figures, which are connected to machinery like springs and cogs, to move. Descartes also compared these motions to those of a clock or a mill, which can be made to move continuously, not just in response to an external push.

Descartes claimed that a similar set of events takes place in the human nervous system when simple reflexes are triggered. The brain, according to Descartes, contained ventricles filled with "animal spirits" or "a very fine air or wind," which reached the ventricles via the blood (Descartes 1664/1985, 100). He believed that the ventricles were connected to networks of nerves, which he thought were mostly hollow save for a set of small fibers running their length.

According to Descartes, the nerves are connected to the brain in such a way that stimulation from the periphery, which tugs on the fibers, is communicated to the brain, triggering a response. Tugging on the fibers opens pores in the nerve, allowing animal spirits to flow from the ventricles through the nerve to the musculature, causing motion, he claimed. Descartes illustrated this with a drawing of a man placing his foot near a flame. He outlined a series of events that supposedly take place in the man's nervous system from the moment his skin contacts the flame to the moment he pulls his foot away. According to Descartes, the "tiny particles" or molecules that comprise the fire cause the area of skin that they touch to move. When the skin moves, a nerve fiber attached to it is pulled, causing a pore at the other end of the nerve to open, in turn allowing animal spirits to flow through the nerve to various muscles, causing the muscles to change shape, and finally pulling the man's foot away from the flame.

Figure 28.1 Descartes' illustration of the man pulling his foot away from the flame. Reproduced from Descartes (1664/1985), out of copyright

Flow of animal spirits down other nerves also causes the man's head and eyes to turn to look at the flame, he says (Descartes 1664/1985, 102) (see Figure 28.1).

From this simple mechanical account of reflexes, Descartes built up a model of the nervous system to explain more complex phenomena like learning and memory. He suggested that associative memory traces—the heat of the flame and how it looks—"are imprinted on the internal part of the brain"; however, he did not have much to say about how that imprinting happens.

If we move ahead to the mid-nineteenth century, the Russian physiologist Ivan Pavlov made the next significant advances in discovering the neural mechanisms of learning. In the process of investigating the alimentary or salivation reflex in dogs, Pavlov discovered that his canine subjects salivate not only in the presence of food, but also in the presence of stimuli that regularly precede the presentation of food, such as a tone, or the experimenter entering the room. He described the first type of reflex (e.g., to food) as "inborn," involving "regular causal connections between definite external stimuli acting on the organism and its necessary reflex actions" (Pavlov 1927/2003, 16). However, he hypothesized that a second type of reflex (e.g., to a tone) involves different "mechanisms" operative in "higher nervous centres" (Pavlov 1927/2003, 25) and is "built up gradually in the course of an animal's own individual existence" (Pavlov 1927/2003, 25). In contrast to Descartes, who thought the mind influences the body through the pineal gland, Pavlov claimed non-physical or psychic causes are not responsible for either innate or conditioned reflexes. Rather, reflexes can be explained solely in terms of neural mechanisms mediating between stimuli and responses. This was in line with the views of mechanist physiologists like Hermann Helmholtz.

To identify "the precise conditions under which new conditioned reflexes are established" (Pavlov 1927/2003, 26), Pavlov and his colleagues ran many rigorously controlled experiments. On the basis of their data, Pavlov concluded that a conditioned reflex can be established if: (1) the presentation of the conditioned stimulus (e.g., a tone) precedes the unconditioned stimulus (e.g., food), (2) the two stimuli overlap in time, (3) the animal is alert and healthy, (4) the conditioned stimulus is an environmentally familiar one to which the animal is otherwise indifferent, and (5) the investigator ensures that the only stimuli operative in the experiment are the conditioned and unconditioned stimuli.

Having reliably produced conditioned reflexes in many canine subjects, Pavlov hypothesized the physiological conditions that allow their formation: "the linking up of impulses in different areas of the brain, by the formation of new nervous connections" is the "nervous mechanism" by which "new conditioned reflexes" are formed (1927/2003, 37). More specifically, Pavlov said "it appears that the cells predominantly excited at a given time" by an unconditioned stimulus (food) "become foci attracting to themselves the nervous impulses aroused by" the conditioned stimulus (tone), and that these impulses "on repetition tend to follow the same path and so to establish conditioned reflexes" (1927/2003, 38). Pavlov illustrated what he had in mind by appeal to "telephonic installation." He explained that he could telephone his laboratory directly, or he could call the operator to connect him to the laboratory. (In those days, operators would manually connect lines by plugging cables into jacks on a switchboard.) Both methods would result in the same outcome. However, "whereas the private line provides a permanent and readily available cable" much like the neural pathway of innate reflexes, "the other line necessitates a preliminary central connection to be established" much like how the neural pathway carrying information about the conditioned stimuli must be connected to the innate pathway. Pavlov did not know precisely the location of the formation of these new connections—he thought that it was possible that it could occur "within the cortex" or "between the cortex and subcortical areas" (Pavlov 1927/2003, 37). He also had no explanation for how such changes in neural connectivity might occur.

Catherine Stinson and Jacqueline Sullivan

An explanation began to emerge at the end of the nineteenth century. Wilhelm His (1886), working with growing nerve cells, August Forel (1887), working on nerve cell degeneration, and the great Spanish histologist Santiago Ramón y Cajal (1888), using Camillo Golgi's (1873) silver nitrate stain on unmyelinated nerve cells, suggested that nerve cells are independent anatomical and functional units (rather than a physically connected web of fibers as previously believed). Golgi's illustrations demonstrated many variations of the nerve cell's typical structure of cell body, single axon, and branching dendrites, in different brain regions. Cajal (1890) showed how growing neurons push their growth code outwards, and gradually form more dendrites and axon collaterals.

In 1891, Cajal discovered that sensory nerves have their dendrites in the periphery and axons projecting toward the brain, while motor cells are the opposite way around. His "law of dynamic polarization" hypothesized that conduction of impulses travel in one direction only, from dendrite to cell body to axon. In the eighteenth century, Luigi Galvani had established that it is electric currents, not corpuscles (i.e., animal spirits), that transmit nerve impulses. However, Cajal and many of his contemporaries believed that neurofibrils contained in nerve cells "underlie the mechanism of neuronal impulse transmission" (Cajal 1890, 95), harkening back to Descartes's account (see Figure 28.2).

These combined discoveries led to speculation in the 1890s that the growth or retraction of dendritic connections and axon collateral branches might account for learning. Cajal (1894a) suggested that genius in a subject such as music might involve increased branching of certain neurons' dendrites and axons. This was pure conjecture, however. Cajal was an anatomist

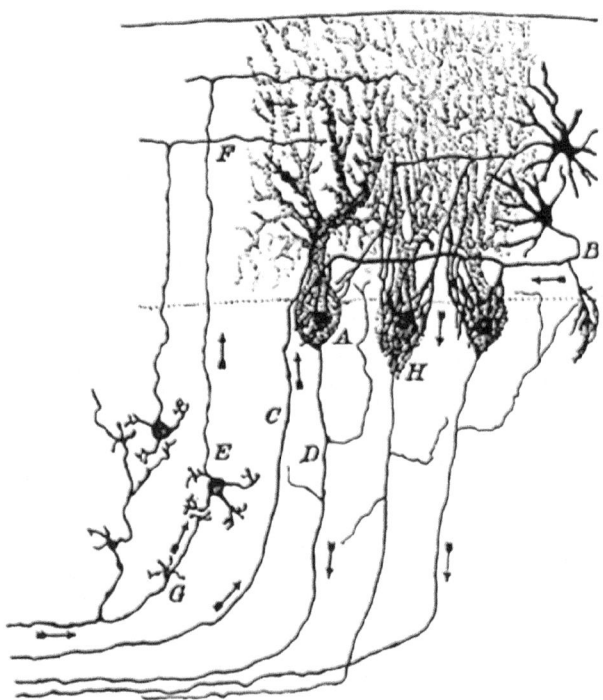

Figure 28.2 Cajal's drawing of Purkinje cells with basket endings. Reproduced from Cajal (1894a), out of copyright

378

working almost exclusively with histological methods (slicing and staining specimens, then examining them under a light microscope), which did not lend themselves well to discovering how learning occurs, nor indeed to discovering much about how nervous impulses are communicated between nerve cells. Physiological methods were required to discover the functional import of Cajal's anatomical findings.

The English physiologist Charles Scott Sherrington first introduced the concept of the synapse in 1897 in a textbook he helped edit (Foster 1897). Based on his work on spinal reflexes, Sherrington had deduced that there is a significant delay in the speed of neural impulses where there are connections made between several nerve cells along the way to or from the spinal cord, rather than single axons traveling the whole distance. He attributed this delay to an "intercellular barrier" or membrane (Bennett 1999). Cajal had convincingly argued that axon collaterals do not directly fuse with the cells they come into contact with, but the hypothesized junction could not be seen under a light microscope. Sherrington's work suggested that the synapse acted as a valve, explaining why conduction occurred in only one direction. He also explored the relationship between inhibitory and excitatory connections. He remarked that the synapse offered "an opportunity for some change in the nature of the nervous impulse as it passes from one cell to the other" (Foster 1897). Physiologists and pharmacologists continued, in the early twentieth century, to uncover the electrical and chemical mechanisms of synaptic transmission.

Advances in *neurophysiological* theorizing and methodology in the first half of the twentieth century were instrumental in connecting this developing knowledge about synaptic transmission to the phenomena of learning and memory. One such advance came in the form of a simple neurophysiological postulate put forward by Donald Hebb in *The Organization of Behavior* (1949). Synthesizing a broad selection of research from psychology (e.g., Pavlov), neuroanatomy (e.g., Cajal), and neurophysiology (e.g. Sherrington, Lorente de Nó), Hebb hypothesized that just as associative learning at the level of behaving organisms required the repetition and contiguity of stimuli or stimuli and responses, so, too, did the permanent metabolic changes or growth processes thought to underlie learning require the contiguous and repetitive excitation of the neurons carrying information about those stimuli and/or responses.

More specifically, Hebb claimed that

> when an axon of [a] cell A is near enough to excite a cell B and repeatedly or persistently takes part in firing it, some growth process or metabolic change takes place in one or both cells such that A's efficiency, as one of the cells that fires B, is increased.
>
> *(Hebb 1949/2002, 62)*

Although the main idea at the heart of Hebb's postulate was not new, as Hebb acknowledged, the postulate provided insight into the kinds of methods that could be employed by physiologists to determine if neurons were plastic in the way that Hebb's predecessors, like Pavlov and Cajal, had claimed.

By the 1960s, neurophysiologists had a working model of the neuron as consisting of (1) an *input* component (the dendrites), (2) an *integrative* component (the axon hillock), (3) a *conductile* component (the axon), and (4) an *output* component—the synaptic terminal from which the neurotransmitter is released (Kandel and Spencer 1968, 69–70). However, what was missing was "a crucial experiment identifying specifically a change occurring in neural tissues as learning takes place" (Hilgard 1956, 481). Such crucial experiments came much later in the form of Nobel Prize winner Eric Kandel and colleagues' development of a simplified preparation for studying the cellular and molecular mechanisms of simple forms of associative and non-associative learning in the invertebrate sea mollusc, *Aplysia californica*.

379

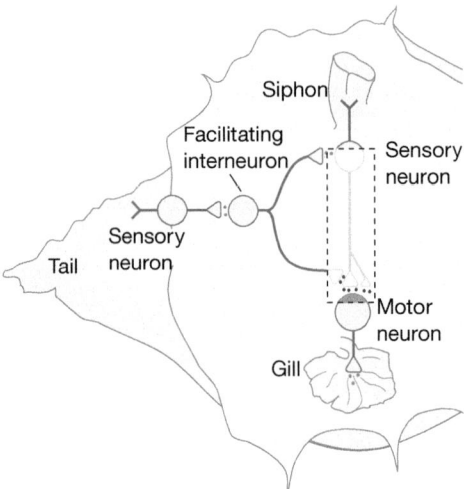

Figure 28.3　The simple neuronal circuit involved in sensitization of the gill-siphon withdrawal reflex in *Aplysia*. (Kandel, E.R., Schwartz, J.H., Jessell, T.M., Siegelbaum, S.A., Hudspeth, A.J. *Principles of Neural Science*, 5th edition, 2012, McGraw-Hill Education. Reproduced with permission of McGraw-Hill Education)

Aplysia has a defensive reflex known as the *gill-siphon withdrawal reflex*. When a tactile stimulus is applied to the animal's siphon—a small fleshy spout located above the gill that expels seawater and waste—it retracts or withdraws the siphon and the gill. In one early set of experiments, Kandel and colleagues experimentally isolated the sensory neurons that carry stimulus information from the siphon, the motor neurons to which these sensory neurons project, and a set of excitatory and inhibitory interneurons that receive input from the sensory neurons and project to the motor neurons. By isolating the neurons that comprise this simple circuit (and other circuits to which it was connected), Kandel and colleagues were able to identify specific cellular and molecular changes that accompany a set of simple forms of associative and non-associative learning in *Aplysia*.

More specifically, they studied a form of learning known as *sensitization*. In a sensitization experiment, an investigator begins by applying a tactile stimulus (e.g., a Q-tip) to an Aplysia's gill or siphon, so as to measure the extent and duration of the withdrawal reflex. The experimenter then delivers a set of noxious shocks to the organism's tail. Following these shocks, she reapplies the tactile stimulus and again measures the extent and duration of the withdrawal reflex. An increase in duration of the withdrawal reflex prior to the tail shocks compared to after the tail shocks is taken as indicative that the animal has learned that there is a noxious stimulus in its environment. Textbooks like *Principles of Neural Science* (5th edition, 2012) reveal in detailed diagrams what we now know about the changes in the strength of the synaptic connections that underlie this form of learning and that they are mediated by specific changes in cellular and molecular activity (see Figure 28.3).

These historical examples should give an idea of some of the ways that scientists discover and explain the mechanisms of the brain. In the following sections we'll revisit these cases, highlighting the fundamental features of mechanistic explanations, and considering some of the methodological issues that arise in neuroscience.

3. Philosophical accounts of mechanisms in neuroscience

One aim of philosophy of science is to understand the structure of science. Another is to account for scientific progress. Up until the latter half of the twentieth century, the examples considered in philosophy of science were taken primarily from the history of physics. This exclusive focus led to an understanding of science that conceived of its history as involving the discovery of laws (e.g., planetary motion, gravitational attraction) and the development of grand unifying theories (e.g., relativity theory). By the middle of the twentieth century, philosophers characterized scientific explanations as arguments, where statements of laws and initial conditions were taken to logically imply the observations to be explained or predicted. Different branches of science were thought to be hierarchically organized with those that studied the most fundamental things (e.g., particles) at the bottom and those that studied the least fundamental things (e.g., societies) at the top. Branches of science were regarded as compartmentalized, and progress within a given branch (e.g., psychology) was not taken to rely on developments within other branches (e.g., neurophysiology). Progress between branches was taken to involve "intertheoretic reduction"—the reduction of theories in "higher-level" sciences like biology to theories in "lower-level" sciences like physics (see Oppenheim and Putnam 1958; Nagel 1961; Chapter 16 in this volume).

The history of scientific research on learning and memory that we described above defies these characterizations in a number of ways. Descartes, Pavlov, Cajal, Hebb, and Kandel were aiming neither to discover large-scale scientific theories nor to reduce those theories to physical ones. The kinds of explanations for learning and memory phenomena they sought combined findings and insights from different areas of science including anatomy, physiology, psychology, and later biochemistry. These diverse branches of science all aimed at understanding learning and memory from different angles and seemed to be making progress interactively rather than independently.

An alternative account of scientific explanation has recently been proposed that provides a more congenial analysis of the discovery strategies and markers of progress described in these historical cases, as well as in the biological sciences more generally. This account focuses on the role of *mechanisms* in scientific explanation. (See, for example, Bechtel and Richardson 1993/2000; Craver 2007; Glennan 1996; Illari and Williamson 2012; Machamer, Darden, and Craver 2000.)

The first crucial step in providing a mechanistic explanation is to identify the phenomenon to be explained (see, for example, Glennan 1996; Bechtel 2008; Craver and Darden 2001). In each case considered above, the phenomenon for which a mechanism is sought is precisely delineated. Consider how Descartes conceives of a reflex; it begins with a stimulus—a man putting his foot into a fire—and ends when the man looks at the fire and retracts his foot from the flame. Similarly, Pavlov's conditioned reflexes begin with repeated and contiguous presentation of the unconditioned stimulus and conditioned stimulus and end with elicitation of the conditioned response (i.e., salivation) to the conditioned stimulus. Sherrington, Hebb, and Kandel postulate a very specific set of inputs—repetition and contiguity in firing of two cells that comprise a synapse—and a very specific output—a change in the way that the two cells communicate. As Peter Machamer, Lindley Darden, and Carl Craver (MDC 2000, 3) claim, the phenomena of interest in mechanistic explanations have clear starting points or set-up conditions and clear endpoints or termination conditions.

Another important feature of mechanistic explanations is that they "account for the behavior of a system in terms of the functions performed by its parts and the interactions between these parts" (Bechtel and Richardson 1993/2000, 17), rather than in terms of general laws or theories. In Descartes' example, *molecules, nerve fibers, pores, animal spirits, and muscles* are all parts or entities of a human organism. *Tugging, opening, flowing, and moving* are all activities in which these parts or entities engage in the production of reflex behaviors. In Cajal's anatomical work, *axon collaterals, dendrites, dendritic spines, and growth cones* are the entities. Sherrington, Hebb, and Kandel later contributed knowledge about the activities of those entities.

William Bechtel and Robert Richardson (1993/2000) emphasize that a central heuristic strategy operative in developing mechanistic explanations is the decomposition of the phenomenon into its component parts and their operations. Decomposing a system in this fashion and explaining its behaviors mechanistically is not something that can be accomplished in a single area of science. As the "New Mechanists" emphasize, it requires input from many different areas of science. In the process, rather than one branch or area of science being reduced to another, input from different areas of science is "integrated into descriptions of multi-level mechanisms" (e.g., Craver 2007).

Cajal's work is a prime example of the decomposition strategy at work. It was critical in convincing anatomists of the "neuron doctrine," which extended cell theory to neural tissues, stating that the brain is made up of anatomically discrete cellular units. Cajal showed how neurons of different types, such as the basket and Purkinje cells of the cerebellum, are connected in organized patterns. He also worked to discover the anatomical properties of the neuron's sub-parts like dendrites, axon collaterals, growth cones, and dendritic spines. Cajal's discoveries of the anatomical properties of neurons and their component parts, in combination with Sherrington and Pavlov's discoveries about how these parts function, shaped the development of Hebb's postulate, which informed Kandel's work in developing simplified preparations that decomposed reflex operations in *Aplysia* to a simple neuronal circuit and its component parts.

Bechtel and Richardson also note that the mechanistic explanatory strategy is often constrained by available technology and that scientists "will appeal analogically to the principles they know to be operative in artificial contrivances as well as in natural systems that are already understood" (1993/2000, 17) to provide mechanistic explanations. Descartes' appeal to the mechanical statues in St. Germain to explain reflex action and Pavlov's appeal to a telephone switchboard to explain how conditioned reflexes come about are clear examples of how the available technology of a time period can shape how investigators conceive of a mechanism.

This raises another important feature of mechanistic explanations detailed by Darden (2002): they are gradually discovered over time (see also Chapter 19). In terms of their empirical support, candidate mechanisms can have the status of "how-possibly," "how-plausibly," or "how-actually" explanations (Craver 2007). In terms of their completeness, mechanistic explanations start out as sketches (MDC 2000) that have gaps in their productive continuity or black boxes left to be filled in with detail. Sketches are revised, filled in, and fit into their surrounding contexts, until they eventually gain the status of adequately complete mechanistic explanations, or are rejected as false starts. The activities and sub-entities that mediate the connections between neurons were black boxes for Cajal before Sherrington developed the concept of the synapse. While Hebb put forward a "how-possibly" mechanism for permanent changes in communication between neurons, this was not yet considered an adequate explanation. Later work by Kandel and colleagues was directed at understanding "how-actually" such changes come about during real learning events.

The process of discovery sometimes requires more substantial revisions to how the phenomenon was originally individuated and circumscribed (see Bechtel and Richardson 1993/2001; Bechtel 2008; Craver 2007, 2009). Experimentation may result in discoveries that prompt a revision to the original taxonomy of kinds of phenomena identified in a given field of research. For example, it may be discovered that what was once considered one phenomenon (e.g., memory) includes at least two forms (e.g., declarative and procedural). Experimentation may reveal that we were looking in the wrong place for a mechanism, or that a single mechanism performs what were originally thought to be two separate functions (e.g., see Eichenbaum and Cohen (2014) on the hippocampus's dual role in memory and navigation). It is well known today that rather than being the conduit between the body and immaterial soul, the pineal gland produces and secretes melatonin, which is involved in the modulation of circadian rhythms in the vertebrate brain.

These historical vignettes as a whole demonstrate another key feature of mechanistic explanations: they are multi-level. From Pavlov to Kandel, for example, we move from observations of mid-scale entities and activities like dogs, bells, and salivation, to micro-scale entities and activities like ions and neurotransmitter release. Mechanistic explanations involve entities and activities at multiple scales, some of which are sub-mechanisms that constitute higher-level components. Even in Descartes' description of the mechanism of the reflex, the behavior of the whole organism is explained by appeal to some of its constituent parts and their sub-parts, like nerves, nerve fibers, pores, and animal spirits.

A related and notable characteristic of mechanistic explanations is that they are not byproducts of a single area of science. Rather, they rely for their development on information emanating from multiple different areas of science that study entities and activities at varying scales (compare Darden and Maull 1977; Craver 2007). Consider Descartes' explanation of the reflex—it combined a corpuscular theory of matter with a rudimentary understanding of the anatomy of the nervous system prevalent in his day, and a theory of animal spirits originating with Galen. Although Pavlov thought that physiology could advance an understanding of the mechanisms of conditioned reflexes without appeal to psychology, he recognized that it could not do so in the absence of advances in anatomy and cell biology. Cajal's histological preparations could not reveal the functional nature of the connections between contiguous neurons without the addition of physiological work, which Sherrington later contributed. Kandel and colleagues' research into the mechanisms of simple forms of non-associative learning in *Aplysia* combines anatomical, electrophysiological, biochemical, behavioral, and pharmacological techniques.

4. Discovering mechanisms: open philosophical problems

The last two features of mechanistic explanation we mentioned—their multi-level nature, and the fact that they integrate results from various branches of science—are very much at odds with traditional thinking about scientific explanation. As mentioned briefly, in the mid-twentieth century, scientific phenomena at different scales, and the fields of science that study them, were thought to be related to one another in terms of reduction. Chemistry, for instance, was supposed to occupy itself with a circumscribed range of chemical phenomena, which the methods of chemistry alone were appropriate for investigating. Furthermore, all of chemistry, it was thought, would eventually prove to be reducible to physics in the way that heat is reducible to the average kinetic energy of physical particles. Higher-level sciences, according to this way of thinking, may serve pragmatic and heuristic purposes along the way to finding the fundamental theory, but eventually should turn out to be superfluous.

Mental phenomena have long posed a challenge to this picture; many philosophers (and others) want to deny that the mind is reducible to more fundamental physical entities and activities. The multi-level nature of mechanistic explanations is meant to provide an alternative to reduction. All of the levels in a mechanism, from low to high, contribute to it performing its function. Going down lower does not provide a more fundamental understanding, even if it might provide finer grained details; in fact, scientists sometimes purposely focus their investigations at higher levels, because that's where the functions they're interested in are performed.

Many questions remain about how exactly this plays out in practice. Craver (2007) describes a picture of "integrative unity" in which a psychological capacity, such as spatial memory, is brought about by anatomically differentiated parts of the brain (area CA1 of the hippocampus), its physiological component parts (neural networks, neurons, synapses), and activities (firing, transmitter release), which in turn are composed of smaller-scale parts (receptors, molecules) and their activities (activation, phosphorylation). If Craver is right, we should be able to fit the results from our historical vignettes into a hierarchy of mechanisms with Pavlov's conditioned reflexes at the top, Hebb's associative synaptic mechanisms slightly below, Cajal's anatomical picture of the neuron and Sherrington's physiological insights into synapses another step down, then finally Kandel's molecular mechanisms of learning at the bottom.

Some of the entities and activities involved do fit together as parts to wholes, such as Kandel's molecular mechanisms, which describe parts of Cajal and Sherrington's neurons and synapses. However, it is not clear that the levels will always connect in such a tidy way, especially at the higher levels. Craver's account seems to presuppose that psychological and neural mechanisms are part of the same ontological hierarchy, yet psychological mechanisms do not necessarily have neural mechanisms as parts (Stinson 2016).

Consider, for example, an information-processing mechanism that explains how an organism learns to respond to stimuli like burning flames or noxious shocks. That mechanism needs to store the relationship between stimulus and response in some memory medium. Reflexes mediated by nerve fibers, as Descartes imagined, can't do the whole job, because we can learn not only to pull our foot away from a flame, but also to do many other things with our limbs in response to many other kinds of stimuli. The nerve fiber doesn't have enough bandwidth to represent all of these learned relationships. This notion of bandwidth is an abstract concept that doesn't appeal specifically to any parts of the stimulus–response system, and yet it provides a psychological-level, mechanistic explanation of why Descartes' fibers can't be the whole story.

Another issue is that what look like the natural boundaries of a phenomenon from the perspective of one science (including start and finish conditions, and the way components are picked out) might not match up with what look like the natural boundaries of the same phenomenon from the perspective of another science (see Chapter 9). In Stinson (2016) one of us argues that the science of memory has this problem. From the perspective of psychology, it seems clear that memory encoding, storage, and recall are distinct processes, for example. Yet from the perspective of neuroscience, there do not appear to be clear distinctions between these memory processes. When you try to integrate the mechanisms of memory studied in these two sciences, you do not find a neat relationship where neural mechanisms turn out to be the parts of psychological mechanisms. At the neural level, encoding, storage, and recall are all intertwined, so neural mechanisms of memory don't turn out to be related to psychological mechanisms of memory as parts to wholes.

Thus, rather than different areas of science like psychology and neuroscience being seamlessly integrated into unified mechanistic explanations, we often find explanations that cross levels in the mind–brain sciences to be messy and partial. Craver illustrates his mosaic unity with images like the one on the left of Figure 28.4. Instead, we suggest that the inter-field relationships we should expect will look more like the more complex image on the right.

Scientists working in different fields conceive of their phenomena of interest in different ways, experimentally investigate phenomena in different ways, ask different research questions, use

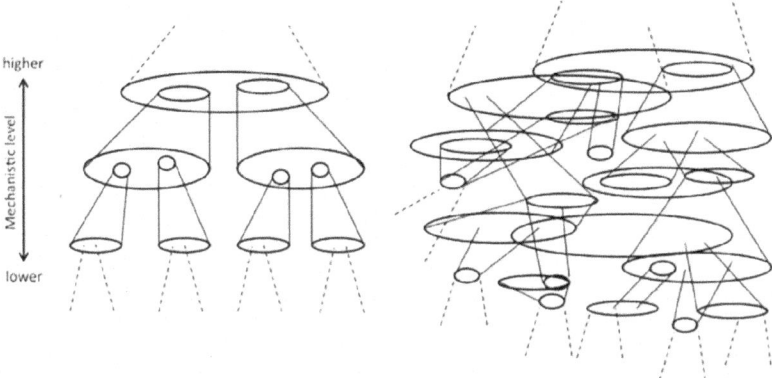

Figure 28.4 A comparison of Craver's (2007) view of inter-level relations [left] between mechanisms, and Stinson's [right]. Copyright 2016 Catherine Stinson. Used with permission

different methods aimed at vastly different scales, and investigate these phenomena in different species. For example, psychologists have historically been characterized as interested in providing explanations of cognitive capacities by functional analysis (e.g., Cummins 1983; Fodor 1968). Many psychologists believe this requires a clear specification and decomposition of abstract cognitive processes (like learning) involved in psychological tasks. When tasks are regarded as inappropriate for individuating a discrete function, they are often refined. However, neurobiologists investigating learning in invertebrates (like Kandel) and physiologists who investigate learning in non-human mammals (rodents, dogs (like Pavlov)) are often not interested in individuating the cognitive processes engaged during training in learning paradigms. While not worrying about abstract cognitive processes makes good sense in the case of *Aplysia*, which has a simple nervous system and can be studied using reduced preparations, ignoring the component cognitive processes that may be involved when rodents are trained in learning paradigms will render the connections investigators would like to make between cellular and molecular mechanisms and cognitive capacities tenuous at best (see Sullivan 2009, 2010, 2016).

A related problem with this unity picture is the issue of comparing mechanisms across species. Kandel's sea slugs are vastly different from humans, yet it is assumed that the results of experiments undertaken in one species are generalizable to others. Pavlov and Kandel are interested in mechanisms of human learning, but perform their experiments on canines and invertebrates. For both ethical and practical reasons, the systems scientists have historically and continue to use are model organisms like dogs, frogs, birds, rodents, sea slugs, and fruit flies. While certain cellular and molecular mechanisms are conserved across species, there are obvious differences between sea slugs and humans that prohibit direct inference from one to the other.

Although neuropsychological research on patients with localized brain damage and fMRI experiments involving human beings have shed some light on the loci of specific types of learning and memory in the human brain, we continue to lack a mechanistic understanding of human learning; the explanations we currently have are patchy at best. It is supposed that advances in imaging technologies will eventually enable a visualization of the loci and mechanisms of human learning. However, before such discoveries are to be feasible, scientists require better methods for individuating learning phenomena in human beings and non-human mammals. It is not simply that scientists lack the available imaging technologies; it is that many experiments in the cognitive neurosciences lack the rigor of work with model organisms that have smaller repertoires of behaviors. It is more difficult to design tasks that tease apart discrete kinds of learning in human beings than in *Aplysia*. One reason for this difficulty is that human beings might use multiple strategies to perform a cognitive task, and it's not always possible to predict the range of strategies that might be used, or to detect whether subjects are using the expected one.

Despite these difficulties in drawing connections between experimental findings using protocols from different fields, and in phylogenetically distant species, it is necessary for mechanistic explanations in neuroscience to find ways of bridging these gaps. As we mentioned earlier, Cajal's histological experiments could only go so far. He was able to get a fairly accurate picture of the anatomy of the neuron, but the structure alone couldn't reveal how neurons communicate. Pavlov could only figure out the functional characteristics of reinforcement learning using his experimental methods. His functional picture could not reveal what sorts of structures might give rise to reinforcement learning. As a general rule, neither bottom-up (from structure to function) nor top-down (from function to structure) methods in isolation can get us all the way to understanding mechanisms. Instead what is needed is a multi-level approach, with researchers simultaneously using many strategies to investigate different phenomena, alongside some efforts at linking the results of these together, i.e. something very much like how neuroscientific research is in fact pursued.

Scientists approach the problem of understanding the brain at various levels because there are robust regularities at various levels, both in neuroscience and in the life sciences more generally. There are some phenomena that we feel compelled to think of as real or natural kinds, like molecules, cells, organs, organisms, and species, even when we can't give them tidy definitions in terms of their component parts. We think that neurons are a genuine kind of thing despite the fact that (contra Cajal) nerve cells sometimes do fuse together in ways that challenge their anatomical and physiological independence. Organisms often end up in symbiotic relationships with other organisms, like our gut microbiota, without which we couldn't live, challenging the independence of organisms. The action potential depends on a membrane, ion channels, and extra and intra-cellular ions, but it only exists within a narrow range of conditions.

A complex biological system like the brain will likely prove impossible to fit into a neat hierarchy of nested parts, because the borders of mechanisms are fuzzy. This does not mean that we can't ever have an integrated science that links together different levels. There are connections to be made between the results from different experimental paradigms, experiments on different species, models of different phenomena, and different models of the same phenomenon. Many of these connections will be partial, and the integrated picture will be patchy (see Schaffner 2006; Stinson 2016).

5. Conclusion

The historical case studies we considered span several centuries, but a common aim in each case was discovering mechanisms. Constructing multi-level mechanistic explanations involves intensive collaboration across different branches of science, and involves many challenges, both pragmatic and methodological. Available technologies, training in experimental methods, choice of model organisms, levels of investigation, and inter-field collaborators all can either ensure success or act as barriers to progress. Integrating the discoveries from various fields where the phenomena are circumscribed in different ways requires piecing together results in complex ways, and carefully considering when and how results can be generalized to different contexts.

Note

1 The authors would like to thank Stuart Glennan and Phyllis Illari for very helpful comments on an earlier draft of this chapter. Co-authors had equivalent input and are listed in alphabetical order.

References

Bechtel, W. (2008). *Mental Mechanisms: Philosophical Perspectives on Cognitive Neuroscience.* New York: Taylor and Francis.

Bechtel, W. and Richardson, R. (1993/2000). *Discovering Complexity: Decomposition and Localization as Strategies in Scientific Research.* Princeton, NJ: Princeton University Press.

Bennett, M. R. (1999). The early history of the synapse: From Plato to Sherrington. *Brain Research Bulletin,* 50(2), 95–118.

Craver, C. (2007). *Explaining the Brain: Mechanisms and the Mosaic Unity of Neuroscience.* Oxford: Oxford University Press.

Craver, C. (2009). Mechanisms and natural kinds. *Philosophical Psychology,* 22(5), 575–594.

Craver, C. and Darden, L. (2001). Discovering mechanisms in neurobiology: The case of spatial memory, in P.K. Machamer, R. Grush, and P. McLaughlin (eds.), *Theory and Method in the Neurosciences* (pp. 112–137). Pittsburgh: University of Pittsburgh Press.

Cummins, Robert. (1983). *The Nature of Psychological Explanation.* Cambridge, MA: MIT Press.

Darden, L. (2002). Strategies for discovering mechanisms: Schema instantiation, modular subassembly, forward/backward chaining. *Philosophy of Science,* 69(3), S354–S365.

Darden, L. and Maull, N. (1977). Interfield theories. *Philosophy of Science,* 43, 44–64.

Descartes, R. (1664/1985). *Treatise on Man* in The Philosophical Writings of Descartes Volume I, translated by John Cottingham, Robert Stoothoff, and Dugald Murdoch. Cambridge: Cambridge University Press.

Eichenbaum, H. and Cohen, N. (2014). Can we reconcile the declarative memory and spatial navigation views on hippocampal function? *Neuron,* 83(4):764–770.

Fodor, J. (1968). *Psychological Explanation: An Introduction to the Philosophy of Psychology.* New York: Random House.

Forel, A. (1887/1991) Out of my life and work. In G. M. Shepherd (Ed.), *Foundations of the Neuron Doctrine* (pp. 115–116). New York: Oxford University Press.

Foster, M. with Sherrington, C. S. (1897). *A Textbook of Physiology, Part Three: The Central Nervous System,* 7th ed. London: Macmillan and Co. Ltd.

Glennan, S. (1996). "Mechanisms and the nature of causation." *Erkenntnis,* 44: 49–71.

Golgi, C. (1873/1991). On the structure of the gray matter of the brain. *Foundations of the Neuron Doctrine,* Gordon M. Shepherd (Trans.) (pp. 84–88). Oxford: Oxford University Press.

Hebb, D. (1949/2002). *The Organization of Behavior: A Neuropsychological Theory.* Mahwah, NJ: Lawrence Erlbaum.

Hilgard, E. R. (1956). *Theories of Learning,* 2nd ed. New York: Appleton-Century-Crofts.

His, W. (1886/1991). On the structure of the human spinal cord and nerve roots. In G. M. Shepherd (Ed.), *Foundations of the Neuron Doctrine* (pp. 106–110). New York: Oxford University Press.

Illari, P. and Williamson, J. (2012). What is a mechanism? Thinking about mechanisms across the sciences. *European Journal for the Philosophy of Science,* 2(1), 119–135.

Kandel, E. R. and W. A. Spencer. (1968). Cellular neurophysiological approaches in the study of learning. *Physiological Review,* 48(1), 65–134.

Kandel, E. R., Schwartz, J. H., Jessell, T. M., Siegelbaum, S. A., and Hudspeth, A. J. (Eds.). (2012). *Principles of Neural Science* (Vol. 5). New York: McGraw-Hill.

Nagel, E. (1961). *The Structure of Science: Problems in the Logic of Scientific Explanation.* London: Harcourt, Brace & World.

Oppenheim, Paul and Putnam, Hilary. (1958). The unity of science as a working hypothesis. In Herbert Feigl, Grover Maxwell, and Michael Scriven (Eds.) *Minnesota Studies in the Philosophy of Science* (pp. 3–36). Minneapolis: Minnesota University Press.

Pavlov, I. (1927/60). *Conditioned Reflexes.* Mineola, NY: Dover Publications.

Machamer, P., Darden, L., and Craver, C. (2000). Thinking about mechanisms. *Philosophy of Science,* 67(1), 1–25.

Ramón y Cajal, S. (1888/1991). Structure of the nervous system of birds. In G. M. Shepherd (Ed.), *Foundations of the Neuron Doctrine* (pp. 141–148). New York: Oxford University Press.

Ramón y Cajal, S. (1890/1988). On the structure of the cerebral cortex of certain mammals. In J. DeFelipe and E. G. Jones (Eds.), *Cajal on the Cerebral Cortex: An Annotated Translation of the Complete Writings* (pp. 23–54). New York: Oxford University Press.

Ramón y Cajal, S. (1894a/1991). The fine structure of the nervous centers (Croonian lecture). In G. M. Shepherd (Ed.), *Foundations of the Neuron Doctrine* (pp. 239–253). New York: Oxford University Press.

Ramón y Cajal, S. (1894b/1990). *New Ideas on the Structure of the Nervous System in Man and Vertebrates.* Neely Swanson and Larry W. Swanson (Trans.). Cambridge, MA: The MIT Press.

Schaffner, K. F. (2006). Reduction: The Cheshire cat problem and a return to roots. *Synthese*, 151(3), 377–402.

Shepherd, G. M. (1991). *Foundations of the Neuron Doctrine.* New York: Oxford University Press.

Stinson, Catherine. (2016). Mechanisms in neuroscience: Ripping nature at its seams. *Synthese*, 193, 1585–1614.

Sullivan, Jacqueline. (2009). The multiplicity of experimental protocols: A challenge to reductionist and nonreductionist models of the unity of science. *Synthese*, 167, 511–539.

Sullivan, Jacqueline. (2010). Reconsidering spatial memory and the Morris water maze. *Synthese*, 177(2), 261–283.

Sullivan, Jacqueline. (2016). Construct Stabilization and the unity of the mind-brain sciences. *Philosophy of Science*, 83, 662–673.

29

MECHANISMS IN COGNITIVE SCIENCE[1]

Carlos Zednik

1. Introduction

A principal goal of cognitive scientific research is to explain behavioral and cognitive phenomena such as perception, action, categorization, memory, learning, language, and attention. The most influential account of explanation in cognitive science is due to David Marr (1982). On Marr's account, cognitive scientists answer questions at three distinct *levels of analysis*: the *computational* level, which concerns questions about a particular system's computational goals and about the appropriateness thereof; the *algorithmic* level, which is driven by questions about the representations and algorithms that are used by the system to achieve these goals; and the *implementational* level, which addresses questions about the way in which these representations and algorithms are physically realized (Marr 1982: 24ff). According to Marr, questions at all three levels of analysis must be answered to "completely understand" a cognitive system, and to thereby explain its behavior.[2]

Although Marr's account remains influential to this day, there are reasons to be unsatisfied about its clarity and scope. First, although it is relatively clear which questions should be asked, it is not quite as clear how these questions might be answered. Among others, it is unclear what it actually takes to show that a computational goal is appropriate, and what it means for an algorithm to be physically realized. Second, because Marr's account is formulated in terms co-opted from computer science, rather than in terms endemic to philosophical discussions of scientific explanation, it is difficult to know how explanations in cognitive science compare to explanations in other disciplines that center on, for example, subsumption under law, the development of models, and/or the discovery of mechanisms. Finally, many of the terms that play a central role in Marr's account are far less prominent in cognitive science today than they were at the time of Marr's writing. In particular, the *computationalist* research program which predominated in the 1970s and 1980s (Pylyshyn 1980) now competes for attention and resources with alternative research programs such as *connectionism* (Rumelhart et al. 1986), *dynamicism* (van Gelder 1998), and the *Bayesian approach* (Zednik & Jäkel 2016)—some of which may not rely on notions of "computation," "algorithm," and "implementation" at all.

Despite these reasons to be unsatisfied with Marr's account of explanation in cognitive science, its lasting influence recommends it as a productive starting point for discussion. Indeed, the aim of this chapter is to show that many of the ambiguities in Marr's account can be resolved, and that its scope can be extended, by considering Marr to have been an early advocate of *mechanistic explanation* (Chapter 1). Although Marr's account was originally designed to capture explanations in computationalist cognitive science, the questions it identifies at each level of analysis are in fact variations on the types of questions that are asked in any research program that aims to discover and describe (cognitive) mechanisms. Specifically, questions at the computational level can be construed as questions about *what* a mechanism is doing and *why*: questions that concern the mechanism's behavior and containing environment. Questions at the algorithmic level, in contrast, are questions about *how* a mechanism does what it does, and concern its component operations. Finally, questions at the implementational level of analysis can be understood as questions about *where* a particular mechanism's component operations are carried out—that is, questions about the component parts in which such operations might be localized. Although Marr showed how each one of these questions might be answered using the concepts and methods of computationalist cognitive science, they are also answered in the context of other research programs such as connectionism, dynamicism, and the Bayesian approach. In other words, there is reason to believe that all of these research programs seek three-level explanations, and moreover, that all of them aspire to discover and describe mechanisms.

This chapter will outline and defend a mechanistic interpretation of Marr's account of explanation in cognitive science, and thereby attempt to resolve the ambiguities that remain in this account, as well as to extend its scope. Notably, in line with Marr's claim that all three levels of analysis are necessary to "completely understand" a cognitive system, on the current interpretation all three levels must be addressed to provide mechanistic explanations of behavioral or cognitive phenomena.[3] In this way, the present interpretation differs from several previous attempts to relate Marr's account to the framework of mechanistic explanation (see e.g. Bechtel & Shagrir 2015; Bickle 2015; Kaplan 2011; Milkowski 2013; Piccinini & Craver 2011). Moreover, the present interpretation helps to address a number of philosophical debates concerning e.g. the role of idealization in cognitive modeling, the role of abstraction in mechanistic explanation, and the nature of the realization-relationship that obtains between functional processes in the mind and physical structures in the brain.

2. The computational level: "what?" and "why?"

In Marr's original formulation, the computational level of analysis is defined by questions about a cognitive system's computational goals, as well as questions about the appropriateness thereof (Marr 1982: 24ff). More generally, these can be understood as questions about *what* a cognitive system is doing, and questions about *why* it does what it does (see also McClamrock 1991; Shagrir 2010).

What-questions can be answered by describing the relevant system's behavior. Within the computationalist research program, this involves specifying an information-processing function that maps the system's (sensory) inputs onto its (behavioral) outputs. That said, recent contributions by e.g. dynamicist researchers suggest that many behavioral and cognitive capacities—especially those that depend on continuous feedback from the environment—cannot be easily described as mappings between input and output (van Gelder 1995). Rather, these capacities are better described as continuous trajectories through state space, the dimensions of which might correspond to neural activity, a system's bodily position or motion, and/or features of the environment (Kelso 1995). Although it may be terminologically confusing to associate dynamical state-space trajectories with the "computational" level of analysis, they are analogous to information-processing functions in that they too can be used to describe a cognitive system's behavior, and thus to answer questions about *what* the system is doing.

Answers to what-questions play an important role not only in Marr's account of explanation in cognitive science, but also in the framework of mechanistic explanation (see also Chapter 16). Descriptions of a cognitive system's behavior are descriptions of an *explanandum phenomenon* (Cummins 2000). In many scientific disciplines, such descriptions are the starting point of mechanistic explanation: The explanandum phenomenon is identified with the overall behavior of a mechanism, and an attempt is made to describe that mechanism's component parts, operations, and overall organization (Bechtel & Richardson 1993; Darden & Craver 2012; Chapter 19, this volume). Notably, many mechanisms are known whose overall behavior has been described as a form of information-processing (Craver 2013; Kaplan 2011), but also mechanisms whose behavior has been more effectively described as a trajectory through state space (Bechtel & Abrahamsen 2010). Therefore, these ways of answering what-questions in cognitive science are by no means inconsistent with the principles of mechanistic explanation (compare Chemero & Silberstein 2008). That said, merely describing a mechanism's overall behavior is insufficient for the purposes of mechanistic explanation (Kaplan & Craver 2011; Chapter 20, this volume). Indeed, insofar as descriptions of a mechanism's overall behavior often resemble law-like regularities (Chapter 12), they may also feature in other kinds of scientific explanation (Chemero & Silberstein 2008; Zednik 2011). For this reason, determining whether cognitive scientists are in fact in the business of mechanistic explanation requires taking a closer look at the way in which they answer questions beyond "what?"

One such question is the question of "why?" In the context of his celebrated cash register example, Marr deems it important to ask "why the cash register performs addition and not, for instance, multiplication when combining the prices of the purchased items to arrive at a final bill" (Marr 1982: 22). Marr claims that such questions are answered by considering the "appropriateness" of the system's behavior with respect to the "task at hand" (Marr 1982: 24). Whereas many previous discussions of Marr overlook this aspect of the computational level, Shagrir (2010) has provided a compelling analysis according to which a cognitive system's behavior should be deemed appropriate when a mathematical description of that behavior can be mapped onto a relevant (potentially abstract or even counterfactual) property of the system's environment. One analytic technique that is well suited for answering why-questions in this way is *rational analysis* (Anderson 1991; Oaksford & Chater 2007), which lies at the heart of the recently influential Bayesian approach in cognitive science (Zednik & Jäkel 2016). This technique involves formally characterizing a cognitive system's task environment as a form of probabilistic inference, and deriving an optimal solution in the sense prescribed by probability theory. A widely reported—and initially surprising—finding of this approach is that many different kinds of cognition and behavior closely approximate optimal solutions (Pouget et al. 2013). Whenever this is the case, why-questions can be answered because the optimal solutions describe a cognitive system's behavior while simultaneously reflecting a mathematical property of the system's environment—namely, the Bayes-optimal solution to a particular task within that environment. Intuitively, the method of rational analysis allows researchers to show that cognitive systems behave as they do *because* that way of behaving is optimal in the sense prescribed by probability theory (but compare Danks 2008).

Shagrir's analysis of "appropriateness" implies that answers to why-questions cannot be found by looking solely at the cognitive system whose behavior is being explained; it also involves looking at the environment in which that behavior unfolds. It is far from obvious how looking at properties external to a cognitive system might be conducive to revealing the mechanisms internal to it. Moreover, it is tempting to think of why-questions as pertaining to a teleological approach to scientific explanation which is traditionally contrasted with the mechanistic approach.[4] Nevertheless, philosophical proponents of mechanistic explanation have recently argued that environmental and teleological considerations play a significant role in mechanistic explanation. In particular, Carl Craver argues that understanding the role a mechanism is supposed to play in a containing environment can greatly facilitate the task of describing that mechanism's actual behavior, and that many investigators for this reason "search for a higher-level mechanism within which it has a role" (Craver 2013: 153). Similarly, William Bechtel (2009) argues that mechanistic explanation involves "looking up" at the environment in which a mechanism is naturally embedded, because doing so greatly facilitates the task of characterizing the mechanism's actual behavior: It might reveal complexities that the mechanism must accommodate, as well as regularities that it can exploit.

These considerations suggest that answers to why-questions do in fact play a role in mechanistic explanation (see also Chapter 8), and indeed, that they may be instrumental for coming up with answers to what-questions. Notably, the Bayesian approach illustrates this kind of dependence of the *what* on the *why*. Recall that many different kinds of behavior and cognition have been found to approximate optimal solutions within a particular task environment. Although initially surprising, this finding is not accidental. Investigators regularly find discrepancies between an optimal solution and the observed behavioral data, but then often go on to tweak the specification of the task environment until the optimal solution is closely approximated by the data (Anderson 1991). Although some commentators denounce this kind of tweaking as a post-hoc model-fitting exercise of limited explanatory value (see e.g. Bowers & Davis 2012), others take it to be an efficient means of deriving mathematical formalisms that simultaneously answer what- and why-questions at the computational level of analysis, and that even facilitate the search for answers at lower levels (Zednik & Jäkel 2016).

In short, Marr's computational level of analysis concerns two types of questions: questions about *what* a cognitive system is doing, and questions about *why*. It is a mistake to consider either kind of question to be antithetical to the principles of mechanistic explanation. On the contrary, both kinds of questions play an important role in the discovery and description of mechanisms in several different research programs. But although the computational level of analysis can therefore be seen to play a critical role in mechanistic explanation, it is by no means sufficient. Just as Marr deems it necessary to answer questions below the computational level for the purposes of "completely understanding" a cognitive system and its behavior, mechanistic explanation also involves describing the internal features of a mechanism—its component parts, operations, and organization.

3. The algorithmic level: "how?"

Marr's algorithmic level of analysis is defined by questions about "the representation for the input and output," and about the "algorithm by which the transformation [between them] may actually be accomplished" (Marr 1982: 23). Questions of this kind are traditionally answered through *cognitive modeling*, which involves *functionally analyzing* the complex behavioral or cognitive capacity being explained into an organized collection of simpler capacities (Cummins 1983), and subsequently describing this collection in formal mathematical or computational terms (Busemeyer & Diederich 2010; Luce 1995). In the computationalist research program, cognitive models often consist of lists of production rules with which to manipulate symbolic expressions (Pylyshyn 1980). These models are quite naturally viewed as descriptions of algorithms for transforming representations so as to achieve a particular computational goal. That said, cognitive models are a commonplace even in non-computationalist research programs. Indeed, some of the most influential cognitive models today consist of mathematical equations that determine the numerical values of output variables (Rumelhart et al. 1986; Nosofsky 1986), the evolution of state variables over time (Busemeyer & Townsend 1993), or the probability distribution over a hypothesis space (Pouget et al. 2013). It may not always be useful or even possible to think of these models as describing representation-transforming algorithms (Ramsey 2007; van Gelder 1995). Nevertheless, insofar as they can be used to reproduce a particular input-output transformation or a series of state-changes, they can still be said to compute a particular information-processing function or state-space trajectory. Insofar as the function or trajectory being computed accurately describes a particular cognitive system's behavior, the relevant cognitive model is a possible answer to a question about *how* that cognitive system does what it does (see also Cummins 2000; McClamrock 1991).

Notably, there are often many different ways to compute a particular information-processing function or state-space trajectory. For this reason, many different cognitive models may be developed to answer a particular how-question, and investigators need a way of distinguishing good answers from bad ones. Interestingly, Robert Cummins is sometimes interpreted as being unwilling or unable to make such a distinction. Consider the following oft-quoted passage:

> Any way of interpreting the transactions causally mediating the input-output connection as steps in a program for doing ϕ will, provided it is systematic and not *ad hoc*, make the capacity to ϕ intelligible. Alternative interpretations, provided they are possible, are not competitors; the availability of one in no way undermines the explanatory force of another.
>
> *(Cummins 1983: 43)*

Because Cummins attributes the same degree of explanatory force to a (potentially) wide variety of models, he has been accused of being unable to distinguish between answers to how-questions that capture the way a cognitive system *actually* does what it does and answers that merely capture the way it might *possibly* do so (Kaplan 2011; Piccinini & Craver 2011). This accusation seems unwarranted, however; it fails to acknowledge Cummins' demand that cognitive models accurately reflect the "transactions causally mediating the input-output connection." Moreover, Cummins goes on to acknowledge that the elements of some cognitive models are more likely than others to be instantiated in a particular cognitive system—and that these models are for this reason to be preferred (Cummins 1983: 44; see also Feest 2003). Thus, Cummins does in fact outline a criterion for distinguishing good answers to how-questions from bad ones: Good answers are provided by cognitive models whose elements are in fact instantiated by the cognitive system being investigated, and that actually reflect the causally relevant factors that contribute to that system's behavior.

This criterion can be fleshed out by aligning it with the idea that cognitive models should accurately describe the internal features of the mechanism responsible for the explanandum phenomenon: its component parts, operations, and overall organization. More precisely, cognitive models should satisfy what has come to be known as the *model-to-mechanism mapping constraint* (3M):

> In successful explanatory models . . . (*a*) the variables in the model correspond to components, activities, properties, and organizational features of the target mechanism that produces, maintains, or underlies the phenomenon, and (*b*) the (perhaps mathematical) dependencies posited among these variables in the model correspond to the (perhaps quantifiable) causal relations among the components of the target mechanism.
>
> *(Kaplan & Craver 2011: 611, see also Chapter 17, this volume)*

3M allows for the possibility that many different models have equal explanatory force—as long as the rules, symbols, equations, and variables posited by each one of these models correspond to the features of the target mechanism. Moreover, although 3M requires that the features of a mechanism be described correctly, it does not require that they be described completely. In this spirit, Piccinini and Craver (2011) argue that cognitive models are designed to provide *mechanism sketches*, elliptical descriptions that correctly, albeit incompletely, describe a mechanism's component parts, operations, and/or organization. Although mechanism sketches satisfy 3M by correctly describing some of a mechanism's features, additional "filling in" is required to transform these sketches into full-fledged mechanistic explanations (Craver 2007; Machamer et al. 2000, see also section 4 of this chapter).

Several commentators have argued that cognitive models can in fact be viewed as mechanism sketches. To this end, they have considered examples from a variety of research programs (for discussion see e.g. Abrahamsen & Bechtel 2006; Milkowski 2013; Zednik 2011). That said, other commentators have identified counterexamples in the form of models that contain *idealizations*: constructs that do not correspond to any one of a particular mechanism's features (see also Chapter 17). For example, Weiskopf (2011) introduces Hummel and Biederman's (1992) model of object-recognition, which contains *Fast Enabling Links* (FELs) that "possess physically impossible characteristics such as infinite speed" (Weiskopf 2011: 331). Similarly, Buckner (2015) considers the role of backpropagation learning in connectionist networks, the execution of which depends on the biologically implausible "backwards transmission of information across neural synapses, the need for prior knowledge of correct output, and the distinct, individualized error signals used to adjust the thresholds and weights of each node and link in the network" (Buckner 2015: 3925). Because FELs and backpropagation learning are psychologically, biologically, and/or physically implausible, cognitive models that incorporate these constructs can be known with relative certainty to *not* satisfy 3M.

Although there are counterexamples to the claim that cognitive models generally *satisfy* 3M, there is nevertheless reason to believe that these models are *constrained* by 3M. That is, there is reason to believe that FELs and backpropagation learning are subject to replacement by less idealized constructs as the explanatory demands on the relevant models increase. Notably, Weiskopf denies this claim, arguing that FELs are "not clearly intended to be eliminated by any better construct in later iterations" (Weiskopf 2011: 331). The key premise in Weiskopf's argument is that FELs allow Hummel and Biederman to understand the unique contribution of synchronous firing to object-recognition, independent of other factors such as mutual excitation and/or inhibition. However, it is important not to conflate intelligibility with explanation: Although FELs may allow investigators to understand the unique contribution of one causally relevant factor, the use of such idealizations often comes at the cost of obscuring or altogether neglecting the contribution of other factors. Thus, although models that contain idealizations may reproduce gross behavioral trends, they are unlikely to capture precise quantitative detail such as reaction-times and learning curves that may be of significant explanatory interest. Indeed, as Buckner goes on to argue, it is in order to capture just this kind of detail that connectionist researchers often seek to replace backpropagation learning with "more biologically plausible training rules" (Buckner 2015: 3925). In general, as the explanatory demands on a particular model increase, it seems likely that idealizations will eventually be replaced by constructs that more accurately reflect the causal structure of a mechanism. Even if the cognitive models being used today do not satisfy 3M, future iterations of these models will presumably strive to do so. For this reason, the requirement that cognitive models correctly describe mechanisms in the sense of 3M is useful for determining whether these models provide good or bad answers to questions about *how* cognitive systems do what they do.

4. The implementational level: "where?"

How-questions at the algorithmic level of analysis are often the primary concern of research programs in cognitive science. Nevertheless, it would be a mistake to think that an explanation has been provided just as soon as these questions have been answered. As has already been stressed repeatedly, on Marr's account all three levels of analysis are needed to "completely understand" a cognitive system and explain its behavior. Accordingly, investigations at the computational and algorithmic levels must be supplemented by investigations at the implementational level, which are driven by questions about the way in which the constructs of a cognitive model are "realized physically" (Marr 1982: 25; see also Polger 2004). Unfortunately, it remains unclear what it actually takes to show that the production rules, symbols, equations, and/or variables specified by a cognitive model are realized by complex and potentially unruly physical systems such as the brain. The aim of this section is to show that this lack of clarity can be remedied by considering what it takes to "fill in" a sketchy cognitive model so as to deliver a full-fledged mechanistic explanation.[5]

In general, "filling in" is a matter of describing a mechanism in greater detail than before (Craver 2007; Machamer et al. 2000). Although the present discussion embraces the view that cognitive models are mechanism sketches which leave out certain details, it has not yet been discussed exactly what kinds of detail are typically left out. Because most cognitive models specify mathematical constructs such as lists of production rules or systems of equations, it is tempting to think of them as specifying details about a mechanism's abstract mathematical or computational properties, rather than about its concrete physical properties. Indeed, this view is quite widespread in the literature, despite often being left implicit (see e.g. Bechtel 2008; Bechtel & Shagrir 2015; Chemero & Silberstein 2008; Craver 2007; Danks 2008; McClamrock 1991; Piccinini & Craver 2011; Shagrir 2010; Stinson 2016). Notably, on this view the details that are included may pertain to any one or more of a mechanism's internal features. For example, on this view cognitive models may describe a mechanism's component operations in terms of e.g. their informational properties, rather than in terms of neuronal spike trains (Pylyshyn 1980; Weiskopf 2011). They may also specify a mechanism's component parts as e.g. filters, without identifying these with any particular neural structures (Stinson 2016). Finally, they might characterize a mechanism's overall organization as a graph, abstracting over anatomical detail (Bechtel & Shagrir 2015; Levy & Bechtel 2013).

Thus understood, the realization-relationship that obtains between the algorithmic and implementational levels is one of *instantiation*; "filling in" involves specifying the concrete physical properties that instantiate a particular set of abstract mathematical properties. Thus, whereas investigations at both levels describe the same component parts, operations, and/or organization of a particular mechanism, they differ with respect to the kinds of properties being described: Abstract mathematical properties on the one hand, and concrete physical properties on the other. Put differently, although the questions being asked at each level are fundamentally the same—they are how-questions in each case—the answers being given differ because they are articulated at different levels of abstraction.

Although this view is relatively widespread in the literature, it faces an important challenge: Why should it be necessary to answer a how-question twice? Since a mechanism's abstract mathematical properties are instantiated by its concrete physical properties, why should the former need to be cited in an explanation, in addition to the latter? Polger (2004) argues that mechanisms can compute information-processing functions or state-space trajectories just in virtue of their physical properties, regardless of whether these properties are also said to instantiate any particular abstract mathematical properties. In the same vein, Bickle (2015) advances a conception of mechanistic explanation in which a characterization of a mechanism's physical properties suffices to explain a wide variety of behavioral and cognitive phenomena; whether or not a cognitive mechanism can also be characterized in abstract mathematical terms at the algorithmic level is explanatorily irrelevant.

There is reason to be wary of this conclusion, however. For one, it contradicts Marr's highly influential claim that all three levels of analysis are needed to "completely understand" a cognitive system's behavior. For another, it calls into question the explanatory relevance of cognitive modeling, a practice that is widely considered to be the centerpiece of cognitive scientific research (see statements to this effect by e.g. Busemeyer & Diederich 2010; Cummins 1983, 2000; Luce 1995; Stinson 2016; Weiskopf 2011). Unfortunately, several previous attempts to avoid this conclusion fall short. Consider, for example, Bechtel and Shagrir's recent claim that cognitive models facilitate the identification of *design principles* that "produce the same results across a wide range of different implementations" (Bechtel & Shagrir 2015: 318). Although Bechtel and Shagrir show that the description of design principles can render a particular mechanism's organization intelligible and generalizable (see also Levy & Bechtel 2013), they do little to argue that these principles actually *produce* anything over and above the physical properties in which they are instantiated. In other words, design principles are subject to Kim's (1993) *causal exclusion* argument, according to which realized properties possess no causal powers over and above their realizers. If the abstract design principles identified at the algorithmic level cannot be thought to possess causal powers over and above the concrete properties that instantiate them, it is unclear why cognitive models are needed to answer questions about *how* a mechanism does what it does. A similarly unsuccessful response is given by Craver, who argues that the algorithmic level describes "realized properties [that] figure in unique causal relevance relations"—relations that, on his manipulationist account of constitutive relevance, "are true of realized properties and are not true of their realizers" (Craver 2007: 220). Recent challenges to Craver's manipulationist account suggest that interventions always affect realized and realizing properties simultaneously (Baumgartner & Gebharter 2015), thereby calling into question the claim that there can actually be any such unique causal relevance relations. Thus, the explanatory relevance of the algorithmic level remains unclear.

These unsuccessful responses to the challenge of explanatory irrelevance are weighed down by the view of cognitive models as descriptions of a mechanism's abstract mathematical properties. But there is an alternative view that has yet to receive serious consideration in the literature: Rather than consider cognitive models as descriptions of a subset of a mechanism's properties, they might instead be thought to describe a subset of its internal features.[6] Specifically, cognitive models can be thought to describe a mechanism's component operations as well as their functional organization, rather than its component parts and their structural organization. Indeed, insofar as cognitive models are developed by functionally analyzing the behavioral or cognitive capacity being explained, it should come as no surprise that they are concerned primarily or even exclusively with a mechanism's component operations. Piccinini and Craver appear to have this in mind when they observe that cognitive models are mechanism sketches "in which *some* structural aspects of a mechanistic explanation are omitted" (Piccinini & Craver 2011: 284, emphasis added). Nevertheless, they do not go far enough: Cognitive models may actually be mechanism sketches in which *all* structural aspects are left out. For sure, many cognitive models contain constructs that are labeled with nouns such as "filter," "channel," or "representation." Nevertheless, these constructs are nearly always defined functionally: A filter is something that *filters*, a channel is something that *channels*, and a representation is something that *represents*.[7] Notably, on this view the fact that a cognitive model's constructs are specified in mathematical or computational terms is irrelevant to its ability to explain; in line with the 3M constraint discussed previously, the production rules, symbols, equations, or variables that typically feature in such a model have just as much explanatory relevance as any other description—abstract or concrete—of the relevant mechanism's component operations. That is, it is misleading to align Marr's levels of analysis with levels of abstraction.

On this alternative view, "filling in" is a matter of *localizing* the production rules, symbols, equations, or variables of a cognitive model by identifying them with a particular mechanism's component parts (Bechtel & Richardson 1993). Thus, implementational-level questions about the way in which certain mathematical or computational constructs are "realized physically" are not questions about *how*, but are in fact questions about *where* a mechanism's component operations are carried out. A cognitive mechanism's component parts may be situated at several different levels of organization—from the level of molecules to the level of organisms (Bechtel 2008; Craver 2007)—and they may be highly distributed within any particular level—spanning whole neural populations, the brain as a whole, but also spanning the physical boundaries between brain, body, and world (Kaplan 2012; Zednik 2011). Like a mechanism's component operations, its component parts can be described in abstract mathematical or concrete physical terms: Molecules, nerve cells, neural populations, bodily limbs, and tools in the environment can all be described in full anatomical or physical detail, but may also be characterized schematically or abstractly, e.g. as lattice-like structures, geometric shapes, or graphical topologies. In general, no matter whether they emphasize abstract mathematical properties or concrete physical ones, descriptions of operations and their functional organization answer how-questions at the algorithmic level of analysis, while descriptions of parts and their structural organization answer where-questions at the implementational level.

In closing, it is worth highlighting an important corollary of this alternative view of the realization-relationship that obtains between the algorithmic and implementational levels: It is wrong to assume that cognitive models at the algorithmic level are *explanatorily autonomous* (compare Feest 2003; Weiskopf 2011). Investigators have recourse to a plethora of analytic and experimental techniques with which to characterize the component operations of a cognitive mechanism, as well as to test the accuracy of any particular characterization. For sure, some of these techniques are driven by purely behavioral methods, and may be independent of the possibility of localizing operations in a mechanism's component parts (Busemeyer & Diederich 2010; Cummins 1983). Nevertheless, the fact that cognitive models are subject to norms that do not involve localization has no bearing on the issue of whether localization—or more generally, the ability to identify the physical structures that are involved in the production of a cognitive or behavioral phenomenon—is *also* important (compare Stinson 2016). Moreover, it is important not to exaggerate the ability to describe a mechanism's component operations independently of its component parts. There are many historical examples in which the successful characterization of a mechanism's component operations was greatly facilitated or even enabled by the prior identification of its component parts (Bechtel 2008; Bechtel & Richardson 1993; Craver 2013). Indeed, as technological advances amplify investigators' ability to individuate structures in the brain and to characterize their functional properties, it seems reasonable to expect that how-questions at the algorithmic level and where-questions at the implementational levels will become increasingly intertwined (see also Boone & Piccinini 2015).

5. Conclusion

The primary aim of this chapter has been to resolve ambiguities in Marr's account of explanation in cognitive science by subsuming it under the framework of mechanistic explanation. To this end, it was argued that Marr's three levels can be individuated by the different types of questions that are typically asked about a cognitive system, and that ambiguities concerning the way in which these questions are posed by Marr can be resolved by aligning each question with a specific aspect of mechanistic explanation. Computational-level questions about a system's computational goals are in fact what-questions that can be answered by describing a mechanism's behavior, and questions about these goals' appropriateness are in fact why-questions that can be answered by situating the mechanism in a containing environment. At the algorithmic level, questions about how a certain computational goal is achieved are in fact questions about a mechanism's component operations. Finally, at the implementational level, questions about the way in which algorithms and representations are physically realized are in fact where-questions that can be answered by identifying the relevant mechanism's component parts. Construed in this way, no single level of analysis bears sole responsibility for delivering mechanistic explanations of behavior and cognition; all three levels are involved in the discovery and description of cognitive mechanisms.

A secondary aim of this chapter has been to show that several different research programs—not just the computationalist approach in which Marr was himself embedded—seek to deliver answers to questions at all three levels of analysis. Thus, the scope of Marr's account is much wider than traditionally assumed, encompassing many different areas and traditions in contemporary cognitive scientific research. Insofar as researchers across cognitive science answer questions about the what, why, how, and where of behavior and cognition, they are all in the business of mechanistic explanation.

Notes

1 The author is indebted to Frank Jäkel, Holger Lyre, the volume editors, and conference audiences in Düsseldorf and Warsaw for helpful feedback on earlier versions of this chapter.

2 Whereas Marr sometimes speaks of explanations at individual levels of analysis, here the term "explanation" will be reserved for a full three-level account, i.e. the kind of account that Marr deems necessary for "complete understanding" (Marr 1982: 4ff). This is not meant to be a commitment to any particular account of the relationship between explanation and understanding, however.

3 Levels of analysis must not be confused with levels of organization within a mechanism (see also Bechtel 2008; Craver 2007). Whereas the former are individuated by the kinds of questions an investigator might ask about a cognitive system, the latter are individuated by constitution-relations within a mechanism. Notably, insofar as many real-world cognitive mechanisms are hierarchical (Bechtel & Richardson 1993; Craver 2007), it may often be profitable to apply all three levels of analysis at several different levels of organization within a single mechanism.

4 Indeed, some proponents of the Bayesian approach take themselves to be delivering teleological explanations *rather than* mechanistic explanations (see e.g. Oaksford & Chater 2007). The accuracy of this characterization has been questioned, however, with some claiming that the Bayesian approach does not provide explanations at all (Bowers & Davis 2012; Danks 2008), and others arguing that it does in fact explain in a way that centers on the discovery of mechanisms (Zednik & Jäkel 2016).

5 The present discussion concerns only the realization-relationship that is relevant to Marr's account of explanation in cognitive science. No commitment will here be made regarding the realization-relationship in other contexts (see e.g. Kim 1993). Moreover, the present contribution will not explore the question of whether the realization-relationship that obtains between the implementational and algorithmic levels also obtains between the algorithmic and computational levels.

6 Intriguingly, Glennan (2010) has called it a "category mistake" to attribute a mechanism's causal powers to its properties, instead of to its component parts, operations, and/or overall organization. The view advanced here of the relationship that obtains between the algorithmic and implementational levels is consistent with Glennan's.

7 The view outlined here is consistent with the proposal that what a particular representation represents may be secondary to the causal-functional role it plays in a cognitive system (see e.g. Fodor 1980). A particular representation's causal-functional role is akin to a mechanism's operation, rather than, for example, to a part.

References

Abrahamsen, A. & Bechtel, W. (2006). Phenomena and mechanisms: Putting the symbolic, connectionist, and dynamical systems debate in broader perspective. In R. Stainton (Ed.), *Contemporary debates in cognitive science*. Oxford: Basil Blackwell, 159–186.

Anderson, J. R. (1991). Is human cognition adaptive? *Behavioral and Brain Sciences* 14: 471–517.

Baumgartner, M. & Gebharter, A. (2015). Constitutive relevance, mutual manipulability, and fat-handedness. *The British Journal for the Philosophy of Science*. doi: 10.1093/bjps/axv003.

Bechtel, W. (2008). *Mental mechanisms: Philosophical perspectives on cognitive neuroscience*. London: Routledge.

Bechtel, W. (2009). Looking down, around, and up: Mechanistic explanation in psychology. *Philosophical Psychology* 22: 543–564.

Bechtel, W. & Abrahamsen, A. (2010). Dynamic mechanistic explanation: Computational modeling of circadian rhythms as an exemplar for cognitive science. *Studies in History and Philosophy of Science Part A* 1: 321–333.

Bechtel, W. & Richardson, R. C. (1993). *Discovering complexity: Decomposition and localization as strategies in scientific research*. Princeton, NJ: Princeton University Press.

Bechtel, W. & Shagrir, O. (2015). The non-redundant contributions of Marr's three levels of analysis for explaining information processing mechanisms. *Topics in Cognitive Science* 7: 312–322.

Bickle, J. (2015). Marr and reductionism. *Topics in Cognitive Science* 7: 299–311.

Boone, W. & Piccinini, G. (2015). The cognitive neuroscience revolution. *Synthese* 193: 1509. doi: 10.1007/s11229-015-0783-4.

Bowers, J. S. & Davis, C. J. (2012). Bayesian just-so stories in psychology and neuroscience. *Psychological Bulletin* 138: 389–414.

Buckner, C. (2015). Functional kinds: A skeptical look. *Synthese* 192: 3915. doi: 10.1007/s11229-014-0606-z.

Busemeyer, J. R. & Diederich, A. (2010). *Cognitive modeling*. Newbury Park, CA: SAGE.

Busemeyer, J. R. & Townsend, J. (1993). Decision field theory: A dynamic-cognitive approach to decision making in an uncertain environment. *Psychological Review* 100: 432–459.

Chemero, A. & Silberstein, M. (2008). After the philosophy of mind: Replacing scholasticism with science. *Philosophy of Science* 75: 1–27.

Craver, C. F. (2007). *Explaining the brain*. Oxford: Oxford University Press.

Craver, C. F. (2013). Functions and mechanisms: A perspectivalist view. In P. Huneman (ed.), *Functions: Selection and mechanisms*. Dordrecht: Springer, 133–158.

Cummins, R. (1983). *The nature of psychological explanation*. Cambridge, MA: MIT Press.

Cummins, R. (2000). 'How does it work?' versus 'what are the laws?': Two conceptions of psychological explanation. In F. Keil & R. Wilson (eds.), *Explanation and cognition*. Cambridge, MA: MIT Press, 117–144.

Danks, D. (2008). Rational analyses, instrumentalism, and implementation. In N. Chater & M. Oaksford (eds.), *The probabilistic mind: Prospects for Bayesian cognitive science*. Oxford: Oxford University Press, 59–75.

Darden, L. & Craver, C. F. (2012). *In search of mechanisms*. Chicago, IL: The University of Chicago Press.

Feest, U. (2003). Functional analysis and the autonomy of psychology. *Philosophy of Science* 70: 937–948.

Fodor, J. A. (1980). Methodological solipsism considered as a research strategy in cognitive psychology. *Behavioral and Brain Sciences* 3: 63–73.

Glennan, S. (2010). Mechanisms, causes, and the layered model of the world. *Philosophy and Phenomenological Research* LXXXI: 362–381.

Hummel, J. E. & Biederman, I. (1992). Dynamic binding in a neural network for shape recognition. *Psychological Review* 99: 480–517.

Kaplan, D. M. (2011). Explanation and description in computational neuroscience. *Synthese* 183: 339–373.

Kaplan, D. M. (2012). How to demarcate the boundaries of cognition. *Biology and Philosophy* 27: 545–570.

Kaplan, D. M. & Craver, C. F. (2011). The explanatory force of dynamical and mathematical models in neuroscience: A mechanistic perspective. *Philosophy of Science* 78: 601–627.

Kelso, J. A. S. (1995). *Dynamic patterns: The self-organization of brain and behavior*. Cambridge, MA: MIT Press.

Kim, J. (1993). *Supervenience and mind: Selected philosophical essays*. Cambridge: Cambridge University Press.

Levy, A. & Bechtel, W. (2013). Abstraction and the organization of mechanisms. *Philosophy of Science* 80: 241–261.

Luce, R. D. (1995). Four tensions concerning mathematical modeling in psychology. *Annual Review of Psychology* 46: 1–26.

Machamer, P., Darden, L. & Craver, C. F. (2000). Thinking about mechanisms. *Philosophy of Science* 67: 1–25.

Marr, D. (1982). *Vision: A computational investigation into the human representation and processing of visual information*. San Francisco: W. H. Freeman.

McClamrock, R. (1991). Marr's three levels: A re-evaluation. *Minds and Machines* 1: 185–196.

Milkowski, M. (2013). *Explaining the computational mind*. Cambridge, MA: MIT Press.

Nosofsky, R. M. (1986). Attention, similarity, and the identification-categorization relationship. *Journal of Experimental Psychology* 115: 39–57.

Oaksford, M. & Chater, N. (2007). *Bayesian rationality: The probabilistic approach to human reasoning*. Oxford: Oxford University Press.

Piccinini, G. & Craver, C. F. (2011). Integrating psychology and neuroscience: Functional analyses as mechanism sketches. *Synthese* 183: 283–311.

Polger, T. (2004). Neural machinery and realization. *Philosophy of Science* 71: 997–1006.

Pouget, A., Beck, J. M., Ma, W. J., & Latham, P. E. (2013). Probabilistic brains: Knowns and unknowns. *Nature Neuroscience* 16: 1170–1178.

Pylyshyn, Z. (1980). Computation and cognition: Issues in the foundations of cognitive science. *Behavioral and Brain Sciences* 3: 111–169.

Ramsey, W. M. (2007). *Representation reconsidered*. Cambridge: Cambridge University Press.

Rumelhart, D. E., McClelland, J. E. & the PDP Research Group (1986). *Parallel distributed processing: Explorations in the microstructure of cognition*. Cambridge, MA: MIT Press.

Shagrir, O. (2010). Marr on computational-level theories. *Philosophy of Science* 77: 477–500.

Stinson, C. (2016). Mechanisms in psychology: Ripping nature at its seams. *Synthese* 193: 1585. doi: 10.1007/s11229-015-0871-5.

Strevens, M. (2008). *Depth*. Cambridge, MA: Harvard University Press.

van Gelder, T. J. (1995). What might cognition be, if not computation? *Journal of Philosophy* 91: 345–381.

van Gelder, T. J. (1998). The dynamical hypothesis in cognitive science. *Behavioral and Brain Sciences* 21: 1–14.

Weiskopf, D. A. (2011). Models and mechanisms in psychological explanation. *Synthese* 183: 313. doi: 10.1007/s11229-011-9958-9.

Zednik, C. (2011). The nature of dynamical explanation. *Philosophy of Science* 78: 238–263.

Zednik, C. & Jäkel, F. (2016). Bayesian reverse-engineering considered as a research strategy for cognitive science. *Synthese*. doi: 10.1007/s11229-016-1180-3.

30

SOCIAL MECHANISMS[1]

Petri Ylikoski

Social mechanisms and mechanism-based explanation have attracted considerable attention in the social sciences and the philosophy of science during the past two decades. The idea of mechanistic explanation has proved to be a useful tool for criticizing existing research practices and meta-theoretical views on the nature of the social-scientific enterprise. Many definitions of social mechanisms have been articulated, and have been used to support a wide variety of methodological and theoretical claims. It is impossible to cover all of these in one chapter, so I will merely highlight some of the most prominent and philosophically interesting ideas.

1. Mechanisms in the social sciences

As in other sciences, mechanism as a notion belongs to the everyday casual vocabulary of many social scientists. In this context the word "mechanism" could refer to a cause, a causal pathway, or an explanation without explicit theorizing about the nature of mechanisms. This casual and occasional mechanistic way of talking is probably as old as the social sciences. Equally old are certain negative connotations of the word "mechanical" implying simple, rigid, and reductionist. Being part of everyday casual vocabulary explains much of the intuitive appeal of mechanistic language, although the negative connotations have made some social scientists suspicious of the mechanistic turn.

Theorizing about mechanisms has multiple origins in the social sciences. Among the most prominent early sources is Rom Harré's (1970) philosophy of science. Although his later work has been highly influential in social psychology, the biggest impact of his philosophy of science was through the so-called critical realist movement. The key thinker in this movement is Roy Bhaskar (1978, 1979), in whose philosophy Harré's ideas about causation and mechanisms have a central role. Despite Bhaskar's transcendental argumentation, his layered account of reality, and his ideas about essences and internal relations that have raised philosophical suspicions and doubts about their relevance to the social sciences, critical realism is still one of the most influential meta-theoretical movements in the social sciences (Lawson 2003; Elder-Vass 2011). Philosophers of mechanisms might be unfamiliar with critical realism, but in the view of many social scientists the idea of causal mechanisms is strongly associated with this movement.

Another influential early advocate of mechanism-based thinking is Jon Elster, his many books containing excellent examples of mechanistic thinking in action. His early definition

(Elster 1989), according to which a mechanism explains by providing a continuous and contiguous chain of causal or intentional links between the *explanans* and the *explanandum*, is quite in line with the general mechanistic perspective. Although Elster's mechanisms tend to be psychological rather than social, his work has inspired many social scientists to open up black boxes and show the cogs and wheels of the internal machinery of social processes. Elster has also strongly associated the mechanistic attitude with the intellectual virtues of clarity and precision: he sees mechanism-based theorizing as clearheaded causal thinking about social processes and for a large group of social scientists this is the core of the mechanistic perspective. Note that there are few connections between social scientists inspired by Bhaskar's critical realism and those supporting Elster's approach. The idea of mechanism-based thinking has multiple interpretations in the social sciences.

Mechanistic thinking started to become mainstream in the 1990s. A work of particular significance in this respect was *Social Mechanisms* (1998), edited by Peter Hedström and Richard Swedberg. Among contributors to this volume are many scholars who have been influential in the development of mechanism-based thinking in the social sciences. Later this approach developed into the analytical sociology movement (Hedström & Bearman 2009; Hedström & Ylikoski 2010) that has most systematically developed the program of mechanism-based social science. At the same time, philosophers of the social sciences such as Daniel Little (1991) and Mario Bunge (1997) started to talk about mechanisms. In political science mechanisms became a focus of attention a little bit later. There causal mechanisms played an important role in debates concerning research methodology and causal inference. Especially noteworthy has been the development of process-tracing methodology for case-study research (Bennett & Checkel 2015). In general, the debates in political science and sociology have been quite similar: advocates of mechanisms have criticized the simplistic use of statistical methodology and the downplaying of the importance of causal process assumptions in causal inference. However, there are also some important differences, as the later discussion will show.

In the following discussion I focus, first, on what is known as Coleman's diagram, which helps to identify the core challenges of mechanism-based theorizing in the social sciences. It also provides a context in which to discuss the appeal of mechanism-based explanation among social scientists. In an attempt to make sense of this debate I will introduce a distinction between causal scenarios and causal mechanism schemes. By way of illustrating social-scientific thinking about mechanisms, I discuss the use of agent-based simulation as a tool for mechanism-based theorizing, and introduce the idea of metamechanism. In the concluding discussion I show how mechanistic ideas have been used in social-scientific debates about causal inference.

2. Coleman's diagram

A useful starting point for discussing social mechanisms is a diagram known as Coleman's boat (Coleman 1987, 1990; see Figure 30.1). It is commonly used in discussions of social mechanisms, and one could say that it has become an emblem of mechanism-based thinking in the social sciences. Coleman himself did not employ mechanistic vocabulary, but in many ways the diagram exemplifies mechanistic thinking in the social sciences (Ylikoski 2016).

The recommended starting point for unpacking the diagram is node D, which represents the macro-social *explanandum* that the sociologist finds interesting. Node A represents some macro-social variable that is associated with D. For example, let us assume that A is the implementation of a government job-training program and D is the decrease in the level of youth unemployment. Did the job-training program cause the decline in unemployment? In other words, can arrow 4 be interpreted as causal?

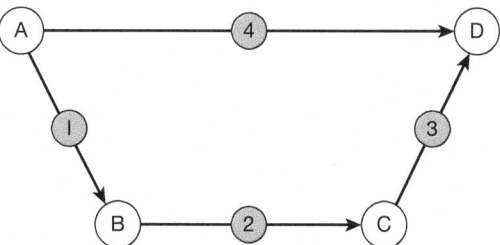

Figure 30.1 Coleman's diagram

Coleman's main point is the following. To justify causal claims like this, it is necessary to understand *how* the suggested cause brings about the effect in question. This idea has two components. The first of these concerns the justification of causal claims: the statistical data on the relevant macro variables is usually so sparse as to be insufficient for establishing such a claim. Coleman's suggestion is to test the causal claim by finding out whether there is an empirically supported mechanism by which A brings about D: if there is, the claim is supported, and if not, then it should be scrapped. The second component concerns explanation. Coleman and other supporters of mechanism-based thinking stress the importance of understanding how the effect was brought about. Thus, even if it were possible to show that variable A is a causal difference-maker for variable D, it would not suffice for a theoretically satisfactory explanation: the explanation has to spell out the mechanism. These two ideas are at the core of social-scientific debates about causal mechanisms. I will discuss both in detail later, but first I will consider other elements of the diagram.

The key point in the diagram is that macro variables have to be connected to activities by agents. In most cases these agents are human persons, but Coleman also allows for various sorts of corporate agents to take this role. Thus arrow 1 describes the way in which changes in macro conditions influence the relevant agents. The change in A may bring about changes in the beliefs, desires, or other mental attributes of these agents, or it might change the opportunities or incentives they are facing. Hedström and Swedberg (1998) call these influences *situational mechanisms*. They cover the ways in which social structures constrain and enable individuals' opportunities for action, and how the cultural and social contexts influence individuals' goals, beliefs, habits, or cognitive frames.

Arrow 2 in the diagram covers the role of the theory of action in sociological explanation. The purpose of theory of action is to connect changes in agents' opportunities and mental states to changes in their behaviors or actions. Coleman used rational choice theory for this purpose, but it is also possible to use other theories of action (Hedström & Ylikoski 2014). The important point here is that mechanical explanations in the social sciences bottom out (Craver 2007) at the level of individual action (Coleman 1990: 4). The behaviors of individual persons are the basic components of social mechanisms and social scientists do not look for explanations for these *action-formation mechanisms*: this is a job for cognitive scientists and psychologists.

Although both situational and action-formation mechanisms certainly pose their own challenges, Coleman argues that the transformational mechanisms (arrow 3) are the biggest bottleneck in sociological theory. Social scientists know a great deal about how individuals' desires, beliefs, and opportunities, for example, are influenced by the social contexts in which they are embedded (situational mechanisms), and about how these desires, beliefs, and opportunities influence actions (action-transformation mechanisms), but when it comes to the link between individual actions and social outcomes, they are often forced to resort to hand-waving. This is something Coleman aimed to highlight with his diagram: macro-level patterns are often difficult to predict from individual-level descriptions, and the way in which individual actions produce social patterns is rarely a simple process of aggregation. This is not a problem that is specific to individualistic theories: although holistic theories tend to highlight the contextuality and complexity of everything, in practice their micro-macro assumptions tend to be rather simplistic. It is frequently assumed that macro facts simply reflect the relevant micro facts, and vice versa. Schelling's (1978) well-known checkerboard model shows how wrong this assumption is even in very simple settings: it shows how residents who do not favor segregation may still end up in a highly segregated neighborhood. It is clear even from this very simple model that one cannot assume that macro facts will reflect individual (or average individual) preferences. It is necessary to understand how the macro outcome is brought about by the interdependent actions of individuals; in other words, one has to understand "the rules of the game" (Coleman 1990: 19).

Coleman was never able to fully unpack the metaphor of the rules of the game, but his discussion and examples were enough to interest sociologists in the theoretical challenge of the micro-macro link. Consequently, whereas others might talk loosely about emergent macro-scale properties, sociologists following Coleman set themselves the much harder task of explaining how that emergence comes about. The key mechanistic idea captured in the diagram is that only when we have understood the whole chain of situational, action-formation, and transformational mechanisms have we understood the relation between the macro-scale social facts. Underlying each A–D association or causal relation is a combination of these mechanisms. All three are required, otherwise the explanation would not cover all elements of the micro-macro link.

Users of the diagram have been criticized (Jepperson & Meyer 2011; Little 2012) for being committed to a reductive ideal of methodological individualism. I believe this criticism is misplaced, at least to some extent. The diagram does not describe the reduction of macro facts to facts about individuals and their relations. The mistake is the assimilation of the Coleman diagram into the supervenience/realization diagrams used in the philosophical debate on mental causation. Coleman's arrows represent explanatory rather than reductive relations, and if one considers the details it is evident that various sorts of structural assumptions play a central role in situational and transformational mechanisms. Coleman's main point is that the structural facts are not explanatory in themselves. It is necessary to understand *how* they bring about their effects via the activities and cognitions of individuals, and a full understanding of how the social whole works requires an understanding of how its behavior is *generated* by the activities of its members. In Coleman's view, individual agents are the basic building blocks of social mechanisms, and therefore have to be *included* in any mechanistic social explanation. However, this does not imply that the explanation entails the reduction of macro-social facts to facts about individuals.

3. Mechanism-based explanations

In the social sciences, as in the philosophy of science, the mechanism-based accounts of explanations have been developed as alternatives to the once dominant covering-law account of explanation (Hempel 1965). While social scientists have been familiar with famous counterexamples to that theory and philosophical problems raised by them (such as problems of the explanatory relevance, the symmetry between explanation and prediction, the asymmetry of explanation, and many problems in analyzing the notion of law), social scientists have mainly been concerned with the apparent implausibility of the theory as a model for the social sciences. There are very few laws in the social sciences, and even those are better described as *explananda* rather than *explanatia*. Most social-scientific generalizations that are not truisms are quite limited in domain, and include exceptions that neat *ceteris paribus* conditions cannot cover.

Furthermore, the strategy of formulating social-scientific theories in terms of axioms and laws has turned out to be very unproductive, resulting in incomplete collections of sterile generalizations marred with unclear domains of application and high levels of conceptual indeterminacy. Finally, the basic idea of a covering-law theory is counterintuitive for social scientists. The best social-scientific explanations seem to do more than simply subsume the phenomenon under more general empirical regularities: a good explanation shows *how* the suggested cause brings about the effect to be explained. The initial appeal of mechanistic ideas lies in this generative notion of explanation.

The dissatisfaction with covering-law theory has not led to general agreement on the definition of a mechanism, however. The intuitive idea can be developed in multiple directions, especially when the people have different applications in mind. As a result, the literature on social mechanisms notoriously abounds with apparently incompatible definitions of mechanisms. Mahoney (2001), for example, lists 24, and subsequent contributions have added to the number. Some critics (e.g., Norkus 2005) regard this multiplicity as a serious problem for mechanism-based explanation. This could be an overstatement in that, although the definitions are formally incompatible, most of them could be considered attempts to capture the same basic ideas. The absence of a generally agreed definition for basic concepts such as gene and species has not stalled the development of the biological sciences. As long as mechanism-oriented social scientists agree on the central exemplars of mechanism-based explanations and share a similar understanding of their general characteristics (Hedström & Ylikoski 2010), they should manage well without a general definition. Thus the main challenge is not to provide a general definition of mechanism-based explanations; it is rather to arrive at a consensus about prototypical examples of mechanistic explanation.

A more serious problem is that much of the mechanism discourse in the social sciences is quite loose. As noted above, as a notion, mechanism belongs to the general causal vocabulary of social scientists. Some advocates of social mechanisms seem to go along with this loose talk, and it is not uncommon to name processes that produce certain kinds of outcome as mechanisms and leave it at that. In such usage, the mechanism is just a label for a black box, a name for an effect, not an explanation. The invisible hand, cumulative advantage, and democratization are not mechanisms in themselves; they are processes that produce specific kinds of outcomes. At most they are names for families of mechanisms. However, if one does not distinguish at least some family members, all one has is a placeholder for something substantial. The key point here is that if one looks into *how* the invisible hand or cumulative advantage work in practice, one finds that there may be multiple mechanisms underlying these effects, which sometimes work separately, and sometimes act together in various combinations. The point here is not about the proper use of the word "mechanism": it is about dangerous ambiguity. Confusion between naming an effect and providing a mechanistic explanation for it may give rise to an illusion of depth of understanding (Ylikoski 2009). Furthermore, the ambiguity makes one blind to the possibility of there being multiple mechanisms that are responsible for similar kinds of effect.

4. Causal scenarios and causal mechanism schemes

Social scientists are interested in explaining particular causal outcomes and in developing general theories about social mechanisms. In both contexts they refer to mechanisms, which sometimes causes unnecessary confusion. To avoid this, it is useful to distinguish between causal scenarios and causal mechanism schemes (Ylikoski & Aydinonat 2014). *Causal scenarios* are (selective) representations of particular causal processes responsible for some concrete event or phenomenon. Used thus, causal mechanism refers to a causal narrative that describes the process that is responsible for the *explanandum*. This narrative may be highly detailed or a mere sketch, but in any case it does more than cite a cause that had an effect on the event (see Chapter 31): it describes the crucial elements in the relevant causal chain. In other words, the causal scenario describes how the *explanandum* event came about. The distinction between *how-possibly* and *how-actually* explanations applies here. Usually there are many different ways in which the outcome *could* have come about, which competing how-possibly scenarios describe. The challenge for researchers is to find evidence that could discriminate between these alternatives and enable them to make a judgment about which scenario is the true explanation.

When political scientists talk about process tracing, they are concerned with causal scenarios. On the other hand, when analytical sociologists discuss causal mechanisms, they tend to refer to *causal mechanism schemes* (compare Chapter 19). For example, sociologists talking about self-fulfilling prophecies or vacancy chains are referring to causal mechanism schemes, which are abstract representations of mechanisms that could bring about effects of a certain kind. The *explanandum* of a causal mechanism scheme tends to be quite abstract, or stylized, reflecting the fact that such schemes are not primarily explanations of particular facts, but schemes for constructing them. Thus it is useful to think of them as abstract building blocks that can be adapted and filled in to serve a role in causal scenarios that explain particular facts. A single causal scenario might be a combination of many different causal mechanism schemes, and might even contain mechanism schemes that have opposite causal effects.

Causal mechanism schemes are at the core of analytical sociology's account of growth of theoretical knowledge, according to which social-scientific knowledge accumulates through the development of middle-range theories (Hedström & Udéhn 2009) about social mechanisms. In this view the core theoretical knowledge comprises a collection of causal mechanism schemes that can be adapted to particular situations and explanatory tasks. According to this toolbox view (Hedström & Ylikoski 2010; Elster 2015), social-scientific knowledge is not integrated into highly abstract general theory, but consists of a growing collection of causal mechanism schemes that are mutually compatible. The understanding of the social world accumulates as the knowledge of the mechanism schemes becomes more detailed and the number of known mechanisms increases. Understanding of more complicated phenomena requires combining different mechanism schemes, hence knowledge also expands through learning how to create these "molecular" mechanisms.

The similarity between this view of the architecture of theoretical knowledge in the social sciences and the mechanistic view of the biological sciences is obvious (see Chapter 19), and offers a fruitful opportunity to compare two domains of knowledge. However, the relevance of the toolbox vision is not limited to meta-theory: it also gives new tools to counter the fragmentation of the social sciences. Causal mechanism schemes can be shared among the different subfields, which would allow for a novel type of integration: various subfields employ and develop the same theoretical toolbox and thereby benefit from each other's work. The toolbox vision for sociological theory shows that in the social sciences mechanistic ideas are not confined to discussions about explanation and causation, but they also play an important role in how social scientists think about the nature of social-scientific knowledge.

5. Agent-based simulation and generative explanation

One of the main points about mechanism-based explanation is that it should describe how the properties, activities, and relations of components bring about the phenomenon to be explained (see Chapters 1 and 19). This concern with generative processes makes *generative sufficiency* a central concern in the evaluation of explanations: the suggested explanation should, at least in principle, be capable of bringing about the outcome in the specified circumstances. However, demonstrating generative sufficiency is not easy. The social outcomes of interest typically result from numerous individuals acting and interacting with one another over extended periods of time. Furthermore, the aggregate behavior of these complex dynamic systems is extremely difficult to understand and to predict without the aid of analytical tools. *Agent-based simulation modeling* (ABM) provides such a tool (Miller & Page 2007; Squazzoni 2012; Ylikoski 2014). There is a natural affinity between the components of mechanism-based explanations and ABM. Like any society, social-scientific ABMs comprise agents with goals and beliefs. These agents possess resources and influence each other. It is easy to see how the macro-patterns they create come to be regarded as analogical to the processes that social scientists study. An ABM sheds light on how the phenomenon to be explained could have been generated, and how changes in agents' attributes or relational structures change the macro outcomes.

Joshua Epstein (2006) offers the strongest ABM-based formulation of the idea of generative explanation. According to Epstein, "If you didn't grow it, you didn't explain it." He meant that producing the macro-level outcome by means of ABM is a necessary condition for its explanation. However, it could be argued that simply "growing" the phenomenon of interest is not sufficient to engender a proper understanding. The crucial challenge is to understand *how* the specified micro-configuration produces the phenomenon. Thus, in reformulating Epstein's slogan Macy and Flache (2009: 263) make a crucial point: "If you don't know how you grew it, you didn't explain it." In any case, social scientists using ABM are not satisfied with mechanistic explanation as mere storytelling. They do not think it is enough to provide a qualitative narrative about the process and its components. They also consider it important to demonstrate that the suggested mechanism can, in fact, bring about the effect to be explained. Having a detailed *how-possibly* scenario is also a precondition for the real empirical testing of it. Only if competing causal scenarios are clearly articulated is it possible to look for crucial empirical evidence that discriminates between them. Still, ABM has a long way to go to become a mainstream tool in sociological research: it is basically a tool for working with causal mechanism schemes rather than concrete causal scenarios (Ylikoski 2014). However, the connection between ABM and mechanism-based thinking is strong.

6. Metamechanisms

Jeremy Freese and Karen Lutfey's (2011) *metamechanism* is an interesting addition to the conceptual toolbox of mechanical philosophy. Behind this idea is Link and Phelan's (1995) suggestion that socioeconomic status is a *fundamental cause* of health differences. They refer to an extremely robust empirical finding: socioeconomic status and health are strongly correlated. The lower-status people are, the sooner they die, and the worse health they have while alive. This association holds for virtually any society and historical period for which there is adequate empirical data. The puzzling thing is that the causes of death and disease have changed a lot over the past hundred years. In other words, the proximate mechanisms of death and ill-health are highly variable. What explains this puzzling pattern?

Freese and Lutfey suggest that the idea of fundamental cause lends itself to a mechanistic interpretation: socioeconomic status (SES) is associated with a metamechanism, in other words a general mechanism that explains the generation of multiple proximate mechanisms that reproduce a particular relationship in different places and at different times. This helps to make sense of Link and Phelan's findings. People with a higher SES have more resources and education, which makes it easier for them to use new medical services and health-improving inventions. Even if healthcare is universal and free, the higher-SES people are better placed to make use of it.

Other metamechanisms may have a similar influence. There are spillover effects, for example, even among individuals who do not especially care about their health. Higher-SES individuals will have better health because they tend to gain more benefits from the purposive actions of others in their social networks (that are partly based on SES). Similarly, health-related institutions might be biased toward higher-SES people: they are given better service and better understand the instructions they get. The proximate mechanisms vary, but as long as the metamechanisms remain in place, there will be health inequalities. The usefulness of such mechanisms is not limited to the sociology of health, and could extend to some biological contexts, for example.

7. Mechanisms and statistical methodology

Apart from explanation, mechanisms are also important in the context of justification of causal claims. Especially in non-experimental contexts that are common in the social sciences, they are said to play a crucial role in distinguishing true causal relations from spurious correlations. Knowing that there is a mechanism through which X could influence Y supports the inference that X is a cause of Y. In addition, the absence of a plausible mechanism linking X to Y gives good reason to be suspicious of any straightforward causal interpretation of the association. Knowledge of mechanisms is also applied in extrapolating causal findings. The assumption of similarity among causal mechanisms is a crucial element in making inferences from one setting or population to another (Steel 2008; see also Chapter 32). However, there is much ambiguity in these mechanistic slogans. It is impossible to ascertain that knowledge of mechanisms is *necessary* for justifying causal claims without a clear idea of what such knowledge consists of, and how much of it is needed. All causal inference presupposes some causal background assumptions, but do all such assumptions concern causal mechanisms? It should also be recognized that mechanisms are not a magic wand for causal inference in the social sciences. The problem in many cases is not the absence of a possible mechanism, but insufficient evidence to discriminate between competing mechanistic hypotheses. Similarly, lazy mechanism-based storytelling is a constant threat: having a good story is no substitute for real statistical evidence. It is not rare for a good story about a (possible) mechanism to make people forget how important it is to test whether such a mechanism really is in place and whether it can really account for the intended *explanandum*.

Thinking in terms of mechanisms is often set against statistical methodology in the social-scientific debates on causal inference. This opposition takes different forms. Many critical realists rely on theorizing about mechanisms as an alternative to using statistics and causal-modeling techniques. Causal modeling is said to embody problematic Humean ideas about causation that make it suspect and of limited value, which is why critical realists tend to use statistics for descriptive purposes only and prefer qualitative evidence and theoretical argumentation.

An alternative is not to give up statistical and causal modeling, but to object to their use without consideration of the relevant causal mechanisms. Here the claim is that statistical techniques have replaced substantial theorizing. Peter Hedström conveys a fairly common sentiment in the following:

> Although most causal modelers refer to sociological theories in their work, they rarely pay it any serious attention. More often than not, they simply use theories to justify the inclusion of certain variables taken from a data set that has often been collected for entirely different purposes than the one to hand. Theoretical statements have become synonymous with hypotheses about relationships between variables, and variables have replaced actors as the active subjects with causal powers.
>
> *(Hedström 2005: 105)*

What he is calling for is a fuller incorporation of theory (about causal mechanisms) into the research design and the interpretation of statistical data. According to this view, sociological research should not be limited to the measurement of causal effects among conventional variables, but should focus on how the social world works. Given this background, it is understandable that advocates of mechanisms have been resistant to the common assumption that mechanisms are just intervening variables. Although the existence of a causal process or mechanism implies that there are intervening variables, not all of them are necessarily of the right sort. They may tell more about other effects of the mechanism than about the mechanism itself, for example. Thus, mere intervening variables do not guarantee the explanatory depth that is the main concern among mechanists.

The promise of mechanisms is to provide something more to causal inference, but what is that additional element? It is now generally recognized that the uses of statistical tools such as regression analysis presupposes substantial assumptions about the causal relations that are modeled (Kincaid 2012). A substantial proportion of these background assumptions concern possible causal mechanisms. There have also been attempts to incorporate mechanistic thinking into causal modeling (e.g., Knight & Winship 2014; see also Imai et al. 2011) that defines mechanisms as *modular sets of entities connected by relations of counterfactual dependence.* (Philosophers will easily recognize the influence of Woodward 2002 here.) According to Knight and Winship, as long as the mechanisms studied satisfy the requirement of modularity, Judea Pearl's DAG (Directed Acyclic Graph) calculus is a powerful tool in terms of facilitating the rigorous consideration of mechanisms in causal analysis. Their main argument is that mechanisms and causal analysis can be combined fruitfully in a way that could help in identifying causal effects even when traditional techniques fail. It is to be expected that attempts like this to combine mechanistic thinking with causal modeling will become more frequent. It appears that the issue is not really about statistical methods, but concerns the way they are used.

Causal graphs are useful when it comes to thinking about social mechanisms, and mechanists should welcome them. However, it should be borne in mind that there are serious limitations in terms of what can be represented as DAGs. Coleman's diagram discussed above cannot be interpreted as a DAG, for example. One problematic spot is arrow 1 between A and B. Consider how the demographic change caused by war affects the number of potential marriage partners available to women, how laws allowing same-sex marriage change the opportunities of same-sex couples to arrange their legal relationship, and how the improved educational level of society is related to the education of individuals. It cannot be said in any of these cases that the relations are strictly causal, whereas it is plausible to say in these cases the cited A-facts partially consist of the mentioned B-facts. However, some of the relevant consequences of changes in A might well be causal at the same time. Thus it seems that A–B explanatory dependency is based on (various) mixtures of causal and constitutive relations (Ylikoski 2013, 2016). This makes sense on the theoretical level, but also makes it impossible to interpret arrow 1 as causal or the diagram as a DAG.

In political science debates, especially in international relations, mechanistic thinking has also been set against statistical methods. However, the context has been that of qualitative case studies, the point being that "the standard quantitative template" is ill adapted for such research. The alternative methodology for causal inference is called *process tracing*, which Bennett and Checkel define as "the analysis of evidence on processes, sequences, and conjunctures of events within a case for the purposes of either developing or testing hypotheses about causal mechanisms that might causally explain the case" (Bennett & Checkel 2015: 7). As this definition implies, mechanisms—causal scenarios—play a central role in process tracing. Causation is understood here as a continuous process and the task is to explain a singular event. A central concern in process tracing is with the sequence of events and mechanisms involved in the unfolding of the process. The researcher looks for diagnostic evidence that can be used to discriminate between alternative causal scenarios that could explain the event.

Process tracing is often presented as a method for "within-a-case causal inference," but its functions remain somewhat unclear: it has been presented as a tool for theory testing, theory development, and for explaining singular outcomes (Beach & Pedersen 2013). These roles are naturally interlocking and not so easy to distinguish. It is rare to begin a case study with the goal of developing a theory, but the search for explanation might generate novel theoretical ideas. The competing hypotheses are competing explanations, and hence it does not make much sense to distinguish hypothesis testing and explanation as separate activities. Furthermore, it is doubtful that process tracing captures something that is unique to qualitative case-study research given that similar case-based *causal process observations* play an important role in the evaluation of evidence in experimental, comparative, and statistical studies. Consideration of the causal processes that produce the data to be analyzed is a major concern in all research. Thus there is much room for building new bridges between different research methodologies based on mechanistic ideas.

An interesting contribution in the literature on process tracing is the taxonomy of tests for causal hypotheses. Originally presented by Van Evera (1997), but later adapted by others (Beach & Pedersen 2013; Bennett & Checkel 2015), this taxonomy describes different kinds of tests that people struggling with the problem of multiple competing causal scenarios could look for. Passing a *Smoking-Gun test* gives strong support to the hypothesis and substantially weakens its rivals, but failure does not imply that the hypothesis is eliminated: it is only weakened and the rivals gain some additional support. The very name of the test is illustrative: finding a person with a smoking gun straight after a shooting makes him or her a strong suspect, but the lack of such evidence does not eliminate this person from the list of suspects. In the case of the *Hoop test* the implications are the opposite: passing the test affirms the relevance of the hypothesis, but does not confirm it. However, if the hypothesis fails the test, it is eliminated. Here the illustration is the familiar idea of an alibi: giving a speech to an audience of dozens of people at the time of the crime provides strong grounds for elimination from the list of suspects. However, the mere lack of an alibi does not yet provide positive evidence of guilt. *Doubly Decisive tests* are rare, but they are the strongest. A hypothesis that passes the test is confirmed and competing hypotheses are eliminated. The consequences of failure are also drastic, but only for the failed hypothesis: it is eliminated. Having a clear video recording of the crime with the shooter's face clearly visible is an example of a doubly decisive test: the video evidence demonstrates that the particular suspect is responsible, and also shows that other suspects took no part in the shooting. Finally, there are *Straw-in-the-Wind tests*. Passing such a test gives the hypothesis some support, but failure does not mean that it is eliminated. The evidence is weak or circumstantial and cannot in itself prove or disprove the suspect's guilt. However, in favorable circumstances enough accumulated evidence of this type may convince the jury. In the context of the social sciences most evidence is inherently of the *Straw-in-the-Wind* kind, and social scientists rarely encounter evidence that could be considered *Doubly Decisive*, or that would constitute a *Smoking-Gun* test. However, the taxonomy is useful in highlighting the fact that the value of evidence depends on the set of alternative hypotheses, not on some intrinsic relationship between a single hypothesis and the empirical material.

8. Conclusion

In the above review I have covered some prominent and interesting themes in the social-scientific debate on mechanisms. I have left many things out, some of which are discussed in other chapters of this book. I have attempted to show that mechanism-based thinking is a strong and expanding meta-theoretical idea in the social sciences, and that some of the ideas, such as the distinction between causal scenarios and causal mechanism schemes, and the notion of metamechanisms, might also be of interest in other disciplines.

Note

1 This research received funding from the European Research Council under the European Union's Seventh Framework Programme (FP7/2007-2013)/ERC grant agreement no. 324233, Riksbankens Jubileumsfond (DNR M12-0301:1), and the Swedish Research Council (DNR 445-2013-7681 and DNR 340-2013-5460).

References

Beach D & Pedersen RB (2013). *Process-Tracing Methods. Foundations and Guidelines.* Ann Arbor, MI: The University of Michigan Press.

Bennett A & Checkel JT (2015). *Process Tracing. From Metaphor to Analytic Tool.* Cambridge: Cambridge University Press.

Bhaskar R (1978). *A Realist Theory of Science.* 2nd ed. Brighton: Harvester.

Bhaskar R (1979). *The Possibility of Naturalism.* Brighton: Harvester.

Bunge M (1997). Mechanism and Explanation. *Philosophy of the Social Sciences* 27:410–65.

Coleman JS (1987). Microfoundations and Macrosocial Behavior, in Alexander, Giesen, Münch & Smelser (eds.): *The Micro-Macro Link.* Berkeley, CA: University of California Press: 153–73.

Coleman JS (1990). *Foundations of Social Theory.* Cambridge, MA: The Belknap Press.

Craver CF (2007). *Explaining the Brain: Mechanisms and the Mosaic Unity of Neuroscience.* New York: Oxford University Press.

Elder-Vass D (2011). *The Causal Power of Social Structures: Emergence, Structure and Agency.* Cambridge: Cambridge University Press.

Elster J (1989). *Nuts and Bolts for the Social Sciences.* Cambridge: Cambridge University Press.

Elster J (2015). *Explaining Social Behavior: More Nuts and Bolts for the Social Sciences.* 2nd ed. Cambridge: Cambridge University Press.

Epstein J (2006). *Generative Social Science: Studies in Agent-Based Computational Modeling.* Princeton, NJ: Princeton University Press.

Freese J & Lutfey K (2011). Fundamental Causality: Challenges of an Animating Concept for Medical Sociology, in Pescosolido et al. (eds.): *Handbook of the Sociology of Health, Illness, and Healing.* Dordrecht: Springer: 67–81.

Harré R (1970). *The Principles of Scientific Thinking.* London: Macmillan.

Hedström, Peter (2005). *Dissecting the Social: On the Principles of Analytical Sociology.* Cambridge: Cambridge University Press.

Hedström P & Bearman P (eds.) (2009). *The Oxford Handbook of Analytical Sociology.* Oxford: Oxford University Press.

Hedström P & Swedberg R (eds.) (1998). *Social Mechanisms: An Analytical Approach to Social Theory.* Cambridge: Cambridge University Press.

Hedström P & Udéhn L (2009). Analytical Sociology and Theories of the Middle Range, in Hedström & Bearman (eds.): *The Oxford Handbook of Analytical Sociology.* Oxford: Oxford University Press: 25–49.

Hedström P & Ylikoski P (2010). Causal Mechanisms in the Social Sciences. *Annual Review of Sociology* 36, 49–67.

Hedström P & Ylikoski P (2014). Analytical Sociology and Rational Choice Theory, in Gianluca Manzo (ed.): *Analytical Sociology: Norms, Actions and Networks.* New York: Wiley: 57–70.

Hempel C (1965). *Aspects of Scientific Explanation.* New York: Free Press.

Imai K, Keele L, Tingley D, & Yamamoto T (2011). Unpacking the Black Box of Causality: Learning about Causal Mechanisms from Experimental and Observational Studies. *American Political Science Review* 105, 765–89.

Jepperson R & Meyer JW (2011). Multiple Levels of Analysis and the Limitations of Methodological Individualisms. *Sociological Theory* 29, 54–73.

Kincaid H (2012). Mechanisms, Causal Modeling, and the Limitations of Traditional Multiple Regression, in Kincaid (ed.): *The Oxford Handbook of Philosophy of Social Science.* Oxford: Oxford University Press: 46–64.

Knight CR & Winship C (2014). The Causal Implications of Mechanistic Thinking: Identification Using Directed Acyclic Graphs (DAGs), in Morgan (ed.): *Handbook of Causal Analysis for Social Research*, Dordrecht: Springer: 275–300.

Lawson T (2003). *Reorienting Economics*. London and New York: Routledge.

Link BG & Phelan J (1995). Social Conditions as Fundamental Causes of Disease. *Journal of Health and Social Behavior* 35 (Extra Issue), 80–94.

Little D (1991). *Varieties of Social Explanation: An Introduction to the Philosophy of Social Science*. Boulder, CO: Westview.

Little D (2012). Explanatory Autonomy and Coleman's Boat. *Theoria* 74, 137–51.

Macy M & Flache A (2009). Social Dynamics from the Bottom Up: Agent-Based Models of Social Interaction. In Hedström and Bearman (eds.): *The Oxford Handbook of Analytical Sociology*. Oxford: Oxford University Press: 245–268.

Mahoney J (2001). Review Essay: Beyond Correlational Analysis: Recent Innovations in Theory and Method. *Sociological Forum* 16, 575–93.

Miller JH & Page SE (2007). *Complex Adaptive Systems: An Introduction to Computational Models of Social Life*. Princeton, NJ: Princeton University Press.

Norkus Z (2005). Mechanisms as Miracle Makers? The Rise and Inconsistencies of the "Mechanismic Approach" in Social Science and History. *History and Theory* 44: 348–72.

Schelling T (1978). *Micromotives and Macrobehavior*. New York: W.W. Norton.

Squazzoni F (2012). *Agent-Based Computational Sociology*. Chichester: John Wiley & Sons.

Steel D (2008). *Across the Boundaries: Extrapolation in Biology and Social Sciences*. Oxford: Oxford University Press.

Van Evera S (1997). *Guide to Methods for Students of Political Science*. Ithaca, NY: Cornell University Press.

Woodward J (2002). What Is a Mechanism? A Counterfactual Account. *Philosophy of Science* 69, S366–S377.

Ylikoski P (2009). The Illusion of Depth of Understanding in Science. In De Regt, Leonelli & Eigner (eds.): *Scientific Understanding: Philosophical Perspectives*. Pittsburgh: Pittsburgh University Press: 100–119.

Ylikoski P (2013). Causal and Constitutive Explanation Compared. *Erkenntnis* 78, 277–97.

Ylikoski P (2014). Agent-Based Simulation and Sociological Understanding. *Perspectives on Science* 22, 318–35.

Ylikoski P (2016). *Thinking with the Coleman Boat*. The IAS Working Paper Series, 1, http://urn.kb.se/resolve?urn=urn:nbn:se:liu:diva-132711.

Ylikoski P & Aydinonat E (2014). Understanding with Theoretical Models. *Journal of Economic Methodology* 21, 19–36.

31

DISAGGREGATING HISTORICAL EXPLANATION

The move to social mechanisms

Daniel Little

1. The covering-law model

Carl Hempel published his chief contribution to the philosophy of history in 1942, nearly 75 years ago. The article is "The Function of General Laws in History" (Hempel 1942), and it set the stage for several fruitless decades of debate within analytic philosophy about the nature of historical explanation. Hempel argued that all scientific explanations have the same logical structure: a deductive (or probabilistic) derivation of the explanandum from one or more general laws and one or more statements of fact. Explanation, in Hempel's view, simply is derivation of the explanandum from general laws. He is emphatic, moreover, in insisting that valid explanations in history *must* have this form:

> We have tried to show that in history no less than in any other branch of empirical inquiry, scientific explanation can be achieved only by means of suitable general hypotheses, or by theories, which are bodies of systematically related hypotheses.
>
> *(1942: 44)*

What kinds of general laws does Hempel think that historians have in the back of their minds when they offer elliptical explanations? He refers to regularities of individual or social psychology (40), regularities of collective behavior ("groups migrate to regions which offer better living conditions"), or at the macro level, regularities linking growing discontent to the outbreak of revolution (41).

Hempel concedes the point that few existing historical explanations actually look like this, with explicit law statements embedded in a deductive argument; but he argues that this shows only that existing explanations are elliptical, incomplete, or invalid. And often, he finds, what is offered as a historical explanation is in fact no more than an "explanation sketch" (42), with placeholders for the general laws.

This set of assumptions leads to insolvable problems for historical explanation if we accept Hempel's account, however, because it is hard to think of a real historical research question where there might be a set of social or individual regularities sufficient to deductively entail the outcome. Bluntly, the social and behavioral sciences have never produced theories of individual or collective behavior that issue in statements of general laws that could be the foundation for a covering-law explanation of an historical event or pattern. And given that social phenomena are formed by actors with a range of features of agency and decision-making, we have very good reason to think that this lack of regularities is inherent in the social world. The social world is simply not governed by a set of social or individual laws.

So the strong governing laws that would be needed for a covering-law explanation do not exist in the social domain. There are regularities and generalizations, but they are small in scope and do not aggregate through deduction to the derivation of regularities among large events or structures. Social regularities are phenomenal, not governing; they reflect characteristics of the actors rather than governing the behavior of the ensembles (Little 1993). This does not this mean that historical explanation is impossible. But we need to turn our attention from large social regularities to causal mechanisms and powers to see what a good historical explanation looks like. (Philosophers of social science who continue to maintain that social explanation depends upon the discovery of generalizations include (Kincaid 1990, 1996; McIntyre 1996).) More generally, the dominant quantitative paradigm for sociological research depends crucially on the idea that social explanation requires the discovery of robust correlations and associations (Abbott 1998).

Instead of imagining the social world in analogy with the law-governed world of natural phenomena, some philosophers of history propose an approach to social science theorizing that emphasizes agency, contingency, and plasticity in the makeup of social facts (Little 2010). This approach recognizes that there is a degree of pattern in social life—but emphasizes that these patterns fall far short of the regularities associated with laws of nature. It emphasizes contingency of social processes and outcomes. It insists upon the importance and legitimacy of eclectic use of social theories: the processes are heterogeneous, and therefore it is appropriate to appeal to different types of social theories as we explain social processes. It emphasizes the importance of path-dependence in social outcomes. It suggests that the most valid scientific statements in the social sciences have to do with the discovery of concrete social-causal mechanisms, through which some types of social outcomes come about.

McAdam, Tarrow, and Tilly illustrate this approach in their treatment of contentious politics in *Dynamics of Contention* (McAdam, Tarrow, and Tilly 2001), and the field of contentious politics is in fact highly suitable to the mechanisms approach. There are numerous clear examples of social processes instantiated in groups and organizations that play into a wide range of episodes of contention and resistance—the mechanics of mobilization, the processes that lead to escalation, the communications mechanisms through which information and calls for action are broadcast, the workings of organizations. So when we are interested in discovering explanations of the emergence and course of various episodes of contention and resistance, it is both plausible and helpful to seek out the specific mechanisms of mobilization and resistance that can be discerned in the historical record.

These considerations suggest that a good historical explanation identifies a number of independent mechanisms and processes that are at work in a particular circumstance, and then demonstrates how these mechanisms, and the actions of the actors involved, lead to the outcome. Historical explanations typically take the form of causal narratives linking actions, circumstances, and consequences leading up to the outcome to be explained.

2. Social mechanisms

The social mechanisms approach to explanation (SM) has filled a very important gap in the theory of social explanation in the past twenty years, between the covering-law model and merely particularistic accounts of specific events. The SM approach is particularly prominent in the emerging program of analytical sociology (Hedström and Swedberg 1996) and critical realism (Bhaskar 1975), but has made its mark in comparative historical sociology and other areas of the social sciences as well (Steinmetz 2004; Gorski 2013; Gross 2009; McAdam et al. 2001).

The core of this approach is the idea that social explanation requires discovery of the under-lying causal mechanisms that give rise to outcomes of interest (Glennan 1996; Cartwright 1999; Cummins 2000; Machamer et al. 2000; Woodward 2003; Hedström 2005; Craver and Darden 2013). On this perspective, explanation of a phenomenon or regularity involves identifying the specific causal processes and causal relations that underlie this phenomenon or regularity, and causal mechanisms are heterogeneous (see Chapter 12). Social mechanisms are concrete social processes in which a set of social conditions, constraints, or circumstances combine to bring about a given outcome (see Chapter 30). A strong example of this approach is to be found in *Dynamics of Contention* where McAdam, Tarrow, and Tilly place primary emphasis on causal mechanisms and processes as the underlying factors that serve to explain complex episodes of social contention (McAdam et al. 2001: 30). On this approach, social explanation does not take the form of inductive discovery of laws; rather, the generalizations that may be discovered in the course of social science research are subordinate to the more fundamental search for causal mechanisms and pathways in individual outcomes and sets of outcomes. This approach casts doubt on the search for generalizable theories across numerous societies. It looks instead for specific causal influence and variation. The approach emphasizes variety, contingency, and the availability of alternative pathways leading to an outcome, rather than expecting to find a small number of common patterns of development or change. The contingency of particular path-ways derives from several factors, including the local circumstances of individual agency and the across-case variation in the specifics of institutional arrangements—giving rise to significant variation in higher-level processes and outcomes.

There is an important ontological assumption underlying the concept of a mechanism—the idea that there is a substrate that makes the mechanism work. By referring to a nexus between I and O as a "mechanism," we presume that there is some underlying domain of things and powers that makes the observed regularity a "necessary" one: given how the world works, the input I brings about events that lead to output O. In evolutionary biology it is the specifics of organisms within an ecology conjoined with natural selection. Crucial to a valid understand-ing of social mechanisms is a general answer to the question: what ontological features of the social world facilitate and empower the causal connection from one end of a social mechanism to another? What stands in the place of "causal powers of natural entities" in the social world?

In the social world it is the concrete situation of the actor and the social and natural envi-ronment in which he/she acts. The causal substrate of social mechanisms is the domain of facts about intentional agents, socially situated in embodied social relations, that constitute the motive power of social causation. This corresponds to the ontology of "methodological localism" and actor-centered sociology I have developed elsewhere (Little 2006, 2009, 2014).

On this approach, the causal capacities of social entities are to be explained in terms of the structuring of preferences, worldviews, information, emotions, incentives, and opportunities for agents. The causal properties of a social entity inhere in its power to affect individuals' behavior through incentives, preference-formation, belief-acquisition, or powers and oppor-tunities. The micro-mechanism that conveys cause to effect is supplied by an account of the actions of agents with specific goals, beliefs, and powers, within a specified set of institutions, rules, and normative constraints.

The framework of social mechanisms as a basis for social explanation raises an important question about the role and scope of generalizability that we can expect from a social explana-tion. Briefly, the mechanisms identified by McAdam, Tarrow, and Tilly show a limited but real degree of generalizability; as they assert, social mechanisms can be expected to recur in other circumstances and times. But the event itself is one of a kind. This is a familiar feature of Tilly's way of thinking about contentious episodes as well: the American Civil War was a singular historical event. But a good explanation will invoke mechanisms that recur in other contentious settings as well. We should not expect to find general theories of civil wars; but our explanations of particular civil wars can invoke quasi-general theories of mid-level mechanisms of conflict, mobilization, and escalation.

Are there any social mechanisms? In fact, a cursory survey of comparative sociology, political science, and the new institutionalism provides a very large body of explanations that identify common social mechanisms available for historical explanations—for example, "collective action problems often cause strikes to fail," "increasing demand for a good causes prices to rise for the good in a competitive market," "transportation systems cause shifts of social activity and habitation." Here is a short but suggestive list: public goods problems (Hardin 1982); political entrepreneurship (Bates 1981); principal-agent problems (Ensminger 1992); features of ethnic or religious group mobilization (Hardin 1995); market mechanisms and failures (Akerlof 1970); rent-seeking behavior (Seligson and Passé-Smith 1993); mechanisms of corruption (Klitgaard 1988); the social psychology of race (Steele and Aronson 1995). The literature now includes examples of dozens of well-developed models of social mechanisms that can be theorized and observed in concrete social settings.

These are all mechanisms that work at the level of socially situated actors within extended institutions and organizations. They characterize one or more of the features of agency and structure in concrete social circumstances. We understand how they work; individuals with specified motivational and cognitive characteristics, placed within the context of the social settings identified by the mechanisms, will behave in ways that bring about the outcome. And they are abstract enough that they can be identified in a wide range of settings: the feudal manor, the collective farm, the Wall Street law firm. In fact, we might say (along with Robert Merton (1967)) that the most fundamental value of theories in the social sciences is the formulation of models of mechanisms at this level—theories of the middle range.

We might attempt to classify mechanisms according to a variety of criteria, as Carl Craver and Lindley Darden have done for biology (Craver and Darden 2013) and Andrew Bennett has begun to do for international relations theory (Bennett 2014; see also Chapter 7, this volume). One approach is to classify mechanisms according to what they do—the level of action and process within which they operate and the kinds of things they influence or bring about (Little 2015). Some mechanisms influence individual behavior directly; others influence the formation of the individual; others provide pathways of aggregation from individual to collective; and so forth. These questions imply a handful of regions of activity for social mechanisms: for example, mechanisms affecting the neuro-cognitive system; mechanisms affecting action and deliberation; mechanisms affecting identity formation; mechanisms of institutional influence on individuals; mechanisms of aggregation from individual to social; mechanisms occurring within the domain of social action and collective action; mechanisms of hierarchy and control; and mechanisms of higher-level causal influences. The key point is that mechanisms are heterogeneous, deriving from different causal properties and powers, and playing a causal role in different parts of the social whole. It is worthwhile to try to provide a preliminary way of classifying the mechanisms that arise in social and historical explanations.

Emphasis on causal mechanisms for adequate social explanation has several beneficial effects for historians. It provides a credible alternative to the covering-law model of explanation. It takes us away from easy reliance on uncritical statistical models. It makes it apparent why historians and social scientists benefit from collaboration. And it also leads us away from excessive emphasis on large-scale classification of events into revolutions, democracies, or religions, and toward more specific analysis of the processes and features that serve to discriminate among instances of large social categories (Tilly 1995). This view parallels arguments offered by Stuart Glennan on the "ephemeral" nature of historical causes (Glennan 2010).

3. Causal narratives

The idea of causal mechanisms fits very well into a characteristic mode of historical writing, the form of a causal narrative. Essentially the idea is that a causal narrative of a complicated outcome or occurrence is an orderly analysis of the sequence of events and the causal processes that connected them, leading from a set of initial conditions to the outcome in question. The narrative pulls together our best understanding of the causal relations, mechanisms, and conditions that were involved in the process and arranges them in an appropriate temporal order. It is a series of answers to "why and how did X occur?" designed to give us an understanding of the full unfolding of the process. A narrative is more than an explanation; it is an attempt to "tell the story" of a complicated outcome. So a causal narrative will include a number of causal claims, intersecting in such a way as to explain the complex event or process that is of interest. A key part of the justification of a causal narrative is the existence of well-developed empirical theories of the social mechanisms that are invoked, and these theories emerge from several of the social sciences. (There is a more extensive discussion of causal narratives in *New Contributions to the Philosophy of History* (Little 2010).)[1]

We might illustrate this idea by looking at the approach taken to contentious episodes and periods by McAdam, Tarrow, and Tilly (2001). In their treatment of various contentious periods, they break the given complex period of contention into a number of mechanisms and processes, conjoined with contingent and conjunctural occurrences that played a significant causal role in the outcome. The explanatory work that their account provides occurs at two levels: the discovery of a relatively small number of social mechanisms of contention that recur across multiple cases, and the construction of complex narratives for particular episodes that bring together their understanding of the mechanisms and processes that were in play in this particular case.

The narrative for a particular case (the Mau Mau uprising, for example) takes the form of a chronologically structured account of the mechanisms that their analysis identifies as having been relevant in the unfolding of the insurgent movement and the government's responses. McAdam, Tarrow, and Tilly give attention to "episodes" within larger processes, with the clear implication that the episodes are to some degree independent from each other and are amenable to a mechanisms analysis themselves. So a narrative is both a concatenated series of episodes and a nested set of mechanisms and processes. (Here again the idea of ephemeral mechanisms is pertinent to the social case (Glennan 2010).)

In short, a narrative describes a particular process or event; but it does so by identifying recurring processes, mechanisms, and forces that can be discerned within the unfolding of the case. So generalizability comes into the story at the level of the components of the narrative—the discovery of common social processes within the historically unique sequence of events.

4. Social variation and mechanisms

Variation within a social or historical phenomenon is all but ubiquitous in history and in the contemporary world. Think of the Cultural Revolution in China, demographic transition in early modern Europe, the ideology of a market society, or the experience of being black in the

United States. We have the noun—"Cultural Revolution"—which can be explained or defined in a sentence or two as an extended social phenomenon of mobilization and conflict that took place in China from 1966–76; and we have the complex underlying social realities to which it refers, spread out over many cities, villages, and communes across China (Esherick et al. 2006). The Cultural Revolution consisted of a myriad of local and regional moments of political mobilization and conflict, with complex relationships to national party organizations as well. So the Cultural Revolution was in fact a highly heterogeneous social and political phenomenon.

This description focuses on locational variation in processes—village to village, country to country. But social scientists often also highlight variations across social segments *within* a given location: class, race, gender, religion, occupation. Do poor sharecroppers have a different fertility profile over time than the wealthy in a particular region at a particular time? Are there significant differences in survival strategies for distinct groups defined by race or ethnicity in a city or a group of cities? Detailed investigations almost always reveal substantial variations across social phenomena like these.

Careful study of the causal and social mechanisms giving rise to historical phenomena in particular settings is a crucial means for understanding and explaining the facts of historical variation. How did the activism and ideology of Cultural Revolution spread from Beijing to Nanjing and other locations? How did activism spread from city to rural locations? How did youth organizations mobilize their followers? How did local circumstances cause changes and variations in the political movement? How much path dependency existed in the spread of revolutionary ideas and strategies? The answers to all these questions most naturally take the form of a sketch of one or more underlying social mechanisms.

This is another place where the appeal to social mechanisms can be seen once more to be highly relevant and helpful to historical research. If we work on the assumption that any large social process—the dispersed locations of contention associated with the Cultural Revolution, say—is the compound result of a set of underlying causal social mechanisms, and if we hypothesize that many of these mechanisms are in play in some places but not in others, then we can explain both similarity and difference in the occurrence of the phenomenon across time and place. Now the work of historical investigation can be put in these terms: identify some of the social mechanisms that evidently recur in various locations; identify some of the mechanisms that lead to significantly different results in some places; and identify some of the cross-location mechanisms that are at work to secure a degree of synchrony and parallel in the developments observed in different locations (communication systems, networks of leaders, dissemination of activists). Case studies and comparative research permit both a degree of generalization and an explanation of variation.

In other words, the intellectual strategy here is to disaggregate the large social factor into the results of a larger number of underlying mechanisms; and then to attempt to discover how these mechanisms played out differently in different settings throughout the range of the Cultural Revolution, protoindustrialization, or ethnic conflict in South Asia. Significantly, as we have seen above, this is exactly the strategy of research and explanation that Charles Tilly and his colleagues were led to in their emphasis on discovering the component social mechanisms that underlie social contention (McAdam et al. 2001).

5. Mechanisms and predictions?

To what extent is it possible to predict the course of large-scale history—the rise and fall of empires, the occurrence of revolution, the crises of capitalism, or the ultimate failure of twentieth-century Communism (see also Chapter 13)? One possible basis for predictions is the availability of theories of underlying processes. To arrive at a supportable prediction about a state of affairs, we might possess a theory of the dynamics of the situation, the mechanisms and processes that interact to bring about subsequent states, and we might be able to model the future effects of those mechanisms and processes. A biologist's projection of the spread of a disease through an isolated population of birds is an example. Or, second, predictions might derive from the discovery of robust trends of change in a given system, along with an argument about how these trends will aggregate in the future. For example, we might observe that the population density is rising in water-poor southern Arizona, and we might predict that there will be severe water shortages in the region in a few decades. However, neither approach is promising when it comes to large historical change. There are strong reasons for doubting the availability of long-term predictions in history.

One reason for the failure of large-scale predictions about social systems is the complexity of causal influences and interactions within the domain of social causation. We may be confident that X causes Z when it occurs in isolated circumstances. But it may be that when U, V, and W are present, the effect of X is unpredictable, because of the complex interactions and causal dynamics of these other influences. This is one of the central findings of complexity studies—the unpredictability of the interactions of multiple causal powers whose effects are non-linear. It is the feature of social life that Roy Bhaskar designates as "open" rather than "closed" (Bhaskar 1975).

Another difficulty is the typical fact of path dependency of social processes. Outcomes are importantly influenced by the particulars of the initial conditions, so simply having a good idea of the forces and influences the system will experience over time does not tell us where it will wind up. There are too many contingent causal influences along the way to allow us to argue that a given future scenario is substantially more likely than dozens of others.

Third, social processes are sensitive to occurrences that are singular and idiosyncratic and not themselves governed by systemic properties. If the winter of 1812 had not been exceptionally cold, perhaps Napoleon's march on Moscow might have succeeded, and the future political course of Europe might have been substantially different. But variations in the weather are not themselves systemically explicable—or at least not within the parameters of the social sciences.

Fourth, social events and outcomes are influenced by the actions of purposive actors. So it is possible for a social group to undertake actions that avert the outcomes that are otherwise predicted. Take climate change and rising ocean levels as an example. We may be able to predict a substantial rise in ocean levels in the next fifty years, rendering existing coastal cities largely uninhabitable. But what should we predict as a consequence of this fact? Societies may pursue different strategies for evading the bad consequences of these climate changes—retreat, massive water control projects, efforts at atmospheric engineering to reverse warming. And the social consequences of each of these strategies are widely different. So the acknowledged fact of global warming and rising ocean levels does not allow clear predictions about social development.

For these and other reasons, it is difficult to have any substantial confidence in predictions of the large course of change that a society, cluster of institutions, or population will experience. History is contingent, and there are always alternative pathways that might have been taken, and history has no general plan. So, no grand predictions in history.

But then we have to ask a different question. What kinds of predictions or projections *are* possible in history? Analysis of existing causal mechanisms does in fact provide a modest basis for limited predictions about the future. Consider a few examples: labor unrest will intensify in China in the next ten years; large technology disasters will occur in Europe; conflicts over land use in East Jerusalem will continue to deepen in the coming decade. Each of these predictions is credible because we can identify salient mechanisms leading to the outcome, and we can be reasonably sure that they will play out as expected.

Several things are apparent when we consider these predictions. First, they are limited in scope; they have to do with small-scale features of the historical drama. Second, they depend on specific and identifiable social circumstances, along with clear ideas about social mechanisms connecting the present to the future. Third, they are at least by implication probabilistic; they indicate likelihoods rather than inevitabilities. Fourth, they imply the existence of *ceteris paribus* conditions: "Absent intervening factors, such-and-so is likely to occur." But, finally, they all appear to be intellectually justifiable. They may not be true, but they can be grounded in an empirically and historically justified analysis of the mechanisms that produce social change, and a model projecting the future effects of those mechanisms in combination.

The heart of prediction is our ability to identify dynamic processes and mechanisms that are at work in the present, and our ability to project their effects into the future. *Modest* predictions are those that single out fairly humdrum current processes in specific detail, and derive some expectations about how these processes will play out in the relatively short run. *Grand* predictions, on the other hand, purport to discover wide and encompassing patterns of development and then to extrapolate their civilizational consequences over a very long period. A modest prediction about China is the expectation that labor protest will intensify over the next ten years. A grand prediction about China is that it will become the dominant economic and military superpower of the late twenty-first century. We can have a fair degree of confidence in the first type of prediction, whereas there are vastly too many possible branches in history, too many countervailing tendencies, too many accidents and contingencies that may occur to give us any confidence in the latter prediction.

Ceteris paribus conditions are unavoidable in formulating historical expectations about the future, because social change is inherently complex and multi-causal. So even if it is the case that a given process, accurately described in the present, creates a tendency for a certain kind of result, it remains the case that there may well be other processes at work that will offset this result. The tendency of powerful agents to seize opportunities for enhancing their wealth through processes of urban development implies a certain kind of urban geography in the future; but this outcome might be offset by a genuinely robust and sustained citizens' movement at the level of local politics.

The idea that historical predictions are generally *probabilistic* is partly a consequence of the fact of the existence of unknown *ceteris paribus* conditions. But it is also, more fundamentally, a consequence of the fact that social causation itself is almost always probabilistic. If we say that rising conflict over important resources (X) is a cause of inter-group violence (Y), we don't mean that X is necessarily followed by Y; instead, we mean that X raises the likelihood of the occurrence of Y (see Chapter 13).

So two conclusions seem justified. First, there is a perfectly valid intellectual role for making historical predictions. Indeed, this is the basis of all policy reasoning. But these need to be modest predictions: limited in scope, closely tied to theories of existing social mechanisms, and accompanied by *ceteris paribus* conditions. And second, grand predictions should be treated with great suspicion. At their best, they depend on identifying a few existing mechanisms and processes; but the fact of multi-causal historical change, the fact of the compounding of uncertainties, and the fact of the unpredictability of complex systems should all make us dubious about large and immodest claims about the future.

6. Conclusion

Seventy years after Hempel's classic article, the covering-law theory is now generally regarded as a fundamentally wrong-headed way of thinking about historical (and social) explanation. Logical positivism is not a convenient lens through which to examine the social and historical sciences. There is too much contingency in the social world. Rather than being the result of law-governed processes, social outcomes proceed from the contingent and historically variable features of the actors and situations who make them. So the attention of many researchers interested in specifying the nature of historical and social explanation has focused on social mechanisms constituted and driven by common features of agency. This results in a different kind of explanation: accounts of particular episodes that shed light on the causal processes that appear to have been involved in their production, but no general accounts of large-scale historical patterns or outcomes.

Analysis of concrete causal mechanisms provides a basis for a degree of generality in historical inquiry. This is true in several ways. First, explanations based on social mechanisms can take place in both a generalizing and a particular context. We can explain a group of similar social outcomes by hypothesizing the workings of a common causal mechanism giving rise to them; and we can explain a unique event by identifying the mechanisms that produced it in the given unique circumstances. Second, a social-mechanism explanation relies on a degree of lawfulness; but it refrains from the strong commitments of the deductive-nomological method. Mechanistic causation is inherently conjunctural causation (Little 2000), in that a given outcome is almost invariably the result of multiple causal influences and mechanisms. There are no high-level social regularities. Third, we can refer both to particular individual mechanisms and a class of similar mechanisms. For example, the situation of "easy access to valuable items along with low probability of detection" constitutes a mechanism leading to pilferage and corruption. We can invoke this mechanism to explain a particular instance of corrupt behavior—a specific group of agents in a business who conspire to issue false invoices—or a mid-level general fact—the logistics function of a large military organization which shows itself to be prone to repeated corruption. So mechanistic explanations support a degree of generalization across instances of social activity while equally reflecting the facts of historical uniqueness and contingency.

Note

1 Robert Bates introduces a similar idea under the rubric of "analytic narrative" (1998). The chief difference between his notion and mine is that his account is limited to the use of game theory and rational choice theory to provide the linkages within the chronological account, whereas I want to allow a pluralistic understanding of the kinds and levels of causes that are relevant to social processes.

References

Abbott, A. 1998. "The causal devolution." *Sociological Methods and Research* 27(2): 148–181.

Akerlof, G. 1970. "The market for 'lemons': quality, uncertainty and the market mechanism." *Quarterly Journal of Economics* 84: 488–500.

Bates, R. H. 1981. *Markets and states in tropical Africa: the political basis of agricultural policies*. Berkeley, CA: University of California Press.

Bates, R. H. 1998. *Analytic narratives*. Princeton, NJ: Princeton University Press.

Bennett, A. 2014. "The mother of all isms: causal mechanisms and structured pluralism in international relations theory." *European Journal of International Relations* 9(3): 459–481.

Bhaskar, R. 1975. *A realist theory of science*. Leeds: Leeds Books.

Cartwright, N. 1999. *The dappled world: a study of the boundaries of science*. Cambridge: Cambridge University Press.

Craver, C. F. and L. Darden. 2013. *In search of mechanisms: discoveries across the life sciences*. Chicago: The University of Chicago Press.

Cummins, R. 2000. "'How does it work?' vs. 'What are the laws?': Two conceptions of psychological explanation." In *Explanation and cognition*. F. Keil and R. Wilson. Cambridge, MA: MIT Press: 117–144.

Ensminger, J. 1992. *Making a market: the institutional transformation of an African society*. Cambridge: Cambridge University Press.

Esherick, J., P. Pickowicz, and A. G. Walder. 2006. *The Chinese cultural revolution as history: studies of the Walter H. Shorenstein Asia-Pacific Research Center*. Stanford, CA: Stanford University Press.

Glennan, S. 1996. "Mechanisms and the nature of causation." *Erkenntnis* 44: 49–71.

Glennan, S. 2010. "Ephemeral mechanisms and historical explanation." *Erkenntnis* 72: 251–266.

Gorski, P. 2013. "What is critical realism? And why should you care?" *Contemporary Sociology* 42(5): 658–670.

Gross, N. 2009. "A pragmatist theory of social mechanisms." *American Sociological Review* 74(3): 358–379.

Hardin, R. 1982. *Collective action*. Baltimore: The Johns Hopkins University Press.

Hardin, R. 1995. *One for all: the logic of group conflict*. Princeton, NJ: Princeton University Press.

Hedström, P. 2005. *Dissecting the social: on the principles of analytical sociology*. Cambridge: Cambridge University Press.

Hedström, P. and R. Swedberg. 1996 "Social mechanisms." *Acta Sociologica* 39: 281–308.

Hempel, C. 1942. "The function of general laws in history." *Journal of Philosophy* 39(2): 35–48.

Kincaid, H. 1990. "Defending laws in the social sciences." *Philosophy of the Social Sciences* 20(1): 56–83.

Kincaid, H. 1996. *Philosophical foundations of the social sciences: analyzing controversies in social research*. Cambridge: Cambridge University Press.

Klitgaard, R. E. 1988. *Controlling corruption*. Berkeley, CA: University of California Press.

Little, D. 1993. "On the scope and limits of generalizations in the social sciences." *Synthese* 97: 183–207.

——. 2000. "Explaining large-scale historical change." *Philosophy of the Social Sciences* 30(1): 89–112.

——. 2006. "Levels of the social." In *handbook for philosophy of anthropology and sociology*. S. Turner and M. Risjord. Amsterdam: Elsevier Publishing: 343–371.

——. 2009. "The heterogeneous social." In *Philosophy of the social sciences: philosophical theory and scientific practice*. C. Mantzavinos. Cambridge: Cambridge University Press: 154–178.

——. 2010. *New contributions to the philosophy of history*. Dordrecht: Springer Science.

——. 2011. "Causal mechanisms in the social realm." In *Causality in the sciences*. P. Illari, F. Russo and J. Williamson. Oxford: Oxford University Press. DOI: 10.1093/acprof:oso/9780199574131.003.0013.

——. 2014. "Actor-centered sociology and the new pragmatism." In *Individualism, holism, explanation and emergence*. J. Zahle and F. Collin. Dordrecht, London, New York: Springer: 55–75.

——. 2015. "Classifying mechanisms by location." *Understanding society*. http://understandingsociety.blogspot.com/2014/08/classifying-mechanisms-by-location.html (Accessed 29 March 2017).

Machamer, P., et al. 2000. "Thinking about mechanisms." *Philosophy of Science* 67(1): 1–25.

Mahoney, J. 2001. "Beyond correlational analysis: recent innovations in theory and method." *Sociological Forum* 16(3): 575–593.

Mayntz, R. 2004. "Mechanisms in the analysis of social macro-phenomena." *Philosophy of the Social Sciences* 34(2): 237–259.

McAdam, D., S. G. Tarrow, and C. Tilly. 2001. *Dynamics of contention: Cambridge studies in contentious politics*. New York: Cambridge University Press.

McIntyre, L. C. 1996. *Laws and explanation in the social sciences: defending a science of human behavior*. Boulder, CO: Westview Press.

Merton, R. K. 1967. *On theoretical sociology*. New York: Free Press.

Seligson, M. A. and J. T. Passé-Smith, Eds. 1993. *Development and underdevelopment: the political economy of inequality*. Boulder, CO: L. Rienner Publishers.

Steel, D. 2004. "Social mechanisms and causal inference." *Philosophy of the Social Sciences* 34(1): 55–78.

Steele, C. M. and J. Aronson. 1995. "Stereotype threat and the intellectual test performance of African Americans." *Journal of Personality and Social Psychology* 69(5): 797–811.

Steinmetz, G. 2004. "Odious comparisons: incommensurability, the case study, and 'small N's' in sociology." *Sociological Theory* 22(3): 371–400.

Tilly, C. 1995. "To explain political processes." *American Journal of Sociology* 100: 1594–1610.

Woodward, J. 2003. *Making things happen: a theory of causal explanation*. New York: Oxford University Press.

32

MECHANISMS IN ECONOMICS[1]

Caterina Marchionni

1. Introduction

The market represents the paradigmatic example of an economic mechanism. Adam Smith famously theorized it as functioning as if led by an invisible hand so as to satisfy the needs of market participants. Over the years the market has been variously theorized as a mechanism *for* resource allocation, price discovery, assignment of property rights, and many other things besides (cf. Rosenbaum, 2000; Mirowski, 2007). At the same time, it has also come to be treated more and more abstractedly and transported far from the economic domain to become a mechanism *for* phenomena as diverse as mating behavior in animals, competition between churches, and marriage choices—instances of a wider trend known as *economics imperialism* (Mäki, 2009a). In spite of the centrality of the market mechanism, however, economics is not solely concerned with market-related phenomena. In fact, a recurring theme of this chapter will be that economics is distinct from the other social sciences not so much by virtue of the kind of real-world mechanisms (and phenomena) with which it deals, but because of the way in which mechanisms are identified and analyzed. What I hope to highlight is the role that economists' methodological commitments play in defining what counts as a mechanism in economics.

In economics the term *mechanism* has various uses. Julian Reiss (2013, pp. 104–5) identifies four different notions. The first is the econometricians' notion of mechanisms as individual causal relations. The contrast here is with mere correlation. The second refers to variables that intervene between a cause and an effect and, as Reiss observes, it is often used in the context of causal inference. The third takes mechanisms to be underlying structures or processes (for example, the market), while the fourth takes mechanisms to be pieces of theory (for example, a theoretical hypothesis showing the conditions under which the market clears). It is mainly the last two notions, to which I will simply refer as mechanisms as underlying structures, that come the closest to the conception of mechanisms advanced by current mechanistic philosophers. It is also the one with which I will be mainly concerned in this chapter, even though, as we will see, the notions of mechanisms as underlying structures and as intervening variables are not always kept clearly separate in philosophical discussions about causal inference and extrapolation (cf. Kincaid, 2004).

Mechanisms have been prominent in recent philosophical reflections on economics: they have been claimed to provide justification for methodological individualism, to be necessary for causal inference, and to aid extrapolation of causal claims from one context to another. In what follows, after giving a characterization of how mechanisms are conceived and represented in economics (sections 2 and 3), I discuss the alleged connection between mechanism and methodological individualism (section 4), and the role of mechanisms in causal inference from statistical data and in extrapolation (sections 5 and 6). Section 7 offers some concluding remarks.

2. What is a mechanism in economics?

Let us begin with a minimal definition proposed as a way of capturing the basic features of mechanisms that contemporary mechanistic philosophers would agree on. I also take it to characterize what Reiss calls *mechanisms as underlying structures*.

> [M] A mechanism for a phenomenon consists of entities and activities organized in such a way that they are responsible for the phenomenon.
>
> *(Illari and Williamson, 2012, p. 120; see also Chapter 1 and Glennan, forthcoming)*

Further features can be added to [M] to produce more specific accounts that restrict the scope of what kinds of things qualify as mechanisms. There are two ways in which [M] can be augmented to take into account the specificities of economics. The first concerns the *kind of mechanism* economics deals with, whereas the second concerns *how* economists identify and analyze mechanisms.

Dan Steel defines *social mechanisms* as follows:

> [SM] Social mechanisms are complexes of interacting individuals, usually classified into specific social categories, that generate causal relationships between aggregate-level variables.
>
> *(Steel, 2004, p. 59)*

Compared to [M], [SM] involves individuals as component parts, individuals that are typically classified into social categories, such as buyers and sellers, fathers and daughters, and who engage in certain kinds of activities (such as buying and selling, providing a dowry and marrying) by virtue of the social roles they occupy. Starting from [SM], which is arguably a general description including economic mechanisms as a subset, one obvious way to single out economic mechanisms is by virtue of their being about particular kinds of social roles, namely those pertaining to the market, or the economy more generally. This is, however, at most only a tiny part of the story. Not only is economics concerned with phenomena that do not clearly pertain to the economy, but the other social sciences are also interested in (say) markets and market-related phenomena.

Another way of reformulating [SM] so as to take into account the specificity of economics is to include the *type of assumptions* economists make about the behavior of individuals, namely those assumptions that derive from economists' commitment to rational-choice theory.[2] Although sometimes rational-choice theory is interpreted as being concerned exclusively with individuals and their properties, it often presupposes structural and institutional facts, and its "individuals" can also be firms, households, or organizations (see, for example, Kincaid, 1996; Janssen, 1993). To capture the latter feature, let us replace *individuals* with *rational agents* in [SM]. This gives us the following characterization of mechanisms in economics:

[EM] Mechanisms in economics are complexes of rational agents, usually classified into social categories, whose actions and interactions generate causal relationships between aggregate-level variables.

[EM] defines mechanisms on the basis of the kind of entities that compose them, namely agents, *and* the kind of properties ascribed to them, namely rational behavior. This is a descriptive (not a normative) claim about what economists (typically) take mechanisms to be and does not entail that this is what economic mechanisms really are. In what follows [EM] will be unpacked and related to some of the main debates concerning mechanisms in economics.

3. Theoretical modeling of mechanisms

What distinguishes economics from other social sciences is not only the kind of mechanisms economics deals in but also the way in which these mechanisms are studied. That is, mainly by building and analyzing simple models of mechanisms described at a high level of abstraction. This characteristic is captured by some of the most prominent accounts of models in economics. The connection between theoretical models and mechanisms features in Mäki's account, according to which economic models are means to isolate the operation of a mechanism from the interference of other factors (see, for example, Mäki 1992, 2009b).[3] Similarly, Cartwright (2001, see also her 1989) claims that economic models are "blueprints for socio-economic machines": by theoretical means models create the right conditions for mechanisms to operate unimpeded. Such conditions do not typically occur spontaneously in the real world, implying that the disturbing factors that in the models were isolated away will affect the mechanism's operation. Of course, not all economic models aim at representing mechanisms; some are better thought of as "phenomenological" models (see Chapter 17). Moreover, those models that can be conceptualized as isolating mechanisms might not succeed in actually representing any real-world mechanism.

In their modeling of mechanisms, economists also subscribe to a set of theoretical commitments and desiderata, which contributes to setting the modeling approach of economics apart from that of other sciences (Marchionni, 2013). First, the *mechanistic requirement* holds that the phenomenon to be explained should be shown to result from a mechanism that fits [EM] above. The legitimacy of the mechanistic requirement will be the topic of the next section. In particular, we will see that different interpretations of this requirement have different degrees of plausibility. A second desideratum economists emphasize relates to unification and requires that the mechanism be derived from a unifying theory; that is, rational-choice theory. Economists' commitment to rational-choice theory has been harshly criticized: rational-choice theory has been found to be either empirically wanting, at least as a theory of individual behavior, or empirically vacuous. Its credentials as a unifying theory are also suspicious. As Reiss (2013) points out, its flexibility rather than its content account for its unifying power. This brings us to the third desideratum that holds that, other things being equal, it is a good thing that the same kind of mechanism is shown to account for many phenomena. Since scope is typically a positive function of the level of abstraction at which a mechanism is described, the desideratum of generality leads to a preference for abstract descriptions of mechanisms.

Consider, for example, the application of Hotelling's model—in which firms choose where to locate spatially to maximize their market shares—to political parties choosing where to locate themselves in the political space to maximize the number of votes (Kuorikoski, 2009; see also Reiss, 2013). The main result of Hotelling's model of spatial localization, according to which firms will tend to locate close to one another, is also shown to account for the fact that political parties tend toward the center of the political spectrum. It is the abstract "logic of the situation" that is hypothesized to be similar, and hence to account for the similarity between the economic and political phenomenon (Kuorikoski, 2009).[4] This conception of mechanisms is compatible with the characterization of mechanisms in economics [EM], which in turn is compatible with the minimal definition [M]: in Hotelling's model, the components are the firms (or the political parties), who by virtue of their socio-economic roles perform activities that in interaction bring about the phenomenon to be explained.

The ease with which abstract descriptions of mechanisms can be transferred across domains has drawbacks. Very little might be inferred about the political market for votes on the basis of the set of features it shares in common with standard markets, while at the same time relevant features specific to each domain might be unduly ignored (Kuorikoski, 2009). Furthermore, the similarity between "situations" does not automatically warrant inferences about properties of the agents across domains (Kuorikoski, 2009). For example, if the assumption of maximizing behavior might be justified by the selection pressures the market exerts on firms, it is not necessarily the case that such behavior is legitimately attributed to political parties if similar selection pressures are absent or are counteracted by other institutional mechanisms. Finally, the strategy of building simple models of abstract mechanisms is likely to pose limits on the kind of phenomena economists would succeed in explaining. For some this is not a far-fetched possibility (see, for example, Lawson, 1997; Northcott and Alexandrova, 2015).

4. Methodological individualism

Economists' commitment to the doctrine of methodological individualism emerges with particular clarity from the belief that macroeconomics should be built on microeconomic foundations (Janssen, 1993; Hoover, 2001). The philosophical debate on micro foundations and methodological individualism has mainly concerned whether *individual-level mechanisms* are necessary for economic explanation and/or whether explanations that do include individual-level mechanisms are somehow better than purely macro-level explanations (Kincaid, 1996). To make the discussion more concrete, let us use a stylized example originally presented in Jackson and Pettit (1992), which I also discuss in Marchionni (2008).

Suppose that the phenomenon to be explained is an increase in the crime rate in a particular neighborhood. Such an increase can be explained in two ways. An aggregate-level explanation identifies a recent increase in the level of unemployment as the cause of the increase in crime rate. An individual-level explanation instead would describe the changes in the opportunities and motivations of particular individuals. As a thesis about explanation, methodological individualism would hold either that the aggregate-level explanation alone does not explain or that in any case the individual-level explanation is better.

There is a sense in which the aggregate-level explanation is deficient. What is missing is a description of the mechanism connecting crime and unemployment. But does such a mechanism always need to be at the individual level? Harold Kincaid (1996) has offered both conceptual and empirical arguments against the claim that individual-level mechanisms are necessary for explanation. In its strongest version, methodological individualism holds that underlying mechanisms must only cite individuals and their properties. This is a non-starter, however: as mentioned above, rational-choice theory, the allegedly individualist theory par excellence, is not concerned only with individual behavior and often presupposes social kinds (Janssen, 1993; Kincaid, 1996).

A weaker version of the argument linking methodological individualism, mechanism, and explanation takes it that individual-level explanations are somehow *better*. This idea, too, has been disputed. Compared to an aggregate-level explanation, the individual-level one, describing the changes in opportunities and motivations of particular individuals, misses relevant information, namely that irrespective of the behavior of particular individuals, an increase in unemployment would have brought about an increase in the crime rate (Jackson and Pettit, 1992; Garfinkel, 1981; see also Kincaid, 1996). These arguments show that neither an exclusively individual-level explanation nor an exclusively aggregate-level one is always to be preferred.[5]

Finally, an even weaker version of methodological individualism takes it that the comparison should not be between an explanation that simply relates the level of unemployment and the crime rate and one that only cites individuals, their motivations, and actions (Coleman, 1990; Janssen, 1993, Chapter 30). Instead, the issue is whether the aggregate-level explanation is improved by showing how an increase in the level of unemployment affects individuals and how this in turn causes the crime rate to increase. For example, suppose now that the direction of causality goes from crime rate to unemployment level and that the agents' choice of whether to engage in criminal activities as well as their opportunity to find a job is affected by the social networks in which they interact (Granovetter, 1973). In particular, suppose that agents are more likely to engage in criminal activities the more people around them do so, and are more likely to find jobs the more people around them are actually employed. It follows that an increase in the crime rate in a particular neighborhood makes it more likely for an individual to interact with a criminal than with someone who is employed and therefore can provide information about new jobs (Calvó-Armengol and Zenou, 2003, p. 71). This contributes to increasing the level of unemployment, which in turn contributes to increasing the crime rate. This is a mechanistic explanation, but the mechanism described is not a purely individual-level one. It describes how an aggregate variable (crime rate) affects another aggregate variable (unemployment level) via micro determinants (individuals' job search) (see Figure 32.1 and also Chapter 30).[6]

This style of explanation is compatible with current mechanistic approaches to explanation. Although (constitutive) mechanisms are at a lower level than the phenomenon to be explained, levels of mechanisms do not map onto traditional compositional ones characterized by mereological or aggregative relations (Chapter 14; Bechtel and Hamilton, 2007). Instead, "X's ϕ-ing is at a lower mechanistic level than S's ψ-ing if and only if ϕ-ing is a component in the mechanism for S's ψ-ing" (Craver, 2007, p. 189). This means that philosophical accounts of mechanistic explanation do not require that mechanisms in economics should be at the individual level. For some economic phenomena, the agents interacting in the mechanism can be firms, organizations, whole countries, or even subpersonal entities (Kincaid, 2004). Furthermore, in a mechanistic explanation the level of organization as well as the environment in which mechanisms are embedded are also important—in our example, these are the networks of relations

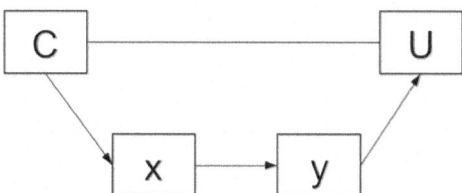

Figure 32.1 Coleman's boat. The mechanism connecting C (crime rate) to U (unemployment level), where x and y represent individual-level variables and the arrows represent causal relations

in which individuals are embedded and changes in the level of unemployment. If economists' mechanistic requirement is interpreted as demanding that economic phenomena be explained by representing the multi-level mechanisms that bring them about, then such a requirement can be justified along the lines proposed by current mechanistic philosophers.

5. Causal inference

Knowledge of mechanisms has been claimed to play a key role in making causal inferences from statistical data more secure. It has even been suggested that to distinguish genuine causal relations from mere correlations, knowledge of mechanisms is *necessary* (Elster, 1983). As an illustration, let us return to the crime-unemployment example. The idea is that knowledge of a connecting mechanism between the aggregate-level variables helps to identify whether the two variables are in fact causally connected, and the direction of causality, or whether they are both effects of a common unmeasured cause. What is under dispute is whether mechanisms are always necessary to identify genuine causal relations in the context of non-experimental research. Note that this claim has two interpretations: the quest for mechanisms can be interpreted as a quest for individual-level mechanisms or for lower-level mechanisms more generally.

Kincaid (1996, pp. 179–82) advances two objections to the first interpretation of this claim; that is, the necessity of *individual-level* mechanisms (see also Hoover, 2001; Reiss, 2008; Steel, 2004). The first is that there is no reason to stop at the level of individuals. If mechanisms are necessary to confirm causal relations, why shouldn't we go down the hierarchy of levels until we reach the rock bottom? Clearly such regress does not help the claim that individual-level mechanisms are necessary for causal inference in economics.[7] The second objection is that causal relationships can be identified with enough confidence through other means such as randomized controlled trials or statistical techniques, and Kincaid offers a few examples of this.

Even if we agree with Kincaid and others that evidence of individual-level mechanisms is not necessary for the confirmation of causal claims, it might still play a useful role. This is the position Steel (2004, 2011) advocates. The key to appreciate how evidence of mechanisms can help causal inference is to distinguish between *direct* and *indirect casual inference* (Steel, 2011). One of Steel's illustrations concerns the postulated causal relationship between the legalization of abortion in 1973 in the US and the decline of the crime rate in the 1990s (Donohue and Levitt, 2001). Information about the micro-level mechanism, from being an unwanted child to criminality, was marshaled as further evidence of the causal link between legalization of abortion and decline of the crime rate. Direct causal inference concerns the causal relationship between legalization of abortion and decline of crime rates, whereas the indirect causal inference concerns the causal relationship between being an unwanted child and criminal behavior. The reason why evidence of the causal relationship between unwanted childhood and criminal behavior is valuable is practical: it concerns the fact that we might be in a position to make a stronger inference about the variables in the mechanism than about the variables in the original relation. In this example, the direct casual inference concerns aggregate variables, whereas the indirect causal inference concerns the individual-level variables, but presumably the same logic applies if the variables in the mechanism were at the same level as those of the primary causal claim. Hence, I agree with Reiss that "it is not necessary that the 'mediating' variable obtains at a lower level than the original cause and effect variables" (2013, p. 104). I suspect the same logic applies more broadly in cases in which a primary causal claim is supported by evidence of different kinds (Claveau, 2012; Staley, 2004). If so, then it is unclear whether the relevant notion of mechanism here involves any form of reduction, not even in the broad sense of underlying or constitutive structures.[8]

The strategy of abstraction and simple models discussed above constitutes the most common source of mechanistic hypotheses in economics (see also Reiss, 2008, pp. 116–17).[9] Assessing the plausibility of these mechanistic hypotheses is ultimately an empirical matter. I agree with Steel that there is no one set of methodologies that uniquely supplies mechanistic evidence, which can be obtained by laboratory and field experimentation, or as in Donohue and Levitt's example, by correlational studies. In Donohue and Levitt's study, evidence in favor of the mechanistic hypothesis also came from studies in countries where for a time abortions had to be approved by the government. In these studies, children born from women who were denied the procedure were found to be more likely to engage in criminal behavior later on. Using such evidence in support of the causal claim about legalization of abortion and decline of criminality in the United States, however, involves a further inferential step—one concerning the relevant similarity between the countries in which the evidence was obtained and the United States. This is known as the problem of extrapolation, to which I now turn.

6. The problem of extrapolation

The problem of extrapolation concerns how to justify transporting causal claims from one context, for example a laboratory experiment or one country, to another, the real world or another country. Steel (2008) offers a comprehensive philosophical treatment of mechanism-based extrapolation. Mechanism-based extrapolation crucially involves the deployment of a methodology he calls *comparative process tracing*. The latter is a matter of comparing mechanisms in the model and in the target by focusing on stages where background knowledge tells us there are likely to be causally relevant differences and/or on those downstream stages where upstream differences are likely to have left a mark (see Figure 32.2).

Steel is optimistic that comparative process tracing can be used to justify extrapolation in biology, but he is more cautious with regard to economics (and social science more generally) for two reasons. The first is that for mechanism-based extrapolation to work, we need mechanisms that are modular, but policy interventions on some part of a mechanism might turn out to affect the mechanism's overall structure. The second is that often there is uncertainty about the mechanisms behind economic phenomena. Let us consider each problem in turn, starting from the latter.

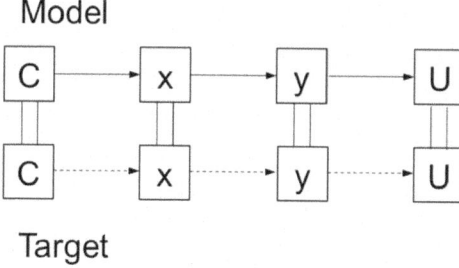

Figure 32.2 Comparative process tracing. Stages of the mechanisms leading from C (crime rate) to U (unemployment) in the model and in the target. The arrows represent causal relations, the dashed arrows represent relations to be inferred about the mechanism in target, and the double lines represent differences and similarities. Adapted from Guala, 2010, p. 1074

First, the viability of mechanism-based extrapolation in economics, Steel (2008) argues, is complicated by the uncertainty concerning what mechanisms are responsible for economic phenomena. For example, in spite of sustained and systematic experimental study since its discovery in the 1970s, preference reversal—"a behavioral tendency for the preference ordering of a pair of alternatives to depend, in a predictable way, on the process used to elicit it" (Starmer, 2008, p. 1)—still lacks a theoretical explanation. Yet, as Guala (2010) observes, uncertainty about the cause of preference reversal only tells us that further work is needed, not that uncertainty is an ineliminable and pervasive feature of economics experiments.

Second, that the effect of policy interventions on causal structure constitutes a problem for economics is captured in the well-known *Lucas critique*, which states that since agents' optimal behavior often changes in response to policy changes, many macroeconomic forecasts, which are based on assumptions about agents' optimal behavior, are bound to fail (Lucas, 1976). More generally, it might be the case that policy interventions on some part of a mechanism affect the mechanism's structure. There are two possible ways of addressing the difficulties posed by such "structure-altering interventions": an experimental and a theoretical one. At least in principle it is possible to design an economic experiment in which the intervention would alter the structure in the same way as the policy intervention, if implemented, would (Guala, 2010, p. 1079). The result of such an experiment could help us overcome the problem of structure-altering interventions and hence deploy mechanism-based extrapolation. Since large-scale interventions are hard to implement in the laboratory or in the field, and experimenting on a smaller scale would still entail uncertainty about the effect of an implementation on a larger scale, Steel's concern with structure-altering interventions remains a practical problem.

The second route, suggested by Steel (2008, p. 158), is to rely on a more fundamental theory to tell what kind of changes in causal structure the intervention is likely to produce. At present the most likely candidate in economics for such a theory is rational-choice theory.[10] Therefore, the issue turns on the appropriateness of rational-choice theory qua *fundamental* theory. In particular, Steel calls upon recent results in experimental economics, which show that small changes in variables that fall outside the domain of the theory have dramatic effects on behavior, casting doubts about whether rational-choice theory can help in anticipating the consequences of structure-altering interventions.

Steel's last conclusion might be too hasty, however. It can be argued that experimental results concerning individual behavior do not suffice to demonstrate that rational-choice theory cannot be relied upon to anticipate changes in causal structure. Rational-choice theory indeed need not be interpreted as a theory about individual behavior as such, but as a theory of individual behavior in settings in which it is supported by the right institutional scaffolding (compare with Satz and Ferejohn, 1994; Ross, 2014). Although the economics experiments that have attracted the most attention are those that demonstrate the existence of behavioral anomalies, these are not the only kind of experiments economists have been engaged with.

Santos (2007), for example, distinguishes between technological and behavioral experiments. Behavioral experiments are aimed at investigating individual behavior and have often been interpreted as yielding results that are at odds with rational-choice theory. By contrast, technological experiments are aimed at investigating institutional (market) mechanisms. Used as complements to the theoretical models developed in the field of mechanism design, technological experiments have guided many of the successful applications of game theory to the design of real-world markets such as the Federal Communications Commission auctions for the allocation of telecommunication licenses (Roth, 2002). Santos (2007) attributes the success of technological experiments to the robustness of the relation between the designed institution and aggregate outcomes to changes in the environment (most notably, preferences). In other words, the market institution is so designed so as to ensure that the resulting actions are rational and income maximizing.

Moreover, if what we need for mechanism-based extrapolation is information about the causal structure in the source and some relevant information about the causal structure in the target, there is no need for the mechanism to include individual-level variables—think, for example, of a causal chain between aggregate-level variables. Hoover (2001, 2009) can be interpreted as making a similar point when he argues that the Lucas critique does not necessarily imply the necessity of individual-level mechanisms. The Lucas critique holds that for estimated relationships to be stable across policy changes, those relationships need to capture the deep parameters in the economy, where deep parameters refer to "the fundamental ontological building blocks of the economy" (Hoover, 2009, p. 393). To go from here to the indispensability of individual-level mechanisms requires the further assumption that those parameters are necessarily micro, which is not obvious. As in the previous discussion about causal inference, it is unclear whether reduction, or underlying constitutive structures, is involved in the context of mechanism-based extrapolation. Rather, it seems that the broader notions of mediating variables and causal chains might be sufficient. This is not a critique of Steel's account of extrapolation—its main insight still stands regardless of the notion of a mechanism sufficient to get it off the ground. At this stage, the relevance of pointing at the possibility that intervening steps need not be at a lower level than the phenomenon to be explained only expands the range of cases in which extrapolation is legitimate.

7. Concluding remarks

Some of the main debates in the philosophy and methodology of economics are intertwined with one or another use of the notion of mechanism. Specifically, we have examined two such notions: mechanisms as (theoretical hypotheses about) underlying structures and mechanisms as intervening variables. I have shown that at least some economic models aim at capturing mechanisms conceived as abstract descriptions of how social-level phenomena result from the actions and interactions of rational agents, paralleling the idea of mechanisms as underlying structures. This modeling strategy is closely linked to economists' commitment to methodological individualism, even though on closer inspection mechanistic explanation does not square well with strong versions of methodological individualism, nor are such strong versions apt descriptions of the actual practice of economics. Furthermore, we have seen that although mechanisms are held to aid both causal inference and extrapolation, it is not always clear whether the relevant notion of mechanism at stake is that of mechanisms as underlying causal structures or as intervening variables.

It might then be that the centrality of the notion of mechanism is a product of the flexibility with which the notion itself is used rather than the role it actually plays in economics. If so, the traction of mechanistic ideas would improve by further clarity about what notion is at stake in a particular case. This is not to claim that the focus on mechanism in the philosophy of economics has had no value. On the contrary, reframing, for example, the age-old issue about methodological individualism in terms of mechanistic explanation contributed to make clear that although explaining by mechanisms is valuable, there is no reason to suppose that such mechanisms should be exclusively at the individual level. Similarly, attention to mechanisms has brought to the fore the function that different kinds of evidence have in both causal inference and extrapolation and, hence, the importance of different methods of generating evidence. This is especially topical now that the toolkit of empirical methodologies at the disposal of economists has been expanding.

Notes

1 I would like to thank the editors, Phyllis Illari and Stuart Glennan, and my colleagues Emrah Aydinonat, Jaakko Kuorikoski, Luis Mireles Flores, Samuli Pöyhönen, and Petri Ylikoski, for their helpful suggestions. All remaining mistakes are obviously mine.

2 According to some, the assumption that rational agents respond to incentives suffices to characterize the core idea behind economics. Landsburg (1993, p. 3), for example, writes: "Most of economics can be summarized in four words: 'People respond to incentives.' The rest is commentary."

3 Some commentators have claimed that economic model building is in fact at odds with a mechanistic world picture (e.g. Lawson, 1997). In rough outline, the criticism is that mainstream economics is committed to a form of deductivism, which presupposes that only events and regularities between them exist, and is unsuitable for the discovery of mechanisms.

4 "The logic of the situation" is part of Popper's method of situational analysis (1994), according to which explanation in social science proceeds by describing the situation the agent is in. Coupled with the rationality principle, the logic of the situation "explains" the agent's behavior.

5 Note that in this case the individual-level explanation involves particular individuals rather than types of individuals (Mäki, 2002).

6 Some wonder whether the explanatory strategy exemplified by Coleman's boat qualifies as a form of methodological individualism in that it allows for the effect of social structure. To distinguish this variety of methodological individualism from its stronger versions, some refer to it as "structural individualism" (see, for example, Udéhn, 2002).

7 Kincaid's objection succeeds only if mechanistic ideas are wedded to reductionism. Recent mechanistic accounts, however, are not reductionistic in this sense. A mechanism is always a mechanism *for* a phenomenon and that sets the boundaries of the relevant mechanism. Whether statistical evidence is sufficient to establish causation, however, is an object of current philosophical debate (see, for example, Russo and Williamson, 2007; Howick, 2011).

8 Kincaid (2004) distinguishes *horizontal* and *vertical* mechanisms, where the former are intervening variables, whereas the latter are the micro-processes that bring about the macro effect. Yet an underlying causal structure need not be a micro-process but can also involve social entities and activities that account for the system having a certain property or exhibiting a certain behavior. Similarly, Craver (2007) distinguishes *etiological* and *constitutive* mechanisms. A complete description of an etiological mechanism represents the entities, their activities and form of organization, a requirement that does not apply in the case of mechanisms as intervening variables. A clarification of the relation between these distinct yet similar notions is a topic for further research.

9 Steel (2008) advocates the method of process tracing, which consists in presenting evidence for the existence of several social practices that, when linked together, produce a chain of causation from one variable to another. The relation between process tracing and the type of mechanistic evidence implicated in indirect causal inference is not fully spelled out. However, since I take mechanistic evidence to be the more general category, in this chapter I only use this terminology.

10 Rational-choice theory is not the only possible fundamental theory for the social sciences. As I interpret it, however, Steel's point is that current alternatives have not yet reached the status currently enjoyed by rational-choice theory. This does not rule out the possibility that an alternative micro-theory may be developed that performs better in predicting the effects of structure-altering interventions (Steel, 2008, p. 160).

References

Bechtel, W & Hamilton, A 2007, "Reduction, integration, and the unity of science: natural, behavioral, and social sciences and the humanities", in TAF Kuipers (ed.), *General philosophy of science: focal issues*, Elsevier, New York, pp. 377–430.

Calvó-Armengol, A & Zenou, Y 2003, "Does crime affect unemployment? The role of social networks", *Annales d'Economie et de Statistique* vol. 71–2, pp. 173–88.

Cartwright, N 1989, *Nature's capacities and their measurement*, Clarendon Press, Oxford.

Cartwright, N 2001, "*Ceteris paribus* laws and socio-economic machines", in U Mäki (ed.), *The economic world view*, Cambridge University Press, Cambridge, pp. 275–92.

Claveau, F 2012, "The Russo-Williamson Theses in the social sciences: Casual inference drawing on two types of evidence", *Studies in History and Philosophy of Biological and Biomedical Sciences* vol. 43, pp. 806–13.

Coleman, JS 1990, *Foundations of social theory*, Harvard University Press, Cambridge, MA.

Craver, C 2007, *Explaining the brain*, Oxford University Press, Oxford.

Donohue, J & Levitt, S 2001, "The impact of legalized abortion on crime", *Quarterly Journal of Economics*, vol. 116, no. 2, pp. 379–420.

Elster, J 1983, *Explaining technical change: a case study in the philosophy of science*. Cambridge University Press, Cambridge.

Garfinkel, A 1981, *Forms of explanation*, Yale University Press, New Haven, CT.

Glennan, S forthcoming, *The new mechanical philosophy*, Oxford University Press, Oxford.

Granovetter, M 1973, "The strength of weak ties", *American Journal of Sociology* vol. 78, no. 6, pp. 1360–80.

Guala, F 2010, "Extrapolation, analogy, and comparative process tracing", *Philosophy of Science* vol. 77, no. 5, pp. 1070–82.

Hoover, K 2001, *Causality in macroeconomics*, Cambridge University Press, Cambridge.

Hoover, K 2009, "Microfoundations and the ontology of macroeconomics", in H Kincaid & D Ross (eds) *The Oxford handbook of philosophy of economics*, Oxford University Press, New York, pp. 386–409.

Howick, J 2011, "Exposing the vanities—and a qualified defense—of mechanistic reasoning in health care decision making", *Philosophy of Science* vol. 78, pp. 926–40.

Illari, P & Williamson, J 2012, "What is a mechanism? Thinking about mechanisms *across* the sciences", *European Journal for Philosophy of Science*, vol. 2, pp. 119–35.

Jackson, F & Pettit, P 1992, "In defense of explanatory ecumenism", *Economics and Philosophy*, vol. 8, 1–21.

Janssen, M 1993, *Micro-foundations: a critical inquiry*, Routledge, London.

Kincaid, H 1996, *Philosophical foundations of the social sciences*, Cambridge University Press, New York.

Kincaid, H 2004, "Contextualism, explanation and the social sciences", *Philosophical Explorations*, vol. 7, no. 3, pp. 201–18.

Kuorikoski, J 2009, "Two concepts of mechanism: componential causal system and abstract form of interaction", *International Studies in the Philosophy of Science*, vol. 23, no. 2, 143–60.

Landsburg, S 1993, *The armchair economist: economics and everyday life*, Free Press, New York.

Lawson, T 1997, *Economics and reality*, Routledge, London.

Lucas, RE 1976, "Econometric policy evaluation: a critique", in K Brunner & A Meltzer (eds.) *The Phillips curve and the labor market*, vol. 1, Carnegie-Rochester Conference on Public Policy, Amsterdam, North Holland, pp. 19–46.

Mäki, U 1992, "On the method of isolation in economics", *Poznan Studies in the Philosophy of the Sciences and the Humanities*, vol. 38, pp. 147–68.

Mäki, U 2002, "Explanatory ecumenism and economics imperialism", *Economics and Philosophy*, vol. 18, pp. 235–57.

Mäki, U 2009a, "Economics imperialism: concept and constraints", *Philosophy of the Social Sciences*, vol. 39, no. 3, pp. 351–80.

Mäki, U 2009b, "MISSing the world: models as isolations and credible surrogate systems", *Erkenntnis* vol. 70, no. 1, pp. 29–43.

Marchionni, C 2008, "Explanatory pluralism and complementarity: from autonomy to integration", *Philosophy of the Social Sciences*, vol. 38, pp. 314–33.

Marchionni, C 2013, "Playing with networks: how economists explain", *European Journal for Philosophy of Science*, vol. 3, no. 3, pp. 331–52.

Mirowski, P 2007, "Markets come to bits: evolution, computation and markomata in economic science", *Journal of Economic Behavior and Organization*, vol. 63, pp. 209–42.

Northcott, R & Alexandrova, A 2015, "Prisoner's dilemma doesn't explain much", in M Peterson (ed.), *The prisoner's dilemma*, Cambridge University Press, Cambridge, pp. 64–84.

Popper, K 1994, *The myth of the framework*, Routledge, London.

Reiss, J 2008, *Error in economics*, Routledge, London.

Reiss, J 2013, *Philosophy of economics: a contemporary introduction*, Routledge, London.

Rosenbaum, E-F 2000, "What is a market? On the methodology of a contested concept", *Review of Social Economy*, vol. 58, no. 4, pp. 455–82.

Ross, D 2014, *Philosophy of economics*, Palgrave Macmillan, New York.

Roth, A 2002, "The economist as engineer: game theory, experimentation, and computation as tools for design economics", *Econometrica*, vol. 70, no. 4, pp. 1341–78.

Russo, F & Williamson, J 2007, "Interpreting causality in the health sciences", *International Studies in the Philosophy of Science*, vol. 21, no. 2, pp. 157–70.

Santos, A 2007, "The 'materials' of experimental economics: technological versus behavioral experiments", *Journal of Economic Methodology*, vol. 14, no. 3, pp. 311–37.

Satz, D & Ferejohn, J 1994, "Rational choice and social theory", *Journal of Philosophy*, vol. 91, no. 2, pp. 71–87.

Staley, K 2004, "Robust evidence and secure evidence claims", *Philosophy of Science*, vol. 71, no. 4, pp. 467–88.

Starmer, C 2008, "Preference reversals", in SN Durlauf & LE Blume (eds.), *The new Palgrave dictionary of economics online*, 2nd edn, Palgrave Macmillan, London.

Steel, D 2004, "Social mechanisms and causal inference", *Philosophy of the Social Sciences*, vol. 34, pp. 55–78.

Steel, D 2008, *Across the boundaries: extrapolation in biology and the social sciences*, Oxford University Press, New York.

Steel, D 2011, "Causality, causal models, and social mechanisms", in IC Jarvie & J Zamora-Bonilla (eds), *The SAGE handbook of the philosophy of the social sciences*, SAGE Publications, London, pp. 288–304.

Steel, D 2013, "Mechanisms and extrapolation in the abortion-crime controversy", in H-K Chao, S-T Chen & R Millstein (eds.), *Mechanisms and causality in biology and economics*, Springer, New York, pp. 185–206.

Udéhn, L 2002, "The changing face of methodological individualism", *Annual Review of Sociology*, vol. 28, pp. 479–507.

Ylikoski, P 2012, "Micro, macro, and mechanisms", in H Kincaid (ed.), *The Oxford handbook of philosophy of the social sciences*, Oxford University Press, Oxford, pp. 21–45.

33

COMPUTATIONAL MECHANISMS[1]

Gualtiero Piccinini

For a long time, computation was seen as making the notion of mechanism precise. Computation explicates mechanism. More recently, the notion of mechanism articulated by philosophers of science has been applied to shed light on computational systems. Mechanism explicates computation. This chapter recounts how this shift came about.

1. The mathematical theory of computation

The notion of computation comes from mathematics. To a first approximation, mathematical computation is the solving of mathematical problems by following an algorithm. A classic example is Euclid's algorithm for finding the greatest common divisor to two numbers. Because algorithms solve problems automatically within finitely many steps, they are sometimes called "mechanical" procedures. This is a first hint that computation has something to do with mechanisms.

In the early twentieth century, mathematicians and logicians developed formal logic to answer questions about the foundations of mathematics. One of these questions was the decision problem for first-order logic; namely, the question of whether an algorithm can determine, for any given statement written in a first-order logical system, whether that statement is provable within the system.[2] In 1936, Alonzo Church and Alan Turing proved that the answer is negative—there is no algorithm solving the decision problem for first-order logic (Church 1936; Turing 1936–7).

Turing's proof is especially relevant because it appeals to machines. To show that no algorithm can solve the decision problem for first-order logic, Turing needed to make precise what counts as an algorithm. He did so in terms of simple machines for manipulating symbols on an unbounded tape divided into squares. These *Turing machines*, as they are now called, have a mobile control device that travels along the tape and reads, writes, and erases symbols on each square of the tape—one square at a time. The control device can be in one of a finite number of states. It determines what to do based on its state as well as what's on the tape, in accordance with a finite set of instructions.

Turing established three striking conclusions. First, he argued persuasively that any algorithm, suitably encoded, can be followed by one of his machines—this conclusion is now known as the *Church–Turing thesis*. Second, Turing showed how to construct *universal* Turing machines—special

Turing machines that can be used to simulate any other Turing machine by executing the relevant instructions, which are written on their tape along with data for the computation. Finally, Turing proved that none of his machines solves the decision problem for first-order logic.

Turing's last result may be generalized as follows. There are denumerably many Turing machines (and therefore algorithms). That is to say, you can list all the Turing machines one after the other and put a natural number next to each. You can do this because you can enumerate all the finite lists of instructions that determine the behavior of each Turing machine. Each Turing machine computes a function from a denumerable domain to a denumerable range of values—for instance, from (numerals representing) natural numbers to (numerals representing) natural numbers. It turns out that there are undenumerably many such functions. Since an undenumerable infinity is much larger than a denumerable one, there are many more functions than Turing machines. Therefore, most functions are not computable by Turing machines. The decision problem for first-order logic is just one of these functions, which cannot be computed by Turing machines. The functions computable by Turing machines are called *Turing-computable*; the rest are called *Turing-uncomputable*. The existence of Turing-uncomputable functions over denumerable domains is theoretically very important. We will run into it again.

2. Mechanisms as Turing machines?

A few years after Turing's paper came out, Warren McCulloch and Walter Pitts argued that brains were a kind of Turing machine (without tapes). More specifically, McCulloch and Pitts defined circuits of simplified neurons for performing logical operations such as AND, OR, and NOT on digital inputs and outputs (strings of "ones" and "zeros"). They showed how to build neural networks out of such circuits and argued that their networks are equivalent to Turing machines; they also suggested that brains work similarly enough to their artificial neural networks, and therefore to Turing machines (McCulloch and Pitts 1943).[3]

Shortly after that, John von Neumann offered a sweeping interpretation of McCulloch and Pitts's work:

> The McCulloch–Pitts result . . . proves that anything that can be exhaustively and unambiguously described, anything that can be completely and unambiguously put into words, is ipso facto realizable by a suitable finite neural network.
>
> *(von Neumann 1951: 22)*

Since McCulloch–Pitts networks are computationally equivalent to Turing machines (with bounded tapes), von Neumann's statement entails that anything that can be exhaustively and unambiguously described is realizable by a Turing machine. It expands the scope of the Church–Turing thesis from covering mathematical algorithms to covering anything that can be described "exhaustively and unambiguously."

Thus was born the idea that Turing machines are not just a special kind of machine for performing computations. According to von Neumann's broader interpretation, Turing machines are a model of anything exhaustively and unambiguously described. Insofar as mechanisms can be exhaustively and unambiguously described, Turing machines are a model of mechanisms. Insofar as physical systems can be exhaustively and unambiguously described, Turing machines are a model of physical systems. (Insofar as physical systems are mechanisms, the previous two statements are equivalent.)

Around the time that von Neumann was making his bold statement, the first digital computers were being designed, built, and put to use. Digital computers are computationally equivalent to *universal* Turing machines (until they run out of memory). The main use of early computers was to run computational simulations of physical systems whose dynamics was too complex to be solved analytically. This sort of computational model became ubiquitous in many sciences. Computers became a tool for simulating just about any physical system—provided, of course, that their behavior could be described—in von Neumann's words—"exhaustively and unambiguously."

Aided by the spread and popularity of computational models, views along the lines of von Neumann's hype about what McCulloch and Pitts had allegedly shown became influential.

One development was the widespread impression that minds must be computable in Turing's sense. After all, McCulloch and Pitts had allegedly shown that brains are basically Turing machines. The theory that minds are computer programs running on neural software was soon to emerge from this milieu (Putnam 1967; Chapter 6, this volume).

Closely related is the idea that explaining clearly how a system produces a behavior requires providing a computer program for that behavior or, at least, some kind of computational model. Sometimes, this view was explicitly framed in terms of mechanisms—to the effect that explaining a behavior mechanistically requires providing a computer program for that behavior (e.g., Dennett 1978: 83). Conversely, some theorists argued that, if human beings can behave in a way that is Turing-*un*computable, then the mind is not a machine (e.g., Lucas 1961).

Other theorists made a similar point using different terminology. They distinguished between the functions being computed and the mechanisms implementing the computation. They argued that the functions performed by a system ought to be explained by functional analysis—a kind of explanation that describes the functions and sub-functions being performed by a system while abstracting away from the implementing mechanisms. For example, multiplying numbers (by the method of partial products) is functionally analyzed into performing single-digit multiplications, writing the results so as to form partial products, and then adding the partial products. Since functional analysis abstracts away from mechanisms, allegedly it is distinct and autonomous from mechanistic explanation. These theorists often maintained that psychology provides functional analyses, while neuroscience studies the implementing mechanisms—therefore, psychological explanations are distinct and autonomous from neuroscientific explanations (see Chapter 29). These same theorists also maintained that functional analyses consist in a computer program or other computational model (Fodor 1968; Cummins 1983). As a result, again, explaining a behavior requires providing a computer program or other computational model for that behavior.

Even more grandly, von Neumann's statement prefigured pancomputationalism—the view that *every* physical system is computational (e.g., Putnam 1967), that the entire universe is a giant computing machine (Zuse 1970), or that computation is somehow the building block of the physical world (e.g., Fredkin 1990; Wheeler 1982). Pancomputationalism took on a life of its own within certain physics circles, where it became known as digital physics or "it from bit" (Wheeler 1990) and—after the birth of quantum computation—transmuted itself into the view that the universe is a quantum computer (Lloyd 2006).[4]

3. Physical computation

All this talk of computation, especially in psychology and philosophy of mind, raised the question of what it takes for a physical system to perform a computation. Is the brain a computing system? Is the mind the software of the brain? What about other physical systems: which of them perform computations? For these questions to be answerable, some account must be given of what it takes for a brain or other physical system to perform a computation.

One early and influential account of physical computation asserted that a physical system performs a computation just in case there is a mapping between the computational state transitions that constitute that computation and the physical state transitions that the system undergoes (Putnam 1967). The main appeal of this *simple mapping account* is that it's simple and intuitive. Eventually it became clear that it's far too weak.

The main problem with the simple mapping account is that, without further constraints, it's too easy to map a computation to a series of physical state transitions. Ordinary physical descriptions define trajectories with undenumerably many state transitions, whereas classical computational descriptions such as Turing machine programs define trajectories with denumerably many state transitions. If nothing more than a mapping is required, any denumerable series of state transition can be mapped onto any undenumerable series of state transitions. Thus, by the simple mapping account, any computation is implemented by any ordinary physical system (Putnam 1988; Searle 1992). This result is now known as *unlimited pancomputationalism*; it is widely seen as a reductio ad absurdum of the simple mapping account of physical computation.

To avoid unlimited pancomputationalism, several theorists introduced constraints on which mappings are acceptable. Perhaps the most popular constraint is a causal one. According to the *causal account* of physical computation, only mappings that respect the causal relations between the computational state transitions—including those that are not instantiated in a given computation—are acceptable (Chalmers 2011; Chrisley 1994; Scheutz 1999). In other words, mapping individual computations onto individual physical state space trajectories is insufficient. What is also needed is that the mapping be set up so that the physical states that correspond to the computational states stand in appropriate causal relations with one another, mirroring the causal relations implicitly defined by the computational description.

The main advantage of sophisticated mapping accounts, such as the causal account, is that they remain close to the simplicity of the mapping account while avoiding unlimited pancomputationalism. The main disadvantage of these accounts is that they remain committed to *limited* pancomputationalism—the view that everything performs some computation or another. Limited pancomputationalism is in line with von Neumann's sweeping interpretation of the Church–Turing thesis, but it is in tension with a common way of understanding the computational theory of mind.

According to many theorists (following Fodor 1968, 1975), the mind (or at least cognition) is explained computationally in a way that other physical processes are not. Specifically, the computational explanation of mind has to do with the manipulation of representations. Another popular family of accounts of physical computation—*semantic accounts*—is tailored to the perceived needs of this representational version of the computational theory of mind. According to semantic accounts, physical computation is the manipulation of representations—or, more precisely, the manipulation of certain kinds of representation in appropriate ways (Cummins 1983; Fodor 1975; Pylyshyn 1984; Shagrir 2006).

The greatest advantage of semantic accounts is that they can avoid all forms of pancomputationalism. In fact, semantic accounts restrict physical computation to those physical systems that manipulate the right kinds of representation in the right way—presumably, only minds and artificial computing systems qualify. Semantic accounts also come with a disadvantage, though: they make no sense of any (alleged) computational system that does not manipulate representations. It turns out that, following computability theory, there is no difficulty in defining computations that manipulate meaningless letters. It is just as easy to define a Turing machine that manipulates uninterpreted marks as it is to define one that manipulates meaningful symbols. It is equally easy to program a digital computer to sort meaningless marks into meaningless orders as it is to program it to manipulate meaningful symbols in meaningful ways. Semantic accounts cannot account for this. Later we will see that the difficulties of both mapping and semantic accounts led to the most recent accounts of physical computation: the mechanistic accounts. But first we need to prepare the terrain by questioning some of the assumptions that are built into the traditional understanding of physical computation.

4. The scope of the Church–Turing thesis

Given the hype about computability that we reviewed, eventually scholars began to investigate its history and proper scope. One of the earliest, rigorous investigations of this sort was Guglielmo Tamburrini's PhD dissertation (Tamburrini 1988), which was soon followed by a series of incisive articles by his advisor, Wilfried Sieg (e.g., Sieg 1994, 1999). Such investigations reconstructed the intellectual context of Church and Turing's work—which, as I mentioned above, was the foundations of mathematics. They also clarified the interpretation and conceptual foundation of Turing's results. They showed that the Church–Turing thesis, properly understood, is about what mathematical problems can be solved by following algorithms—not about what the mind in particular or physical systems in general can do. The latter questions are independent of the original Church–Turing thesis and must be answered by other means.

With this more rigorous and restrictive understanding of the Church–Turing thesis in the background, Jack Copeland (2000) argued that we should distinguish between a mechanism in the broad, generic sense and a mechanism in the sense of a procedure that computes a Turing-computable function. A mechanism in the broad, generic sense is a system of organized components—this is the notion behind the new mechanism in philosophy of science, which we will discuss in the next section. Whether a mechanism in the broad, generic sense computes a Turing-computable function or a Turing-uncomputable function has nothing to do with the original Church–Turing thesis.

Copeland pointed out that, at least in mathematics, there are hypothetical machines that are more powerful than Turing machines—they compute Turing-uncomputable functions (Turing 1939). Copeland called a machine that is computationally more powerful than a Turing machine a *hypercomputer*. Finally, Copeland suggested that whether the brain or another physical system is a regular computing system or a hypercomputer is an empirical question—it is possible that brains are hypercomputers or that artificial hypercomputers can be built. Given Copeland's argument, the mind may be mechanistic in the generic sense even if it does something that is Turing-uncomputable.

What about the view that something is mechanical if and only if it is computable by some Turing machine? This kind of view relies on the notion of "mechanical" used in logic, whereby algorithms are said to be mechanical procedures. Since Turing machines are formal counterparts to algorithms, any procedure that is not computable by Turing machines is ipso facto not mechanical in that sense. As we've seen, this reasonable conclusion was fallaciously used to imply that the human mind is mechanical or mechanistic (in an unspecified sense) if and only if it is equivalent to a Turing machine. Copeland dubbed this kind of inference—from the Church–Turing thesis to conclusions about whether minds are mechanistic or Turing-computable—the Church–Turing fallacy (Copeland 2000, Piccinini 2007a).

Parallel to and independently of Copeland's argument, a literature on physical hypercomputation began to emerge within the foundations of physics. Various philosophers and physicists proposed hypothetical means by which exotic physical systems may be able to compute Turing-uncomputable functions. A prominent example involves a receiver system traveling toward a huge, rotating black hole while receiving signals on the results of a Turing machine's computation while the Turing machine orbits the black hole. Because of the way the black hole distorts space-time, the Turing machine may be able to complete an infinite number of operations while the receiver system that launched the Turing machine traverses a finite time-like trajectory; this allows the receiver to compute a Turing-uncomputable function, at least for one value of the function (Hogarth 1994; Etesi and Németi 2002). Even though there is little indication that hypercomputation is feasible, this literature as well as Copeland's argument undermine the idea that Turing machines are a general model of mechanisms. Still, none of this directly challenges (limited) pancomputationalism in a broader sense. It may still be that every physical system is computational—only now some physical systems may compute functions that no Turing machine can compute.

5. The rise of the new mechanism

Around the same time that Copeland and others were defending the physical possibility of hypercomputation, another group of philosophers of science revived the view that constitutive explanation—explanation of what a system does in terms of what its organized subsystems do—is provided by mechanisms (Chapter 1). These philosophers became known as *the new mechanists*. They argued that many special sciences, including biology and neuroscience, explain phenomena by uncovering mechanisms. What they meant by "mechanism" is what Copeland meant: an organized system of components and activities such that the components and activities, organized the way they are, produce the phenomenon (Bechtel and Richardson 1993; Glennan 1996; Machamer, Darden, and Craver 2000).

A few points about the new mechanism are relevant here.

First, the notion of mechanism articulated by the new mechanists is grounded in the explanations provided by the special sciences. The new mechanists scrutinized scientific practices in sciences such as molecular biology and neuroscience. They argued that such sciences explain phenomena mechanistically. This raises the question of whether other sciences—including, say, computer science and computer engineering—explain mechanistically.

Second, the new mechanists differed slightly in the way they conceptualized mechanisms. For example, some talked about mechanisms performing operations, others of mechanisms performing activities, yet others of mechanisms performing functions. Those differences don't matter for our purposes. What matters is that none of these theorists conceptualized mechanisms in terms of computation. The view examined in previous sections, whereby a mechanism is something whose behavior can be modeled by a Turing machine, was absent from the new mechanism. Mechanisms were systems of concrete components organized to perform operations, activities, or functions. (Elsewhere, I argue that these formulations are essentially equivalent anyway; Piccinini forthcoming, cf. Illari and Williamson 2012.)

Third, even when mechanists talked about functions, they were not talking about the kind of mathematical functions of a denumerable domain that may or may not be Turing-computable. Instead, they were talking about functions as a kind of activity. In this sense of function, the function of the heart is to pump blood, and the function of a drill is to make holes. Even with respect to functions, the new mechanist notion of mechanism is independent of the notion of computation.

Fourth, many new mechanists were hostile to teleology. When they talked about functions, they typically understood them in a nonteleological way, as something a mechanism *does* rather than something a mechanism is *for* (cf. Craver 2001). But the functions of mechanisms can also be understood teleologically, as what something is *for*. There is a large literature, independent of the new mechanist literature, devoted to explicating teleological functions in nonteleological terms; it may be combined with the new mechanist notion of mechanism to yield *functional* mechanisms; that is, mechanisms with teleological functions (Garson 2013). Functional mechanisms in this teleological sense are a subset of all mechanisms (Chapter 8).

6. The mechanistic account of digital computation

The new mechanists argued forcefully that many special sciences explain by uncovering mechanisms. The most obvious candidate sciences were molecular biology, physiology, neuroscience, and perhaps the study of engineered artifacts.

If all of these sciences explain mechanistically, the question arises of how computer science and computer engineering explain. A closely related question is how computational explanation works—not only in computer science and engineering but also in psychology and neuroscience (Chapter 29). A mechanistic answer to these questions is the starting point of the mechanistic account of physical computation.

I began defending a version of the mechanistic account in the early 2000s (Piccinini 2003; although at that time I called it the "functional account") and spent much of the subsequent decade developing and extending it (2015). Others followed suit with their own versions, sometimes tailoring it to specific sciences or varieties of computation (Kaplan 2011; Fresco 2014; Milkowski 2013). The mechanistic account explicates physical computation in terms of mechanisms and has become a prominent account of physical computation. In explicating the mechanistic account, I will focus on the version I know best.

A first observation is that computer science and computer engineering have their own special domain. They study *computational* systems. By contrast, other sciences—with the exception of cognitive science and neuroscience—generally do not focus on computational systems. Thus, if computing systems are mechanisms, there seems to be something distinctive about them—something that makes them the proper domain of a specific science. Accordingly, an important task of the mechanistic account is to say what distinguishes computing mechanisms from other mechanisms.

This task is made difficult by the contested boundaries between computing systems and non-computing systems. Some theorists take a very restrictive attitude: only systems that manipulate representations according to rules and such that their manipulations are caused by representations of the rules count as genuinely computing systems (Fodor 1975; Pylyshyn 1984). Other theorists take a very liberal attitude: every physical system performs some computation or another (Chalmers 2011; Scheutz 1999; Shagrir 2006). Yet other theorists fall somewhere between these two extremes.

To establish reasonable boundaries, I used the following strategy. First, identify a set of paradigmatic examples of physical computing systems (digital computers, calculators, etc.) and a set of relatively uncontroversial examples of non-computing systems (digestive systems, hurricanes, etc.). Second, identify what distinguishes the computing systems from the non-computing systems. Finally, use the resulting account to decide any boundary cases. My first stab at a mechanistic account covered only digital computation (including hypothetical hypercomputers), precisely because digital computing systems such as computers and calculators are generally accepted as computational and because they fall under Turing's mathematical theory of computation (Piccinini, 2007b, 2008a).

To construct a mechanistic account of digital computation, the key notion was that of a digit, which is a concrete counterpart of the abstract notion of letter (or symbol) employed by computability theorists. As we saw above, a Turing machine is said to manipulate strings of *letters* (symbols) according to a set of instructions. A digit is a physical macrostate that corresponds to the mathematical notion of letter (or symbol). It comes in finitely many types that can be reliably distinguished and manipulated by a physical system. Like a set of letters, it can be concatenated into strings such that the physical system that manipulates the digits can differentiate between digits based on their place within a string and treat each digit differently depending on its place within the string. Given these preliminaries, I argued that physical computing systems are physical systems whose function is manipulating strings of digits according to a rule defined over the digits. The rule is just the mapping from input strings of digits to output strings of digits.

This account has several virtues.

First, it makes computation an objective property of physical systems. Whether a physical system has functions, whether it manipulates strings of digits, and whether it does so in accordance with rules defined over the digits are objective properties of the system, which can be empirically investigated. Unlimited pancomputationalism is thus avoided.

Second, only a fairly restricted class of physical systems counts as computational. Thus, limited pancomputationalism is avoided.

Third, the mechanistic account does not assume that computation requires representation. It shares this virtue with mapping accounts. This is a virtue because many paradigmatic examples of computing systems do not manipulate representations in any interesting sense. In addition, the notion of representation is somewhat murky and in need of a clear foundation, so we should avoid representation, if possible, in explicating computation (Chalmers 2011).

Fourth, the mechanistic account comes with a clear and compelling notion of computational explanation, which fits the practices of the relevant sciences. That is, computational explanation is a species of mechanistic explanation. It is the species of mechanistic explanation that appeals to components that manipulate strings of digits in accordance with rules defined over the digits.

Fifth, the mechanistic account is the first to explain miscomputation. Computer scientists and engineers work hard to test the reliability of computer circuits, debug computer programs, and fix computers. In other words, they fight miscomputation. Mapping accounts have a hard time making any sense of that—after all, there is always a mapping between any physical state transition and some computation or another. Semantic accounts are not much better off. In any case, no one attempted to make sense of miscomputation until mechanistic accounts came along. According to mechanistic accounts, there are several possible sources of miscomputation. A system may miscompute because it was poorly designed (relative to the goals of the designers), or poorly built (relative to the design plan), or misused. If a system is programmable, it can also miscompute because it was programmed incorrectly (relative to the programmer's intentions).[5]

Sixth, the mechanistic account can be used to provide an illuminating taxonomy of computing systems (Piccinini 2008b, 2008c).

Based on this initial mechanistic account, I argued that so-called analog computers and neural networks that have no digital inputs and outputs are not computing mechanisms properly so called (Piccinini 2008b, 2008c). This turned out to be overly restrictive.

7. The mechanistic account of generic computation

Analog computers in the strict sense (Pour-El 1974) are physical systems that are used to solve systems of differential equations. They solve systems of equations by embodying the mathematical relations between the variables in the equations within the physical variables in the computer and by physically manipulating the physical variables in a way that corresponds to the relations defined by the equations. Perhaps the most important operation performed by analog computers is integration, which allows the computer to output a value corresponding to the integral of a function over a time interval. Notice that systems of differential equations are defined over real (continuous) variables, which can take any real number as a value. To embody such variables, analog computers must manipulate physical variables that (are assumed to) take any real values, at least within reasonable intervals. (Outside of such reasonable intervals, an analog computer would malfunction; for example, overly high voltage levels within a circuit would fry the circuit.) Of course, physical variables that (are assumed to) take any real value can only be measured and manipulated to a finite degree of approximation, which means that analog computers produce outputs that are approximations of the desired result.

Clearly, there are important differences between digital computers and analog computers. Whereas digital computers manipulate strings of digits, analog computers manipulate continuous variables. Whereas digital computers can follow precise instructions defined over digits, analog computers follow systems of differential equations. Whereas digital computers compute functions defined over denumerable domains, analog computers solve systems of differential equations. Whereas digital computers can be universal in Turing's sense, Turing's notion of universality does not apply to analog computers. Whereas digital computers produce exact results, analog computers produce approximate results. These differences may be seen as evidence that analog "computers" are so different from digital computers that they shouldn't even be considered computers at all (Piccinini 2008b).

And yet, those same differences also point at deep similarities. Both digital and analog computers are used to solve mathematical problems. Both digital and analog computers solve mathematical problems encoded in the values of physical variables. Both digital and analog computers solve mathematical problems by following some sort of procedure or rule defined over the variables they manipulate. Both digital and analog computers used to compete within the same market—the market for machines that solve systems of equations. (Digital computers eventually won that competition, in part because they do much more than solving systems of equations.) These similarities are serious enough that many scientific communities found it appropriate to use the same word—"computer"—for both classes of systems. Maybe they are onto something; philosophers may be better off figuring out what that is rather than asserting that scientists are misguided.

A similar point applies to other unconventional models of computation, such as DNA computing and quantum computing. In recent decades, computer scientists have investigated models of computation that depart in various ways from mainstream electronic digital computing. These models employ chemical reactions, DNA molecules, or quantum systems to perform computations—or so the people who study them assert. It would be valuable to have an account of physical computation general enough to cover them all. Notice that neither supporters of mapping accounts nor supporters of semantic accounts ever addressed this sort of problem.

In light of the above, we need a notion of physical computation broader than that of digital computation—a notion of computation in a generic sense that covers digital computation, analog computation, and other unconventional kinds of computation. One option would be to declare that everything computes (limited pancomputationalism). But that erases all boundaries between the domain of computer science and every other scientific domain. It would be better if there were a way to characterize what digital computing systems have in common with analog computers and other unconventional computing systems as well as what distinguishes all of them from other physical systems.

A helpful notion is *medium independence*, which was used by Justin Garson (2003) to characterize the notion of information introduced by Edgar Adrian (1928) in neurobiology. Garson argued that information is medium-independent in the sense that it can be carried by a physical variable (such as spike trains) regardless of the physical origin of the information (which may be light, sound waves, pressure, etc.).

Medium independence in this sense is stronger than multiple realizability. If a property is medium-independent, it is also multiply realizable, because it's realizable in different media. The converse does not hold: a multiply realizable property may or may not be medium-independent. For example, the property of being a mousetrap is multiply realizable but not medium-independent: any mousetrap must handle the same medium—mice!

A similar notion of medium independence can be used to characterize all the vehicles of computation, whether digital or not. Specifically, all physical computing systems—digital, analog, or what have you—can be implemented using physically different variables so long as the variables possess the right degrees of freedom and the implementing mechanisms can manipulate those degrees of freedom in the right way. In other words, computational vehicles are macroscopic variables defined in a medium-independent way. This, then, became my mechanistic account of computation in the generic sense: a physical computing system is a mechanism whose function is to manipulate medium-independent variables in accordance with a rule defined over the variables (Piccinini and Scarantino 2011; Piccinini 2015).

This more general mechanistic account of physical computation inherits the virtues of the mechanistic account of digital computation. In addition, it covers analog computers and other unconventional models of computation. In virtue of its generality, the account covers the notion of computation used by computational neuroscientists without assuming that it must be digital—or analog, for that matter. In fact, by relying on this generalized mechanistic account, Sonya Bahar and I have argued on empirical grounds that neural computation is neither digital nor analog—it is a third, *sui generis*, notion of computation (Piccinini and Bahar 2013).

Perhaps the biggest point of contention surrounding the mechanistic account of physical computation is whether it adequately subsumes all forms of computational explanation. Carl Craver and I (2011) argued that it does because functional analysis—the traditional alternative to mechanistic explanation—is just a sketch of a mechanism. Others responded that, at least in cognitive science or cognitive neuroscience, there are forms of computational explanation that are not mechanistic (Chirimuuta 2014; Shagrir 2010; Shapiro 2016; Weiskopf 2011). The most common reason they give is that such forms of computational explanation abstract away from some aspects of the mechanism that carries out the computation. I believe that when the proper roles of abstraction within mechanistic explanation are appreciated, constitutive explanation—including computational explanation—remains mechanistic (Boone and Piccinini 2016, forthcoming). This debate is likely to continue (see Chapters 16, 17, and 20).

Notes

1 Thanks to Ken Aizawa, Giovanni Camardi, Joe Dewhurst, Anne Jacobson, Jack Mallah, Marcin Milkowski, Alessio Plebe, Charles Rathkopf, Oron Shagrir, Wilfried Sieg, Guglielmo Tamburrini, my audience at IACAP 2016, and especially Stuart Glennan and Phyllis Illari for helpful comments on previous versions. This project was supported in part by International Studies and Programs at the University of Missouri—St. Louis.

2 This formulation is due to Church (1936). Hilbert and Ackerman (1928) had asked whether some algorithm could determine whether any given statement written in first-order logic is universally valid—valid in every structure satisfying the axioms. In more modern terms, Hilbert and Ackerman's question is whether any given statement written in first-order logic is logically valid—true under every possible interpretation. Church's formulation is equivalent to Hilbert and Ackerman's because of the completeness of first-order logic (Church 1936, fn. 6).

3 Stephen Kleene (1956) confirmed their assertions by proving that McCulloch–Pitts networks are computationally equivalent to finite state automata, which are computationally equivalent to Turing machines without tapes or with *bounded* tapes (strictly speaking, Turing machines have an *unbounded* tape, which increases their computational power).

4 Independently, some logicians and mathematicians showed that large classes of physical or computational systems could be simulated more or less exactly by Turing machines (e.g., Gandy 1980; Rubel 1989); others looked for possible exceptions (e.g., Pour-El and Richards 1989).

5 For more on miscomputation, see Piccinini 2015: 148–150.

References

Adrian, E. D. (1928). *The Basis of Sensation: The Action of the Sense Organs*. New York: Norton.

Bechtel, W. and R. C. Richardson (1993). *Discovering Complexity: Decomposition and Localization as Scientific Research Strategies*. Princeton, NJ: Princeton University Press.

Boone, W. and G. Piccinini (2016). "The Cognitive Neuroscience Revolution," *Synthese* (in press). 10.1007/s11229-015-0783-4

Boone, W. and G. Piccinini (forthcoming). "Mechanistic Abstraction," *Philosophy of Science*.

Chalmers, D. J. (2011). "A Computational Foundation for the Study of Cognition," *Journal of Cognitive Science* 12(4): 323–57.

Chirimuuta, M. (2014). "Minimal Models and Canonical Neural Computations: The Distinctness of Computational Explanation in Neuroscience," *Synthese* 191(2): 127–154.

Chrisley, R. L. (1994). "Why Everything Doesn't Realize Every Computation," *Minds and Machines* 4: 403–430.

Church, A. (1936). "An Unsolvable Problem in Elementary Number Theory," *The American Journal of Mathematics* 58: 345–363.

Copeland, B. J. (2000). "Narrow Versus Wide Mechanism: Including a Re-Examination of Turing's Views on the Mind-Machine Issue." *The Journal of Philosophy* XCVI(1): 5–33.

Craver, C. (2001). "Role Functions, Mechanisms, and Hierarchy," *Philosophy of Science* 68: 53–74.

Cummins, R. (1983). *The Nature of Psychological Explanation*. Cambridge, MA: MIT Press.

Dennett, D. C. (1978). *Brainstorms*. Cambridge, MA: MIT Press.

Etesi, G. and I. Németi (2002). "Non-Turing Computations via Malament-Hogarth Spacetimes," *International Journal of Theoretical Physics* 41: 342–370.

Fodor, J. A. (1968). *Psychological Explanation*. New York: Random House.

Fodor, J. A. (1975). *The Language of Thought*. Cambridge, MA: Harvard University Press.

Fredkin, E. (1990). "Digital Mechanics: An Information Process Based on Reversible Universal Cellular Automata," *Physica D* 45: 254–270.

Fresco, N. (2014). *Physical Computation and Cognitive Science*. New York: Springer.

Gandy, R. (1980). "Church's Thesis and Principles for Mechanism." In J. Barwise, H. J. Keisler, and K. Kuhnen (Eds.), *The Kleene Symposium* (123–148). Amsterdam: North-Holland.

Garson, J. (2003). "The Introduction of Information into Neurobiology," *Philosophy of Science* 70: 926–936.

Garson, J. (2013). "The Functional Sense of Mechanism," *Philosophy of Science* 80: 317–333.

Glennan, S. (1996). "Mechanisms and the Nature of Causation," *Erkenntnis* 44: 49–71.

Hilbert, D. and W. Ackermann (1928). *Grundzüge der theoretischen Logik*. Berlin: Springer.

Hogarth, M. L. (1994). "Non-Turing Computers and Non-Turing Computability," *PSA* 1: 126–138.

Illari, P. M. and J. Williamson (2012). "What Is a Mechanism? Thinking about Mechanisms across the Sciences," *European Journal of Philosophy of Science* 2: 119–135.

Kaplan, D. M. (2011). "Explanation and Description in Computational Neuroscience," *Synthese* 183 (3): 339–373.

Kleene, S. C. (1956). "Representation of Events in Nerve Nets and Finite Automata." In C. E. Shannon and J. McCarthy (Eds.), *Automata Studies* (3–42). Princeton, NJ: Princeton University Press.

Lloyd, S. (2006). *Programming the Universe: A Quantum Computer Scientist Takes on the Cosmos*. New York: Knopf.

Lucas, J. R. (1961). "Minds, Machines, and Gödel," *Philosophy* 36: 112–137.

Machamer, P. K., Darden, L., and C. Craver (2000). "Thinking about Mechanisms," *Philosophy of Science* 67: 1–25.

McCulloch, W. S. and W. H. Pitts (1943). "A Logical Calculus of the Ideas Immanent in Nervous Activity," *Bulletin of Mathematical Biophysics* 7: 115–33.

Milkowski, M. (2013). *Explaining the Computational Mind*. Cambridge, MA: MIT Press.

Piccinini, G. (2003). Computations and Computers in the Sciences of Mind and Brain.

Doctoral Dissertation, Pittsburgh, PA: University of Pittsburgh. URL = <http://etd.library. pitt.edu/ETD/available/etd-08132003-155121/>.

Piccinini, G. (2007a). "Computationalism, the Church-Turing Thesis, and the Church-Turing Fallacy," *Synthese* 154(1): 97–120.

Piccinini, G. (2007b). "Computing Mechanisms," *Philosophy of Science* 74(4): 501–526.

Piccinini, G. (2008a). "Computation without Representation," *Philosophical Studies* 137(2): 205–241.

Piccinini, G. (2008b). "Computers," *Pacific Philosophical Quarterly* 89(1): 32–73.

Piccinini, G. (2008c). "Some Neural Networks Compute, Others Don't," *Neural Networks* 21(2–3): 311–321.

Piccinini, G. (2015). *Physical Computation: A Mechanistic Account*. Oxford: Oxford University Press.

Piccinini, G. (forthcoming). "Activities Are Manifestations of Causal Powers." In M. Adams, Z. Biener, U. Feest, and J. Sullivan (Eds.), *Eppur Si Muove: Doing History and Philosophy of Science with Peter Machamer*. Berlin: Springer.

Piccinini, G. and S. Bahar (2013). "Neural Computation and the Computational Theory of Cognition," *Cognitive Science* 34: 453–488.

Piccinini, G. and C. Craver (2011). "Integrating Psychology and Neuroscience: Functional Analyses as Mechanism Sketches," *Synthese* 183: 283–311.

Piccinini, G. and A. Scarantino (2011). "Information Processing, Computation, and Cognition," *Journal of Biological Physics* 37(1): 1–38.

Pour-El, M. B. (1974). "Abstract Computability and Its Relation to the General Purpose Analog Computer (Some Connections Between Logic, Differential Equations and Analog Computers)," *Transactions of the American Mathematical Society* 199: 1–28.

Pour-El, M. B. and J. I. Richards (1989). *Computability in Analysis and Physics*. Berlin: Springer-Verlag.

Putnam, H. (1967). "Psychological Predicates." In W. H. Capitan and D. D. Merrill (Eds.), *Art, Mind, and Religion* (37–48). Pittsburgh, PA: University of Pittsburgh Press.

Putnam, H. (1988). *Representation and Reality*. Cambridge, MA: MIT Press.

Pylyshyn, Z. W. (1984). *Computation and Cognition*. Cambridge, MA: MIT Press.

Rubel, L. A. (1989). "Digital Simulation of Analog Computation and Church's Thesis," *Journal of Symbolic Logic* 54(3): 1011–1017.

Scheutz, M. (1999). "When Physical Systems Realize Functions . . . ," *Minds and Machines* 9(2): 161–196.

Searle, J. R. (1992). *The Rediscovery of the Mind*. Cambridge, MA: MIT Press.

Shagrir, O. (2006). "Why We View the Brain as a Computer," *Synthese* 153(3): 393–416.

Shagrir, O. (2010). "Brains as Analog-Model Computers," *Studies in History and Philosophy of Science* 41: 271–279.

Shapiro, L. A. 2016. "Mechanism or Bust? Explanation in Psychology." *British Journal for the Philosophy of Science* 10.1093/bjps/axv062.

Sieg, W. (1994). "Mechanical Procedures and Mathematical Experience." In A. George (Ed.), *Mathematics and Mind* (71–117). New York: Oxford University Press.

Sieg, W. (1999). "On Computability." In A. Irvine (Ed.), *Philosophy of Mathematics* (Handbook of the Philosophy of Science) (535–630). Amsterdam: North-Holland.

Tamburrini, G. (1988). *Reflections on Mechanism*, unpublished Ph.D. dissertation, Columbia University.

Turing, A. M. (1936–7). "On Computable Numbers, with an Application to the Entscheidungsproblem," *Proceeding of the London Mathematical Society* 42(1): 230–265.

Turing, A. M. (1939). "Systems of Logic Based on Ordinals," *Proceedings of the London Mathematical Society*, Ser. 2 45: 161–228.

von Neumann, J. (1951). "The General and Logical Theory of Automata." In L. A. Jeffress (Ed.), *Cerebral Mechanisms in Behavior* (1–41). New York: Wiley.

Weiskopf, D. (2011). "Models and Mechanisms in Psychological Explanation," *Synthese* 183(3): 313–338.

Wheeler, J. A. (1982). "The Computer and the Universe," *International Journal of Theoretical Physics* 21(6–7): 557–572.

Wheeler, J. A. (1990). "Information, Physics, Quantum: The Search for Links." In W. H. Zurek (Ed.), *Complexity, Entropy, and the Physics of Information* (3–28). Redwood City, CA: Addison-Wesley.

Zuse, K. (1970). *Calculating Space*. Cambridge, MA: MIT Press.

34

MECHANISMS AND ENGINEERING SCIENCE[1]

Dingmar van Eck

1. Introduction

Use of "mechanism talk" is ubiquitous in engineering science (e.g., Chandrasekaran and Josephson 2000; Goel 2013). Philosophical discussions of mechanisms also frequently invoke engineered systems, such as pumps, car engines, mouse traps, toilets, soda vending machines, and the like in illustrating various aspects of mechanisms and mechanistic explanation (see Levy 2014). Nevertheless, focused philosophical analyses of the structure of mechanistic explanations in engineering science are scarce (see van Eck 2015). Reference to engineered systems in discussions of mechanisms and mechanistic explanation is often a loose metaphor, not a conceptualization that offers sophisticated understanding of what mechanistic explanation looks like in engineering practice. Moreover, philosophical work that aims to elucidate the connection(s) between engineering and systems biology—connections that practicing engineers and biologists have been stressing for more than a decade (e.g. Csete and Doyle 2002)—is also few and far between, in particular with respect to the use of engineering principles in the construction of mechanistic explanations in systems biology (see Braillard 2015). In this chapter I address both these issues.

In this chapter I give an outline of the structure of mechanistic explanation in engineering science, and organize this discussion around two features that extend the mechanistic program toward explanation when applied to engineering science. First, in section 2, I show that in engineering, two distinct sub-types of role function—"behavior function" and "effect function"—are employed in the functional individuation of mechanisms, rather than role function *simpliciter*. Empirically informed understanding of mechanistic explanation in engineering science requires sensitivity to this distinction (van Eck 2015). I illustrate this point in terms of reverse engineering and malfunction explanations in engineering science.

Second, in section 3, I discuss connections between (control) engineering and systems biology, focusing on the usage of engineering principles in the construction of mechanistic explanations in systems biology. Systems biology has adopted engineering tools and principles, in particular from control engineering, to model and explain complex biological systems. These tools are often in the service of characterizing the organization of mechanisms in abstract, truncated fashion. I discuss a case of heat shock response in *Escherichia coli* to illustrate the role of engineering principles in mechanistic explanation in systems biology (see El-Samad et al. 2005; Braillard 2015).

This case again shows the relevance of distinguishing behavior from effect function and, moreover, gives means to elaborate a key issue in a recent and general discussion on the explanatory power of mechanistic explanations, viz. to flesh out the distinctions between the explanatory desiderata of "completeness and specificity" (Craver 2007) and "abstraction" (Levy and Bechtel 2013). Rather than being in competition, as some authors have it, I argue that these desiderata are suitable for different explanation-seeking contexts.

2. Mechanistic explanation in engineering science

By now, quite a few accounts of mechanistic explanation are on offer in the literature. Although they come in different flavors, there is broad consensus on a number of key features:

> All mechanistic explanations begin with (a) the identification of a phenomenon or some phenomena to be explained, (b) proceed by decomposition into the entities and activities relevant to the phenomenon, and (c) give the organization of entities and activities by which they produce the phenomenon.
>
> *(Illari and Williamson 2012: 123; see also Chapter 1)*

Mechanistic explanations thus explain how mechanisms, i.e. organized collections of entities and activities, produce phenomena (Machamer et al. 2000; Glennan 2005; Bechtel and Abrahamsen 2005; Craver 2007).

Role function ascription is considered crucial for (b) decomposition and (c) the elucidation of mechanisms' organization (Machamer et al. 2000; Craver 2001; Illari and Williamson 2010; see Chapter 8). As Machamer et al. (2000) write:

> Mechanisms are identified and individuated by the activities and entities that constitute them, by their start and finish conditions, and by their functional roles. Functions are the roles played by entities and activities in a mechanism. To see an activity as a function is to see it as a component in some mechanism, that is, to see it in a context that is taken to be important, vital, or otherwise significant.
>
> *(Machamer et al. 2000: 6)*

Mechanistic role functions thus refer to activities that make a contribution to the workings of mechanisms of which they are a part, and mechanistic organization is key for the ascription of functions. For instance, in the context of explaining the circulatory system's activity of "delivering goods to tissues," the heart's "pumping blood through the circulatory system" is ascribed a function relative to organizational features such as the availability of blood, and the manner in which veins and arteries are spatially organized (Craver 2001: 64).

This perspective on the general structure of mechanistic explanation, and the importance of (role) functional individuation of mechanisms, finds widespread support in the literature on mechanistic explanation in the life sciences. Frequently, in this literature, mechanisms of technical artifacts, such as clocks, mousetraps, and car engines, are invoked as metaphors to elucidate features of biological mechanisms (Craver 2001) and features of mechanisms in general (Glennan 2005; Darden 2006; Illari and Williamson 2012). The mechanistic concept of role function, and its utility in the functional individuation of mechanisms, has likewise been explicated in terms of mechanisms of technical artifacts such as car engines (Craver 2001).

However, as mentioned in section 1, reference to such technical mechanisms must not be understood as providing insight into mechanistic explanation in engineering science per se since engineers use multiple notions of function in the functional individuation of technical mechanisms, rather than the concept of role function *simpliciter* (van Eck 2015). (And, as we will see, the distinction between "contextual" and "isolated" descriptions of an entity's activity (Craver 2001) also does not capture the distinction in engineering concepts of function.)

Function is a key term in engineering and an ambiguous one (e.g. Chandrasekaran and Josephson 2000). A variety of function notions are used in mechanism individuation and explanation in engineering science, and the precise notion of function invoked depends on the explanatory and design task at hand. Contrary to explanation in other sciences, explanation in engineering science cannot be seen in isolation from design. For instance, failure analysis—malfunction explanation—is an important type of explanation (Bell et al. 2007) that has as its ultimate aim the improvement of technical systems; also reverse engineering explanation is not "merely" mechanistic explanation since its ultimate aim is the redesign and subsequent improvement of extant technical systems (Otto and Wood 2001). Similarly, explanation is in the service of conceptual design in knowledge base-assisted designing in which (mechanistic) explanations of the workings of extant technical systems and their components are archived and put to use to develop novel design specifications (Stone and Wood 2000). Below I zoom in on two such contexts and the relevance of different notions of engineering function for the functional individuation of technical mechanisms in these contexts, viz. reverse engineering explanation and malfunction explanation. As we will see, specific notions of function are optimally "engineered" for specific explanatory settings (I focus here on explanatory contexts, not the design contexts to which they are related).

Function has no uniform meaning in engineering: different approaches advance different conceptualizations (Erden et al. 2008), and some researchers use the term with more than one meaning simultaneously (Chandrasekaran and Josephson 2000). This ambiguity led to philosophical analysis of the precise meanings of function involved. Vermaas (2009) regimented the spectrum of available function meanings into three "archetypical" engineering conceptualizations of function: *behavior function*—function as the desired behavior of a technical artifact; *effect function*—function as the desired effect of behavior of a technical artifact; and *purpose function*—function as the purpose for which a technical artifact is designed.[2] In the ensuing discussion on reverse engineering explanation and malfunction explanation, the notions of behavior function and effect function are most relevant.

Behavior functions are typically modeled as conversions of flows of materials, energy, and signals, where input flows and output flows in the conversion (are assumed to) match in terms of physical conservation laws (Stone and Wood 2000; Otto and Wood 2001). For instance, the function "loosen/tighten screws" of an electric screwdriver is then represented as a conversion of input flows of "screws" and "electricity" into corresponding output flows of "screws," "torque," "heat," and "noise" (see Stone and Wood 2000: 364). Since these descriptions of functions are specified such that input and output flows match in terms of physical conservation laws—here, the conservation of energy through the conversion of electrical energy into rotational, thermal, and acoustic energy—they are taken to refer to specific physical behaviors of technical artifacts (Vermaas 2009).

Effect function descriptions refer to only the technologically relevant *effects* of the physical behaviors of technical artifacts: the requirements are dropped that descriptions of these effects meet conservation laws and that matching input and output flows are specified (Vermaas 2009). The function of an electric screwdriver is then described simply as, say, "loosen/tighten screws," leaving the physical antecedents of this effect unmentioned. Behavior function descriptions thus refer to the "complete" behaviors involved, including features like thermal and acoustic energy flows, whereas effect functions refer to subsets of these behaviors, i.e. desired effects.[3]

Engineering descriptions and explanations of the workings of extant technical artifacts and artifact designs are often constructed by functionally decomposing functions into a number of sub-functions. The relationships between functions and sets of their sub-functions are often graphically represented in functional decomposition models. Like the concept of function, such models come in a variety of "archetypical" flavors (van Eck 2011). In the context of reverse engineering explanation and malfunction explanation, the relevant ones are *behavior functional decomposition*—a model of an organized set of behavior functions, and *effect functional decomposition*—a model of an organized set of effect functions.

In reverse engineering explanations, elaborate behavior functions and functional decompositions are used; in malfunction explanations, less detailed effect functions and functional decompositions are employed.

In engineering science, reverse engineering and engineering design go hand in glove (e.g. Otto and Wood 2001; Stone and Wood 2000). Consider Otto and Wood's (2001) reverse engineering and redesign method, in which a reverse engineering phase in which reverse engineering explanations are developed for existing artifacts precedes and drives a subsequent redesign phase of those artifacts. The goal of the reverse engineering phase is to explain how existing artifacts produce their overall (behavior) functions in terms of underlying mechanisms, i.e. organized components and sub-functions (behaviors) by which overall (behavior) functions are produced. These explanations are subsequently used in the redesign phase to identify components that function sub-optimally and to either improve them or replace them with better-functioning ones.

In the reverse engineering phase, an artifact is first broken down component-by-component, and hypotheses are formulated concerning the functions of those components. In this method, functions are behavior functions and are represented by conversions of flows of materials, energy, and signals. Since the aim of the reverse engineering phase is to understand in detail the manner in which an extant technical system operates, elaborate behavior function descriptions are used. Descriptions of input–output conversions give more relevant details than descriptions of effects. After this analysis, a different reverse engineering analysis commences in which components are removed, one at a time, and the effects are assessed of removing single components on the overall functioning of the artifact. Such single-component removals are used to detail the behavior functions of the (removed) components further. The idea behind this latter analysis is to compare the results from the first and second reverse engineering analysis to gain a potentially more nuanced understanding of the functions of the components of the (reverse engineered) artifact. Using these two reverse engineering analyses, a behavior functional decomposition of the artifact is then constructed in which the behavior functions of the components are specified and interconnected by their input and output flows of materials, energy, and signals

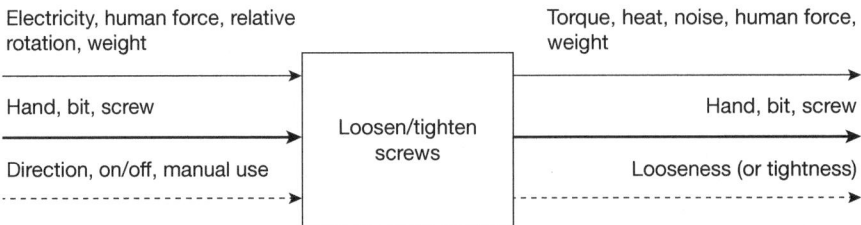

Figure 34.1　Overall behavior function of an electric power screwdriver. Thin arrows represent energy flows; thick arrows represent material flows, dashed arrows represent signal flows (adapted from Stone and Wood 2000: 363, figure 2)

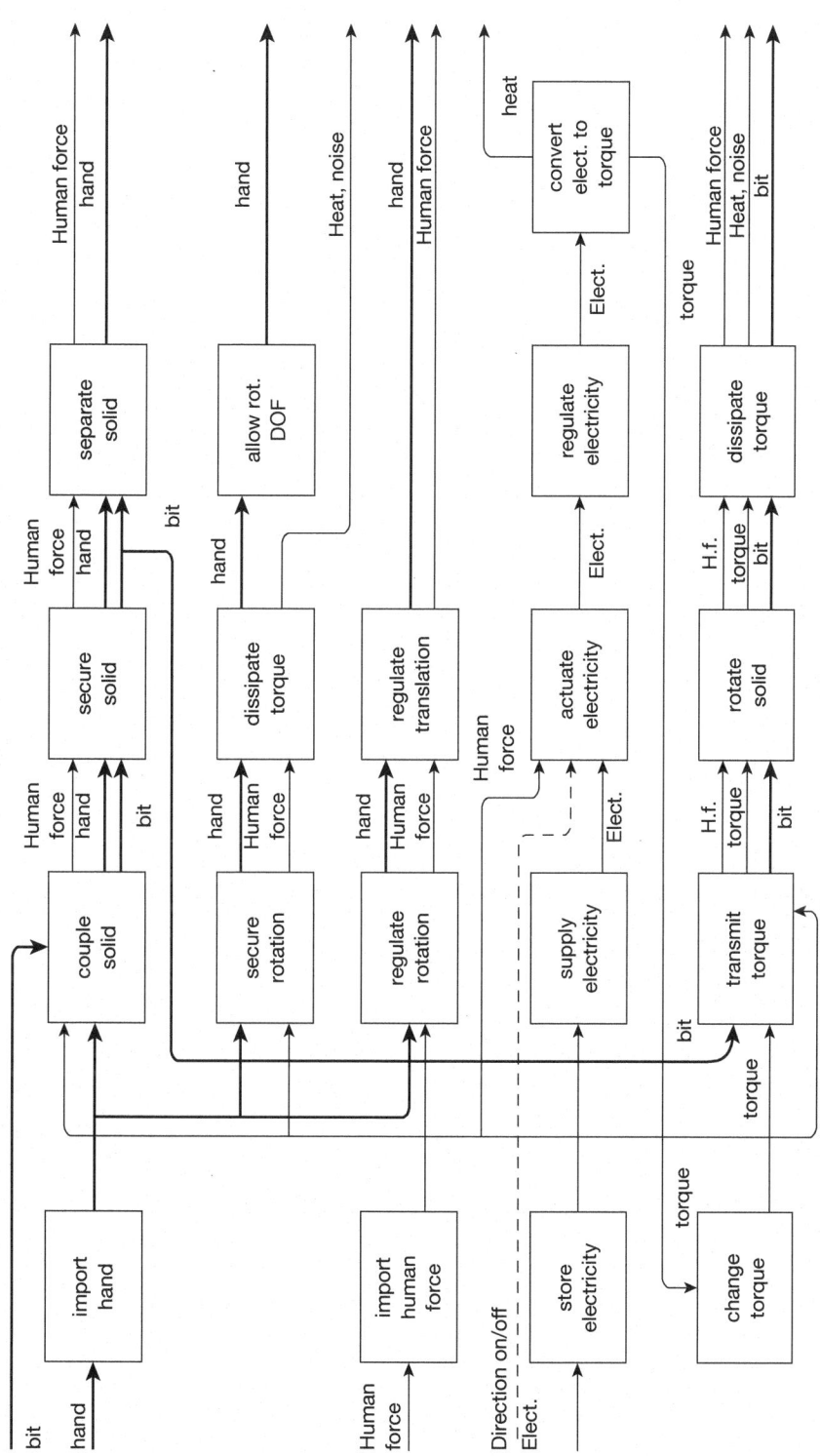

Figure 34.2 Behavior functional decomposition of an electric power screwdriver. Thin arrows represent energy flows; thick arrows represent material flows, dashed arrows represent signal flows (adapted from Stone and Wood 2000: 364, figure 4)

(Otto and Wood 2001). Such models represent parts of the mechanisms by which technical systems operate, to wit: causally connected behaviors of components. Examples of an overall behavior function and behavior functional decomposition of a reverse engineered electric screwdriver are given in Figures 34.1 and 34.2, respectively.

In the model in Figure 34.2, temporally organized and interconnected behaviors are described. Components of artifacts are described in Otto and Wood's method in tables, what in engineering are called "bills of materials," together with a model, called an "exploded view," of the components composing the artifacts. Taken together, these component and behavior functional decomposition models provide functional individuations and representations of mechanisms of artifacts.

Such (behavior functional decomposition) models are subsequently used to identify suboptimally functioning components and so drive succeeding redesign phases (Otto and Wood 2001). The focus here is on the reverse engineering explanation part of the methodology.

In malfunction explanation, this detail in mechanistic models is, however, not required: engineers take it that less detailed effect functions and functional decompositions there do a better explanatory job.

In malfunction analysis, explanation-seeking questions of the following format arise:

Why does artifact x not serve the expected function to ϕ?

Such questions are *contrastive*: why malfunction, rather than normal function? In the engineering literature, malfunction explanations that answer contrastive questions list different and fewer mechanistic features than reverse engineering explanations which answer questions about normal behavior or function.[4] Such explanations are constructed using effect functions and functional decompositions.

Malfunction explanations in engineering pick out only a few features of mechanisms, i.e. those causal factors—failing components or sub-mechanisms—that are taken to make a difference to the occurrence of a specific malfunction, as well as some coarse-grained details of the containing mechanism to understand where the fault is located. Yet most information about structural and behavioral specifics of malfunctioning components/sub-mechanisms, and their containing mechanisms, is left out (Hawkins and Woollons 1998; Bell et al. 2007).[5,6]

Consider, by way of example, the Functional Interpretation Language (FIL) methodology for malfunction analysis and explanation (Bell et al. 2007). In FIL, functions are effect functions and represented in terms of their *triggers* and *effects*. Triggers describe input states that actuate physical behaviors which result in certain (expected) effects. So triggers are the input conditions for effects, i.e. functions, to be achieved but they are not the immediate physical antecedents of these effects. These physical antecedents, i.e. input flows, are not referred to in trigger-effect descriptions and neither do these descriptions meet conservation laws. For instance, consider the function description "depress_brake_pedal"—"red_stop_lamps_lit" of a car's stop light (Bell et al. 2007: 400), in which electromagnetic radiation ("light") is "created" rather than being converted from other energies. This description, rather, is a summary of some salient features of (manipulating) such artifacts; depressing the brake pedal will, if the system functions properly, result in the lighting of the stop lamps, i.e. the effect function.

According to Bell et al. (2007), such trigger and effect representations serve two explanatory ends in malfunction analyses: first, they *highlight* relevant behavioral features of a given artifact, i.e. effects, and, simultaneously, provide the means to *ignore* less relevant or irrelevant behavioral features, i.e. physical behaviors underlying these effects; second, they support assessing which components are malfunctioning (Bell et al. 2007: 400–1).

For instance, the trigger-effect representation "depress_brake_pedal"—"red_stop_lamps_lit" highlights the input condition of a pedal being depressed, and the resulting desired effect of lighted lamps, yet ignores the structural and behavioral specifics of the brake pedal and stop lamps, such as the pedal lever and electrical circuit mechanisms, as well as the energy conversions—e.g., mechanical energy conversions into electricity—that are needed to achieve this effect. Such representations only highlight those features that are considered explanatorily relevant to assess malfunctioning systems, and omit reference to physical behaviors/energy conversions by which the desired effects are achieved.

Second, such trigger-effect descriptions support comparing normally functioning technical systems with malfunctioning ones (Bell et al. 2007). Trigger-effect descriptions support assessing whether the expected effects in fact obtain, and, if not, which and how components are malfunctioning (Bell et al. 2007). A normally functioning artifact, say the car's stop lights, has both a trigger and an effect occurring; the brake pedal is depressed and the stop lights are lit. Trigger-effect descriptions support analysis of two varieties of malfunction. First, a trigger may occur, yet fail to result in the intended effect. Say, the brake pedal is depressed, yet the stoplights are not on. Second, a trigger may not be occurring, yet the effect is nevertheless present. Say, the brake pedal is not depressed, yet the stoplights are on (see Bell et al. 2007). Such analysis of the actual states of triggers and effects allows one to focus on the most likely causes of failure (Bell et al. 2007). Say, if the pedal is depressed and the lights fail to ignite, the first likely causes to investigate may be whether the electrical circuits in the lights are broken or the "on/off" connection between the brake and electrical circuitry (connected to the lamp) is damaged. On the other hand, if the pedal is not depressed and the lights are lit, a first likely cause to investigate may be whether the "on/off "connection between the brake and the electrical circuitry is damaged. To support more detailed malfunction analyses, functions are often decomposed into sub-functions in FIL. An example of a functional decomposition of a two-ring cooking hotplate is given in Figure 34.3.

The usage of effect functions and functional decompositions in FIL is the optimal choice given that function descriptions are used to black-box or suppress reference to unwanted behavioral

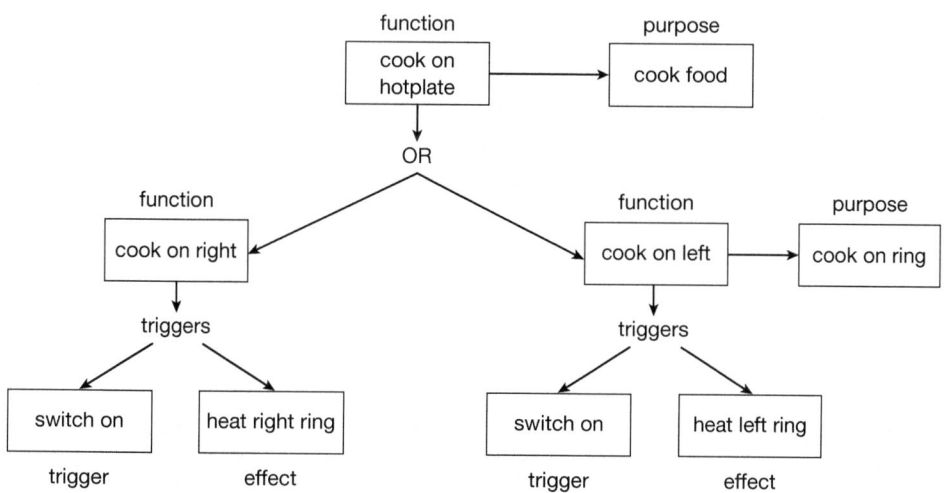

Figure 34.3 Effect functional decomposition of a two-ring cooking hotplate (adapted from Bell et al. 2007)

and structural details. Effect function descriptions only highlight the relevant difference-making properties with respect to malfunctioning artifacts, whereas more elaborate behavior function descriptions include irrelevant details such as, say, the thermal energy generated when lamps are lit.

The upshot of these two cases is that explanations in engineering are furnished relative to explanatory objectives and, importantly, the level of detail included in these explanations hinges on specific concepts of technical function. Engineering scientists simplify or increase the details of explanations—functional decompositions—depending on the explanatory purpose at hand, and these adjustments are made using specific concepts of technical function (compare e.g. Figures 34.2 and 34.3). In reverse engineering explanation, elaborate or "complete" descriptions of mechanisms are provided, in terms of behavior functions and functional decompositions, to answer the question of how a technical system exhibits a given overall behavior. In malfunction explanation, less elaborate "sketches" of mechanisms are provided in terms of effect functions and functional decompositions, referring only to some mechanistic features, namely those difference-making factors that mark the *contrast* between normal functioning and malfunctioning technical systems. So, depending upon explanatory context, mechanisms are individuated in different ways using different conceptualizations of function in engineering science. Neither function conceptualization in itself accommodates both ways in which mechanisms are functionally individuated in engineering science. Behavior and effect function ascriptions are invoked to individuate mechanisms in different ways depending on the task at hand.

However, this distinction in functional individuation, and its reliance on different function concepts, remains opaque when seen from a perspective that conceives of mechanism individuation and mechanistic explanation in terms of mechanistic role function ascription *simpliciter*. The concept of mechanistic role function, an activity that makes a contribution to the workings of a mechanism of which it is a part, admits of two interpretations in the context of engineering science: behavior function on the one hand and effect function on the other. Note that behavior and effect descriptions of function describe, in different ways, the contributions of components to mechanisms of which they are a part. The distinction between behavior and effect function thus is not to be conflated with the distinction between a mechanism description and a description of a mechanism's overall activity. Neither is the behavior-effect function distinction to be conflated with the distinction between "isolated" and "contextual" descriptions of an entity's activity (Craver 2001): isolated descriptions describe activities without taking into account the mechanisms in which they are situated; contextual descriptions describe activities in terms of the mechanistic contexts in which they are situated and to which they contribute. Both behavior and effect functions are of the contextual variety, describing contributions of components to the mechanisms of which they are a part. So to arrive at an empirically informed understanding of explanatory practices in engineering, and at consistency of the general structure of mechanistic explanation with these practices, regimenting the concept of role function into domain-specific engineering concepts of behavior and effect function, i.e. sub-types of role function, is needed.

I now turn to another facet of the relationship between mechanistic explanation and engineering that has received little sustained analysis: the usage of engineering principles in the construction of mechanistic explanations in systems biology. Here we see again the relevance of knowing the ways in which mechanisms are functionally individuated in engineering: the manner in which biological mechanisms are individuated in engineering terms also hinges on specific *engineering* conceptualizations of function. To better understand the specifics of mechanism individuation along engineering lines in systems biology, sensitivity to the varieties of engineering function and functional individuation is thus called for. In the case I discuss below, mechanism individuation hinges on the use of *effect* function descriptions and ascriptions.

3. Explanation and systems thinking: where engineering and systems biology meet

Although philosophy, it seems, is only recently picking up on the fruitful cross-talk between engineering and systems biology (see Braillard 2015), with Wimsatt's (e.g. 2006) work being a notable exception, engineers and systems biologists alike have been stressing the conceptual ties for more than a decade (Hartwell et al. 1999; Lazebnik 2002). With biological data about complex biological systems exploding during the last 20 years or so because of (functional) genomics projects and the like, opportunities to understand complex biological systems in far greater detail became available. Yet cashing out that promise also signaled the need for new tools that enabled massive data analysis and integration to build explanatory models of these complex systems with a scale and complexity hitherto unknown (see Chapter 27). Here is where, amongst others, engineering tools came in.

As Hartwell et al. (1999), for instance, commented with respect to engineering representational schemes:

> In our opinion, the most effective language to describe functional modules and their interactions [in systems biology] will be derived from the synthetic sciences, such as computer science or engineering, in which function appears naturally.
>
> *(Hartwell et al. 1999: C49)*

Or as Csete and Doyle (2002) commented with respect to systemic organization:

> Advanced technologies [like cars and airplanes] and biology have extremely different physical implementations, but they are far more alike in systems-level organization than is widely appreciated
>
> *(Csete and Doyle 2002: 1664)*

Functional engineering parlance and systemic organization come together in the connection between control engineering and systems biology, in which (effect functional) decomposition and control principles governing (the construction of) engineering systems are used to characterize complex biological systems:

> Some insightful recent papers advocate a similar modular decomposition of biological systems according to the well defined functional parts used in engineering and, specifically, engineering control theory.
>
> *(Tomlin and Axelrod 2005: 4219)*

A case in point is research by El-Samad et al. on the mechanism(s) to counter heat shock in *Escherichia coli* (El-Samad et al. 2005; see Tomlin and Axelrod 2005; Braillard 2015). Heat shock response is a widely conserved response of cells to cope with environmental stress brought about by unusual increases in temperature, involving the induced expression of heat shock proteins. Such temperature increases can damage proteins by breaking down their tertiary structures. Heat shock proteins come in two varieties and mitigate this effect in two different ways: molecular chaperones do so by refolding denatured proteins and proteases by degrading denatured proteins. If the response is sufficiently swift and massive, cell death can be prevented by protein repair and/or removal of damaged proteins. The response needs to be tightly controlled in the sense that it is only activated in the case of heat shock, since the response is highly energy-consuming and would make too high energy demands if heat shock proteins would be produced all the time. Cells thus must maintain a delicate balance between the protective effect of heat shock protein production and the metabolic cost of overproducing these proteins. In *E. coli*, the RNA polymerase cofactor σ^{32} promotes the transcription of heat shock proteins. After heat shock stress—temperature increase—σ^{32} activity increases, resulting in the transcription of specific heat shock gene promoters which initiate the transcription of genes that in turn encode specific heat shock proteins—chaperones and proteases. This heat shock protein expression, when appropriate, prevents cell death. This mechanism uses both feed-forward and feedback loops that process information about temperature and the folding state of proteins in the cell. σ^{32} activity is crucial in all this and depends on a feed-forward mechanism that senses temperature and controls σ^{32} transcription, and feedback regulatory mechanisms that register the folding levels of proteins (levels of denatured cellular protein) and degrade σ^{32}. These regulatory feedback mechanisms are crucial to ensure that σ^{32} synthesis, activity, and stability are brought back to normal levels after a sufficient number of heat shock proteins have been produced and the threat to cell death is averted.

The above qualitative information on the heat shock response system is well known. El-Samad and his group (2005) went further and constructed a quantitative, mathematical model of the heat shock response to "use this description to pose questions about the regulatory architecture of the system" (El-Samad et al. 2005: 2737), i.e. the dynamical, mechanistic organization that sustains the heat shock response. They came up with an elaborate mathematical model consisting of 31 equations and seven parameters. Now, to make the model computationally tractable and pose and answer questions about the dynamical, mechanistic organization of the system, the original model had to be trimmed down. This model reduction was effected by various simplifications and, importantly, the salient modularity of the system which made it possible to decompose the system into functional modules, described in terms of effect functions, along control engineering lines. The resulting model was taken to be a "simplified yet reasonably accurate version of the original model" (El-Samad et al. 2005: 2737).

As Braillard (2015) stressed, control engineering principles played an important heuristic role in this model reduction, and thus in the discovery of the mechanism's core organizational features that sub-serve its overall regulatory behavior. The close analogy between engineered systems and biological ones with respect to functional modular organizations sub-serving regulatory processes makes this possible. As El-Samad et al. (2005) explain:

> Control and dynamical systems theory is a discipline that uses modular decompositions extensively to make modeling and model reduction more tractable. Because biological networks are themselves complex regulation systems, it is reasonable to expect that seeking similarities with the functional modules traditionally identified in engineering schemes can be particularly useful.
>
> *(El-Samad et al. 2005: 2737)*

In control engineering, decomposition into functional modules (modules defined in terms of their effect-role functions) often begins with identification of the process to be regulated called the "plant" (see Lind 1994), for instance altitude regulation of an airplane or temperature regulation of a thermostat. Modules of the system that contribute to the regulation are described in terms of their contributing functions, the most common of which are "sensors," "detectors," "controllers," "actuators," and "feed-forward" and "feedback" signals. For instance, in a simple heating system, the plant is the temperature regulation process, which is achieved, inter alia, by a sensor module which measures ambient temperature, calculates the deviation from the desired temperature, and feeds this information into the thermostat (controller). The thermostat then outputs signals that are sent to an actuator (heat fuel valve) that generates an actuation signal (e.g. fuel to furnace) that corrects deviation from the desired temperature. The sensor module again measures the ambient temperature and, if needed, feeds back information on temperature deviations to the controller, and so on.

El-Samad et al. (2005) applied this control engineering perspective to the *E. coli* heat shock response system. In this application, the protein folding task (the refolding of denatured proteins) is taken to be the process to be regulated (plant), the feed-forward signal (sent by a sensor) is the temperature-dependent translational efficiency of σ^{32} synthesis, the controller is the level of σ^{32} activity, chaperones function as actuators of the plant (the actuated plant input is the number of molecular chaperones), and sensors measure plant output (the amount of denatured protein), which in turn is fed back to the controller.

This decomposition allowed El-Samad et al. (2005) to construct a simplified model consisting of just six equations and 11 parameters in which each equation describes the behavior of a module. They remark:

> This model provides useful insight into the heat shock system design architecture. It also suggests a mathematical and conceptual modular decomposition that defines the functional blocks or submodules of the heat shock system. This decomposition is drawn by analogy to manmade control systems and is found to constitute a canonical blueprint representation for the heat shock network.
>
> *(El-Samad et al. 2005: 2736)*

What we here thus see is that analogical reasoning with respect to regulation processes, and the functional architecture sub-serving these processes in engineered and biological systems, led to a functional modular decomposition of a biological system that laid bare core organizational features of the system by which it produces regulatory behavior. Engineering tools here serve as a discovery heuristic for a mechanism's core organizational features that sub-serve its overall regulatory behavior (see Braillard 2015; Chapter 19, this volume). This usefulness of engineering concepts, i.e. modular decompositions in terms of effect functions, is not specific to the *E. coli* case, and generalizes to a variety of cases (see Tomlin and Axelrod 2005) and suggests a general discovery heuristic:

> If the heat shock mechanism can be described and understood in terms of engineering control principles, it will surely be informative to apply these principles to a broad array of cellular regulatory mechanisms and thereby reveal the control architecture under which they operate.
>
> *(Tomlin and Axelrod 2005: 4220)*[7]

In concluding this chapter, I suggest that this case gives relevant insights into a general discussion on explanatory power in the recent mechanisms literature by providing an empirical illustration of the complementarity of two allegedly competing perspectives on the explanatory power of mechanistic explanations.

I have argued elsewhere that differences between two main (allegedly) competing perspectives on the explanatory power of mechanistic explanations, "completeness and specificity" (Craver 2007) and "abstraction" (Levy and Bechtel 2013), essentially boil down to differences in the notions of difference making endorsed in these accounts and that they are in fact not in competition (van Eck 2015). They are rather suitable for different explanation-seeking contexts. Whereas abstraction dictates that mechanistic explanations should only list the "primary factors" responsible for the occurrence of system function, "completeness and specificity" prescribes that in addition to primary ones, "higher-order factors" should also be described, which concern factors influencing the precise manner in which a system function occurs or those sub-serving the primary factors. The *E. coli* case gives an empirical illustration of this view.

The notion of "robustness" looms large in the *E. coli* case, as well as in systems biology and engineering in general. Robust systems—ones resilient to perturbations to parts of the mechanism or the environment in which it functions—require complex sub-systems dedicated to counteracting perturbations (Kitano 2004). This holds both for complex biological systems and (most) engineered systems. Think, for instance, of all the sub-systems of an airplane dedicated to counteracting changes to make it fly in the appropriate manner, or the sub-systems in *E. coli* that play a role in counteracting the effects of heat shock on protein deformation—chaperons and proteases. As El-Samad et al. (2005) elaborate:

> The modular decomposition of the hsr [heat shock response] shows a level of complexity not justified by the basic functionality demanded from an operational heat shock system. A simple and operational heat-shock system would consist solely of a temperature sensor . . . and a transcriptional/translational apparatus that responds appropriately to temperature changes.
>
> *(El-Samad 2015: 2738)*

Why, then, is additional complexity present? Computational modeling indicated that "complexity is indeed necessary to achieve robustness, noise rejection, speed of response, and economical use of cellular resources, much like engineering systems" (El-Samad 2015: 2738).

Complexity and robustness provide an interesting slant on "abstraction" and "completeness and specificity." Depending on the questions one asks with respect to complex, robust systems, either "completeness and specificity" or "abstraction" are better suited. For instance, one may address the question: "How does the mechanism execute its regulatory function?", or the joint questions: "How does the mechanism execute its regulatory function?" and "Why does it (execute its) function in a robust manner?" If one is interested in the key organizational details that enable complex systems to function, abstract description suffices. If, on the other hand, one is also interested in the mechanistic details that enable a mechanism to function in a robust fashion, more specific and elaborate descriptions are called for. In the latter case, one is not only interested in the "primary factors" responsible for the occurrence of system function, but also in the "higher-order factors" influencing the precise manner in which it occurs or those sub-serving the primary factors (see Weisberg 2007).

To round up, the functional individuation of mechanisms—in terms of behavior and effect function ascription and decomposition—proceeds differently in engineering science than the manner in which it is taken to work in the life sciences. Understanding these specifics is required to understand the structure of mechanistic explanation in engineering science and, moreover, adds to our understanding of the ways in which tools and insights from engineering are used in mechanism individuation and explanation in systems biology. Finally, cases from engineering and systems biology give general insights into the explanatory power of mechanistic models in specific explanation-seeking contexts.

Notes

1 It is a pleasure thanking Phyllis Illari and Stuart Glennan for the opportunity to write this chapter and for their helpful comments on a previous version.

2 The term "archetypical" here refers to "most common"; the three conceptualizations of function are not meant to be exhaustive. For instance, some engineers use "function" to refer to intentional behaviors of agents (see van Eck 2010). In reverse engineering analyses, "function" refers to actual or expected behavior, without the normative connotation "desired."

3 Behavior and effect functions thus have a partly common semantic structure: certain aspects or features of behaviors that they both refer to. They are dissimilar in the sense that behavior function descriptions refer to additional behavioral aspects, not referred to in effect function descriptions, so as to make these descriptions accord with physical conservation laws. The relation between behavior and effect function is asymmetrical in the sense that effects, being subsets of behaviors, are straightforwardly derivable from behaviors, but not vice versa. From a given effect one cannot automatically derive the behavior of which the effect is a part. Cars that run on gasoline operate by means of different energy conversions than cars that run on electricity, yet both display the same effects; say, delivering acceleration. The semantic structure that they partly have in common creates the possibility and need to be pluralist about mechanistic role functions, i.e. different ways to conceive of the role functions of mechanisms, in the context of engineering science. I defend this pluralism about mechanistic role functions later on in this section. To be sure, I am thus not advocating a pluralist view about functions of mechanisms with a completely dissimilar semantic structure.

4 Reverse engineering explanation, like mechanistic explanation in general, is not contrastive, whereas malfunction explanation is. The role of contrasts, essential to counterfactual accounts of explanation, seems not vital to most accounts of mechanistic explanation. Mechanistic explanations are often taken to track mechanisms that exhibit productive continuity, and are typically not construed in counterfactual fashion. Counterfactual reasoning, rather, is often invoked in analyses of mechanism discovery and in explanatory relevance assessments where interventions on putative components are stressed (Craver 2007). Malfunction explanations thus make for an interesting extension of mechanistic conceptions of explanation, since they are both mechanistic and contrastive.

5 That is, structural and behavioral characteristics are considered irrelevant in a first-round functional analysis of malfunction. After this analysis, more detailed behavioral models of components and their behaviors are used for identifying specific explanatorily relevant structural and behavioral characteristics of malfunctioning components/sub-mechanisms (Bell et al. 2007). However, immediately specifying these details in functional models is taken to result in listing a lot of irrelevant details.

6 Malfunction explanations in engineering thus exemplify Garson's (2013) "functional sense of mechanism" (see Chapter 8); a malfunction is seen as a breakdown of a mechanism, not as the result of a specific mechanism for malfunction.

7 The analysis I gave in this section illustrates what Glennan and Illari call "methodological mechanism" (see Chapter 1): seen from an epistemological and methodological perspective, biology and engineering have much in common and tools and insights from the latter can help address explanatory issues in the former (and vice versa). I thus do not address the metaphysical nature of mechanisms and how differences play out in this regard between biology and engineering. So, for instance, whether biological mechanisms are modular like (most) technical ones or whether we impose modularity on them to understand how they work is, although a very intriguing question, not one I am concerned with here. Neither do I focus here on possible differences that may emerge between the life sciences and engineering when we consider normative conceptions of function. It might transpire, however, that these differences are not so great as some might suspect, at least in the context of explanation: of course, engineers design and build technical systems-mechanisms with desired ("proper") functions, yet a "functional sense of mechanism" (Garson 2013; see Chapter 8) suffices to account for a token mechanism with a function that it fails to perform, both in engineering and biological contexts. Furthermore, in engineering contexts of explanation, engineering sub-types of role function do the explanatory work, not an etiological conception of function (van Eck and Weber 2014).

References

Bechtel, W., and A. Abrahamsen. (2005). "Explanation: A Mechanist Alternative", *Studies in History and Philosophy of Biological and Biomedical Sciences* 36:421–41.

Bell, J., Snooke, N., and C. Price. (2007). "A Language for Functional Interpretation of Model Based Simulation", *Advanced Engineering Informatics* 21:398–409.

Braillard, P.A. (2015). "Prospects and Limits of Explaining Biological Systems in Engineering Terms". In: *Explanation in Biology* (Eds. P.A. Braillard & C. Malaterre), Dordrecht: Springer, 319–44.

Chandrasekaran, B., and J.R. Josephson. (2000). "Function in Device Representation", *Engineering with Computers* 16:162–77.

Craver, C.F. (2001). "Role Functions, Mechanisms, and Hierarchy", *Philosophy of Science* 68:53–74.

Craver, C.F. (2007). *Explaining the Brain: Mechanisms and the Mosaic Unity of Neuroscience*. New York: Oxford University Press.

Csete, M.E., and J.C. Doyle. (2002). "Reverse Engineering of Biological Complexity", *Science* 295: 1664–9.

Darden, L. (2006). *Reasoning in Biological Discoveries*. Cambridge: Cambridge University Press.

El-Samad, H., Kurata, H., Doyle, J.C., Gross, C.A., and M. Khammash. (2005). "Surviving Heat Shock: Control Strategies for Robustness and Performance", *PNAS* 102(8):736–41.

Erden, M.S., Komoto, H., Van Beek, T.J., D'Amelio, V., Echavarria, E., and T. Tomiyama. (2008). "A Review of Function Modeling: Approaches and Applications", *Artificial Intelligence for Engineering Design, Analysis, and Manufacturing* 22:147–69.

Garson, J. (2013). "The Functional Sense of Mechanism", *Philosophy of Science* 80:317–33.

Glennan, S. (2005). "Modeling Mechanisms", *Studies in the History and Philosophy of the Biological and Biomedical Sciences* 36(2):375–88.

Goel, A.K. (2013). "A 30-Year Case Study and 15 Principles: Implications of an Artificial Intelligence Methodology for Functional Modeling", *AIEDAM* 27(3):203–15.

Hartwell, L.H., Hopfield, J.J., Leibner, S., and A.W. Murray. (1999). "From Molecular to Modular Cell Biology", *Nature* 402:C47–C52.

Hawkins, P.G., and D.J. Woollons. (1998). "Failure Modes and Effects Analysis of Complex Engineering Systems using Functional Models", *Artificial Intelligence in Engineering* 12(4):375–97.

Illari, P., and J. Williamson. (2010). "Function and Organization: Comparing the Mechanisms of Protein Synthesis and Natural Selection", *Studies in History and Philosophy of Biological and Biomedical Sciences* 41:279–91.

Illari, P., and J. Williamson. (2012). "What Is a Mechanism? Thinking about Mechanisms Across the Sciences", *European Journal for Philosophy of Science* 2:119–35.

Kitano, H. (2004). "Biological Robustness", *Nature* 5:826–37.

Lazebnik, Y. (2002). "Can a Biologist Fix a Radio?—Or, What I Learned While Studying Apoptosis", *Cancer Cell* 2:179–82.

Levy, A. (2014). "Machine-Likeness and Explanation by Decomposition", *Philosopher's Imprint* 6:1–15.

Levy, A. and W. Bechtel. (2013). "Abstraction and the Organization of Mechanisms," *Philosophy of Science* 80:241–61.

Lind, M. (1994). "Modeling Goals and Functions of Complex Industrial Plants", *Applied Artificial Intelligence* 8:259–83.

Machamer, P.K., Darden, L., and C.F. Craver. (2000). "Thinking about Mechanisms", *Philosophy of Science* 57:1–25.

Otto, K.N., and K.L. Wood. (2001). *Product Design: Techniques in Reverse Engineering and New Product Development*. Upper Saddle River, NJ: Prentice Hall.

Stone, R.B., and K.L. Wood. (2000). "Development of a Functional Basis for Design", *Journal of Mechanical Design* 122:359–70.

Tomlin, C.J., and Axelrod, J.D. (2005). "Understanding Biology by Reverse Engineering the Control", *PNAS* 102(12):4219–20.

Van Eck, D. (2010). "On the Conversion of Functional Models: Bridging Differences Between Functional Taxonomies in the Modeling of User Actions", *Research in Engineering Design* 21(2):99–111.

Van Eck, D. (2011). "Supporting Design Knowledge Exchange by Converting Models of Functional Decomposition", *Journal of Engineering Design* 22(11–12):839–58.

Van Eck, D. (2015). "Mechanistic Explanation in Engineering Science", *European Journal for Philosophy of Science* 5(3):349–75.

Van Eck, D., and E. Weber. (2014). "Function Ascription and Explanation: Elaborating an Explanatory Utility Desideratum for Ascriptions of Technical Functions", *Erkenntnis* 79:1367–89.

Vermaas, P.E. (2009). "The Flexible Meaning of Function in Engineering", *Proceedings of the 17th International Conference on Engineering Design (ICED 09)* 2:113–24.

Weisberg, M. (2007). "Three Kinds of Idealization", *The Journal of Philosophy* 104(12):639–59.

Wimsatt, W. (2006). *Re-Engineering Philosophy for Limited Beings: Piecewise Approximations to Reality.* Cambridge, MA: Harvard University Press.

INDEX